Progress in Mathematics

Volume 254

Series Editors

H. Bass
J. Oesterlé
A. Weinstein

Xiaonan Ma
George Marinescu

QA
331
M257
2007
web

Holomorphic Morse Inequalities and Bergman Kernels

Birkhäuser
Basel · Boston · Berlin

Authors:

Xiaonan Ma
Centre de Mathématiques
Laurent Schwartz (C.M.L.S.)
École Polytechnique
91128 Palaiseau Cedex
France
e-mail: ma@math.polytechnique.fr

George Marinescu
Mathematisches Institut
Universität zu Köln
Weyertal 86–90
50931 Köln
Germany
e-mail: gmarines@math.uni-koeln.de

2000 Mathematics Subject Classification: 23J, 58J, 53C, 53D, 32F, 32L, 32Q

Library of Congress Control Number : 2007922259

Bibliographic information published by Die Deutsche Bibliothek
Die Deutsche Bibliothek lists this publication in the Deutsche Nationalbibliografie;
detailed bibliographic data is available in the Internet at <http://dnb.ddb.de>.

ISBN 978-3-7643-8096-0 Birkhäuser Verlag AG, Basel - Boston - Berlin

This work is subject to copyright. All rights are reserved, whether the whole or part
of the material is concerned, specifically the rights of translation, reprinting, re-use of
illustrations, broadcasting, reproduction on microfilms or in other ways, and storage in
data banks. For any kind of use whatsoever, permission from the copyright owner must
be obtained.

© 2007 Birkhäuser Verlag AG
Basel · Boston · Berlin
P.O. Box 133, CH-4010 Basel, Switzerland
Part of Springer Science+Business Media
Printed on acid-free paper produced of chlorine-free pulp. TCF ∞
Printed in Germany

ISBN 978-3-7643-8096-0   e-ISBN 978-3-7643-8115-8

9 8 7 6 5 4 3 2 1   www.birkhauser.ch

Ferran Sunyer i Balaguer (1912–1967) was a self-taught Catalan mathematician who, in spite of a serious physical disability, was very active in research in classical mathematical analysis, an area in which he acquired international recognition. His heirs created the Fundació Ferran Sunyer i Balaguer inside the Institut d'Estudis Catalans to honor the memory of Ferran Sunyer i Balaguer and to promote mathematical research.

Each year, the Fundació Ferran Sunyer i Balaguer and the Institut d'Estudis Catalans award an international research prize for a mathematical monograph of expository nature. The prize-winning monographs are published in this series. Details about the prize and the Fundació Ferran Sunyer i Balaguer can be found at

http://ffsb.iec.cat

**This book has been awarded the
Ferran Sunyer i Balaguer 2006 prize.**

The members of the scientific committee of the 2006 prize were:

Antonio Córdoba
  Universidad Autónoma de Madrid

Paul Malliavin
  Université de Paris VI

Joseph Oesterlé
  Université de Paris VI

Oriol Serra
  Universitat Politècnica de Catalunya, Barcelona

Alan Weinstein
  University of California at Berkeley

Ferran Sunyer i Balaguer Prize winners since 1997:

To Ling and Cristina

Contents

Introduction

Let X be a compact complex manifold and L be a holomorphic line bundle on X. We denote by $H^q(X, L)$ the qth cohomology group of the sheaf of holomorphic sections of L on X.

Many important results in algebraic and complex geometry are derived by combining a vanishing property with an index theorem, or from the asymptotic results on the tensor powers L^p when $p \to \infty$. One of the most famous examples is the Kodaira–Serre vanishing theorem which asserts that if L is positive, then $H^q(X, L^p)$ vanish for $q \geqslant 1$ and large p. The key remark is that the spectrum of the Kodaira–Laplace operator $\Box_p$ acting on $(0, q)$-forms, $q \geqslant 1$, with values in the tensor powers L^p, shifts to the right linearly in the tensor power p. As a consequence the kernel of $\Box_p$ is trivial on forms of higher degree and the vanishing theorem follows by the Hodge theory and the Dolbeault isomorphism. Moreover, the Riemann–Roch–Hirzebruch theorem implies that L^p has a lot of holomorphic sections on X for large p, which indeed embed the manifold X in a projective space.

An important generalization which we will emphasize is the asymptotic holomorphic Morse inequalities of Demailly. They give asymptotic bounds on the Morse sums of the $\overline{\partial}$-Betti numbers $\dim H^q(X, L^p)$ in terms of certain integrals of the curvature form of L. The holomorphic Morse inequalities provide a useful tool in complex geometry. They are again based on the asymptotic spectral behavior of the Kodaira–Laplace operator $\Box_p$ for large p.

The applications of these vanishing theorems and holomorphic Morse inequalities are numerous. Let us mention here only the Kodaira embedding theorem, the classical Lefschetz hyperplane theorem for projective manifolds, the computation of the asymptotics of the Ray-Singer analytic torsion by Bismut and Vasserot, as well as the solution of the Grauert–Riemenschneider conjecture by Siu and Demailly or the compactification of complete Kähler manifolds of negative Ricci curvature by Nadel and Tsuji. Donaldson's work on the existence of symplectic submanifolds was inspired by the same circle of ideas.

The holomorphic Morse inequalities are global statements which can be deduced from local information such as the behavior of the heat or Bergman kernels. In this refined form we can establish the asymptotic expansion of the Bergman kernel associated to L^p as $p \to \infty$, which have had a tremendous impact on research in the last years. Especially, let's single out its applications in Donaldson's approach to the existence of Kähler metrics with constant scalar curvature in relation to the Mumford–Chow stability which was mainly motivated by a conjecture of Yau. Other applications include the convergence of the induced Fubini–Study metrics, the distribution of zeroes of random sections, the Berezin–Toeplitz quantization and sampling problems.

Another important operator which we will study, also in view of the generalization to symplectic manifolds, is the Dirac operator acting on high tensor powers of L on symplectic manifolds. For a Kähler manifold the square of the Dirac operator is twice the Kodaira Laplacian.

In the present book we will give for the first time a self-contained and unified treatment to the holomorphic Morse inequalities and the asymptotic expansion of the Bergman kernel by using heat kernels, and we present also various applications. Our point of view comes from the local index theory, especially from the analytic localization techniques developed by Bismut–Lebeau. Basically, the holomorphic Morse inequalities are a consequence of the small time asymptotic expansion of the heat kernel. The Bergman kernel corresponds to the limit of the heat kernel when the time parameter goes to infinity, and the asymptotic is more sophisticated. A simple principle in this book is that the existence of the spectral gap of the operators implies the existence of the asymptotic expansion of the corresponding Bergman kernel, no matter if the manifold X is compact or not, or singular, or with boundary. Moreover, we will present a general and algorithmic way to compute the coefficients of the expansion.

Let us now give a rapid account of the main results discussed in this book.

In the first chapter we introduce the basic material. After giving a self-contained presentation of the connections on the tangent bundle, Dirac operator and Lichnerowicz formula, we specify them for the Kodaira Laplacian, especially we study in detail the Bochner–Kodaira–Nakano formula without and with boundary term. These various formulas are fundamental and have a lot of applications. We will use them repeatedly throughout the text. As a direct application, we establish immediately classical vanishing results and the spectral gap property for Kodaira Laplacians and modified Dirac operators. The latter will play an essential role in our approach to the asymptotic expansion of Bergman kernel.

The last two sections of this chapter are dedicated to Demailly's holomorphic Morse inequalities. They originally arose in connection with the generalization of the Kodaira vanishing theorem for Moishezon manifolds proposed by Grauert and Riemenschneider, who conjectured that a compact connected complex manifold X possessing a semi-positive line bundle L, which is positive at at least one point, is Moishezon. The conjecture was solved by Siu and Demailly. The solution of Demailly involves the following strong Morse inequalities:

$$\sum_{j=0}^{q}(-1)^{q-j} \dim H^j(X, L^p) \leqslant \frac{p^n}{n!} \int_{X(\leqslant q)} (-1)^q \left(\tfrac{\sqrt{-1}}{2\pi} R^L \right)^n + o(p^n) \qquad (1)$$

as $p \longrightarrow \infty$, where R^L is the curvature of L (cf. (1.5.15)), and $X(\leqslant q)$ is the set of points where $\dot{R}^L \in \mathrm{End}(T^{(1,0)}X)$, defined by $R^L(u, \overline{v}) = g^{TX}(\dot{R}^L u, \overline{v})$ for $u, v \in T^{(1,0)}X$ and a Riemannian metric g^{TX} on TX, is non-degenerate and has at most q negative eigenvalues. For $q = n$ we have equality, so we obtain an asymptotic Riemann–Roch–Hirzebruch formula.

Demailly's discovery was triggered by Witten's influential analytic proof of the standard Morse inequalities. Witten analyzes the spectrum of the Schrödinger operator $\Delta_t = \Delta + t^2|df|^2 + tV$, where $t > 0$ is a real parameter, Δ is the Bochner Laplacian acting on forms on X, f is a Morse function on X and V is a 0-order operator. For $t \longrightarrow \infty$, the spectrum of Δ_t approaches the spectrum of a sum of harmonic oscillators attached to the critical points of f. In Demailly's holomorphic Morse inequalities, the role of the Morse function is played by the Hermitian metric on the line bundle and the Hessian of the Morse function becomes the curvature of the bundle. The original proof was based on the study of the semi-classical behavior as $p \to \infty$ of the spectral counting functions of the Kodaira Laplacians $\Box_p$ on L^p. Subsequently, Bismut gave a heat kernel proof which involves probability theory, and then Demailly and Bouche were able to replace the probability technique by a classical heat kernel argument.

We present here a new approach based on the asymptotic of the heat kernel of the Kodaira Laplacian, $\exp(-\frac{u}{p}\Box_p)$. The analytic core follows in Section 1.6 where, inspired by the work of Bismut–Lebeau, we present a new proof for the asymptotic of the heat kernel. In Section 1.7 we apply these results to obtain a heat equation proof of the holomorphic Morse inequalities following Bismut.

In Chapter 2 we study the properties of the field of meromorphic functions. We establish further two fundamental results about Moishezon manifolds. Then we give the proof of the Siu–Demailly criterion which answers the Grauert–Riemenschneider conjecture. For $q = 1$, the Morse inequalities (1) give

$$\dim H^0(X, L^p) \geqslant \frac{p^n}{n!} \int_{X(\leqslant 1)} \left(\frac{\sqrt{-1}}{2\pi} R^L\right)^n + o(p^n), \quad p \longrightarrow \infty. \tag{2}$$

Therefore if L satisfies

$$\int_{X(\leqslant 1)} \left(\frac{\sqrt{-1}}{2\pi} R^L\right)^n > 0, \tag{3}$$

(in particular, if L is semi-positive and positive at at least one point), there are a lot of sections in $H^0(X, L^p)$, which by taking quotients deliver n independent meromorphic functions, i.e., X is Moishezon.

In Section 2.4 we present an algebraic reformulation of the holomorphic Morse inequalities.

In Chapter 3 we prove the Morse inequalities for the Dolbeault L^2-cohomology spaces for a non-compact manifold satisfying the fundamental estimate (Poincaré inequality) at infinity. Using this more abstract formulation of the Morse inequalities, we can find a lower bound for the growth of the holomorphic section space for uniformly positive line bundles (Theorem 3.3.5) and an extension of the Siu–Demailly criterion for compact complex spaces with isolated singularities.

We end the chapter with a study of a class of manifolds satisfying pseudoconvexity conditions in the sense of Andreotti–Grauert, namely q-convex and weakly 1-complete manifolds and also covering manifolds. Pseudoconvex manifolds are very important in complex geometry and analysis.

In Chapter 4, we study the asymptotic expansion of the Bergman kernel. We assume now that L is positive, equivalently, there exists a Hermitian metric h^L on L, such that $\omega = \frac{\sqrt{-1}}{2\pi} R^L$ defines a Kähler form on X, where R^L is the curvature of the holomorphic Hermitian connection ∇^L on (L, h^L). In the rest of the Introduction we denote by g^{TX} the associated Kähler metric to ω on TX. We also let E be a holomorphic vector bundle on X with a Hermitian metric h^E.

Since L is positive, the Kodaira–Serre vanishing theorem shows that

$$H^q(X, L^p \otimes E) = 0 \tag{4}$$

for p large enough and $q \geqslant 1$. Thus the whole cohomology of $L^p \otimes E$ concentrates in degree zero.

The Bergman kernel $P_p(x, x')$ associated to $L^p \otimes E$ for p large enough, is the smooth kernel of the orthogonal projection P_p from $\mathscr{C}^\infty(X, L^p \otimes E)$, the space of smooth sections of tensor powers $L^p \otimes E$, on the space of holomorphic sections of $L^p \otimes E$, or, equivalently, on the kernel of the Kodaira Laplacian $\square_p$ on $L^p \otimes E$. More precisely, let $\{S_i^p\}_{i=1}^{d_p}$ be any orthonormal basis of $H^0(X, L^p \otimes E)$ with respect to the global inner product induced by g^{TX}, h^L and h^E (cf. (1.3.14)). Then for p large enough,

$$P_p(x, x') = \sum_{i=1}^{d_p} S_i^p(x) \otimes (S_i^p(x'))^* \in (L^p \otimes E)_x \otimes (L^p \otimes E)_{x'}^*. \tag{5}$$

Especially,

$$P_p(x, x) = \sum_{i=1}^{d_p} |S_i^p(x)|^2, \quad \text{if } E = \mathbb{C}. \tag{6}$$

The Bergman kernel has been studied by Tian, Yau, Bouche, Ruan, Catlin, Zelditch, Lu, Wang, and many others, in various generalities, establishing the asymptotic expansion for high powers of L. Moreover, it was discovered that the coefficients in the asymptotic expansion encode geometric information about the underlying complex projective manifolds.

Our approach to the study of the asymptotic expansion continues the method applied in Chapter 1. We treat both the Dirac operator and the Kodaira Laplacian in the same time by means of the modified Dirac operator. The key point of our method is that the spectrum $\mathrm{Spec}(\square_p)$ of $\square_p$ (or of the half of the square of the

Dirac operator) has a spectral gap, cf. Section 1.5. This means that there exists $C > 0$ such that for $p \geqslant 1$,

$$\mathrm{Spec}(\square_p) \subset \{0\} \cup \,]2\pi\, p - C, +\infty[. \tag{7}$$

We can divide our approach in three steps. The first step is to establish the spectral gap property (7). The second is the localization: the spectral gap property (7) and the finite propagation speed of solutions of hyperbolic equations allow us first to localize the asymptotic of $P_p(x_0, x')$ in the neighborhood of x_0. We pull-back and extend the operator to $T_{x_0}X \cong \mathbb{R}^{2n}$, and verify that it inherits also the spectral gap property. The third step is to work on $\mathbb{R}^{2n}$. Here we combine the spectral gap property, the rescaling of the coordinates and functional analysis techniques, to conclude the proof of our final result. Moreover, by using a formal power series trick, we get a general and algorithmic way to compute the coefficients in the expansion. Certainly, for the last two steps it makes no difference whether the manifold X is compact or not. Thus in various new situations, we only need to verify the spectral gap property (cf. Chapters 5, 6, 8).

We obtain finally the following asymptotic expansion (cf. Theorem 4.1.2):

$$P_p(x, x) \sim \sum_{r=0}^{\infty} \boldsymbol{b}_r(x) p^{n-r}, \tag{8}$$

where $\boldsymbol{b}_r(x) \in \mathrm{End}(E)_x$ are smooth coefficients, which are polynomials in R^{TX}, R^E and their derivatives with order $\leqslant 2r - 2$. Moreover

$$\boldsymbol{b}_0 = \mathrm{Id}_E, \quad \boldsymbol{b}_1 = \frac{1}{4\pi} \Big[2R^E(w_j, \overline{w}_j) + \frac{1}{2} r^X \, \mathrm{Id}_E \Big], \tag{9}$$

where r^X is the scalar curvature of (TX, g^{TX}) and $\{w_j\}_{j=1}^n$ is an orthonormal basis of $T^{(1,0)}X$. In the case of trivial bundle E the term $\boldsymbol{b}_1$ was calculated by Lu and used by Donaldson in his work on the existence of Kähler metrics with constant scalar curvature.

We also find the full off-diagonal expansion of the Bergman kernel $P_p(x, x')$ with the help of the heat kernel.

In Chapter 5, we study in detail the metric aspect of the Kodaira map as an application of the asymptotic expansion of the Bergman kernel. First, we present an analytic proof of the Kodaira embedding theorem following an original idea of Bouche, and we study the convergence of the induced Fubini–Study metric. Then the Kodaira map $\Phi_p : X \longrightarrow \mathbb{P}(H^0(X, L^p)^*)$, defined by $\Phi_p(x) = \{s \in H^0(X, L^p) : s(x) = 0\}$ for $x \in X$, is an embedding for p large enough and for any $l \in \mathbb{N}$, there exists $C_l > 0$ such that

$$\Big| \frac{1}{p}\, \Phi_p^*(\omega_{FS}) - \omega \Big|_{\mathscr{C}^l(X)} \leqslant \frac{C_l}{p^2}, \tag{10}$$

where ω_{FS} is the Fubini–Study form on $\mathbb{P}(H^0(X, L^p)^*)$.

By using the Kodaira embedding, we also discuss briefly the relation of the Bergman kernel and the existence of Kähler metrics with constant scalar curvature. Then, as an easy consequence of our approach, we describe the asymptotic expansion of the Bergman kernel on complex orbifolds, and the metric aspect of the Kodaira map.

Finally, we give an introduction to the Ray-Singer analytic torsion and study its asymptotic behavior. The analytic torsions have a lot of applications, especially in Arakelov geometry. This seems to be quite independent of our subject, but in fact, Donaldson has used the analytic torsion in his study of the existence of Kähler metrics with constant scalar curvature.

In Chapter 6 we establish the existence of the expansion on compact sets of a non-compact manifold, as long as the spectral gap exists. One interesting situation is the case of Zariski open sets in compact complex spaces endowed with the generalized Poincaré metric. The expansion of the Bergman kernel implies a new proof of the Shiffman–Ji–Bonavero–Takayama criterion for a Moishezon manifold. Then we obtain again Morse inequalities which are suitable for the study of the compactification of complete Kähler manifolds with pinched negative curvature.

In Chapter 7, using the full off-diagonal expansion of the Bergman kernel, we study the properties of Toeplitz operators and the Berezin–Toeplitz quantization. For $f \in \mathscr{C}^\infty(X, \mathrm{End}(E))$, we define the Toeplitz operator $\{T_{f,p}\}$ as the family of linear operators

$$T_{f,p} : L^2(X, L^p \otimes E) \longrightarrow L^2(X, L^p \otimes E), \quad T_{f,p} = P_p\, f\, P_p\,. \qquad (11)$$

One of our main goals is to show that the set of Toeplitz operators is closed under the composition of operators, so they form an algebra. More precisely, let $f, g \in \mathscr{C}^\infty(X, \mathrm{End}(E))$, then there exist $C_r(f,g) \in \mathscr{C}^\infty(X, \mathrm{End}(E))$ with

$$T_{f,p}\, T_{g,p} = \sum_{r=0}^{\infty} p^{-r} T_{C_r(f,g),p} + \mathcal{O}(p^{-\infty}), \qquad (12)$$

where C_r are differential operators. In particular $C_0(f,g) = fg$.

If $f, g \in \mathscr{C}^\infty(X)$, then

$$[T_{f,p}, T_{g,p}] = \frac{\sqrt{-1}}{p} T_{\{f,g\},p} + \mathcal{O}(p^{-2}), \qquad (13)$$

here $\{f, g\}$ is the Poisson bracket of f, g on $(X, 2\pi\omega)$.

In Chapter 8, we find the asymptotic expansion of the Bergman kernel associated to the modified Dirac operator and the renormalized Bochner Laplacian, as well as their applications.

We hope the material of this book can also be used by graduate students. To help the readers, we add five appendices. In Appendix A, we recall the Sobolev embedding theorems and basic elliptic estimates. In Appendix B, we present useful material from Hermitian geometry. We also introduce the basics of Chern–Weil and Chern–Simons theories. In Appendix C, we collect some facts about self-adjoint operators. In Appendix D, we explain in detail the relation of the heat kernel and the finite propagation speed of solutions of hyperbolic equations. Finally, in Appendix E, we explain the basic facts about the harmonic oscillator.

The book should also serve as an analytic introduction to the applications to algebraic geometry of the holomorphic Morse inequalities as developed by Demailly and his school, as well as to Donaldson's approach to the existence of Kähler metrics of constant scalar curvature.

To keep the book within reasonable size, we list several classical results without proofs, and we indicate the corresponding references in the bibliographic notes of each chapter. The literature concerning the various themes we treat is quite vast and contains many important contributions. We could not include them all in the Bibliography, and restrained to the references which directly influenced our work.

Prerequisites for this book are a course on differentiable manifolds and vector bundles. This book is not necessarily meant to be read sequentially. The reader is encouraged to go directly to the chapter of interest. Basically, Chapters 1 and 4 introduce the main technical ideas, and other chapters are various generalizations and applications. Here is a roadmap for our book.

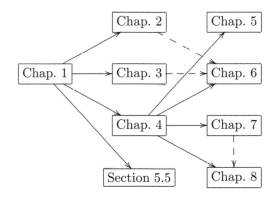

Notation

We denote by $\mathbb{C}, \mathbb{N}, \mathbb{Q}, \mathbb{R}, \mathbb{Z}$ the complex, natural, rational, real, integer numbers, and $\mathbb{C}^* = \mathbb{C} \setminus \{0\}$, $\mathbb{N}^* = \mathbb{N} \setminus \{0\}$, $\mathbb{R}^* = \mathbb{R} \setminus \{0\}$, $\mathbb{R}_+ = [0, \infty[$, $\mathbb{R}_+^* =]0, \infty[$, $\mathbb{Q}_+ = \mathbb{Q} \cap \mathbb{R}_+$. For $u \in \mathbb{R}$, we denote by $\lfloor u \rfloor$ the integer part of u.

For $\alpha = (\alpha_1, \ldots, \alpha_m) \in \mathbb{N}^m$, $B = (B_1, \ldots, B_m) \in \mathbb{C}^m$, we write by

$$|\alpha| = \sum_{j=1}^m \alpha_j, \quad \alpha! = \prod_j (\alpha_j!), \quad B^\alpha = \prod_j B_j^{\alpha_j}.$$

$SL(n, \mathbb{C})$ is the space of $\mathbb{C}$-valued $n \times n$ matrices with determinant 1. $O(n)$ is the orthogonal group of degree n over $\mathbb{R}$. $U(n)$ is the unitary group of degree n over $\mathbb{C}$.

We denote by dim or $\dim_\mathbb{C}$ the complex dimension of a complex (vector) space. We denote also by $\dim_\mathbb{R}$ the real dimension of a space.

For a complex vector bundle E on a manifold X, $\mathrm{rk}(E)$ denotes its rank, and Id_E the identity morphism. Also, $\det(E) := \Lambda^{\mathrm{rk}(E)}(E)$ is its determinant line bundle, E^* its dual bundle and $\mathrm{End}(E) := E \otimes E^*$. The space of smooth sections of E over X is denoted by $\mathscr{C}^\infty(X, E)$.

If Q is an operator, we denote by $\mathrm{Ker}(Q)$ its kernel, $\mathrm{Im}(Q)$ its image set.

If U is a subset of V, we write $U \subset V$. If U is a relatively compact subset of V, we write $U \Subset V$. The characteristic function 1_U of U is defined as 1 on U and 0 on the complement of U.

In the whole book, if there is no other specific notification, when in a formula a subscript index appears two times, then we sum up with this index.

Acknowledgments

We are happy to acknowledge our debt to Professor Jean-Michel Bismut. His impact can be felt throughout the book. The first-named author in particular – having been a student of Professor Jean-Michel Bismut – wishes to express his hearty thanks for discussions on various subjects and his enlightened support since 1994.

Over the years, the second-named author benefited from many inspirational discussions with Jean-Pierre Demailly about holomorphic Morse inequalities.

The subject of Bergman kernels was tremendously influenced by the work of Louis Boutet de Monvel and Johannes Sjöstrand, with whom we had the privilege of discussing the topics.

We thank Xianzhe Dai, Kefeng Liu and Weiping Zhang who helped us understand better the Bergman kernel. We thank also Christophe Margerin, Pierre Milman, Richard Thomas and Xiaowei Wang for several useful conversations and the staff of Birkhäuser for its patience and dedication.

We express our gratitude to the institutions which welcomed us, Humboldt Universität zu Berlin, Centre de Mathématiques Laurent Schwartz de l'École Polytechnique (Palaiseau), Johann Wolfgang Goethe Universität (Frankfurt am Main) and Universität zu Köln.

In particular, our collaboration started at the Humboldt Universität zu Berlin, where, in his Thesis from 1922, Stefan Bergman discovered the kernel function in response to a question of Erhard Schmidt. We reproduce here the account of Menahem Max Schiffer from 1981[1]: "Bergman participated in Schmidt's seminar and was charged to give a lecture on the development of arbitrary functions with finite square integrals in terms of an orthogonal set. As he told me, he misunderstood the task and instead of dealing with real functions over a real interval, he attacked the problem for analytic functions over a complex domain. He found the task hard but attacked it courageously and carried it through. This was the genesis of his famous theory of the kernel function."

---

[1]M.M. Schiffer, *Stefan Bergman* (1885–1977) *in memoriam*, Ann. Pol. Math. 39 (1981), 5–9.

Chapter 1

Demailly's Holomorphic Morse Inequalities

The first aim of this chapter is to provide the background material on differential geometry for the whole book. Then, in the last two sections, we present a heat kernel proof of Demailly's holomorphic Morse inequalities, Theorem 1.7.1.

This chapter is organized as follows. In Section 1.1 we review the theory of connections on vector bundles. In Section 1.2, we explain different connections on the tangent bundle and their relations. In Section 1.3, we define the modified Dirac operator for an almost complex manifold and prove the related Lichnerowicz formula. We explain also the Atiyah–Singer index theorem for the modified Dirac operator. In Section 1.4, we show that the operator $\overline{\partial}^E + \overline{\partial}^{E,*}$ is a modified Dirac operator, and we establish the Lichnerowicz and Bochner–Kodaira–Nakano formulas for the Kodaira Laplacian. In Section 1.5, we deal with vanishing theorems for positive line bundles and the spectral gap property for the modified Dirac operator and the Kodaira Laplacian. In Section 1.6, we establish the asymptotic of the heat kernel which is the analytic core result of this chapter. Finally, in Section 1.7, we prove Demailly's holomorphic Morse inequalities.

1.1 Connections on vector bundles

In this section, we review the definition on connections and the associated curvatures. Section 1.1.1 reviews some general facts on connections on vector bundles, and we specify them to the holomorphic case in Section 1.1.2.

1.1.1 Hermitian connection

Let E be a complex vector bundle over a smooth manifold X. Let TX be the tangent bundle and T^*X be the cotangent bundle. Let $\mathscr{C}^\infty(X, E)$ be the space of smooth sections of E on X. Let $\Omega^r(X, E)$ be the spaces of smooth r-forms on X with values in E, and set $\mathscr{C}^\infty(X) := \mathscr{C}^\infty(X, \mathbb{C})$, $\Omega^\bullet(X) := \Omega^\bullet(X, \mathbb{C})$.

Let $d : \Omega^\bullet(X) \to \Omega^{\bullet+1}(X)$ be the exterior differential. It is characterized by

a) $d^2 = 0$;

b) for $\varphi \in \mathscr{C}^\infty(X)$, $d\varphi$ is the one form such that $(d\varphi)(U) = U(\varphi)$ for a vector field U;

c) (Leibniz rule) for any $\alpha \in \Omega^q(X), \beta \in \Omega(X)$, then

$$d(\alpha \wedge \beta) = d\alpha \wedge \beta + (-1)^q \alpha \wedge d\beta. \tag{1.1.1}$$

Then we verify that for any 1-form α, vector fields U, V on X, we have

$$d\alpha(U, V) = U(\alpha(V)) - V(\alpha(U)) - \alpha([U, V]), \tag{1.1.2}$$

here $[U, V]$ is the Lie bracket of U and V.

A linear map $\nabla^E : \mathscr{C}^\infty(X, E) \to \mathscr{C}^\infty(X, T^*X \otimes E)$ is called a *connection* on E if for any $\varphi \in \mathscr{C}^\infty(X)$, $s \in \mathscr{C}^\infty(X, E)$ and $U \in TX$, we have

$$\nabla^E_U(\varphi s) = U(\varphi)\, s + \varphi \nabla^E_U s. \tag{1.1.3}$$

Connections on E always exist. Indeed, let $\{V_k\}_k$ an open covering of X such that $E|_{V_k}$ is trivial. If $\{\eta_{kl}\}_l$ is a local frame of $E|_{V_k}$, any section $s \in \mathscr{C}^\infty(V_k, E)$ has the form $s = \sum_l s_l \eta_{kl}$ with uniquely determined $s_l \in \mathscr{C}^\infty(V_k)$. We define a connection on $E|_{V_k}$ by $\nabla^E_k s := \sum_l ds_l \otimes \eta_{kl}$. Consider now a partition of unity $\{\psi_k\}_k$ subordinated to $\{V_k\}_k$. Then $\nabla^E s := \sum_k \nabla^E_k(\psi_k s)$, $s \in \mathscr{C}^\infty(X, E)$, defines a connection on E.

If ∇^E_1 is another connection on E, then by (1.1.3), $\nabla^E_1 - \nabla^E \in \Omega^1(X, \operatorname{End}(E))$.

If ∇^E is a connection on E, then there exists a unique extension $\nabla^E : \Omega^\bullet(X, E) \to \Omega^{\bullet+1}(X, E)$ verifying the Leibniz rule: for $\alpha \in \Omega^q(X)$, $s \in \Omega^r(X, E)$, we have

$$\nabla^E(\alpha \wedge s) = d\alpha \wedge s + (-1)^q \alpha \wedge \nabla^E s. \tag{1.1.4}$$

From (1.1.2), for $s \in \mathscr{C}^\infty(X, E)$ and vector fields U, V on X, we have

$$(\nabla^E)^2(U, V)s = \nabla^E_U \nabla^E_V s - \nabla^E_V \nabla^E_U s - \nabla^E_{[U,V]} s. \tag{1.1.5}$$

Then $(\nabla^E)^2(U, V)(\varphi s) = (\nabla^E)^2(U, \varphi V)s = (\nabla^E)^2(\varphi U, V)s = \varphi(\nabla^E)^2(U, V)s$ for any $\varphi \in \mathscr{C}^\infty(X)$. We deduce that:

Definition and Theorem 1.1.1. The operator $(\nabla^E)^2$ defines a bundle morphism $(\nabla^E)^2 : E \to \Lambda^2(T^*X) \otimes E$, called the *curvature operator*. Therefore, there exists $R^E \in \Omega^2(X, \operatorname{End}(E))$, called the *curvature* of ∇^E, such that $(\nabla^E)^2$ is given by multiplication with R^E, i.e., $(\nabla^E)^2 s = R^E s \in \Omega^2(X, E)$ for $s \in \mathscr{C}^\infty(X, E)$.

Let h^E be a *Hermitian metric* on E, i.e., a smooth family $\{h^E_x\}_{x \in X}$ of sesquilinear maps $h^E_x : E_x \times E_x \to \mathbb{C}$ such that $h^E_x(\xi, \xi) > 0$ for any $\xi \in E_x \setminus \{0\}$. We call (E, h^E) a Hermitian vector bundle on X. There always exist Hermitian metrics on E by using the partition of unity argument as above.

Definition 1.1.2. A connection ∇^E is said to be a *Hermitian connection* on (E, h^E) if for any $s_1, s_2 \in \mathscr{C}^\infty(X, E)$,

$$d\langle s_1, s_2 \rangle_{h^E} = \langle \nabla^E s_1, s_2 \rangle_{h^E} + \langle s_1, \nabla^E s_2 \rangle_{h^E}. \tag{1.1.6}$$

There always exist Hermitian connections. In fact, let ∇_0^E be a connection on E, then $\langle \nabla_1^E s_1, s_2 \rangle_{h^E} = d\langle s_1, s_2 \rangle_{h^E} - \langle s_1, \nabla_0^E s_2 \rangle_{h^E}$ defines a connection ∇_1^E on E. Now $\nabla^E = \frac{1}{2}(\nabla_0^E + \nabla_1^E)$ is a Hermitian connection on (E, h^E).

Let $\{\xi_l\}_{l=1}^m$ be a local frame of E. Denote by $h = (h_{lk} = \langle \xi_k, \xi_l \rangle_{h^E})$ the matrix of h^E with respect to $\{\xi_l\}_{l=1}^m$. The *connection form* $\theta = (\theta_k^l)$ of ∇^E with respect to $\{\xi_l\}_{l=1}^m$ is defined by, with local 1-forms θ_k^l,

$$\nabla^E \xi_k = \theta_k^l \xi_l. \tag{1.1.7}$$

Remark 1.1.3. If E is a real vector bundle on X, certainly, everything still holds, especially, a connection ∇^E is said to be an Euclidean connection on (E, h^E) if it preserves the Euclidean metric h^E.

1.1.2 Chern connection

Let E be a holomorphic vector bundle over a complex manifold X. Let h^E be a Hermitian metric on E. We call (E, h^E) a holomorphic Hermitian vector bundle.

The almost complex structure J induces a splitting $TX \otimes_{\mathbb{R}} \mathbb{C} = T^{(1,0)}X \oplus T^{(0,1)}X$, where $T^{(1,0)}X$ and $T^{(0,1)}X$ are the eigenbundles of J corresponding to the eigenvalues $\sqrt{-1}$ and $-\sqrt{-1}$, respectively. Let $T^{*(1,0)}X$ and $T^{*(0,1)}X$ be the corresponding dual bundles. Let

$$\Omega^{r,q}(X, E) := \mathscr{C}^\infty(X, \Lambda^r(T^{*(1,0)}X) \otimes \Lambda^q(T^{*(0,1)}X) \otimes E)$$

be the spaces of smooth (r, q)-forms on X with values in E.

The operator $\overline{\partial}^E : \mathscr{C}^\infty(X, E) \to \Omega^{0,1}(X, E)$ is well defined. Any section $s \in \mathscr{C}^\infty(X, E)$ has the local form $s = \sum_l \varphi_l \xi_l$ where $\{\xi_l\}_{l=1}^m$ is a local holomorphic frame of E and φ_l are smooth functions. We set $\overline{\partial}^E s = \sum_l (\overline{\partial} \varphi_l) \xi_l$, here $\overline{\partial} \varphi_l = \sum_j d\overline{z}_j \frac{\partial}{\partial \overline{z}_j} \varphi_l$ in holomorphic coordinates $(z_1, \cdots, z_n)$.

Definition 1.1.4. A connection ∇^E on E is said to be a *holomorphic connection* if $\nabla_U^E s = i_U(\overline{\partial}^E s)$ for any $U \in T^{(0,1)}X$ and $s \in \mathscr{C}^\infty(X, E)$.

Theorem 1.1.5. *There exists a unique holomorphic Hermitian connection ∇^E on (E, h^E), called the* Chern connection*. With respect to a local holomorphic frame, the connection matrix is given by* $\theta = h^{-1} \cdot \partial h$.

Proof. By Definition 1.1.4, we have to define ∇_U^E just for $U \in T^{(1,0)}X$. Relation (1.1.6) implies for $U \in T^{(1,0)}X$, $s_1, s_2 \in \mathscr{C}^\infty(X, E)$,

$$U\langle s_1, s_2 \rangle_{h^E} = \langle \nabla_U^E s_1, s_2 \rangle_{h^E} + \langle s_1, \nabla_{\overline{U}}^E s_2 \rangle_{h^E}. \tag{1.1.8}$$

Since $\nabla^E_{\overline{U}} s_2 = i_{\overline{U}}(\overline{\partial}^E s_2)$, the above equation defines $\nabla^E_{\overline{U}}$ uniquely. Moreover, if $\{\xi_l\}_{l=1}^m$ is a local holomorphic frame, from (1.1.6) we deduce that $\theta = h^{-1} \cdot \partial h$. $\square$

Since E is holomorphic, similar to (1.1.4), the operator $\overline{\partial}^E$ extends naturally to $\overline{\partial}^E : \Omega^{\bullet,\bullet}(X, E) \longrightarrow \Omega^{\bullet,\bullet+1}(X, E)$ and $(\overline{\partial}^E)^2 = 0$.

Let ∇^E be the holomorphic Hermitian connection on (E, h^E). Then we have a decomposition of ∇^E after bidegree

$$
\begin{aligned}
\nabla^E &= (\nabla^E)^{1,0} + (\nabla^E)^{0,1}, \quad (\nabla^E)^{0,1} = \overline{\partial}^E, \\
(\nabla^E)^{1,0} &: \Omega^{\bullet,\bullet}(X, E) \longrightarrow \Omega^{\bullet+1,\bullet}(X, E).
\end{aligned}
\tag{1.1.9}
$$

By (1.1.8), (1.1.9) and $(\overline{\partial}^E)^2 = 0$ we have

$$
(\overline{\partial}^E)^2 = \left((\nabla^E)^{1,0}\right)^2 = 0, \quad (\nabla^E)^2 = \overline{\partial}^E(\nabla^E)^{1,0} + (\nabla^E)^{1,0}\overline{\partial}^E.
\tag{1.1.10}
$$

Thus the curvature $R^E \in \Omega^{1,1}(X, \mathrm{End}(E))$. If $\mathrm{rk}(E) = 1$, $\mathrm{End}(E)$ is trivial and R^E is canonically identified to a $(1,1)$-form on X, such that $\sqrt{-1}R^E$ is real.

In general, let us introduce an auxiliary Riemannian g^{TX} metric on X, compatible with the complex structure J (i.e., $g^{TX}(\cdot, \cdot) = g^{TX}(J \cdot, J \cdot)$). Then R^E induces a Hermitian matrix $\dot{R}^E \in \mathrm{End}(T^{(1,0)}X \otimes E)$ such that for $u, v \in T_x^{(1,0)}X$, $\xi, \eta \in E_x$, and $x \in X$,

$$
\langle R^E(u, \overline{v})\xi, \eta \rangle_{h^E} = \langle \dot{R}^E(u \otimes \xi), v \otimes \eta \rangle.
\tag{1.1.11}
$$

Definition 1.1.6. We say that (E, h^E) is *Nakano positive* (*resp. semi-positive*) if $\dot{R}^E \in \mathrm{End}(T^{(1,0)}X \otimes E)$ is positive definite (resp. semi-definite), and *Griffiths positive* (*resp. semi-positive*) if $\langle R^E(v, \overline{v})\xi, \xi \rangle_{h^E} = \langle \dot{R}^E(v \otimes \xi), v \otimes \xi \rangle > 0$ (resp. $\geqslant 0$) for all non-zero $v \in T_x^{(1,0)}X$ and all non-zero $\xi \in E_x$. Certainly, these definitions do not depend on the choice of g^{TX}.

1.2 Connections on the tangent bundle

On the tangent bundle of a complex manifold, we can define several connections: the Levi–Civita connection, the holomorphic Hermitian (i.e., Chern) connection and Bismut connection. In this section, we explain the relation between them. We shall see that these three connections coincide, if X is a Kähler manifold.

We start by recalling in Section 1.2.1 some facts about the Levi–Civita connection. In Section 1.2.2, we study in detail the holomorphic Hermitian connection on the tangent bundle. In Section 1.2.3, we define the Bismut connection.

Let (X, J) be a complex manifold with complex structure J and $\dim_{\mathbb{C}} X = n$. Let $T_h X$ be the holomorphic tangent bundle on X, and let TX be the corresponding real tangent bundle. Let g^{TX} be any Riemannian metric on TX compatible with J, i.e., $g^{TX}(Ju, Jv) = g^{TX}(u, v)$ for any $u, v \in T_x X$, $x \in X$. We will shortly express this relation by $g^{TX}(J \cdot, J \cdot) = g^{TX}(\cdot, \cdot)$.

1.2.1 Levi–Civita connection

The results of this section apply for any Riemannian manifold (X, g^{TX}). We denote by $\langle \cdot, \cdot \rangle$ the $\mathbb{C}$-bilinear form on $TX \otimes_{\mathbb{R}} \mathbb{C}$ induced by the metric g^{TX}. Let ∇^{TX} be the Levi–Civita connection on (TX, g^{TX}). By the explicit equation for $\langle \nabla^{TX} \cdot, \cdot \rangle$, for any U, V, W, Y vector fields on X,

$$
\begin{aligned}
2 \langle \nabla_U^{TX} V, W \rangle = U \langle V, W \rangle + V \langle U, W \rangle - W \langle U, V \rangle \\
- \langle U, [V, W] \rangle - \langle V, [U, W] \rangle + \langle W, [U, V] \rangle .
\end{aligned} \tag{1.2.1}
$$

∇^{TX} is the unique connection on TX which preserves the metric (satisfies (1.1.6)) and is torsion free, i.e.,

$$
\nabla_U^{TX} V - \nabla_V^{TX} U = [U, V]. \tag{1.2.2}
$$

The curvature $R^{TX} \in \Lambda^2(T^* X) \otimes \operatorname{End}(TX)$ of ∇^{TX} is defined by

$$
R^{TX}(U, V) = \nabla_U^{TX} \nabla_V^{TX} - \nabla_V^{TX} \nabla_U^{TX} - \nabla_{[U,V]}^{TX}. \tag{1.2.3}
$$

Then we have the following well-known facts

$$
\begin{aligned}
R^{TX}(U, V)W + R^{TX}(V, W)U + R^{TX}(W, U)V = 0, \\
\langle R^{TX}(U, V)W, Y \rangle = \langle R^{TX}(W, Y)U, V \rangle.
\end{aligned} \tag{1.2.4}
$$

Let $\{e_i\}_{i=1}^{2n}$ be an orthonormal frame of TX and $\{e^i\}_{i=1}^{2n}$ its dual basis in $T^* X$. The *Ricci curvature* Ric and *scalar curvature* r^X of (TX, g^{TX}) are defined by

$$
\operatorname{Ric} = - \sum_j \langle R^{TX}(\cdot, e_j) \cdot, e_j \rangle, \quad r^X = - \sum_{ij} \langle R^{TX}(e_i, e_j)e_i, e_j \rangle. \tag{1.2.5}
$$

The Riemannian volume form dv_X of (TX, g^{TX}) has the form $dv_X = e^1 \wedge \cdots \wedge e^{2n}$ if the orthonormal frame $\{e_i\}$ is oriented.

If α is a 1-form on X, the function $\operatorname{Tr}(\nabla \alpha)$ is given by the formula

$$
\operatorname{Tr}(\nabla \alpha) = \sum_i e_i(\alpha(e_i)) - \alpha(\nabla_{e_i}^{TX} e_i). \tag{1.2.6}
$$

The following formula is quite useful.

Proposition 1.2.1. *For any $\mathscr{C}^1$ 1-form α with compact support, we have*

$$
\int_X \operatorname{Tr}(\nabla \alpha) dv_X = 0. \tag{1.2.7}
$$

Proof. Let W be the vector field on X corresponding to α under the Riemannian metric g^{TX}, so that $\langle W, Y \rangle = (\alpha, Y)$ for any $Y \in TX$.

We denote by L_W the Lie derivative of the vector field W. Recall that for any vector field Y on X,

$$
L_W Y = [W, Y] = \nabla_W^{TX} Y - \nabla_Y^{TX} W. \tag{1.2.8}
$$

Thus by (1.2.8) and $\langle \nabla^{TX}_W e_j, e_j \rangle = 0$, we get

$$
\begin{aligned}
L_W dv_X = \langle L_W e^j, e_j \rangle dv_X &= -\langle e_j, L_W e_j \rangle dv_X \\
&= \left\langle \nabla^{TX}_{e_j} W, e_j \right\rangle dv_X = \left(e_j \langle W, e_j \rangle - \left\langle W, \nabla^{TX}_{e_j} e_j \right\rangle \right) dv_X \\
&= \mathrm{Tr}(\nabla \alpha) dv_X. \quad (1.2.9)
\end{aligned}
$$

We will denote by $\wedge$ and i the exterior and interior product respectively. E. Cartan's homotopy formula tells us that on the bundle of exterior differentials $\Lambda(T^*X)$,

$$
L_W = d \cdot i_W + i_W \cdot d. \quad (1.2.10)
$$

From (1.2.9) and (1.2.10), we get

$$
0 = \int_X L_W dv_X = \int_X \mathrm{Tr}(\nabla \alpha) dv_X. \quad (1.2.11)
$$

The proof of Proposition 1.2.1 is complete. $\qquad \square$

For $x_0 \in X$, $W \in T_{x_0} X$, let $\mathbb{R} \ni u \to x_u = \exp^X_{x_0}(uW)$ be the geodesic in X such that $x_u|_{u=0} = x_0$, $\frac{dx_u}{du}|_{u=0} = W$. For $\varepsilon > 0$, we denote by $B^X(x_0, \varepsilon)$ and $B^{T_{x_0}X}(0, \varepsilon)$ the open balls in X and $T_{x_0}X$ with center x_0 and radius ε, respectively. Then the map $T_{x_0}X \ni Z \to \exp^X_{x_0}(Z) \in X$ is a diffeomorphism from $B^{T_{x_0}X}(0, \varepsilon)$ onto $B^X(x_0, \varepsilon)$ for ε small enough; by identifying $Z = \sum Z_i e_i \in T_{x_0}X$ with $(Z_1, \ldots, Z_{2n}) \in \mathbb{R}^{2n}$, it yields a local chart for X around x_0, called normal coordinate system at x_0. We will identify $B^{T_{x_0}X}(0, \varepsilon)$ with $B^X(x_0, \varepsilon)$ by this map.

Let $\{e_i\}_i$ be an oriented orthonormal basis of $T_{x_0}X$. We also denote by $\{e^i\}_i$ the dual basis of $\{e_i\}$. Let $\tilde{e}_i(Z)$ be the parallel transport of e_i with respect to ∇^{TX} along the curve $[0, 1] \ni u \to uZ$. Then $e_j = \frac{\partial}{\partial Z_j}$.

The *radial vector field* $\mathcal{R}$ is the vector field defined by $\mathcal{R} = \sum_i Z_i e_i$ with $(Z_1, \ldots, Z_{2n})$ the coordinate functions.

Proposition 1.2.2. *The following identities hold:*

$$
\mathcal{R} = \sum_j Z_j e_j = \sum_j Z_j \tilde{e}_j(Z),
$$
$$
\langle \mathcal{R}, e_j \rangle = Z_j. \quad (1.2.12)
$$

Proof. Note that $x_u : [0, 1] \ni u \to uZ$ is a geodesic, and $\mathcal{R}(x_u) = u\frac{dx_u}{du}$, thus by the geodesic equation $\nabla^{TX}_{\frac{dx_u}{du}} \frac{dx_u}{du} = 0$, we get

$$
\nabla^{TX}_{\mathcal{R}} \mathcal{R} = u \nabla^{TX}_{\frac{dx_u}{du}} (u\frac{dx_u}{du}) = u\frac{dx_u}{du} = \mathcal{R}. \quad (1.2.13)
$$

Thus we have

$$\mathcal{R}\langle \mathcal{R}, \widetilde{e}_j \rangle = \langle \nabla_{\mathcal{R}}^{TX} \mathcal{R}, \widetilde{e}_j \rangle + \langle \mathcal{R}, \nabla_{\mathcal{R}}^{TX} \widetilde{e}_j \rangle = \langle \mathcal{R}, \widetilde{e}_j \rangle. \qquad (1.2.14)$$

This means that $\langle \mathcal{R}, \widetilde{e}_j \rangle$ is homogeneous of order 1. But

$$\langle \mathcal{R}, \widetilde{e}_j \rangle = \sum_k Z_k \langle e_k, \widetilde{e}_j \rangle = Z_j + \mathscr{O}(|Z|^2). \qquad (1.2.15)$$

Thus from (1.2.14) and (1.2.15), we infer the first equation of (1.2.12).

Since the Levi–Civita connection ∇^{TX} is torsion free and $[\mathcal{R}, e_i] = -e_i$, we have

$$\langle \mathcal{R}, \nabla_{\mathcal{R}}^{TX} e_i \rangle = \langle \mathcal{R}, \nabla_{e_i}^{TX} \mathcal{R} \rangle + \langle \mathcal{R}, [\mathcal{R}, e_i] \rangle = \frac{1}{2} e_i \langle \mathcal{R}, \mathcal{R} \rangle - \langle \mathcal{R}, e_i \rangle. \qquad (1.2.16)$$

From (1.2.13) and (1.2.16), we obtain

$$\mathcal{R}\langle \mathcal{R}, e_i \rangle = \langle \nabla_{\mathcal{R}}^{TX} \mathcal{R}, e_i \rangle + \langle \mathcal{R}, \nabla_{\mathcal{R}}^{TX} e_i \rangle = \frac{1}{2} e_i \langle \mathcal{R}, \mathcal{R} \rangle = Z_i. \qquad (1.2.17)$$

But $\langle \mathcal{R}, e_i \rangle = \sum_j Z_j \langle e_j, e_i \rangle = Z_i + \mathscr{O}(|Z|^2)$. Thus we get the second equation of (1.2.12). $\qquad \square$

For $\alpha = (\alpha_1, \ldots, \alpha_{2n}) \in \mathbb{N}^{2n}$, set $Z^{\alpha} = Z_1^{\alpha_1} \ldots Z_{2n}^{\alpha_{2n}}$.

Lemma 1.2.3. *If $\widetilde{e}_i(Z)$ is written in the basis $\{e_i\}$, its Taylor expansion up to order r is determined by the Taylor expansion up to order $r - 2$ of $R_{mqkl} = \langle R^{TX}(e_q, e_m) e_k, e_l \rangle_Z$. Moreover we have*

$$\widetilde{e}_i(Z) = e_i - \frac{1}{6} \sum_j \langle R_{x_0}^{TX}(\mathcal{R}, e_i) \mathcal{R}, e_j \rangle_{x_0} e_j + \sum_{|\alpha| \geqslant 3} \left(\frac{\partial^{\alpha}}{\partial Z^{\alpha}} \widetilde{e}_i \right)(0) \frac{Z^{\alpha}}{\alpha!}. \qquad (1.2.18)$$

Thus the Taylor expansion up to order r of $g_{ij}(Z) = g^{TX}(e_i, e_j)(Z) = \langle e_i, e_j \rangle_Z$ is a polynomial of the Taylor expansion up to order $r - 2$ of R_{mqkl}; moreover

$$g_{ij}(Z) = \delta_{ij} + \frac{1}{3} \langle R_{x_0}^{TX}(\mathcal{R}, e_i) \mathcal{R}, e_j \rangle_{x_0} + \mathscr{O}(|Z|^3). \qquad (1.2.19)$$

Proof. Let Γ^{TX} be the connection form of ∇^{TX} with respect to the frame $\{\widetilde{e}_i\}$ of TX. Then $\nabla^{TX} = d + \Gamma^{TX}$. Let $\partial_i = \nabla_{e_i}$ be the partial derivatives along e_i. By the definition of our fixed frame, we have $i_{\mathcal{R}} \Gamma^{TX} = 0$. Thus

$$L_{\mathcal{R}} \Gamma^{TX} = [i_{\mathcal{R}}, d] \Gamma^{TX} = i_{\mathcal{R}}(d\Gamma^{TX} + \Gamma^{TX} \wedge \Gamma^{TX}) = i_{\mathcal{R}} R^{TX}. \qquad (1.2.20)$$

Let $\widetilde{\theta}(Z) = (\theta_j^i(Z))_{i,j=1}^{2n}$ be the $2n \times 2n$-matrix such that

$$e_i = \sum_j \theta_i^j(Z) \widetilde{e}_j(Z), \quad \widetilde{e}_j(Z) = (\widetilde{\theta}(Z)^{-1})_j^k e_k. \qquad (1.2.21)$$

Set $\theta^j(Z) = \sum_i \theta_i^j(Z)e^i$ and

$$\theta = \sum_j e^j \otimes e_j = \sum_j \theta^j \widetilde{e}_j \in T^*X \otimes TX. \tag{1.2.22}$$

As ∇^{TX} is torsion free, $\nabla^{TX}\theta = 0$, thus the $\mathbb{R}^{2n}$-valued 1-form $\theta = (\theta^j(Z))$ satisfies the structure equation,

$$d\theta + \Gamma^{TX} \wedge \theta = 0. \tag{1.2.23}$$

Observe first that under our trivialization by $\{\widetilde{e}_i\}$, by (1.2.12), for the $\mathbb{R}^{2n}$-valued function $i_{\mathcal{R}}\theta$,

$$i_{\mathcal{R}}\theta = \sum_j Z_j e_j = (Z_1, \ldots, Z_{2n}) =: Z. \tag{1.2.24}$$

Substituting (1.2.12), (1.2.24) and $(L_{\mathcal{R}} - 1)Z = 0$, into the identity $i_{\mathcal{R}}(d\theta + \Gamma^{TX} \wedge \theta) = 0$, from (1.2.20), we obtain

$$(L_{\mathcal{R}} - 1)L_{\mathcal{R}}\theta = (L_{\mathcal{R}} - 1)(dZ + \Gamma^{TX}Z) = (L_{\mathcal{R}}\Gamma^{TX})Z = (i_{\mathcal{R}}R^{TX})Z. \tag{1.2.25}$$

Where we consider the curvature R^{TX} as a matrix of two-forms and θ is a $\mathbb{R}^{2n}$-valued one-form. The ith component of $R^{TX}Z$, θ is $\langle R^{TX}\mathcal{R}, \widetilde{e}_i \rangle$, θ^i, from (1.2.25), we get

$$i_{e_j}(L_{\mathcal{R}} - 1)L_{\mathcal{R}}\theta^i(Z) = \langle R^{TX}(\mathcal{R}, e_j)\mathcal{R}, \widetilde{e}_i \rangle(Z). \tag{1.2.26}$$

By (1.2.12), $L_{\mathcal{R}}e^j = e^j$. Thus from the Taylor expansion of $\theta_j^i(Z)$, we get

$$\sum_{|\alpha| \geqslant 1} (|\alpha|^2 + |\alpha|)(\partial^\alpha \theta_j^i)(0)\frac{Z^\alpha}{\alpha!} = \langle R^{TX}(\mathcal{R}, e_j)\mathcal{R}, \widetilde{e}_i \rangle(Z). \tag{1.2.27}$$

Now by (1.2.21) and $\theta_j^i(x_0) = \delta_{ij}$, (1.2.27) determines the Taylor expansion of $\theta_j^i(Z)$ up to order m in terms of the Taylor expansion of the coefficients of R^{TX} up to order $m - 2$. And

$$(\widetilde{\theta}^{-1})_j^i = \delta_{ij} - \frac{1}{6}\langle R_{x_0}^{TX}(\mathcal{R}, e_i)\mathcal{R}, e_j \rangle_{x_0} + \mathscr{O}(|Z|^3). \tag{1.2.28}$$

By (1.2.21), (1.2.27), we infer (1.2.18).
    From (1.2.21),

$$g_{ij}(Z) = \theta_i^k(Z)\theta_j^k(Z). \tag{1.2.29}$$

Thus the rest of Lemma 1.2.3 follows from (1.2.28) and (1.2.29). The proof of Lemma 1.2.3 is complete. $\square$

Let E be a complex vector bundle on X, and let ∇^E be a connection on E with curvature $R^E := (\nabla^E)^2$. Let $(\mathcal{U}, Z_1, \ldots, Z_{2n})$ be a local chart of X such that $0 \in \mathcal{U}$ represents $x_0 \in X$. Set $\mathcal{R} = \sum_i Z_i \frac{\partial}{\partial Z_i}$. Now we identify E_Z to E_{x_0} by parallel transport with respect to the connection ∇^E along the curve $[0,1] \ni u \to uZ$; this gives a trivialization of E near 0. We denote by Γ^E the connection form with respect to this trivialization of E near 0. Then in the frame $e_j = \frac{\partial}{\partial Z_j}$, Γ^E becomes a function with values in $\mathbb{R}^{2n} \otimes \mathrm{End}(\mathbb{C}^{\mathrm{rk}(E)})$ and $\nabla^E = d + \Gamma^E$.

Lemma 1.2.4. *The Taylor coefficients of $\Gamma^E(e_j)(Z)$ at x_0 up to order r are determined by Taylor coefficients of R^E up to order $r-1$. More precisely,*

$$\sum_{|\alpha|=r} (\partial^\alpha \Gamma^E)_{x_0}(e_j) \frac{Z^\alpha}{\alpha!} = \frac{1}{r+1} \sum_{|\alpha|=r-1} (\partial^\alpha R^E)_{x_0}(\mathcal{R}, e_j) \frac{Z^\alpha}{\alpha!}. \tag{1.2.30}$$

Especially,

$$\Gamma_Z^E(e_j) = \frac{1}{2} R_{x_0}^E(\mathcal{R}, e_j) + \mathcal{O}(|Z|^2). \tag{1.2.31}$$

Proof. By the definition of our fixed frame, we have $R^E = d\Gamma^E + \Gamma^E \wedge \Gamma^E$ and

$$i_{\mathcal{R}} \Gamma^E = 0, \qquad L_{\mathcal{R}} \Gamma^E = [i_{\mathcal{R}}, d]\Gamma^E = i_{\mathcal{R}}(d\Gamma^E + \Gamma^E \wedge \Gamma^E) = i_{\mathcal{R}} R^E. \tag{1.2.32}$$

Using $L_{\mathcal{R}} dZ^j = dZ^j$ and expanding both sides of the second equation of (1.2.32) in Taylor's series of at $Z = 0$, we obtain

$$\sum_\alpha (|\alpha| + 1)(\partial^\alpha \Gamma^E)_{x_0}(e_j) \frac{Z^\alpha}{\alpha!} = \sum_\alpha (\partial^\alpha R^E)_{x_0}(\mathcal{R}, e_j) \frac{Z^\alpha}{\alpha!}. \tag{1.2.33}$$

By equating coefficients of Z^α of both sides, we get Lemma 1.2.4. $\qquad\square$

1.2.2 Chern connection

Recall that $T^{(1,0)}X$ is a holomorphic vector bundle with Hermitian metric $h^{T^{(1,0)}X}$ induced by g^{TX}. The map $T_h X \ni Y \to \frac{1}{2}(Y - \sqrt{-1}JY) \in T^{(1,0)}X$ induces the natural identification of $T_h X$ and $T^{(1,0)}X$.

We will denote by $\langle \cdot, \cdot \rangle$ the $\mathbb{C}$-bilinear form on $TX \otimes_{\mathbb{R}} \mathbb{C}$ induced by g^{TX}. Note that $\langle \cdot, \cdot \rangle$ vanishes on $T^{(1,0)}X \times T^{(1,0)}X$ and on $T^{(0,1)}X \times T^{(0,1)}X$.

For $U \in TX \otimes_{\mathbb{R}} \mathbb{C}$, we will denote by $U^{(1,0)}, U^{(0,1)}$ its components in $T^{(1,0)}X$ and $T^{(0,1)}X$. Let $\{w_j\}_{j=1}^n$ be a local orthonormal frame of $T^{(1,0)}X$ with dual frame $\{w^j\}_{j=1}^n$. Then

$$e_{2j-1} = \tfrac{1}{\sqrt{2}}(w_j + \overline{w}_j) \quad \text{and} \quad e_{2j} = \tfrac{\sqrt{-1}}{\sqrt{2}}(w_j - \overline{w}_j), \quad j = 1, \ldots, n, \tag{1.2.34}$$

form an orthonormal frame of TX. We fix this notation throughout the book and use it without further notice.

Let $\nabla^{T^{(1,0)}X}$ be the holomorphic Hermitian connection on $(T^{(1,0)}X, h^{T^{(1,0)}X})$ with curvature $R^{T^{(1,0)}X}$. For $v \in \mathscr{C}^\infty(X, T^{(0,1)}X)$, we define

$$\nabla^{T^{(0,1)}X} v := \overline{\nabla^{T^{(1,0)}X} \overline{v}}.$$

Then $\nabla^{T^{(0,1)}X}$ defines a connection on $T^{(0,1)}X$. Set

$$\widetilde{\nabla}^{TX} = \nabla^{T^{(1,0)}X} \oplus \nabla^{T^{(0,1)}X}. \tag{1.2.35}$$

Then $\widetilde{\nabla}^{TX}$ is a connection on $TX \otimes_{\mathbb{R}} \mathbb{C}$ and it preserves TX; we still denote by $\widetilde{\nabla}^{TX}$ the induced connection on TX. Then $\widetilde{\nabla}^{TX}$ preserves the metric g^{TX}.

Let T be the torsion of the connection $\widetilde{\nabla}^{TX}$. Then $T \in \Lambda^2(T^*X) \otimes TX$ is defined by

$$T(U, V) = \widetilde{\nabla}_U^{TX} V - \widetilde{\nabla}_V^{TX} U - [U, V], \tag{1.2.36}$$

for vector fields U and V on X. Hence

T maps $T^{(1,0)}X \otimes T^{(1,0)}X$ (resp. $T^{(0,1)}X \otimes T^{(0,1)}X$) into $T^{(1,0)}X$
(resp. $T^{(0,1)}X$) and vanishes on $T^{(1,0)}X \otimes T^{(0,1)}X$. $\tag{1.2.37}$

Set

$$S = \widetilde{\nabla}^{TX} - \nabla^{TX}, \quad \mathcal{S} = \sum_i S(e_i)e_i. \tag{1.2.38}$$

Then S is a real 1-form on X taking values in the skew-adjoint endomorphisms of TX. Since ∇^{TX} is torsion free, we have for $U, V \in TX$,

$$T(U, V) = S(U)V - S(V)U. \tag{1.2.39}$$

Moreover, from (1.2.1), (1.2.36), (1.2.38) and since $\widetilde{\nabla}^{TX}$ preserves g^{TX} we obtain directly

$$2\langle S(U)V, W \rangle - \langle T(U, V), W \rangle - \langle T(W, U), V \rangle + \langle T(V, W), U \rangle = 0. \tag{1.2.40}$$

By (1.2.37), (1.2.39) and (1.2.40), we get

$$\langle S(w_i)w_k, w_j \rangle = 0,$$
$$2\langle S(w_i)\overline{w}_k, w_j \rangle = 2\langle S(\overline{w}_k)w_i, w_j \rangle = -\langle T(w_i, w_j), \overline{w}_k \rangle. \tag{1.2.41}$$

Since $T(w_i, \overline{w}_j) = 0$, $S(\overline{w}_j)w_i = S(w_i)\overline{w}_j$, and so

$$\mathcal{S} = 2S(w_j)\overline{w}_j = \langle T(w_i, w_j), \overline{w}_j \rangle \overline{w}_i + \langle T(\overline{w}_i, \overline{w}_j), w_j \rangle w_i$$
$$= \langle T(e_i, e_j), e_j \rangle e_i, \tag{1.2.42}$$
$$2\langle S(\cdot)w_j, \overline{w}_j \rangle = \langle T(w_i, w_j), \overline{w}_j \rangle w^i - \langle T(\overline{w}_i, \overline{w}_j), w_j \rangle \overline{w}^i.$$

The connection $\widetilde{\nabla}^{TX}$ on TX induces naturally a covariant derivative on the exterior bundle $\Lambda(T^*X)$ and we still denote it by $\widetilde{\nabla}^{TX}$. For any differential forms α, β and vector field Y, it satisfies

$$\widetilde{\nabla}_Y^{TX}(\alpha \wedge \beta) = (\widetilde{\nabla}_Y^{TX}\alpha) \wedge \beta + \alpha \wedge \widetilde{\nabla}_Y^{TX}\beta. \tag{1.2.43}$$

For a 1-form α and vector fields U, V, we have $(\widetilde{\nabla}_U^{TX}\alpha, V) = U(\alpha, V) - (\alpha, \widetilde{\nabla}_U^{TX}V)$. Likewise, ∇^{TX} induces naturally a connection ∇^{TX} on $\Lambda(T^*X)$. We denote by ε the exterior product $T^*X \otimes \Lambda^\bullet(T^*X) \to \Lambda^{\bullet+1}(T^*X)$.

Lemma 1.2.5. *For the exterior differentiation operator d acting on smooth sections of $\Lambda(T^*X)$, we have*

$$d = \varepsilon \circ \widetilde{\nabla}^{TX} + i_T, \qquad d = \varepsilon \circ \nabla^{TX}. \tag{1.2.44}$$

Proof. We write $\mathbf{d} := \varepsilon \circ \widetilde{\nabla}^{TX} + i_T$. Then by using (1.2.43), we know that for any homogeneous differential forms α, β, we have

$$\mathbf{d}(\alpha \wedge \beta) = (\mathbf{d}\alpha) \wedge \beta + (-1)^{\deg \alpha}\alpha \wedge \mathbf{d}\beta. \tag{1.2.45}$$

From Leibniz's rule (1.2.45), it suffices to show that $\mathbf{d}$ agrees with d on functions (which is clear) and 1-forms. Now, for any smooth function f on X, we have

$$\begin{aligned}
\varepsilon \circ \widetilde{\nabla}^{TX} df &= e^i \wedge e^j \langle \widetilde{\nabla}_{e_i}^{TX} df, e_j \rangle = e^i \wedge e^j \left(e_i(e_j(f)) - \langle df, \widetilde{\nabla}_{e_i}^{TX} e_j \rangle \right) \\
&= \frac{1}{2} e^i \wedge e^j \left(e_i(e_j(f)) - \langle df, \widetilde{\nabla}_{e_i}^{TX} e_j \rangle - \left(e_j(e_i(f)) - \langle df, \widetilde{\nabla}_{e_j}^{TX} e_i \rangle \right) \right) \quad (1.2.46) \\
&= -\frac{1}{2} e^i \wedge e^j \langle df, T(e_i, e_j) \rangle = -i_T df.
\end{aligned}$$

Thus $\mathbf{d}$ coincides also d on 1-forms. Thus we get the first equation of (1.2.44). As ∇^{TX} is torsion free, from the above argument, we obtain the second equation of (1.2.44). $\qquad\square$

If $B \in \Lambda^2(T^*X) \otimes TX$ we will denote by B_{as} the anti-symmetrization of the tensor $V, W, Y \to \langle B(V, W), Y \rangle$. Then

$$B_{as}(V, W, Y) = \langle B(V, W), Y \rangle + \langle B(W, Y), V \rangle + \langle B(Y, V), W \rangle. \tag{1.2.47}$$

Especially from (1.2.37), we infer

$$\begin{aligned}
T_{as} &= \frac{1}{2} \langle T(e_i, e_j), e_k \rangle e^i \wedge e^j \wedge e^k \\
&= \frac{1}{2} \langle T(w_i, w_j), \overline{w}_k \rangle w^i \wedge w^j \wedge \overline{w}^k + \frac{1}{2} \langle T(\overline{w}_i, \overline{w}_j), w_k \rangle \overline{w}^i \wedge \overline{w}^j \wedge w^k \quad (1.2.48) \\
&=: T_{as}^{(1,0)} + T_{as}^{(0,1)}.
\end{aligned}$$

Here $T_{as}^{(1,0)}$, $T_{as}^{(0,1)}$ are the anti-symmetrizations of the components $T^{(1,0)}$, $T^{(0,1)}$ of T in $T^{(1,0)}X$ and $T^{(0,1)}X$.

Let Θ be the real $(1, 1)$-form defined by

$$\Theta(X, Y) = g^{TX}(JX, Y). \tag{1.2.49}$$

Note that the exterior differentiation operator d acting on smooth sections of $\Lambda(T^*X)$ has the decomposition

$$d = \partial + \overline{\partial}. \tag{1.2.50}$$

Proposition 1.2.6. *We have the identity of 3-forms on X,*

$$T_{as} = -\sqrt{-1}(\partial - \overline{\partial})\Theta. \tag{1.2.51}$$

Proof. By (1.2.34), we know that $\Theta = \sqrt{-1}\sum_i w^i \wedge \overline{w}^i$. Thus

$$
\begin{aligned}
\widetilde{\nabla}^{TX}\Theta &= \sqrt{-1}((\widetilde{\nabla}^{TX}w^i) \wedge \overline{w}^i + w^i \wedge \widetilde{\nabla}^{TX}\overline{w}^i) \\
&= \sqrt{-1}\left(-\langle\widetilde{\nabla}^{TX}w_i, \overline{w}_j\rangle - \langle w_i, \widetilde{\nabla}^{TX}\overline{w}_j\rangle\right) w^i \wedge \overline{w}^j = 0.
\end{aligned} \tag{1.2.52}
$$

From (1.2.44), (1.2.48) and (1.2.52) we have

$$d\Theta = i_T\Theta = \sqrt{-1}(T_{as}^{(1,0)} - T_{as}^{(0,1)}). \tag{1.2.53}$$

The relations (1.2.48) and (1.2.53) yield

$$\partial\Theta = \sqrt{-1}T_{as}^{(1,0)}, \qquad \overline{\partial}\Theta = -\sqrt{-1}T_{as}^{(0,1)}. \tag{1.2.54}$$

(1.2.54) imply (1.2.51). □

Definition 1.2.7. We call Θ as in (1.2.49) a Hermitian form on X and (X, J, Θ) a complex Hermitian manifold. The metric $g^{TX} = \Theta(\cdot, J\cdot)$ on TX is called a *Kähler metric* if Θ is a closed form, i.e., $d\Theta = 0$. In this case, the form Θ is called a *Kähler form* on X, and the complex manifold (X, J) is called a *Kähler manifold* .

Let $\nabla^X J \in T^*X \otimes \mathrm{End}(TX)$ be the covariant derivative of J induced by the Levi–Civita connection ∇^{TX}.

Theorem 1.2.8. *(X, J, Θ) is Kähler if and only if the bundle $T^{(1,0)}X$ and $T^{(0,1)}X$ are preserved by the Levi–Civita connection ∇^{TX}, or in other words, if and only if $\nabla^X J = 0$. In this case,*

$$\nabla^{TX} = \widetilde{\nabla}^{TX}, \quad S = 0, \quad T = 0. \tag{1.2.55}$$

Proof. As Θ is a $(1,1)$-form, by (1.2.41), (1.2.48) and (1.2.51), $d\Theta = 0$ is equivalent to $T_{as} = 0$ and equivalent to $S(\overline{w}_k)w_i \in T^{(1,0)}X$ for any i, k. But this means that the bundles $T^{(1,0)}X$ and $T^{(0,1)}X$ are preserved by ∇^{TX}. Hence (1.2.55) is equivalent to (X, Θ) being Kähler. Moreover, as J acts by multiplication with $\sqrt{-1}$ on $T^{(1,0)}X$, we get for $U \in TX$,

$$
\begin{aligned}
\langle S(U)w_i, w_j\rangle &= -\langle\nabla_U^{TX}w_i, w_j\rangle = -\frac{1}{2}\langle\nabla_U^{TX}(1 - \sqrt{-1}J)w_i, w_j\rangle \\
&= \frac{\sqrt{-1}}{2}\langle(\nabla_U^X J)w_i, w_j\rangle,
\end{aligned} \tag{1.2.56}
$$

by (1.2.38). Now, from $J^2 = -1$ we deduce

$$J(\nabla^X J) + (\nabla^X J)J = 0. \tag{1.2.57}$$

This means that $(\nabla^X J)$ exchanges $T^{(1,0)}X$ and $T^{(0,1)}X$. By (1.2.44), and (1.2.56), $\nabla^X J = 0$ is equivalent to $S(\overline{w}_k)w_i \in T^{(1,0)}X$ for any i, k. The proof of Theorem 1.2.8 is complete. □

1.2.3   Bismut connection

Let S^B denote the 1-form with values in the antisymmetric elements of $\mathrm{End}(TX)$ which satisfies for $U, V, W \in TX$,

$$\langle S^B(U)V, W \rangle = \frac{\sqrt{-1}}{2}\left((\partial - \bar{\partial})\Theta\right)(U, V, W) = -\frac{1}{2}T_{as}(U, V, W). \tag{1.2.58}$$

By (1.2.40), (1.2.47), (1.2.58), we have for $U, V, W \in TX$,

$$\langle (S^B - S)(U)V, W \rangle = -\langle T(U, V), W \rangle + \langle T(U, W), V \rangle. \tag{1.2.59}$$

Relations (1.2.41), (1.2.48), and (1.2.58) yield

$$
\begin{aligned}
\langle S^B(e_j)\omega_l, \bar{\omega}_m \rangle &= -\frac{1}{2}\langle T(e_j, \omega_l), \bar{\omega}_m \rangle + \frac{1}{2}\langle T(e_j, \bar{\omega}_m), \omega_l \rangle \\
&= -\langle S(e_j)\omega_l, \bar{\omega}_m \rangle, \\
\langle S^B(e_j)\omega_l, \omega_m \rangle &= -\frac{1}{2}\langle T(\omega_l, \omega_m), e_j \rangle = \langle S(e_j)\omega_l, \omega_m \rangle.
\end{aligned}
\tag{1.2.60}
$$

Definition 1.2.9. The *Bismut connection* ∇^B on TX is defined by

$$\nabla^B := \nabla^{TX} + S^B = \widetilde{\nabla}^{TX} + S^B - S. \tag{1.2.61}$$

In view of (1.2.58), the torsion of ∇^B is $2S^B$ which is a skew-symmetric tensor.

The connection ∇^B will be used in the Lichnerowicz formula (1.4.29).

Lemma 1.2.10. *The connection ∇^B preserves the complex structure of TX.*

Proof. Using (1.2.60), we find that for $V, W \in T^{(1,0)}X$, $\langle (S^B - S)(U)V, W \rangle = 0$, for any $U \in TX$. Equivalently, $(S^B - S)(U)$ is a complex endomorphism of TX. Using (1.2.61), we find that ∇^B preserves the complex structure of TX. $\qquad\square$

1.3   Spin^c Dirac operator

This section is organized as follows. In Section 1.3.1, we define the Clifford connection. In Section 1.3.2, we define the spin^c Dirac operator on a complex manifold and prove the related Lichnerowicz formula. In Section 1.3.3, we obtain the Lichnerowicz formula for the modified Dirac operator. In Section 1.3.4, we explain also the Atiyah–Singer index theorem for the modified Dirac operator.

In this section, we work on a smooth manifold with an almost complex structure J.

1.3.1   Clifford connection

Let (X, J) be a smooth manifold with J an almost complex structure on TX. Let g^{TX} be any Riemannian metric on TX compatible with J. Let $h^{\Lambda^{0,\bullet}}$ be the Hermitian metric on $\Lambda(T^{*(0,1)}X)$ induced by g^{TX}.

The fundamental $\mathbb{Z}_2$ spinor bundle induced by J is given by $\Lambda(T^{*(0,1)}X)$, whose $\mathbb{Z}_2$-grading is defined by $\Lambda(T^{*(0,1)}X) = \Lambda^{\mathrm{even}}(T^{*(0,1)}X) \oplus \Lambda^{\mathrm{odd}}(T^{*(0,1)}X)$. For any $v \in TX$ with decomposition $v = v^{(1,0)} + v^{(0,1)} \in T^{(1,0)}X \oplus T^{(0,1)}X$, let $\overline{v}^{(1,0),*} \in T^{*(0,1)}X$ be the metric dual of $v^{(1,0)}$. Then

$$c(v) = \sqrt{2}(\overline{v}^{(1,0),*} \wedge - i_{v^{(0,1)}}) \tag{1.3.1}$$

defines the Clifford action of v on $\Lambda(T^{*(0,1)}X)$, where $\wedge$ and i denote the exterior and interior product, respectively. We verify easily that for $U, V \in TX$,

$$c(U)c(V) + c(V)c(U) = -2\langle U, V \rangle. \tag{1.3.2}$$

For a skew-adjoint endomorphism A of TX, from (1.3.1), using the notation of (1.2.34),

$$
\begin{aligned}
\frac{1}{4}\langle Ae_i, e_j \rangle c(e_i)c(e_j) &= -\frac{1}{2}\langle Aw_j, \overline{w}_j \rangle + \langle Aw_l, \overline{w}_m \rangle \overline{w}^m \wedge i_{\overline{w}_l} \\
&\quad + \frac{1}{2}\langle Aw_l, w_m \rangle i_{\overline{w}_l} i_{\overline{w}_m} + \frac{1}{2}\langle A\overline{w}_l, \overline{w}_m \rangle \overline{w}^l \wedge \overline{w}^m \wedge.
\end{aligned}
\tag{1.3.3}
$$

Let ∇^{det} be a Hermitian connection on $\det(T^{(1,0)}X)$ endowed with metric induced by g^{TX}. Let R^{det} be its curvature. Let $P^{T^{(1,0)}X}$ be the natural projection from $TX \otimes_{\mathbb{R}} \mathbb{C}$ onto $T^{(1,0)}X$. Then the connection $\nabla^{1,0} = P^{T^{(1,0)}X} \nabla^{TX} P^{T^{(1,0)}X}$ on $T^{(1,0)}X$ induces naturally a connection ∇^{det_1} on $\det(T^{(1,0)}X)$.

Let $\Gamma^{TX} \in T^*X \otimes \mathrm{End}(TX)$, Γ^{det} be the connection forms of ∇^{TX}, ∇^{det} associated to the frames $\{e_j\}$, $w_1 \wedge \cdots \wedge w_n$, i.e.,

$$
\begin{aligned}
\nabla^{TX}_{e_i} e_j &= \Gamma^{TX}(e_i)e_j, \quad \nabla^{\mathrm{det}}(w_1 \wedge \cdots \wedge w_n) = \Gamma^{\mathrm{det}} w_1 \wedge \cdots \wedge w_n, \\
\nabla^{\mathrm{det}_1}(w_1 \wedge \cdots \wedge w_n) &= \Big(\sum_j \langle \Gamma^{TX} w_j, \overline{w}_j \rangle\Big) w_1 \wedge \cdots \wedge w_n.
\end{aligned}
\tag{1.3.4}
$$

The *Clifford connection* ∇^{Cl} on $\Lambda(T^{*(0,1)}X)$ is defined for the frame $\{\overline{w}^{j_1} \wedge \cdots \wedge \overline{w}^{j_k}, 1 \leqslant j_1 < \cdots < j_k \leqslant n\}$ by the local formula

$$\nabla^{\mathrm{Cl}} = d + \frac{1}{4}\langle \Gamma^{TX} e_i, e_j \rangle c(e_i)c(e_j) + \frac{1}{2}\Gamma^{\mathrm{det}}. \tag{1.3.5}$$

Proposition 1.3.1. ∇^{Cl} *defines a Hermitian connection on* $\Lambda(T^{*(0,1)}X)$ *and preserves its* $\mathbb{Z}_2$-*grading. For any* V, W *vector fields of* TX *on* X, *we have*

$$[\nabla^{Cl}_V, c(W)] = c(\nabla^{TX}_V W). \tag{1.3.6}$$

Proof. At first, by (1.3.4) and (1.3.5), we have

$$\left[\nabla_V^{\mathrm{Cl}}, c(e_k)\right] = \frac{1}{4}\left[\langle\Gamma^{TX}(V)e_i, e_j\rangle c(e_i)c(e_j), c(e_k)\right]$$
$$= \langle\Gamma^{TX}(V)e_k, e_j\rangle c(e_j) = c(\nabla_V^{TX}e_k). \tag{1.3.7}$$

Thus if ∇^{Cl} is well defined, we get (1.3.6) from (1.3.7).

Now we observe that $c(w_{j_1})\dots c(w_{j_k})1$, $(1 \leqslant j_1 < \cdots < j_k \leqslant n)$ generate a frame of $\Lambda(T^{*(0,1)}X)$. Taking into account (1.3.7), to verify that ∇^{Cl} does not depend on the choice of our frame $\{w_j\}_{j=1}^n$, we only need to verify that $\nabla^{\mathrm{Cl}}1$ is well defined.

Relations (1.2.38), (1.3.3), (1.3.4) and (1.3.5) entail

$$\nabla^{\mathrm{Cl}} = d + \frac{1}{2}(\nabla^{\mathrm{det}} - \nabla^{\mathrm{det}_1}) + \langle\Gamma^{TX}w_l, \overline{w}_m\rangle\,\overline{w}^m \wedge i_{\overline{w}_l}$$
$$- \frac{1}{2}\langle Sw_l, w_m\rangle\, i_{\overline{w}_l}\, i_{\overline{w}_m} - \frac{1}{2}\langle S\overline{w}_l, \overline{w}_m\rangle\,\overline{w}^l \wedge \overline{w}^m \wedge. \tag{1.3.8}$$

From (1.3.8), we know

$$\nabla^{\mathrm{Cl}}1 = \frac{1}{2}(\nabla^{\mathrm{det}} - \nabla^{\mathrm{det}_1}) - \frac{1}{2}\sum_{lm}\langle S\overline{w}_l, \overline{w}_m\rangle\,\overline{w}^l \wedge \overline{w}^m. \tag{1.3.9}$$

Clearly, $\nabla^{\mathrm{det}} - \nabla^{\mathrm{det}_1}$ is a 1-form on X, and the right-hand side of (1.3.9) does not depend on the choice of the frame w_j. Thus $\nabla^{\mathrm{Cl}}1$ is well defined.

Let $c(e_i)^*$ be the adjoint of $c(e_i)$ with respect to the Hermitian product on $\Lambda(T^{*(0,1)}X)$. By (1.3.1), we have

$$c(e_i)^* = -c(e_i). \tag{1.3.10}$$

Using (1.3.5), (1.3.10) and the anti-symmetry of $\langle\Gamma^{TX}e_i, e_j\rangle$ in i, j, we see that ∇^{Cl} preserves the Hermitian metric on $\Lambda(T^{*(0,1)}X)$.

Finally, from (1.3.5), ∇^{Cl} preserves the $\mathbb{Z}_2$-grading on $\Lambda(T^{*(0,1)}X)$. The proof of Proposition 1.3.1 is complete. $\qquad\square$

Let R^{Cl} be the curvature of ∇^{Cl}.

Proposition 1.3.2. *We have the following identity:*

$$R^{\mathrm{Cl}} = \frac{1}{4}\langle R^{TX}e_i, e_j\rangle c(e_i)c(e_j) + \frac{1}{2}R^{\mathrm{det}}. \tag{1.3.11}$$

Proof. At first, observe that if i, j, k, l are different, then $[c(e_i)c(e_j), c(e_k)c(e_l)] = 0$. Thus from (1.3.2),

$$\left[\langle\Gamma^{TX}(W)e_i, e_j\rangle c(e_i)c(e_j), \langle\Gamma^{TX}(V)e_k, e_l\rangle c(e_k)c(e_l)\right]$$
$$= 4\sum_{i\neq j\neq k}\langle\Gamma^{TX}(W)e_i, e_j\rangle\langle\Gamma^{TX}(V)e_k, e_j\rangle\,[c(e_i)c(e_j), c(e_k)c(e_j)]$$
$$= 4\langle\Gamma^{TX}(W)e_i, \Gamma^{TX}(V)e_k\rangle(c(e_i)c(e_k) - c(e_k)c(e_i)) \tag{1.3.12}$$
$$= 4\langle(\Gamma^{TX} \wedge \Gamma^{TX})(W, V)e_i, e_k\rangle c(e_i)c(e_k).$$

Moreover, we have

$$R^{TX} = d\Gamma^{TX} + \Gamma^{TX} \wedge \Gamma^{TX}, \tag{1.3.13}$$
$$R^{\mathrm{Cl}}(e_l, e_m) = \nabla^{\mathrm{Cl}}_{e_l} \nabla^{\mathrm{Cl}}_{e_m} - \nabla^{\mathrm{Cl}}_{e_m} \nabla^{\mathrm{Cl}}_{e_l} - \nabla^{\mathrm{Cl}}_{[e_l, e_m]}.$$

Finally, (1.3.5), (1.3.12) and (1.3.13) yield (1.3.11).   □

1.3.2   Dirac operator and Lichnerowicz formula

Let (E, h^E) be a Hermitian vector bundle on X. Let ∇^E be a Hermitian connection on (E, h^E) with curvature R^E.

Set $\mathbf{E}^q = \Lambda^q(T^{*(0,1)}X) \otimes E$, $\mathbf{E} = \oplus_{q=0}^n \mathbf{E}^q$. We still denote by ∇^{Cl} the connection on $\Lambda(T^{*(0,1)}X) \otimes E$ induced by ∇^{Cl} and ∇^E. Let $\Omega^{0,q}(X, E) := \mathscr{C}^\infty(X, \mathbf{E}^q)$ be the set of smooth sections of $\mathbf{E}^q$ on X.

Along the fibers of $\Lambda(T^{*(0,1)}X) \otimes E$, we consider the pointwise Hermitian product $\langle \cdot, \cdot \rangle_{\Lambda^{0,\bullet} \otimes E}$ induced by g^{TX} and h^E. The L^2-scalar product on $\Omega^{0,\bullet}(X, E)$ is given by

$$\langle s_1, s_2 \rangle = \int_X \langle s_1(x), s_2(x) \rangle_{\Lambda^{0,\bullet} \otimes E} \, dv_X(x). \tag{1.3.14}$$

We denote the corresponding norm with $\|\cdot\|_{L^2}$, and by $L^2(X, \Lambda(T^{*(0,1)}X) \otimes E)$ or $L^2_{0,\bullet}(X, E)$, the L^2 completion of $\Omega^{0,\bullet}_0(X, E)$, which is the subspace of $\Omega^{0,\bullet}(X, E)$ consisting of elements with compact support.

Definition 1.3.3. The *spin^c Dirac operator* D^c is defined by

$$D^c = \sum_{j=1}^{2n} c(e_j) \nabla^{\mathrm{Cl}}_{e_j} : \Omega^{0,\bullet}(X, E) \longrightarrow \Omega^{0,\bullet}(X, E). \tag{1.3.15}$$

By Proposition 1.3.1 and equation (1.3.1), D^c interchanges $\Omega^{0,\mathrm{even}}(X, E)$ and $\Omega^{0,\mathrm{odd}}(X, E)$. We write

$$D^c_+ = D^c|_{\Omega^{0,\mathrm{even}}(X,E)}, \quad D^c_- = D^c|_{\Omega^{0,\mathrm{odd}}(X,E)}. \tag{1.3.16}$$

Lemma 1.3.4. D^c *is a formally self-adjoint, first order elliptic differential operator on* $\Omega^{0,\bullet}(X, E)$.

Proof. Let $s_1, s_2 \in \Omega^{0,\bullet}(X, E)$ with compact support and let α be the 1-form on X given by $\alpha(Y) = \langle c(Y)s_1, s_2 \rangle_{\Lambda^{0,\bullet} \otimes E}$, for any vector field Y on X. Proposition 1.3.1 and (1.3.10) imply that for $x \in X$,

$$\langle s_1, D^c s_2 \rangle_{\Lambda^{0,\bullet} \otimes E, x} = \langle D^c s_1, s_2 \rangle_{\Lambda^{0,\bullet} \otimes E, x} - \mathrm{Tr}(\nabla \alpha)_x. \tag{1.3.17}$$

The integral over X of the last term vanishes by Proposition 1.2.1. Thus D^c is formally self-adjoint.

For $\zeta \in T^*X$, let $\zeta^* \in TX$ be the metric dual of ζ. The principal symbol $\sigma(D^c)$ of D^c is

$$\sigma(D^c)(\zeta) = \sqrt{-1}c(\zeta^*). \tag{1.3.18}$$

By (1.3.2), $(\sigma(D^c)(\zeta))^2 = |\zeta|^2$, which means, that $\sigma(D^c)(\zeta)$ is invertible for any $\zeta \neq 0$. Thus D^c is a first order elliptic differential operator. $\qquad\square$

Let (F, h^F) be a Hermitian vector bundle on X and let ∇^F be a Hermitian connection on F. Then the usual *Bochner Laplacians* Δ^F, Δ are defined by

$$\Delta^F := -\sum_{i=1}^{2n}\left((\nabla_{e_i}^F)^2 - \nabla_{\nabla_{e_i}^{TX}e_i}^F\right), \quad \Delta = \Delta^{\mathbb{C}}. \tag{1.3.19}$$

Let $s_1, s_2 \in \mathscr{C}^\infty(X, F)$, with compact support and let α be the 1-form on X given by $\alpha(Y)(x) = \langle\nabla_Y^F s_1, s_2\rangle(x)$, for any $Y \in T_xX$. Then by (1.2.6), (1.2.7), we get the following useful equation:

$$\int_X \langle\Delta^F s_1, s_2\rangle dv_X = \int_X \langle\nabla^F s_1, \nabla^F s_2\rangle dv_X - \int_X \mathrm{Tr}(\nabla\alpha)dv_X$$
$$= \int_X \langle\nabla^F s_1, \nabla^F s_2\rangle dv_X. \tag{1.3.20}$$

We denote by Δ^{Cl} the Bochner Laplacian on $\Lambda(T^{*(0,1)}X) \otimes E$ associated to ∇^{Cl} as in (1.3.19). Now we prove the Lichnerowicz formula for D^c.

Theorem 1.3.5.

$$(D^c)^2 = \Delta^{\mathrm{Cl}} + \frac{r^X}{4} + \frac{1}{2}\left(R^E + \frac{1}{2}R^{\det}\right)(e_i, e_j)c(e_i)c(e_j). \tag{1.3.21}$$

Proof. By (1.3.2), (1.3.6) and (1.3.15),

$$(D^c)^2 = \frac{1}{2}\sum_{ij}\left\{c(e_i)\nabla_{e_i}^{\mathrm{Cl}}c(e_j)\nabla_{e_j}^{\mathrm{Cl}} + c(e_j)\nabla_{c_j}^{\mathrm{Cl}}c(e_i)\nabla_{e_i}^{\mathrm{Cl}}\right\}$$

$$= \frac{1}{2}\sum_{ij}\left\{(c(e_i)c(e_j) + c(e_j)c(e_i))\nabla_{e_i}^{\mathrm{Cl}}\nabla_{e_j}^{\mathrm{Cl}} + c(e_i)\left[\nabla_{e_i}^{\mathrm{Cl}}, c(e_j)\right]\nabla_{e_j}^{\mathrm{Cl}}\right.$$

$$\left.+c(e_j)[\nabla_{e_j}^{\mathrm{Cl}}, c(e_i)]\nabla_{e_i}^{\mathrm{Cl}} + c(e_j)c(e_i)\left[\nabla_{e_j}^{\mathrm{Cl}}, \nabla_{e_i}^{\mathrm{Cl}}\right]\right\} \tag{1.3.22}$$

$$= -\sum_i(\nabla_{e_i}^{\mathrm{Cl}})^2 + \sum_{ijk}\langle\nabla_{e_i}^{TX}e_j, e_k\rangle c(e_i)c(e_k)\nabla_{e_j}^{\mathrm{Cl}}$$

$$+ \frac{1}{2}\sum_{ij}c(e_j)c(e_i)\left[\nabla_{e_j}^{\mathrm{Cl}}, \nabla_{e_i}^{\mathrm{Cl}}\right].$$

But we have

$$\langle\nabla_{e_i}^{TX}e_j, e_k\rangle = -\langle e_j, \nabla_{e_i}^{TX}e_k\rangle. \tag{1.3.23}$$

In view of (1.3.2), (1.3.23), we obtain

$$\langle \nabla^{TX}_{e_i} e_j, e_k \rangle c(e_i) c(e_k) \nabla^{\mathrm{Cl}}_{e_j} = -c(e_i) c(e_k) \nabla^{\mathrm{Cl}}_{\nabla^{TX}_{e_i} e_k}$$

$$= \nabla^{\mathrm{Cl}}_{\nabla^{TX}_{e_i} e_i} - \frac{1}{2} \sum_{i \neq k} c(e_i) c(e_k) \left(\nabla^{\mathrm{Cl}}_{\nabla^{TX}_{e_i} e_k} - \nabla^{\mathrm{Cl}}_{\nabla^{TX}_{e_k} e_i} \right) \qquad (1.3.24)$$

$$= \nabla^{\mathrm{Cl}}_{\nabla^{TX}_{e_i} e_i} - \frac{1}{2} c(e_i) c(e_k) \nabla^{\mathrm{Cl}}_{[e_i, e_k]}.$$

Comparing to (1.3.13), we have here

$$(R^{\mathrm{Cl}} + R^E)(e_l, e_m) = \nabla^{\mathrm{Cl}}_{e_l} \nabla^{\mathrm{Cl}}_{e_m} - \nabla^{\mathrm{Cl}}_{e_m} \nabla^{\mathrm{Cl}}_{e_l} - \nabla^{\mathrm{Cl}}_{[e_l, e_m]}. \qquad (1.3.25)$$

(1.3.22)–(1.3.25) yield

$$(D^c)^2 = -\sum_i \left((\nabla^{\mathrm{Cl}}_{e_i})^2 - \nabla^{\mathrm{Cl}}_{\nabla^{TX}_{e_i} e_i} \right) + \frac{1}{2} c(e_j) c(e_i) (R^{\mathrm{Cl}} + R^E)(e_j, e_i). \qquad (1.3.26)$$

To simplify the notation, set

$$R_{ijkl} := \langle R^{TX}(e_j, e_i) e_k, e_l \rangle. \qquad (1.3.27)$$

By Proposition 1.3.2, we get

$$c(e_j) c(e_i) R^{\mathrm{Cl}}(e_j, e_i) = -\frac{1}{4} R_{ijkl} c(e_i) c(e_j) c(e_k) c(e_l)$$

$$+ \frac{1}{2} c(e_i) c(e_j) R^{\det}(e_i, e_j). \qquad (1.3.28)$$

By the second equation of (1.2.4) and (1.3.2),

$$\sum_{i \neq k \neq j} R_{ijkl} c(e_i) c(e_j) c(e_k) = 2 \sum_{i < j < k} (R_{ijkl} + R_{jkil} + R_{kijl}) c(e_i) c(e_j) c(e_k) = 0.$$

Thus

$$R_{ijkl} c(e_i) c(e_j) c(e_k) c(e_l) = -R_{ijjl} c(e_i) c(e_l) + R_{ijil} c(e_j) c(e_l)$$

$$= 2 c(e_j) c(e_l) R_{ijil} = -2 R_{ijij}. \qquad (1.3.29)$$

In the last equation of (1.3.29), we use that R_{ijil} is symmetric in j, l (which follows by the first equation of (1.2.4)). By (1.2.5) and (1.3.27), we get the right-hand side of (1.3.29) equals $-2r^X$. Hence (1.3.26)–(1.3.29) imply (1.3.21). $\square$

1.3.3 Modified Dirac operator

For any $\mathbb{Z}_2$-graded vector space $V = V^+ \oplus V^-$, the natural $\mathbb{Z}_2$-grading on $\mathrm{End}(V)$ is defined by

$$\mathrm{End}(V)^+ = \mathrm{End}(V^+) \oplus \mathrm{End}(V^-), \quad \mathrm{End}(V)^- = \mathrm{Hom}(V^+, V^-) \oplus \mathrm{Hom}(V^-, V^+),$$

and we define $\deg B = 0$ for $B \in \mathrm{End}(V)^+$, and $\deg B = 1$ for $B \in \mathrm{End}(V)^-$. For $B, C \in \mathrm{End}(V)$, we define their supercommutator (or graded Lie bracket) by

$$[B, C] = BC - (-1)^{\deg B \cdot \deg C} CB. \tag{1.3.30}$$

For $B, B', C \in \mathrm{End}(V)$, the *Jacobi identity* holds:

$$(-1)^{\deg C \cdot \deg B'} \big[B', [B, C] \big] + (-1)^{\deg B' \cdot \deg B} \big[B, [C, B'] \big]$$
$$+ (-1)^{\deg B \cdot \deg C} \big[C, [B', B] \big] = 0. \tag{1.3.31}$$

We will apply the above notation for spaces $\Lambda(T^{*(0,1)}X)$ and $\Omega^{0,\bullet}(X, E)$ with natural $\mathbb{Z}_2$-grading induced by the parity of the degree.
For $i_1 < \cdots < i_j$, we define

$$^c(e^{i_1} \wedge \cdots \wedge e^{i_j}) = c(e_{i_1}) \ldots c(e_{i_j}). \tag{1.3.32}$$

Then by extending $\mathbb{C}$-linearly, ^cB is defined for any $B \in \Lambda(T^*X \otimes_{\mathbb{R}} \mathbb{C})$.
For $A \in \Lambda^3(T^*X)$, set $|A|^2 = \sum_{i<j<k} |A(e_i, e_j, e_k)|^2$. Now let A be a smooth section of $\Lambda^3(T^*X)$. Let

$$\nabla_U^A = \nabla_U^{\mathrm{Cl}} +^c (i_U A) \quad \text{for } U \in TX \tag{1.3.33}$$

be the Hermitian connection on $\Lambda(T^{*(0,1)}X) \otimes E$ induced by ∇^{Cl} and A. Let Δ^A be the Bochner Laplacian defined by ∇^A as in (1.3.19).

Definition 1.3.6. The *modified Dirac operators* $D^{c,A}$, $D_\pm^{c,A}$ are defined by

$$D^{c,A} := D^c + {}^cA, \quad D_\pm^{c,A} := D_\pm^c + {}^cA. \tag{1.3.34}$$

Theorem 1.3.7. The modified Dirac operator $D^{c,A}$ is formally self-adjoint and

$$(D^{c,A})^2 = \Delta^A + \frac{r^X}{4} + {}^c\Big(R^E + \frac{1}{2}R^{\det}\Big) + {}^c(dA) - 2|A|^2. \tag{1.3.35}$$

Proof. By Lemma 1.3.4 and (1.3.10), the operator $D^c + {}^cA$ is formally self-adjoint.
By (1.3.6), $\nabla_{e_i}^{\mathrm{Cl}} {}^cA = {}^c(\nabla_{e_i}^{TX} A)$. From (1.2.44) and (1.3.2) and since A is odd degree, we have

$$[c(e_i), {}^cA] = -2\,{}^c(i_{e_i} A),$$
$$c(e_i)(\nabla_{e_i}^{\mathrm{Cl}} {}^cA) - (\nabla_{e_i}^{\mathrm{Cl}} {}^cA)c(e_i) = 2\,{}^c(e^i \wedge \nabla_{e_i}^{TX} A) = 2\,{}^c(dA). \tag{1.3.36}$$

By (1.3.19), (1.3.33) and the first equation of (1.3.36),

$$\Delta^A = \Delta^{\mathrm{Cl}} + \frac{1}{2}\Big(\nabla_{e_i}^{\mathrm{Cl}}[c(e_i), {}^cA] + [c(e_i), {}^cA]\nabla_{e_i}^{\mathrm{Cl}} \Big)$$
$$- \frac{1}{2}[c(\nabla_{e_i}^{TX} e_i), {}^cA] - \frac{1}{4}\sum_i [c(e_i), {}^cA]^2 \tag{1.3.37}$$
$$= \Delta^{\mathrm{Cl}} - 2\,{}^c(i_{e_i} A)\nabla_{e_i}^{\mathrm{Cl}} + \frac{1}{2}[c(e_i), \nabla_{e_i}^{\mathrm{Cl}} {}^cA] - \sum_i {}^c(i_{e_i} A)^2.$$

Then Theorem 1.3.5, (1.3.33), (1.3.36) and (1.3.37) imply

$$
\begin{aligned}
(D^c + {}^c A)^2 =&(D^c)^2 + [c(e_i), {}^c A]\nabla^{\mathrm{Cl}}_{e_i} + c(e_i)(\nabla^{\mathrm{Cl}}_{e_i}{}^c A) + ({}^c A)^2 \\
=&\Delta^A + ({}^c A)^2 + \sum_i {}^c(i_{e_i} A)^2 + c(e_i)(\nabla^{\mathrm{Cl}}_{e_i}{}^c A) \\
&- \frac{1}{2}[c(e_i), (\nabla^{\mathrm{Cl}}_{e_i}{}^c A)] + \frac{r^X}{4} + {}^c(R^E + \frac{1}{2}R^{\det}).
\end{aligned}
\tag{1.3.38}
$$

Relations (1.3.36) and (1.3.38) yield

$$
(D^c + {}^c A)^2 =\Delta^A + ({}^c A)^2 + \sum_i {}^c(i_{e_i} A)^2 + {}^c(dA) + \frac{r^X}{4} + {}^c(R^E + \frac{1}{2}R^{\det}).
\tag{1.3.39}
$$

Let $I = \{i_1, \ldots, i_m\}$ be an ordered subset of $\{1, \ldots, 2n\}$, and assume that all $i_j \in I$ are distinct. Let $|I|$ be the cardinal of I. Set ${}^c e_I = c(e_{i_1}) \ldots c(e_{i_m})$. Take $k \leqslant 2n$, and let I, J be two ordered subsets of $\{k+1, \ldots, 2n\}$ such that $I \cap J = \emptyset$. Then

$$
{}^c e_{1 \ldots k}{}^c e_I{}^c e_{1 \ldots k}{}^c e_J = (-1)^{k|I|}({}^c e_{1 \ldots k})^2 {}^c e_I{}^c e_J = (-1)^{k|I| + \frac{k(k+1)}{2}} {}^c e_I{}^c e_J. \tag{1.3.40}
$$

Since A is odd degree, (1.3.40) imply

$$
\begin{aligned}
{}^c(i_{e_i} A)^2 =& \sum_{k=0}^{2} \sum_{i_1 < \cdots < i_k} (-1)^{\frac{k(k-1)}{2}} {}^c((i_{e_{i_1}} \ldots i_{e_{i_k}} i_{e_i} A)^2), \\
{}^c(A)^2 =& \sum_{k=0}^{3} \sum_{i_1 < \cdots < i_k} (-1)^{\frac{k(k+1)}{2}} {}^c((i_{e_{i_1}} \ldots i_{e_{i_k}} A)^2).
\end{aligned}
\tag{1.3.41}
$$

Observe that since $A \in \Lambda^3(T^*X)$, $A^2 = 0$ and $(i_{e_{i_1}} i_{e_{i_2}} A)^2 = 0$. Thus

$$
({}^c A)^2 + \sum_i {}^c(i_{e_i} A)^2 = -2 \sum_{i_1 < i_2 < i_3} (i_{e_{i_1}} i_{e_{i_2}} i_{e_{i_3}} A)^2 = -2|A|^2.
\tag{1.3.42}
$$

From (1.3.39) and (1.3.42), we infer (1.3.35). $\qquad\square$

1.3.4   Atiyah–Singer index theorem

Theorem 1.3.8. *If X is compact, the modified Dirac operator $D^{c,A}$ is an essentially self-adjoint Fredholm operator, thus its kernel $\mathrm{Ker}(D^{c,A})$ is a finite-dimensional complex vector space.*

Proof. At first, if $s_k \in L^2_{0,\bullet}(X, E)$, $D^{c,A} s_k = 0$ and $\lim_{k \to \infty} s_k = s \in L^2_{0,\bullet}(X, E)$, then $D^{c,A} s = 0$ in the sense of distributions. By Theorem A.3.4, $s \in \Omega^{0,\bullet}(X, E)$ and

$s \in \mathrm{Ker}(D^{c,A})$. Thus the space $\mathrm{Ker}(D^{c,A})$ is closed, so a Hilbert space. Since X is compact, Theorems A.3.1, A.3.2 and Lemma 1.3.4 imply that $D^{c,A}$ is essentially self-adjoint and the unit ball

$$B = \{s \in L^2_{0,\bullet}(X, E) : \|s\|_{L^2} \leqslant 1,\ D^{c,A}s = 0\} \subset \mathrm{Ker}(D^{c,A}) \qquad (1.3.43)$$

is compact. Thus $\mathrm{Ker}(D^{c,A})$ is finite-dimensional and $D^{c,A}$ is Fredhlom. $\qquad \Box$

When X is compact, we define the index $\mathrm{Ind}(D^{c,A}_+)$ of $D^{c,A}_+$ as

$$\begin{aligned} \mathrm{Ind}(D^{c,A}_+) &:= \dim \mathrm{Ker}(D^{c,A}_+) - \dim \mathrm{Coker}(D^{c,A}_+) \\ &= \dim \mathrm{Ker}(D^{c,A}_+) - \dim \mathrm{Ker}(D^{c,A}_-). \end{aligned} \qquad (1.3.44)$$

For any Hermitian (complex) vector bundle (F, h^F) with Hermitian connection ∇^F and curvature R^F on X, set

$$\begin{aligned} \mathrm{ch}(F, \nabla^F) &:= \mathrm{Tr}\left[\exp\left(\frac{-R^F}{2\pi\sqrt{-1}}\right)\right], \\ c_1(F, \nabla^F) &:= \mathrm{Tr}\left[\frac{-R^F}{2\pi\sqrt{-1}}\right], \\ \mathrm{Td}(F, \nabla^F) &:= \det\left(\frac{R^F/(2\pi\sqrt{-1})}{\exp(R^F/(2\pi\sqrt{-1})) - 1}\right). \end{aligned} \qquad (1.3.45)$$

By Appendix B.5 these are closed real differential forms on X and their cohomology classes do not depend on the choice of the metric h^F and connection ∇^F. The corresponding cohomology classes are called the *Chern class* of F, the *first Chern class* of F, the *Todd class* of F, respectively, and we denote them by $\mathrm{ch}(F)$, $c_1(F)$, $\mathrm{Td}(F) \in H^*(X, \mathbb{R})$ (see Example B.5.5) .

Theorem 1.3.9 (Atiyah–Singer index theorem). *If X is compact, $\mathrm{Ind}(D^{c,A}_+)$ is a topological invariant given by*

$$\mathrm{Ind}(D^{c,A}_+) = \int_X \mathrm{Td}(T^{(1,0)}X)\, \mathrm{ch}(E). \qquad (1.3.46)$$

1.4 Lichnerowicz formula for $\Box^E$

This section is organized as follows. In Section 1.4.1, we exhibit the relation between the operator $\overline{\partial}^E + \overline{\partial}^{E,*}$ and the Dirac operator D^c. In Section 1.4.2, we prove Bismut's Lichnerowicz formula for the Kodaira Laplacian $\Box^E$. In Section 1.4.3, we establish the Bochner–Kodaira–Nakano formula for $\Box^E$. In Section 1.4.4, we prove the Bochner–Kodaira–Nakano formula with boundary term.

We will use the notation from Sections 1.2, 1.3.

1.4.1   The operator $\overline{\partial}^E + \overline{\partial}^{E,*}$

Let (X, J) be a complex manifold with complex structure J and $\dim_{\mathbb{C}} X = n$, and let g^{TX} be any Riemannian metric on TX compatible with J. We consider a holomorphic Hermitian vector bundle (E, h^E) on X. Let ∇^E be the holomorphic Hermitian (i.e., Chern) connection on (E, h^E) whose curvature is R^E. Let $\overline{\partial}^E$ be the Dolbeault operator acting on $\Omega^{0,\bullet}(X, E) := \oplus_q \Omega^{0,q}(X, E)$. Then

$$(\overline{\partial}^E)^2 = 0. \tag{1.4.1}$$

The complex $(\Omega^{0,\bullet}(X, E), \overline{\partial}^E)$ is called the Dolbeault complex and its cohomology, called Dolbeault cohomology of X with values in E, is denoted by $H^{0,\bullet}(X, E)$.

By the Dolbeault isomorphism (Theorem B.4.4), $H^{0,\bullet}(X, E)$ is canonically isomorphic to the qth cohomology group $H^q(X, \mathscr{O}_X(E))$ of the sheaf $\mathscr{O}_X(E)$ of holomorphic sections of E over X. We shortly denote $H^q(X, E) := H^q(X, \mathscr{O}_X(E))$. Especially for $q = 0$,

$$H^{0,0}(X, E) = H^0(X, \mathscr{O}_X(E)) = H^0(X, E). \tag{1.4.2}$$

Let $\overline{\partial}^{E,*}$ be the formal adjoint of $\overline{\partial}^E$ on the Dolbeault complex $\Omega^{0,\bullet}(X, E)$ with respect to the scalar product $\langle \cdot, \cdot \rangle$ in (1.3.14). Set

$$\begin{aligned} D &= \sqrt{2}\big(\overline{\partial}^E + \overline{\partial}^{E,*}\big), \\ \square^E &= \overline{\partial}^E \overline{\partial}^{E,*} + \overline{\partial}^{E,*} \overline{\partial}^E. \end{aligned} \tag{1.4.3}$$

Then $\square^E$ is called the Kodaira Laplacian and

$$D^2 = 2\square^E. \tag{1.4.4}$$

Thus D^2 preserves the $\mathbb{Z}$-grading of $\Omega^{0,\bullet}(X, E)$. It is a fundamental result, that the elements of $\mathrm{Ker}(\square^E)$, called *harmonic forms*, represent the Dolbeault cohomology. The following theorem follows from the more general Theorem 3.1.8 on noncompact manifolds (cf. Remark 3.1.10).

Theorem 1.4.1 (Hodge theory). *If X is a compact complex manifold, then for any $q \in \mathbb{N}$, we have the following direct sum decomposition:*

$$\begin{aligned} \Omega^{0,q}(X, E) &= \mathrm{Ker}(D|_{\Omega^{0,q}}) \oplus \mathrm{Im}(\square^E|_{\Omega^{0,q}}) \\ &= \mathrm{Ker}(D|_{\Omega^{0,q}}) \oplus \mathrm{Im}(\overline{\partial}^E|_{\Omega^{0,q-1}}) \oplus \mathrm{Im}(\overline{\partial}^{E,*}|_{\Omega^{0,q+1}}). \end{aligned} \tag{1.4.5}$$

Thus for any $q \in \mathbb{N}$, we have the canonical isomorphism,

$$\mathrm{Ker}(D|_{\Omega^{0,q}}) = \mathrm{Ker}(D^2|_{\Omega^{0,q}}) \simeq H^{0,q}(X, E). \tag{1.4.6}$$

Especially, $H^q(X, E) \simeq H^{0,q}(X, E)$ is finite-dimensional.

Definition 1.4.2. The *Bergman kernel* of E is $P(x, x')$, $(x, x' \in X)$, the Schwartz kernel of P, the orthogonal projection from $(L^2(X, \Lambda(T^{*(0,1)}X) \otimes E), \langle \ \rangle)$ onto $\mathrm{Ker}(D)$, the kernel of D acting on $\Omega^{0,\bullet}(X, E) \cap L^2(X, \Lambda(T^{*(0,1)}X) \otimes E)$, with respect to the Riemannian volume form $dv_X(x')$. Especially,

$$P(x, x') \in (\Lambda(T^{*(0,1)}X) \otimes E)_x \otimes (\Lambda(T^{*(0,1)}X) \otimes E)^*_{x'}.$$

Remark 1.4.3. From Theorem 1.4.1, the Bergman kernel $P(x, x')$ is smooth on $x, x' \in X$ when X is compact. In general, by the ellipticity of D and Schwartz kernel theorem, we know $P(x, x')$ is $\mathscr{C}^\infty$ (cf. Problem 1.5).

Recall that the tensors $S, T, \mathcal{S}, T_{as}$ were defined in (1.2.38) and (1.2.48).

Lemma 1.4.4. *For the operators* $\overline{\partial}^E, (\nabla^E)^{1,0}$ *acting on* $\Omega^{\bullet, \bullet}(X, E)$ *in* (1.1.9), *we have*

$$\overline{\partial}^E = \overline{w}^j \wedge \widetilde{\nabla}^{TX}_{\overline{w}_j} + i_{T^{(0,1)}}$$
$$= \overline{w}^j \wedge \widetilde{\nabla}^{TX}_{\overline{w}_j} + \frac{1}{2}\langle T(\overline{w}_j, \overline{w}_k), w_m\rangle \overline{w}^j \wedge \overline{w}^k \wedge i_{\overline{w}_m}, \tag{1.4.7}$$

$$(\nabla^E)^{1,0} = w^j \wedge \widetilde{\nabla}^{TX}_{w_j} + i_{T^{(1,0)}}$$
$$= w^j \wedge \widetilde{\nabla}^{TX}_{w_j} + \frac{1}{2}\langle T(w_j, w_k), \overline{w}_m\rangle w^j \wedge w^k \wedge i_{w_m}. \tag{1.4.8}$$

For the formal adjoints $\overline{\partial}^{E,*}$ *and* $(\nabla^F)^{1,0*}$ *of* $\overline{\partial}^E$ *and* $(\nabla^E)^{1,0}$ *with respect to* (1.3.14), *we have*

$$\overline{\partial}^{E,*} = -i_{\overline{w}_j} \widetilde{\nabla}^{TX}_{w_j} - \langle T(w_j, w_k), \overline{w}_k\rangle i_{\overline{w}_j}$$
$$+ \frac{1}{2}\langle T(w_j, w_k), \overline{w}_m\rangle \overline{w}^m \wedge i_{\overline{w}_k} \wedge i_{\overline{w}_j}, \tag{1.4.9}$$

$$(\nabla^E)^{1,0*} = -i_{w_j} \widetilde{\nabla}^{TX}_{\overline{w}_j} - \langle T(\overline{w}_j, \overline{w}_k), w_k\rangle i_{w_j}$$
$$+ \frac{1}{2}\langle T(\overline{w}_j, \overline{w}_k), w_m\rangle w^m \wedge i_{w_k} i_{w_j}. \tag{1.4.10}$$

Proof. The operator $\overline{\partial}^E$ on E is given by

$$\overline{\partial}^E = \sum_{i=1}^n \overline{w}^i \wedge \nabla^E_{\overline{w}_i}. \tag{1.4.11}$$

We still denote by $\widetilde{\nabla}^{TX}$ the connection $\widetilde{\nabla}^{TX} \otimes 1 + 1 \otimes \nabla^E$ and by i_T the operator $i_T \otimes 1$ on $\Lambda^{\bullet, \bullet}(T^*X) \otimes E$. From (1.2.44), we deduce

$$\nabla^E = \varepsilon \circ \widetilde{\nabla}^{TX} + i_T. \tag{1.4.12}$$

Relations (1.2.37) and (1.4.12) imply (1.4.7) and (1.4.8), by decomposition after bidegree and the definition of T. Observe that from (1.2.38), the $(0,1)$ and $(1,0)$-components of $\mathcal{S}$ are

$$
\begin{aligned}
\mathcal{S}^{(0,1)} &= \left(\left\langle \widetilde{\nabla}^{TX}_{w_i} \overline{w}_i, w_j \right\rangle - \left\langle \nabla^{TX}_{e_k} e_k, w_j \right\rangle \right) \overline{w}_j, \\
\mathcal{S}^{(1,0)} &= \left(\left\langle \widetilde{\nabla}^{TX}_{\overline{w}_i} w_i, \overline{w}_j \right\rangle - \left\langle \nabla^{TX}_{e_k} e_k, \overline{w}_j \right\rangle \right) w_j.
\end{aligned}
\tag{1.4.13}
$$

Let $s_1, s_2 \in \Omega_0^{\bullet,\bullet}(X, E)$ and let α be the $(0,1)$-form on X given for any vector field $U = U^{(1,0)} \oplus U^{(0,1)} \in T^{(1,0)}X \oplus T^{(0,1)}X$ on X, by $\alpha(U) = -\langle i_{U^{(0,1)}} s_1, s_2\rangle_{\Lambda^{\bullet,\bullet} \otimes E}$. Note that from (1.2.6),

$$
\operatorname{Tr}(\nabla \alpha) = w_j \alpha(\overline{w}_j) + \overline{w}_j \alpha(w_j) - \alpha(\nabla^{TX}_{e_k} e_k).
\tag{1.4.14}
$$

Proceeding as in the proof of (1.3.17), (1.4.13) and (1.4.14) entail the following relation between pointwise scalar products:

$$
\langle s_1, \overline{w}^i \widetilde{\nabla}^{TX}_{\overline{w}_i} s_2 \rangle_{\Lambda^{\bullet,\bullet} \otimes E, x} = -\langle i_{\overline{w}_i} \widetilde{\nabla}^{TX}_{w_i} s_1, s_2 \rangle_{\Lambda^{\bullet,\bullet} \otimes E, x} \\
- \operatorname{Tr}(\nabla \alpha)_x + i_{\mathcal{S}^{(0,1)}} \alpha. \tag{1.4.15}
$$

The integral of the last term vanishes by Proposition 1.2.1, so integrating (1.4.15) and (1.2.42) over X, we infer (1.4.9).

Let β be the $(1,0)$ form on X given by $\beta(U) = -\langle i_{U^{(1,0)}} s_1, s_2\rangle_{\Lambda^{\bullet,\bullet} \otimes E}$. Then as in (1.4.15),

$$
\langle s_1, w^j \widetilde{\nabla}^{TX}_{w_j} s_2 \rangle_{\Lambda^{\bullet,\bullet} \otimes E, x} = -\langle i_{w_j} \widetilde{\nabla}^{TX}_{\overline{w}_j} s_1, s_2 \rangle_{\Lambda^{\bullet,\bullet} \otimes E, x} \\
- \operatorname{Tr}(\nabla \beta)_x + i_{\mathcal{S}^{(1,0)}} \beta. \tag{1.4.16}
$$

Integration of (1.4.16) and (1.2.42) gives (1.4.10).   □

In this section, in the definition (1.3.15) of the spin^c Dirac operator D^c, we choose $\nabla^{\det}$ to be the holomorphic Hermitian connection on $\det(T^{(1,0)}X)$. Consequently D is a modified Dirac operator.

Theorem 1.4.5. *We have the following identity:*

$$
D = D^c - \frac{1}{4} c(T_{as}).
\tag{1.4.17}
$$

Proof. In view of (1.3.1), (1.4.7) and (1.4.9), we have

$$
\begin{aligned}
\sqrt{2}\,\overline{\partial}^E &= c(w_i) \widetilde{\nabla}^{TX}_{\overline{w}_i} - \frac{1}{4} c(w_i) c(w_j) c(T(\overline{w}_i, \overline{w}_j)), \\
\sqrt{2}\,\overline{\partial}^{E,*} &= c(\overline{w}_i) \widetilde{\nabla}^{TX}_{w_i} + \frac{\sqrt{2}}{2} \langle T(w_i, w_j), \overline{w}_k \rangle i_{\overline{w}_j} i_{\overline{w}_i} \wedge \overline{w}^k \\
&= c(\overline{w}_i) \widetilde{\nabla}^{TX}_{w_i} + \frac{1}{4} c(\overline{w}_j) c(\overline{w}_i) c(T(w_i, w_j)).
\end{aligned}
\tag{1.4.18}
$$

Taking into account (1.4.3) and (1.4.18), we get

$$
\begin{aligned}
D = {} & c(w_i)\widetilde{\nabla}^{TX}_{\overline{w}_i} + c(\overline{w}_i)\widetilde{\nabla}^{TX}_{w_i} \\
& - \frac{1}{4}c(w_i)c(w_j)c(T(\overline{w}_i,\overline{w}_j)) - \frac{1}{4}c(\overline{w}_i)c(\overline{w}_j)c(T(w_i,w_j)).
\end{aligned}
\tag{1.4.19}
$$

Let $\Gamma^{T^{(1,0)}X} \in T^*X \otimes \operatorname{End}(T^{(1,0)}X)$ be the connection form of $\nabla^{T^{(1,0)}X}$ associated to the frames $\{w_j\}$. Note that for the frame $\{\overline{w}^{j_1}\wedge\cdots\wedge\overline{w}^{j_k}, 1\leqslant j_1 < \cdots < j_k \leqslant n\}$,

$$
\begin{aligned}
\widetilde{\nabla}^{TX} &= d + \langle\Gamma^{T^{(1,0)}X}w_l,\overline{w}_m\rangle\,\overline{w}^m \wedge i_{\overline{w}_l}, \\
\Gamma^{\det} &= \operatorname{Tr}[\Gamma^{T^{(1,0)}X}].
\end{aligned}
\tag{1.4.20}
$$

Comparing with (1.2.38), (1.3.3), (1.3.5), we obtain

$$
\widetilde{\nabla}^{TX} = \nabla^{\mathrm{Cl}} + \frac{1}{4}\sum_{ij}\langle S(\cdot)e_i, e_j\rangle\, c(e_i)c(e_j).
\tag{1.4.21}
$$

Clearly, by (1.2.38),

$$
\frac{1}{4}\Big(\langle S(e_i)e_i, e_j\rangle\,(c(e_i))^2 c(e_j) + \langle S(e_i)e_j, e_i\rangle\, c(e_i)c(e_j)c(e_i)\Big) = -\frac{1}{2}c(S).
\tag{1.4.22}
$$

Thus (1.2.39), (1.4.21), (1.4.22) imply

$$
\begin{aligned}
& c(w_i)\widetilde{\nabla}^{TX}_{\overline{w}_i} + c(\overline{w}_i)\widetilde{\nabla}^{TX}_{w_i} \\
& = D^c - \frac{1}{2}c(S) + \frac{1}{4}\sum_{j\neq i\neq k}\langle S(e_i)e_j, e_k\rangle\, c(e_i)c(e_j)c(e_k) \\
& = D^c - \frac{1}{2}c(S) + \frac{1}{4}{}^c(T_{as}).
\end{aligned}
\tag{1.4.23}
$$

Using (1.2.42), we get

$$
\begin{aligned}
& \frac{1}{4}c(\overline{w}_i)c(\overline{w}_j)c(T(w_i,w_j)) + \frac{1}{4}c(w_i)c(w_j)c(T(\overline{w}_i,\overline{w}_j)) \\
& = \frac{1}{4}\langle T(e_i,e_j),e_k\rangle c(e_i)c(e_j)c(e_k) = \frac{1}{2}{}^c(T_{as}) - \frac{1}{2}c(S).
\end{aligned}
\tag{1.4.24}
$$

Finally (1.4.19), (1.4.23) and (1.4.24) imply (1.4.17). $\qquad\square$

When X is compact, the Euler number $\chi(X, E)$ of the holomorphic vector bundle E is defined by

$$
\chi(X, E) = \sum_{q=0}^{n}(-1)^q \dim H^q(X, E).
\tag{1.4.25}
$$

From Theorems 1.3.9, 1.4.1, 1.4.5, we obtain:

Theorem 1.4.6 (Riemann–Roch–Hirzebruch theorem). *If X is compact, then*

$$
\chi(X, E) = \int_X \operatorname{Td}(T_h X)\, \operatorname{ch}(E).
\tag{1.4.26}
$$

1.4.2   Bismut's Lichnerowicz formula for $\square^E$

Recall that the Bismut connection ∇^B preserves the complex structure on TX by Lemma 1.2.10, thus, as in (1.2.43), it induces a natural connection ∇^B on $\Lambda(T^{*(0,1)}X)$ which preserves its $\mathbb{Z}$-grading. Let $\nabla^{B,\Lambda^{0,\bullet}}$, $\nabla^{B,\Lambda^{0,\bullet}\otimes E}$ be the connections on $\Lambda(T^{*(0,1)}X)$, $\Lambda(T^{*(0,1)}X) \otimes E$ defined by

$$\nabla^{B,\Lambda^{0,\bullet}} = \nabla^B + \langle S(\cdot)w_j, \overline{w}_j \rangle,$$

$$\nabla^{B,\Lambda^{0,\bullet}\otimes E} = \nabla^{B,\Lambda^{0,\bullet}} \otimes 1 + 1 \otimes \nabla^E. \tag{1.4.27}$$

By (1.2.42), $\langle S(\cdot)w_j, \overline{w}_j \rangle$ is a purely imaginary form, thus $\nabla^{B,\Lambda^{0,\bullet}\otimes E}$ is a Hermitian connection on $\Lambda(T^{*(0,1)}X) \otimes E$ which preserves its $\mathbb{Z}$-grading. We denote by $R^{B,\Lambda^{0,\bullet}}$ the curvature of $\nabla^{B,\Lambda^{0,\bullet}}$.

By (1.2.60), (1.3.3) and (1.3.8), as in (1.4.21), we get for $U \in TX$,

$$\nabla^{B,\Lambda^{0,\bullet}\otimes E}_U = \nabla^{\mathrm{Cl}}_U + \frac{1}{2}c(S^B(U)) = \nabla^{\mathrm{Cl}}_U - \frac{1}{4}c(i_U T_{as}). \tag{1.4.28}$$

As in (1.3.19), we denote by $\Delta^{B,\Lambda^{0,\bullet}\otimes E}$ the Bochner Laplacian defined by $\nabla^{B,\Lambda^{0,\bullet}\otimes E}$.

Theorem 1.4.7.

$$D^2 = \Delta^{B,\Lambda^{0,\bullet}\otimes E} + \frac{r^X}{4} + c\left(R^E + \frac{1}{2}\mathrm{Tr}[R^{T^{(1,0)}X}]\right)$$

$$+ \frac{\sqrt{-1}}{2}c(\overline{\partial}\partial\Theta) - \frac{1}{8}|(\partial - \overline{\partial})\Theta|^2. \tag{1.4.29}$$

Proof. Let $R^{\det}$ be the curvature of the holomorphic Hermitian connection on $\det(T^{(1,0)}X)$. Then

$$R^{\det} = \mathrm{Tr}[R^{T^{(1,0)}X}]. \tag{1.4.30}$$

Theorem 1.3.7 and relations (1.2.51), (1.4.17) and (1.4.30) entail (1.4.29). $\square$

Remark 1.4.8. If (X,Θ) is Kähler, then $\nabla^{B,E}$ coincides with $\nabla^{\Lambda(T^{*(0,1)}X)\otimes E}$, the connection on $\Lambda(T^{*(0,1)}X)\otimes E$ induced by the holomorphic Hermitian connections $\nabla^{T^{(1,0)}X}$ and ∇^E. Moreover, $r^X = 2R^{\det}(w_i, \overline{w}_i)$. (1.4.29) reads

$$D^2 = \Delta^{\Lambda(T^{*(0,1)}X)\otimes E} - R^E(w_j, \overline{w}_j)$$

$$+ 2\left(R^E + \frac{1}{2}\mathrm{Tr}[R^{T^{(1,0)}X}]\right)(w_i, \overline{w}_j)\overline{w}^j \wedge i_{\overline{w}_i}. \tag{1.4.31}$$

1.4.3 Bochner–Kodaira–Nakano formula

Let Θ be the real $(1,1)$-form associated to g^{TX} as in (1.2.49). We define the Lefschetz operator $L = (\Theta \wedge) \otimes 1$ on $\Lambda^{\bullet,\bullet}(T^*X) \otimes E$ and its adjoint $\Lambda = i(\Theta)$ with respect to the Hermitian product $\langle \cdot, \cdot \rangle_{\Lambda^{\bullet,\bullet} \otimes E}$ induced by g^{TX} and h^E. For $\{w_j\}_{j=1}^n$ a local orthonormal frame of $T^{(1,0)}X$, we have

$$L = \sqrt{-1} w^j \wedge \overline{w}^j \wedge, \quad \Lambda = -\sqrt{-1} i_{\overline{w}_j} i_{w_j}. \tag{1.4.32}$$

Let us define the formal adjoints $(\nabla^E)^{1,0*}$ of $(\nabla^E)^{1,0}$ and $(\nabla^E)^{0,1*} = \overline{\partial}^{E,*}$ of $(\nabla^E)^{0,1} = \overline{\partial}^E$ with respect to (1.3.14) as in Lemma 1.4.4. We use next the supercommutator defined in (1.3.30), and we apply it on $\Omega^{\bullet,\bullet}(X, E)$ endowed with natural $\mathbb{Z}_2$-grading induced by the parity of degree.

Definition 1.4.9. The *holomorphic* and *anti-holomorphic Kodaira Laplacians* are defined by:

$$\begin{aligned} \overline{\square}^E &= \left[(\nabla^E)^{1,0}, (\nabla^E)^{1,0*} \right], \\ \square^E &= \left[\overline{\partial}^E, \overline{\partial}^{E,*} \right]. \end{aligned} \tag{1.4.33}$$

The *Hermitian torsion operator* is defined by

$$\mathcal{T} := [\Lambda, \partial\Theta] = [i(\Theta), \partial\Theta]. \tag{1.4.34}$$

Let us express now $\mathcal{T}$ in terms of the torsion T of the connection $\widetilde{\nabla}^{TX}$.

Lemma 1.4.10. *We have*

$$\mathcal{T} = \frac{1}{2} \langle T(w_j, w_k), \overline{w}_m \rangle \left[2 w^k \wedge \overline{w}^m \wedge i_{\overline{w}_j} - 2 \delta_{jm} w^k - w^j \wedge w^k \wedge i_{w_m} \right]. \tag{1.4.35}$$

Proof. From (1.2.48), (1.2.54) and (1.4.34), we obtain

$$\begin{aligned} \mathcal{T} = \frac{\sqrt{-1}}{2} \langle T(w_j, w_k), \overline{w}_m \rangle \Big\{ & [\Lambda, \omega^j] \wedge \omega^k \wedge \overline{w}^m \\ & + \omega^j \wedge [\Lambda, \omega^k] \wedge \overline{w}^m + \omega^j \wedge \omega^k \wedge [\Lambda, \overline{w}^m] \Big\}. \end{aligned} \tag{1.4.36}$$

By the formula (1.4.32) for Λ, we easily get

$$[\Lambda, \omega^j] = -\sqrt{-1} i_{\overline{w}_j}, \quad [\Lambda, \overline{w}^m] = \sqrt{-1} i_{w_m}. \tag{1.4.37}$$

Now, (1.4.36), (1.4.37) together with $T(w_j, w_k) = -T(w_k, w_j)$ imply the desired relation (1.4.35). $\square$

We have the following generalization of the usual Kähler identities in the presence of torsion.

Theorem 1.4.11 (generalized Kähler identities).

$$[\overline{\partial}^{E,*}, L] = \sqrt{-1}\left((\nabla^E)^{1,0} + \mathcal{T}\right), \tag{1.4.38a}$$

$$[(\nabla^E)^{1,0*}, L] = -\sqrt{-1}(\overline{\partial}^E + \overline{\mathcal{T}}), \tag{1.4.38b}$$

$$[\Lambda, \overline{\partial}^E] = -\sqrt{-1}\left((\nabla^E)^{1,0*} + \mathcal{T}^*\right), \tag{1.4.38c}$$

$$[\Lambda, (\nabla^E)^{1,0}] = \sqrt{-1}(\overline{\partial}^{E,*} + \overline{\mathcal{T}}^*). \tag{1.4.38d}$$

Proof. Remark that the third and forth formulas are the adjoints of the first two. Thus it suffices to prove (1.4.38a), (1.4.38b). Using (1.4.9) we find

$$\begin{aligned}
\left[\overline{\partial}^{E,*}, L\right] &= [-i_{\overline{w}_i}\widetilde{\nabla}^{TX}_{w_i}, L] - \langle T(w_j, w_k), \overline{w}_k\rangle[i_{\overline{w}_j}, L] \\
&\quad + \frac{1}{2}\langle T(w_j, w_k), \overline{w}_m\rangle[\overline{w}^m \wedge i_{\overline{w}_k} i_{\overline{w}_j}, L].
\end{aligned} \tag{1.4.39}$$

By (1.4.32),

$$[i_{\overline{w}_j}, L] = -\sqrt{-1}\, w^j \wedge, \quad [i_{w_j}, L] = \sqrt{-1}\, \overline{w}^j \wedge. \tag{1.4.40}$$

By (1.2.52), $\widetilde{\nabla}^{TX}_{w_i} L = L \widetilde{\nabla}^{TX}_{w_i}$ so from (1.4.40)

$$[-i_{\overline{w}_j}\widetilde{\nabla}^{TX}_{w_j}, L] = -[i_{\overline{w}_j}, L]\widetilde{\nabla}^{TX}_{w_j} = \sqrt{-1}\, w^j \wedge \widetilde{\nabla}^{TX}_{w_j}. \tag{1.4.41}$$

By (1.4.40), we infer

$$\begin{aligned}
[\overline{w}^m \wedge i_{\overline{w}_k} i_{\overline{w}_j}, L] &= \overline{w}^m \wedge \left([i_{\overline{w}_k}, L]i_{\overline{w}_j} + i_{\overline{w}_k}[i_{\overline{w}_j}, L]\right) \\
&= -\sqrt{-1}\overline{w}^m \wedge (\omega^k \wedge i_{\overline{w}_j} + i_{\overline{w}_k}\omega^j).
\end{aligned} \tag{1.4.42}$$

Relations (1.4.39)–(1.4.42) yield finally

$$\begin{aligned}
\left[\overline{\partial}^{E,*}, L\right] &= \sqrt{-1}\, w^j \wedge \widetilde{\nabla}^{TX}_{w_j} + \sqrt{-1}\, \langle T(w_j, w_k), \overline{w}_k\rangle w^j \\
&\quad + \sqrt{-1}\, \langle T(w_j, w_k), \overline{w}_m\rangle w^k \wedge \overline{w}^m \wedge i_{\overline{w}_j}.
\end{aligned} \tag{1.4.43}$$

Adding (1.4.8) and (1.4.35) shows that $\sqrt{-1}\left((\nabla^E)^{1,0} + \mathcal{T}\right)$ equals the right-hand side of (1.4.43), hence (1.4.38a) holds.

Formula (1.4.38b) can be proved along similar lines as (1.4.38a). Alternatively, as the computation is local, we can choose a local holomorphic frame of E and using (1.4.40), we reduce the proof to the case of a trivial line bundle E. But then (1.4.38b) follows from (1.4.38a) by conjugation. $\square$

Theorem 1.4.12 (Bochner–Kodaira–Nakano formula).

$$\Box^E = \overline{\Box}^E + [\sqrt{-1}R^E, \Lambda] + [(\nabla^E)^{1,0}, \mathcal{T}^*] - [(\nabla^E)^{0,1}, \overline{\mathcal{T}}^*]. \tag{1.4.44}$$

Proof. From (1.4.38d) we deduce that $\overline{\partial}^{E,*} = -\sqrt{-1}\,[\Lambda, (\nabla^E)^{1,0}] - \overline{\mathcal{T}}^*$. Thus

$$\Box^E = [\overline{\partial}^E, \overline{\partial}^{E,*}] = -\sqrt{-1}\,[\overline{\partial}^E, [\Lambda, (\nabla^E)^{1,0}]] - [\overline{\partial}^E, \overline{\mathcal{T}}^*]. \tag{1.4.45}$$

The Jacobi identity (1.3.31) implies

$$\left[\overline{\partial}^E, [\Lambda, (\nabla^E)^{1,0}]\right] = \left[\Lambda, [(\nabla^E)^{1,0}, \overline{\partial}^E]\right] + \left[(\nabla^E)^{1,0}, [\overline{\partial}^E, \Lambda]\right]. \tag{1.4.46}$$

Since $(\overline{\partial}^E)^2 = 0$, $((\nabla^E)^{1,0})^2 = 0$, we have

$$R^E = (\nabla^E)^2 = [(\nabla^E)^{1,0}, \overline{\partial}^E]. \tag{1.4.47}$$

Using the expression of $[\overline{\partial}^E, \Lambda]$ given in (1.4.38c) we find

$$\left[(\nabla^E)^{1,0}, [\overline{\partial}^E, \Lambda]\right] = \sqrt{-1}\left[(\nabla^E)^{1,0}, (\nabla^E)^{1,0*}\right] + \sqrt{-1}\left[(\nabla^E)^{1,0}, \mathcal{T}^*\right]. \tag{1.4.48}$$

Taking into account the definition of $\overline{\square}^E$ (cf. (1.4.33)), we conclude (1.4.44) from (1.4.45)–(1.4.48). $\qquad\square$

Corollary 1.4.13. *Assume that (X, g^{TX}) is Kähler. Then*

$$\square^E = \overline{\square}^E + [\sqrt{-1}R^E, \Lambda], \tag{1.4.49a}$$

$$\Delta = 2\square = 2\overline{\square}. \tag{1.4.49b}$$

Here $\overline{\square} := \overline{\square}^{\mathbb{C}} = \partial\partial^* + \partial^*\partial$; $\square := \square^{\mathbb{C}}$ are usual ∂-Laplacian and $\overline{\partial}$-Laplacian, $\Delta = dd^* + d^*d$ is the Bochner Laplacian on $\Lambda(T^*X)$ and d^* is the adjoint of d.

Therefore, the Hodge decomposition holds for the de Rham cohomology group $H^\bullet(X, \mathbb{C})$:

(a) $H^j(X, \mathbb{C}) \cong \oplus_{p+q=j} H^q(X, \mathscr{O}_X^p) \cong \oplus_{p+q=j} H^{p,q}(X)$,

(b) $H^{p,q}(X) \cong \overline{H^{q,p}(X)}$.

We denote here by $H^{p,q}(X) := H^{p,q}(X, \mathbb{C})$ the Dolbeault cohomology groups.

Proof. Indeed, by Theorem 1.2.8, g^{TX} is Kähler if and only if $\mathcal{T} = 0$, so (1.4.49a) follows trivially from (1.4.44). By taking $E = \mathbb{C}$ with a trivial metric, we obtain $\square = \overline{\square}$. Moreover

$$\Delta = [d, d^*] = [\partial + \overline{\partial}, \partial^* + \overline{\partial}^*] = \square + \overline{\square} + [\partial, \overline{\partial}^*] + [\overline{\partial}, \partial^*], \tag{1.4.50}$$

and the two latter brackets vanish (Problem 1.6). By the real analogue of Theorem 1.4.1 (Hodge theory), $H^\bullet(X, \mathbb{C}) \simeq \mathrm{Ker}(\Delta)$. This completes the proof. $\qquad\square$

Theorem 1.4.14 (Nakano's inequality). *For any $s \in \Omega_0^{\bullet,\bullet}(X, E)$,*

$$\frac{3}{2}\langle\square^E s, s\rangle \geqslant \langle[\sqrt{-1}R^E, \Lambda]s, s\rangle$$

$$-\frac{1}{2}\left(\|\mathcal{T}s\|_{L^2}^2 + \|\mathcal{T}^*s\|_{L^2}^2 + \|\overline{\mathcal{T}}s\|_{L^2}^2 + \|\overline{\mathcal{T}}^*s\|_{L^2}^2\right). \tag{1.4.51}$$

If (X, g^{TX}) is Kähler, then

$$\langle\square^E s, s\rangle \geq \langle[\sqrt{-1}R^E, \Lambda]s, s\rangle. \tag{1.4.52}$$

Proof. Let $s \in \Omega_0^{\bullet,\bullet}(X, E)$. Since

$$
\begin{aligned}
\langle \Box^E s, s \rangle &= \|\overline{\partial}^E s\|_{L^2}^2 + \|\overline{\partial}^{E,*} s\|_{L^2}^2, \\
\langle \overline{\Box}^E s, s \rangle &= \|(\nabla^E)^{1,0} s\|_{L^2}^2 + \|(\nabla^E)^{1,0*} s\|_{L^2}^2,
\end{aligned}
\tag{1.4.53}
$$

we deduce from (1.4.44) that

$$
\begin{aligned}
\|\overline{\partial}^E s\|_{L^2}^2 + \|\overline{\partial}^{E,*} s\|_{L^2}^2 =& \|(\nabla^E)^{1,0} s\|_{L^2}^2 + \|(\nabla^E)^{1,0*} s\|_{L^2}^2 \\
&+ \langle [\sqrt{-1}R^E, \Lambda] s, s \rangle + \langle [(\nabla^E)^{1,0}, \mathcal{T}^*] s, s \rangle - \langle [\overline{\partial}^E, \overline{\mathcal{T}}^*] s, s \rangle.
\end{aligned}
\tag{1.4.54}
$$

By the Cauchy–Schwarz inequality, we find

$$
\left| \langle [(\nabla^E)^{1,0}, \mathcal{T}^*] s, s \rangle \right| \leqslant \frac{1}{2} \left(\|(\nabla^E)^{1,0} s\|_{L^2}^2 + \|(\nabla^E)^{1,0*} s\|_{L^2}^2 + \|\mathcal{T} s\|_{L^2}^2 + \|\mathcal{T}^* s\|_{L^2}^2 \right),
$$

$$
\left| \langle [\overline{\partial}^E, \overline{\mathcal{T}}^*] s, s \rangle \right| \leqslant \frac{1}{2} \left(\|\overline{\partial}^E s\|_{L^2}^2 + \|\overline{\partial}^{E,*} s\|_{L^2}^2 + \|\overline{\mathcal{T}} s\|_{L^2}^2 + \|\overline{\mathcal{T}}^* s\|_{L^2}^2 \right).
$$

Therefore

$$
\begin{aligned}
\frac{3}{2} \left(\|\overline{\partial}^E s\|_{L^2}^2 + \|\overline{\partial}^{E,*} s\|_{L^2}^2 \right) &\geqslant \frac{1}{2} \left(\|(\nabla^E)^{1,0} s\|_{L^2}^2 + \|(\nabla^E)^{1,0*} s\|_{L^2}^2 \right) \\
&+ \langle [\sqrt{-1}R^E, \Lambda] s, s \rangle - \frac{1}{2} \left(\|\mathcal{T} s\|_{L^2}^2 + \|\mathcal{T}^* s\|_{L^2}^2 + \|\overline{\mathcal{T}} s\|_{L^2}^2 + \|\overline{\mathcal{T}}^* s\|_{L^2}^2 \right),
\end{aligned}
\tag{1.4.55}
$$

whereby the conclusion. $\square$

For the purpose of proving vanishing theorems and the spectral gap for forms of bidegree $(0, q)$ with values in a positive bundle (especially on non-compact manifolds or with boundary), we derive sometimes another form of the Bochner–Kodaira–Nakano formula. Set $\widetilde{E} = E \otimes K_X^*$ where

$$
K_X^* = \Lambda^n(T^{(1,0)}X) = \det(T^{(1,0)}X).
$$

Since $K_X \otimes K_X^* \cong \mathbb{C}$, there exists a natural isometry

$$
\begin{aligned}
\Psi &= \sim : \Lambda^{0,q}(T^*X) \otimes E \longrightarrow \Lambda^{n,q}(T^*X) \otimes \widetilde{E}, \\
\Psi s &= \widetilde{s} = (w^1 \wedge \cdots \wedge w^n \wedge s) \otimes (w_1 \wedge \cdots \wedge w_n),
\end{aligned}
\tag{1.4.56}
$$

where $\{w_j\}_{j=1}^n$ a local orthonormal frame of $T^{(1,0)}X$.

Theorem 1.4.15. *For any $s \in \Omega^{0,\bullet}(X, E)$, we have*

$$
\begin{aligned}
\Box^E s =& \Psi^{-1} \overline{\Box}^{\widetilde{E}} \Psi s + R^{E \otimes K_X^*}(w_j, \overline{w}_k) \overline{w}^k \wedge i_{\overline{w}_j} s \\
&+ \Psi^{-1} (\nabla^{\widetilde{E}})^{1,0} \mathcal{T}^* \Psi s - \left[\overline{\partial}^E, \Psi^{-1} \overline{\mathcal{T}}^* \Psi \right] s.
\end{aligned}
\tag{1.4.57}
$$

Proof. We apply (1.4.44) for $\widetilde{s}$:

$$\square^{\widetilde{E}}\widetilde{s} = \overline{\square}^{\widetilde{E}}\widetilde{s} + [\sqrt{-1}R^{\widetilde{E}},\Lambda]\widetilde{s} + [(\nabla^{\widetilde{E}})^{1,0}, \mathcal{T}^*]\widetilde{s} - [\overline{\partial}^{\widetilde{E}}, \overline{\mathcal{T}}^*]\widetilde{s}. \qquad (1.4.58)$$

Since K_X^* is a holomorphic bundle,

$$\overline{\partial}^{\widetilde{E}}\widetilde{s} = (\overline{\partial}^E s)^\sim, \quad \overline{\partial}^{\widetilde{E},*}\widetilde{s} = (\overline{\partial}^{E,*}s)^\sim, \quad \square^{\widetilde{E}}\widetilde{s} = (\square^E s)^\sim. \qquad (1.4.59)$$

Hence $\Psi^{-1}\square^{\widetilde{E}}\widetilde{s} = \square^E s$. Likewise

$$\begin{aligned}
\Psi^{-1}[\overline{\partial}^{\widetilde{E}}, \overline{\mathcal{T}}^*]\widetilde{s} &= [\overline{\partial}^E, \Psi^{-1}\overline{\mathcal{T}}^*\Psi]s, \\
\Psi^{-1}[(\nabla^{\widetilde{E}})^{1,0}, \mathcal{T}^*]\widetilde{s} &= \Psi^{-1}(\nabla^{\widetilde{E}})^{1,0}\mathcal{T}^*\widetilde{s}, \\
\Psi^{-1}\overline{\square}^{\widetilde{E}}\Psi s &= \Psi^{-1}(\nabla^{\widetilde{E}})^{1,0}(\nabla^{\widetilde{E}})^{1,0*}\Psi s.
\end{aligned} \qquad (1.4.60)$$

By (1.4.37) we have

$$[\sqrt{-1}R^{\widetilde{E}},\Lambda] = R^{\widetilde{E}}(w_j, \overline{w}_k)(w^j \wedge i_{w_k} - i_{\overline{w}_j}\overline{w}^k \wedge), \qquad (1.4.61)$$

thus

$$\Psi^{-1}[\sqrt{-1}R^{\widetilde{E}},\Lambda]\widetilde{s} = R^{E\otimes K_X^*}(w_j, \overline{w}_k)\overline{w}^k \wedge i_{\overline{w}_j}s. \qquad (1.4.62)$$

From (1.4.59), (1.4.60) and (1.4.62), we obtain (1.4.57). $\qquad \square$

Remark 1.4.16. Assume now that g^{TX} is Kähler. Then $\mathcal{T} = 0$, and $\widetilde{\nabla}^{TX}$ on $\Lambda(T^{*(0,1)}X) \otimes E$ is induced by the holomorphic Hermitian connections $\nabla^{T^{(1,0)}X}$, ∇^E. On $\Omega^{0,\bullet}(X, F)$, set $\Delta^{0,\bullet} = -\sum_i(\widetilde{\nabla}^{TX}_{w_i}\widetilde{\nabla}^{TX}_{\overline{w}_i} - \widetilde{\nabla}^{TX}_{\nabla^{TX}_{w_i}\overline{w}_i})$. From (1.4.8) and (1.4.10), for $s \in \Omega^{0,\bullet}(X, E)$, we obtain $\Psi^{-1}\overline{\square}^{\widetilde{E}}\Psi s = \Delta^{0,\bullet}s$. We infer from (1.4.57):

$$\square^E s = \Delta^{0,\bullet}s + R^{E\otimes K_X^*}(w_j, \overline{w}_k)\overline{w}^k \wedge i_{\overline{w}_j}s \quad \text{for } s \in \Omega^{0,\bullet}(X, E). \qquad (1.4.63)$$

Corollary 1.4.17. *For any* $s \in \Omega_0^{0,q}(X, E)$,

$$\begin{aligned}
\frac{3}{2}(\|\overline{\partial}^E s\|_{L^2}^2 + \|\overline{\partial}^{E,*}s\|_{L^2}^2) &\geqslant \langle R^{E\otimes K_X^*}(w_j, \overline{w}_k)\overline{w}^k \wedge i_{\overline{w}_j}s, s\rangle \\
&\quad - \frac{1}{2}(\|\mathcal{T}^*\widetilde{s}\|_{L^2}^2 + \|\overline{\mathcal{T}}\widetilde{s}\|_{L^2}^2 + \|\overline{\mathcal{T}}^*\widetilde{s}\|_{L^2}^2).
\end{aligned} \qquad (1.4.64)$$

Proof. By applying (1.4.59) and (1.4.62) to (1.4.51) with $\widetilde{s} \in \Omega^{n,q}(X, \widetilde{E})$, we obtain (1.4.64). Alternatively, we can repeat the proof of Theorem 1.4.14 by replacing (1.4.44) with (1.4.57). $\qquad \square$

1.4.4 Bochner–Kodaira–Nakano formula with boundary term

Keeping the same notations as before, let M be a smooth, relatively compact domain in X. We set $M = \{x \in X : \varrho(x) < 0\}$ where $\varrho \in \mathscr{C}^{\infty}(X)$ satisfies $|d\varrho| = 1$ on ∂M. (This is always possible by replacing ϱ by $\varrho/|d\varrho|$ near ∂M and using a partition of unity argument.) Let $\overline{M}$ be the closure of M.

Let $-e_{\mathrm{n}} \in TM$ be the metric dual of $d\varrho$. Then e_{n} is the inward pointing unit normal at ∂M. We decompose e_{n} as $e_{\mathrm{n}} = e_{\mathrm{n}}^{(1,0)} + e_{\mathrm{n}}^{(0,1)} \in T^{(1,0)}X \oplus T^{(0,1)}X$. Then we have

$$e_{\mathrm{n}}^{(1,0)} = -\overline{w}_j(\varrho)w_j, \quad e_{\mathrm{n}}^{(0,1)} = -w_j(\varrho)\overline{w}_j. \tag{1.4.65}$$

To simplify the notation in the rest of this section, for $s_1, s_2 \in \Omega^{\bullet,\bullet}(\overline{M}, E)$, we will denote by $\langle s_1, s_2 \rangle$ the integral $\int_M \langle s_1, s_2 \rangle_{\Lambda^{\bullet,\bullet} \otimes E, x} dv_X(x)$.

Lemma 1.4.18. For $s_1, s_2 \in \Omega^{\bullet,\bullet}(\overline{M}, E)$, we have

$$\langle \overline{\partial}^E s_1, s_2 \rangle - \langle s_1, \overline{\partial}^{E,*} s_2 \rangle = \int_{\partial M} \langle \overline{\partial}\varrho \wedge s_1, s_2 \rangle_{\Lambda^{\bullet,\bullet} \otimes E} dv_{\partial M},$$
$$\langle s_1, (\nabla^E)^{1,0} s_2 \rangle - \langle (\nabla^E)^{1,0*} s_1, s_2 \rangle = \int_{\partial M} \langle s_1, \partial\varrho \wedge s_2 \rangle_{\Lambda^{\bullet,\bullet} \otimes E} dv_{\partial M}. \tag{1.4.66}$$

Proof. Let γ be a 1-form on X. From (1.2.9), (1.2.10) and Stokes theorem (remark that $dv_X = d\varrho \wedge dv_{\partial M}$ on ∂M), we get

$$\int_M \mathrm{Tr}(\nabla \gamma) dv_X = -\int_{\partial M} \gamma(e_{\mathrm{n}}) dv_{\partial M}. \tag{1.4.67}$$

In view of (1.4.15), (1.4.65) and (1.4.67), we get

$$\langle s_1, \overline{\partial}^E s_2 \rangle - \langle \overline{\partial}^{E,*} s_1, s_2 \rangle = \int_{\partial M} \alpha(e_{\mathrm{n}}) dv_{\partial M}$$
$$= -\int_{\partial M} \langle i_{e_{\mathrm{n}}^{(0,1)}} s_1, s_2 \rangle_{\Lambda^{\bullet,\bullet} \otimes E} dv_{\partial M} = \int_{\partial M} \langle s_1, \overline{\partial}\varrho \wedge s_2 \rangle_{\Lambda^{\bullet,\bullet} \otimes E} dv_{\partial M}. \tag{1.4.68}$$

Similarly, from (1.4.10), (1.4.16) and (1.4.67), we obtain

$$\langle s_1, (\nabla^E)^{1,0} s_2 \rangle - \langle (\nabla^E)^{1,0*} s_1, s_2 \rangle$$
$$= -\int_{\partial M} \langle i_{e_{\mathrm{n}}^{(1,0)}} s_1, s_2 \rangle_{\Lambda^{\bullet,\bullet} \otimes E} dv_{\partial M} = \int_{\partial M} \langle s_1, \partial\varrho \wedge s_2 \rangle_{\Lambda^{\bullet,\bullet} \otimes E} dv_{\partial M}. \tag{1.4.69}$$

The proof of Lemma 1.4.18 is complete. $\qquad\square$

Let $\overline{\partial}_H^{E,*}$ be the Hilbert space adjoint of $\overline{\partial}^E$ on M. By definition, $s \in \mathrm{Dom}(\overline{\partial}_H^{E,*})$ if and only if there exists $s_1 \in L^2(M, \Lambda^{\bullet,\bullet}(T^*X) \otimes E)$ such that for any $s_2 \in \mathrm{Dom}(\overline{\partial}^E)$, $\langle s, \overline{\partial}^E s_2 \rangle = \langle s_1, s_2 \rangle$ and then $\overline{\partial}_H^{E,*} s = s_1$. Let us set

$$B^{0,q}(M, E) = \{s \in \Omega^{0,q}(\overline{M}, E) : i_{e_{\mathrm{n}}^{(0,1)}} s = 0 \text{ on } \partial M\}. \tag{1.4.70}$$

Proposition 1.4.19. *We have* $B^{0,q}(M,E) = \mathrm{Dom}(\overline{\partial}_H^{E,*}) \cap \Omega^{0,q}(\overline{M},E)$ *and* $\overline{\partial}_H^{E,*} = \overline{\partial}^{E,*}$ *on* $B^{0,q}(M,E)$.

Proof. For $s_1 \in \mathrm{Dom}(\overline{\partial}_H^{E,*}) \cap \Omega^{0,q}(\overline{M},E)$, $s_2 \in \Omega^{0,q-1}(\overline{M},E)$,

$$\langle \overline{\partial}_H^{E,*} s_1, s_2 \rangle = \langle s_1, \overline{\partial}^E s_2 \rangle = \langle \overline{\partial}^{E,*} s_1, s_2 \rangle - \int_{\partial M} \langle i_{e_{\mathfrak{n}}^{(0,1)}} s_1, s_2 \rangle_{\Lambda^{\bullet,\bullet} \otimes E} dv_{\partial M}.$$

If $s_2 \in \Omega_0^{0,q-1}(M,E)$, the boundary term vanishes, thus $\langle \overline{\partial}_H^{E,*} s_1, s_2 \rangle = \langle \overline{\partial}^{E,*} s_1, s_2 \rangle$. Since $\Omega_0^{0,q-1}(M,E)$ is dense in $L_{0,q-1}^2(M,E)$, it follows that $\overline{\partial}_H^{E,*} s_1 = \overline{\partial}^{E,*} s_1$. This implies that the boundary term vanishes for all $s_2 \in \Omega^{0,q-1}(\overline{M},E)$, so $i_{e_{\mathfrak{n}}^{(0,1)}} s_1 = 0$ on ∂M. $\qquad\square$

Definition 1.4.20. The *Levi form* of ∂M is the restriction of $\partial \overline{\partial} \varrho$ to the holomorphic tangent bundle of ∂M. For $s \in \Omega^{0,q}(\overline{M},E)$, at $y \in \partial M$, set

$$\mathscr{L}_\varrho(s,s) = (\partial \overline{\partial} \varrho)(w_k, \overline{w}_j) \langle \overline{w}^j \wedge i_{\overline{w}_k} s, s \rangle_{\Lambda^{\bullet,\bullet} \otimes E, y}. \tag{1.4.71}$$

Theorem 1.4.21. *For any* $s \in B^{0,\bullet}(M,E)$, *we have*

$$\|\overline{\partial}^E s\|_{L^2}^2 + \|\overline{\partial}^{E,*} s\|_{L^2}^2 = \|(\nabla^{\widetilde{E}})^{1,0*} \widetilde{s}\|_{L^2}^2 + \langle R^{E \otimes K_X^*}(w_j, \overline{w}_k) \overline{w}^k \wedge i_{\overline{w}_j} s, s \rangle$$
$$- \langle \overline{\partial}^E s, \Psi^{-1} \overline{T} \widetilde{s} \rangle - \langle \Psi^{-1} \overline{T}^* \widetilde{s}, \overline{\partial}^{E,*} s \rangle + \langle T^* \widetilde{s}, (\nabla^{\widetilde{E}})^{1,0*} \widetilde{s} \rangle$$
$$+ \int_{\partial M} \mathscr{L}_\varrho(s,s) \, dv_{\partial M}. \tag{1.4.72}$$

Proof. Since $s \in B^{0,\bullet}(M,E) = \mathrm{Dom}(\overline{\partial}_H^{E,*}) \cap \Omega^{0,q}(\overline{M},E)$, by (1.4.66), we have

$$\|\overline{\partial}^{E,*} s\|_{L^2}^2 = \langle \overline{\partial}^E \overline{\partial}^{E,*} s, s \rangle,$$
$$\|\overline{\partial}^E s\|_{L^2}^2 = \langle \overline{\partial}^{E,*} \overline{\partial}^E s, s \rangle + \int_{\partial M} \langle \overline{\partial}^E s, \overline{\partial} \varrho \wedge s \rangle_{\Lambda^{\bullet,\bullet} \otimes E} dv_{\partial M},$$
$$\langle [\overline{\partial}^E, \Psi^{-1} \overline{T}^* \Psi] s, s \rangle = \langle \overline{\partial}^E s, \Psi^{-1} \overline{T} \widetilde{s} \rangle + \langle \Psi^{-1} \overline{T}^* \widetilde{s}, \overline{\partial}^{E,*} s \rangle, \tag{1.4.73}$$
$$\langle (\nabla^{\widetilde{E}})^{1,0} T^* \widetilde{s}, \widetilde{s} \rangle = \langle T^* \widetilde{s}, (\nabla^{\widetilde{E}})^{1,0*} \widetilde{s} \rangle + \int_{\partial M} \langle \partial \varrho \wedge T^* \widetilde{s}, \widetilde{s} \rangle_{\Lambda^{\bullet,\bullet} \otimes E} dv_{\partial M},$$
$$\langle \Box^{\widetilde{E}} \widetilde{s}, \widetilde{s} \rangle = \|(\nabla^{\widetilde{E}})^{1,0*} \widetilde{s}\|_{L^2}^2 + \int_{\partial M} \langle \partial \varrho \wedge (\nabla^{\widetilde{E}})^{1,0*} \widetilde{s}, \widetilde{s} \rangle_{\Lambda^{\bullet,\bullet} \otimes E} dv_{\partial M}.$$

Thus (1.4.57), (1.4.73) yield

$$\|\overline{\partial}^E s\|_{L^2}^2 + \|\overline{\partial}^{E,*} s\|_{L^2}^2 = \|(\nabla^{\widetilde{E}})^{1,0*} \widetilde{s}\|_{L^2}^2 + \langle R^{E \otimes K_X^*}(w_j, \overline{w}_k) \overline{w}^k \wedge i_{\overline{w}_j} s, s \rangle$$
$$- \langle \overline{\partial}^E s, \Psi^{-1} \overline{T} \widetilde{s} \rangle - \langle \Psi^{-1} \overline{T}^* \widetilde{s}, \overline{\partial}^{E,*} s \rangle + \langle T^* \widetilde{s}, (\nabla^{\widetilde{E}})^{1,0*} \widetilde{s} \rangle \tag{1.4.74}$$
$$+ \int_{\partial M} \left(\langle \overline{\partial}^E s, \overline{\partial} \varrho \wedge s \rangle_{\Lambda^{\bullet,\bullet} \otimes E} + \langle \partial \varrho \wedge ((\nabla^{\widetilde{E}})^{1,0*} + T^*) \widetilde{s}, \widetilde{s} \rangle_{\Lambda^{\bullet,\bullet} \otimes E} \right) dv_{\partial M}.$$

To conclude our theorem, we need to compute the last two terms in (1.4.74). By (1.4.59) and (1.4.65), we infer

$$\left\langle \overline{\partial}^E s, \overline{\partial}\varrho \wedge s \right\rangle_{\Lambda^{\bullet,\bullet}\otimes E} + \left\langle \partial\varrho \wedge ((\nabla^{\widetilde{E}})^{1,0*} + \mathcal{T}^*)\widetilde{s}, \widetilde{s} \right\rangle_{\Lambda^{\bullet,\bullet}\otimes E}$$
$$= \left\langle -i_{e_{\mathfrak{n}}^{(0,1)}}\overline{\partial}^E s + \Psi^{-1}\partial\varrho \wedge ((\nabla^{\widetilde{E}})^{1,0*} + \mathcal{T}^*)\Psi s, s \right\rangle_{\Lambda^{\bullet,\bullet}\otimes E}. \quad (1.4.75)$$

Recall that on $TX \otimes_{\mathbb{R}} \mathbb{C}$, we denote also by $\langle\,,\,\rangle$ the $\mathbb{C}$-bilinear form induced by g^{TX}. As in the proof of Lemma 1.4.4, we denote by $\widetilde{\nabla}^{TX}$ the connection on $\Lambda^{\bullet,\bullet}(T^*X) \otimes E$ induced by ∇^E and $\widetilde{\nabla}^{TX}$. From (1.4.35), we get

$$\mathcal{T}^* = \frac{1}{2}\left\langle T(\overline{w}_j, \overline{w}_k), w_m \right\rangle \left[2\,\overline{w}^j \wedge i_{\overline{w}_m} \wedge i_{w_k} - 2\delta_{jm}i_{w_k} - w^m \wedge i_{w_k}i_{w_j} \right]. \quad (1.4.76)$$

By (1.4.10) and (1.4.76), we obtain

$$(\nabla^{\widetilde{E}})^{1,0*} + \mathcal{T}^* = -i_{w_j}\widetilde{\nabla}^{TX}_{\overline{w}_j} + \left\langle T(\overline{w}_j, \overline{w}_k), w_m \right\rangle \overline{w}^j \wedge i_{\overline{w}_m} \wedge i_{w_k}. \quad (1.4.77)$$

Thus from (1.4.7), (1.4.65), (1.4.77) and $i_{e_{\mathfrak{n}}^{(0,1)}}s = 0$ on ∂M, we have on ∂M,

$$-i_{e_{\mathfrak{n}}^{(0,1)}}\overline{\partial}^E s + \Psi^{-1}\partial\varrho \wedge ((\nabla^{\widetilde{E}})^{1,0*} + \mathcal{T}^*)\Psi s$$
$$= \left\{ -i_{e_{\mathfrak{n}}^{(0,1)}}\overline{w}^j \widetilde{\nabla}^{TX}_{\overline{w}_j} + \langle e_{\mathfrak{n}}^{(0,1)}, w_j \rangle \widetilde{\nabla}^{TX}_{\overline{w}_j} \right.$$
$$\left. + \left\langle T(\overline{w}_j, \overline{w}_k), w_m \right\rangle \left(-\frac{1}{2}i_{e_{\mathfrak{n}}^{(0,1)}}\overline{w}^j \wedge \overline{w}^k \wedge i_{\overline{w}_m} - \langle e_{\mathfrak{n}}^{(0,1)}, w_k \rangle \overline{w}^j \wedge i_{\overline{w}_m} \right) \right\}s$$
$$= \left(-i_{e_{\mathfrak{n}}^{(0,1)}}\overline{w}^j \widetilde{\nabla}^{TX}_{\overline{w}_j} + \langle e_{\mathfrak{n}}^{(0,1)}, w_j \rangle \widetilde{\nabla}^{TX}_{\overline{w}_j} \right)s. \quad (1.4.78)$$

To compute the term in (1.4.78), we use again our boundary condition. Relations (1.4.65) and (1.4.70) yield

$$(i_{\overline{w}_j}s)w_j(\varrho) = -i_{e_{\mathfrak{n}}^{(0,1)}}s = 0 \quad \text{on } \partial M. \quad (1.4.79)$$

Especially, $(i_{\overline{w}_j}s)w_j \in T\partial M \otimes \Lambda^{0,\bullet}(T^*X) \otimes E$ on ∂M. Thus at $y \in \partial M$, we have

$$0 = \langle \widetilde{\nabla}^{TX}_{\overline{w}_j}(i_{e_{\mathfrak{n}}^{(0,1)}}s), i_{\overline{w}_j}s \rangle_{\Lambda^{\bullet,\bullet}\otimes E,y} = \langle \overline{w}^j \widetilde{\nabla}^{TX}_{\overline{w}_j}(i_{e_{\mathfrak{n}}^{(0,1)}}s), s \rangle_{\Lambda^{\bullet,\bullet}\otimes E,y}. \quad (1.4.80)$$

Now

$$\overline{w}^j \widetilde{\nabla}^{TX}_{\overline{w}_j} i_{e_{\mathfrak{n}}^{(0,1)}} = -i_{e_{\mathfrak{n}}^{(0,1)}}\overline{w}^j \widetilde{\nabla}^{TX}_{\overline{w}_j} + \langle e_{\mathfrak{n}}^{(0,1)}, w_j \rangle \widetilde{\nabla}^{TX}_{\overline{w}_j} + \langle \widetilde{\nabla}^{TX}_{\overline{w}_j} e_{\mathfrak{n}}^{(0,1)}, w_k \rangle \overline{w}^j i_{\overline{w}_k}.$$
$$(1.4.81)$$

Moreover, from (1.4.7) and (1.4.65),

$$\langle \widetilde{\nabla}^{TX}_{\overline{w}_j} e_{\mathfrak{n}}^{(0,1)}, w_k \rangle = \overline{w}_j \langle e_{\mathfrak{n}}^{(0,1)}, w_k \rangle - \langle e_{\mathfrak{n}}^{(0,1)}, \widetilde{\nabla}^{TX}_{\overline{w}_j} w_k \rangle$$
$$= -\overline{w}_j(\partial\varrho, w_k) + (\partial\varrho, \widetilde{\nabla}^{TX}_{\overline{w}_j} w_k)$$
$$= -(\widetilde{\nabla}^{TX}_{\overline{w}_j}\partial\varrho)(w_k) = (\overline{\partial}\partial\varrho)(w_k, \overline{w}_j). \quad (1.4.82)$$

Using (1.4.78), (1.4.80), (1.4.81) and (1.4.82), we get at $y \in \partial M$,

$$\Big\langle -i_{e_n^{(0,1)}} \overline{\partial}^E s + \Psi^{-1} \partial \varrho \wedge ((\nabla^{\widetilde{E}})^{1,0*} + \mathcal{T}^*) \Psi s, s \Big\rangle_{\Lambda^{\bullet,\bullet} \otimes E, y}$$
$$= (\partial \overline{\partial} \varrho)(w_k, \overline{w}_j) \langle \overline{w}^j i_{\overline{w}_k} s, s \rangle_{\Lambda^{\bullet,\bullet} \otimes E, y} = \mathscr{L}_\varrho(s, s). \quad (1.4.83)$$

Finally, (1.4.74), (1.4.75) and (1.4.83) imply (1.4.72). $\qquad \square$

Similarly to Corollary 1.4.17, we obtain:

Corollary 1.4.22. *For any* $s \in B^{0,q}(M, E)$,

$$\frac{3}{2} \big(\|\overline{\partial}^E s\|_{L^2}^2 + \|\overline{\partial}^{E,*} s\|_{L^2}^2 \big) \geqslant \frac{1}{2} \|(\nabla^{\widetilde{E}})^{1,0*} \widetilde{s}\|_{L^2}^2 + \big\langle R^{E \otimes K_X^*}(w_j, \overline{w}_k) \overline{w}^k \wedge i_{\overline{w}_j} s, s \big\rangle$$
$$+ \int_{\partial M} \mathscr{L}_\varrho(s, s) \, dv_{\partial M} - \frac{1}{2} \big(\|\mathcal{T}^* \widetilde{s}\|_{L^2}^2 + \|\overline{\mathcal{T}} \widetilde{s}\|_{L^2}^2 + \|\overline{\mathcal{T}}^* \widetilde{s}\|_{L^2}^2 \big). \quad (1.4.84)$$

Our proof of the Bochner–Kodaira–Nakano formula with boundary term (1.4.72) and of the estimate (1.4.84) takes a different route as the usual proof, which consists in integrating by parts starting with the left-hand side and deriving at the end also the Bochner–Kodaira–Nakano formula without boundary. We integrate here directly (1.4.57) and we can easily identify the boundary term. It is remarkable that the curvature and the torsion do not contribute to the boundary integral.

1.5   Spectral gap

As a direct application of the Lichnerowicz formula and Bochner–Kodaira–Nakano formula, we obtain various vanishing theorems and exhibit the spectral gap for the modified Dirac operators. The spectral gap property will play an essential role in our approach to the Bergman kernel.

This section is organized as follows. In Section 1.5.1, we obtain the vanishing theorems and the spectral gap property for the Kodaira Laplacian. In Section 1.5.2 we establish the spectral gap property for a modified Dirac operator on symplectic manifolds.

1.5.1   Vanishing theorem and spectral gap

Lemma 1.5.1 ($\partial \overline{\partial}$-Lemma). *Let* φ *be a smooth, real, d-exact, (q,q)-form on a compact Kähler manifold M; then there exists a smooth, real, $(q-1, q-1)$-form ρ on M such that*

$$\varphi = \sqrt{-1} \partial \overline{\partial} \rho. \quad (1.5.1)$$

Proof. Let $\overline{\partial}^*, \partial^*, d^*$ be the adjoint of $\overline{\partial}, \partial, d$ associated to the Kähler metric g^{TM}. From (1.4.49b) and (1.4.50), $\mathrm{Ker}(\square) = \mathrm{Ker}(d) \cap \mathrm{Ker}(d^*)$.

As φ is d-exact, φ is orthogonal to $\mathrm{Ker}(d^*)$, thus φ is orthogonal to $\mathrm{Ker}(\square)$. By Hodge theory (Theorem 1.4.1) for $E = \Lambda^q(T^{*(1,0)}M)$, there exists a (q,q)-form φ_1 such that

$$\varphi = 2\square\varphi_1 = (d\,d^* + d^*d)\varphi_1. \tag{1.5.2}$$

Again using φ is d-exact and $\mathrm{Im}(d) \cap \mathrm{Im}(d^*) = 0$, we get $\varphi = d\,d^*\varphi_1$.

Let $\psi^{q-1,q}$ (resp. $\psi^{q,q-1}$) be the $(q-1,q)$ (resp. $(q,q-1)$)-component of $d^*\varphi_1$. As φ is a (q,q)-form, we get

$$\varphi = \partial\psi^{q-1,q} + \overline{\partial}\psi^{q,q-1}, \quad \overline{\partial}\psi^{q-1,q} = 0, \quad \partial\psi^{q,q-1} = 0. \tag{1.5.3}$$

(If $q = 1$, we get directly (1.5.3) from the d-exactness of φ).

We claim that if θ is a $(q-1,q)$-form and $\overline{\partial}\theta = 0$, then there exists a $(q-1,q-1)$-form η such that

$$\partial\theta = \partial\overline{\partial}\eta. \tag{1.5.4}$$

By Hodge theory (Theorem 1.4.1) for $E = \Lambda^{q-1}(T^{*(1,0)}M)$, there exists a smooth $(q-1,q-1)$-form η such that

$$\overline{\partial}^*\theta = (\overline{\partial}^*\overline{\partial} + \overline{\partial}\,\overline{\partial}^*)\eta. \tag{1.5.5}$$

(1.4.5) shows that $\mathrm{Im}(\overline{\partial}) \cap \mathrm{Im}(\overline{\partial}^*) = 0$. Thus we get

$$\overline{\partial}^*(\theta - \overline{\partial}\eta) = 0, \qquad \overline{\partial}\,\overline{\partial}^*\eta = 0. \tag{1.5.6}$$

But from $\overline{\partial}(\theta - \overline{\partial}\eta) = 0$, (1.4.49b) and (1.5.6) we know

$$\theta - \overline{\partial}\eta \in \mathrm{Ker}(\overline{\partial}) \cap \mathrm{Ker}(\overline{\partial}^*) = \mathrm{Ker}(\square) = \mathrm{Ker}(\partial) \cap \mathrm{Ker}(\partial^*). \tag{1.5.7}$$

Thus we get (1.5.4) for θ and η.

For $\psi^{q,q-1}$, we will apply (1.5.4) for $\overline{\psi^{q,q-1}}$. Thus there exists ρ such that (1.5.1) holds. As φ is real, we can take ρ as real. $\qquad\square$

For a holomorphic Hermitian line bundle (F, h^F) on a complex manifold M, we will call the curvature R^F associated to the holomorphic Hermitian connection ∇^F on (F, h^F) simply the curvature R^F associated to h^F.

The curvature R^F is a $(1,1)$-form on M and $\sqrt{-1}R^F$ is real. For any holomorphic local frame s of F on an open set U,

$$R^F(x) = \overline{\partial}\partial \log |s(x)|^2_{h^F} \quad \text{on } U. \tag{1.5.8}$$

Definition 1.5.2. A holomorphic line bundle F on a complex manifold M is *positive* (resp. *semi-positive*) if there is a metric h^F on F with associated curvature R^F such that $\sqrt{-1}R^F$ is a positive (resp. semi-positive) $(1,1)$-form on M. F is *negative* if F^* is positive.

Certainly, the notions of positivity (Definition 1.5.2), Griffiths positivity and Nakano positivity (Definition 1.1.6) are equivalent for holomorphic line bundles.

Proposition 1.5.3. *Let F be a holomorphic line bundle on a compact Kähler manifold M. If Ω is a real, closed $(1,1)$-form on M with*

$$[\Omega] = c_1(F) \in H^2(M, \mathbb{R}), \tag{1.5.9}$$

then, up to multiplication by positive constants, there exists a unique metric h^F on F such that $\Omega = \frac{\sqrt{-1}}{2\pi} R^F$, where R^F is the curvature associated to h^F. Thus F is positive if and only if its first Chern class may be represented by a positive form in $H^2(M, \mathbb{R})$.

Proof. Let h_0^F be a Hermitian metric on F and let R_0^F be the curvature associated to h_0^F. Then by (1.5.9), $\Omega - \frac{\sqrt{-1}}{2\pi} R_0^F$ is a real, d-exact, $(1,1)$-form on M. By Lemma 1.5.1, there exists a real function ρ on M such that

$$\Omega = \frac{\sqrt{-1}}{2\pi} R_0^F + \frac{\sqrt{-1}}{2\pi} \overline{\partial} \partial \rho. \tag{1.5.10}$$

From (1.5.8) and (1.5.10), we know $-2\pi\sqrt{-1}\Omega$ is the curvature associated to the metric $e^\rho h_0^F$ on F.

Let h_1^F be another metric on F such that $\Omega = \frac{\sqrt{-1}}{2\pi} R_1^F$. Then there is a real function ρ_1 such that $h_1^F = e^{\rho_1} h^F$. By (1.5.8), we have

$$\overline{\partial} \partial \rho_1 = 0. \tag{1.5.11}$$

Taking the trace of both sides in (1.5.11) and using (1.4.49b), we get $\Delta \rho_1 = 0$. Thus ρ_1 is a constant function on X (cf. Problem 1.9). $\square$

For a variant of Proposition 1.5.3 for singular Hermitian metrics, see Lemma 2.3.5.

Theorem 1.5.4. *Let X be a compact complex manifold of dimension n and F be a positive holomorphic line bundle on X. Then:*

  (a) (Kodaira vanishing theorem) $H^q(X, F \otimes K_X) = 0$, *if* $q > 0$.
  (b) (Nakano vanishing theorem) $H^{r,q}(X, F) = 0$, *if* $r + q > n$.

Proof. Let h^F be a metric on F with associated curvature R^F such that $\omega = \frac{\sqrt{-1}}{2\pi} R^F$ is a positive $(1,1)$-form. Let $g^{TX} := \omega(\cdot, J\cdot)$ be the associated Kähler metric on TX. Then the Hermitian torsion $\mathcal{T} = 0$. Moreover, as $\omega = \sqrt{-1} w^j \wedge \overline{w}^j$, by (1.4.37), we have

$$[\omega, \Lambda] = w^k \wedge i_{w_k} - i_{\overline{w}_k} \overline{w}^k \wedge . \tag{1.5.12}$$

Thus for $s \in \Omega^{r,q}(X, F)$, we have

$$[\omega, \Lambda]s = (r + q - n)s. \tag{1.5.13}$$

Now the Nakano inequality (1.4.51) implies that if s is harmonic, i.e., $\square^F s = 0$, it follows that $s = 0$ wherever $r + q > n$. By Hodge theory (Theorem 1.4.1) for the holomorphic vector bundle $\Lambda^r(T^{*(1,0)}X) \otimes F$, we get (b). (a) is a particular case of (b) for $r = n$. $\square$

Now we will study the spectral gap property for Kodaira Laplacians.

Let (X, J) be a compact complex manifold with complex structure J and $\dim_{\mathbb{C}} X = n$. Consider a holomorphic Hermitian line bundle (L, h^L) on X, and a holomorphic Hermitian vector bundle (E, h^E) on X. Let ∇^E, ∇^L be the holomorphic Hermitian (i.e., Chern) connections on (E, h^E), (L, h^L) with curvatures R^E, R^L. Choose any Riemannian metric g^{TX} on TX, compatible with the complex structure J. Set

$$\omega := \frac{\sqrt{-1}}{2\pi} R^L, \qquad \Theta(\cdot, \cdot) := g^{TX}(J\cdot, \cdot). \qquad (1.5.14)$$

Then ω, Θ are real $(1,1)$-forms on X, and ω is the Chern–Weil representative of the first Chern class $c_1(L)$ of L. Then the Riemannian volume form dv_X of (TX, g^{TX}) is $\Theta^n/n!$.

We will identify the two-form R^L with the Hermitian matrix

$$\dot{R}^L \in \mathrm{End}(T^{(1,0)}X)$$

such that for $W, Y \in T^{(1,0)}X$,

$$R^L(W, \overline{Y}) = \langle \dot{R}^L W, \overline{Y} \rangle. \qquad (1.5.15)$$

Let $\{w_j\}_{j=1}^n$ be a local orthonormal frame of $T^{(1,0)}X$ with dual frame $\{w^j\}_{j=1}^n$. Set

$$\omega_d = -\sum_{l,m} R^L(w_l, \overline{w}_m)\, \overline{w}^m \wedge i_{\overline{w}_l}, \qquad \tau(x) = \sum_j R^L(w_j, \overline{w}_j). \qquad (1.5.16)$$

Then $\omega_d \in \mathrm{End}(\Lambda(T^{*(0,1)}X))$ and R^L acts as the derivative ω_d on $\Lambda(T^{*(0,1)}X)$. By (1.3.32), we have

$$^c(R^L) = \frac{1}{2} \sum_{ij} R^L(e_i, e_j) c(e_i) c(e_j) = -2\omega_d - \tau. \qquad (1.5.17)$$

If we choose $\{w_j\}_{j=1}^n$ to be an orthonormal basis of $T^{(1,0)}X$ such that

$$\dot{R}^L(x) = \mathrm{diag}(a_1(x), \ldots, a_n(x)) \in \mathrm{End}(T_x^{(1,0)}X), \qquad (1.5.18)$$

then

$$\omega_d(x) = -\sum_j a_j(x)\overline{w}^j \wedge i_{\overline{w}_j}, \qquad \tau(x) = \sum_j a_j(x). \qquad (1.5.19)$$

For $p \in \mathbb{N}$, we denote by $L^p := L^{\otimes p}$. By replacing E by $L^p \otimes E$ in (1.4.3), we get

$$D_p = \sqrt{2}\big(\overline{\partial}^{L^p \otimes E} + \overline{\partial}^{L^p \otimes E,*}\big),$$
$$\Box_p = \overline{\partial}^{L^p \otimes E}\, \overline{\partial}^{L^p \otimes E,*} + \overline{\partial}^{L^p \otimes E,*}\, \overline{\partial}^{L^p \otimes E}. \qquad (1.5.20)$$

$D_p^2 = 2\Box_p$ preserves the $\mathbb{Z}$-grading on $\Omega^{0,\bullet}(X, L^p \otimes E)$.

We make the following basic assumption in the rest of this section.

Assumption: $\sqrt{-1}R^L$ is a positive $(1,1)$-form on X, equivalently, for any $0 \neq Y \in T^{(1,0)}X$, we have

$$R^L(Y,\overline{Y}) > 0. \tag{1.5.21}$$

In the notation of (1.5.15)–(1.5.18), the condition (1.5.21) is equivalent to:

$$\dot{R}^L \in \mathrm{End}(T^{(1,0)}X) \text{ is positive-definite, i.e., } a_j(x) > 0 \text{ for any } x \in X, 1 \leqslant j \leqslant n. \tag{1.5.22}$$

Theorem 1.5.5. *There exist $C_0, C_L > 0$ such that for any $p \in \mathbb{N}$ and any $s \in \Omega^{0,>0}(X, L^p \otimes E) = \bigoplus_{q \geqslant 1} \Omega^{0,q}(X, L^p \otimes E)$,*

$$\|D_p s\|_{L^2}^2 \geqslant (2C_0 p - C_L)\|s\|_{L^2}^2. \tag{1.5.23}$$

The spectrum $\mathrm{Spec}(\square_p)$, of the Kodaira Laplacian $\square_p$, is contained in the set $\{0\} \cup]pC_0 - \frac{1}{2}C_L, +\infty[$.

Proof. By (1.4.64) and (1.5.16), we get for any $s \in \Omega^{0,\bullet}(X, L^p \otimes E)$,

$$\|D_p s\|_{L^2}^2 = 2\big(\|\overline{\partial}^{L^p \otimes E} s\|_{L^2}^2 + \|\overline{\partial}^{L^p \otimes E,*} s\|_{L^2}^2\big) \geqslant \frac{4}{3}\langle -\omega_d s, s\rangle\, p - C\|s\|_{L^2}^2. \tag{1.5.24}$$

Hence (1.5.18) and (1.5.22) yield (1.5.23). If $s \in \mathscr{C}^\infty(X, L^p \otimes E)$ is an eigensection of D_p^2 with $D_p^2 s = \lambda s$ and $\lambda \neq 0$, then $0 \neq D_p s \in \Omega^{0,1}(X, L^p \otimes E)$, and $D_p^2 D_p s = \lambda D_p s$. Thus $\lambda \geqslant 2C_0 p - C_L$. This finishes the last part of Theorem 1.5.5. $\square$

By Theorems 1.4.1, 1.5.5, we conclude:

Theorem 1.5.6 (Kodaira–Serre vanishing theorem). *If L is a positive line bundle, then there exists $p_0 > 0$ such that for any $p \geqslant p_0$,*

$$H^q(X, L^p \otimes E) = 0 \quad \text{for any } q > 0. \tag{1.5.25}$$

1.5.2 Spectral gap of modified Dirac operators

Let (X, J) be a compact manifold with almost complex structure J and $\dim_{\mathbb{R}} X = 2n$. Let (L, h^L) be a Hermitian line bundle on X, and let (E, h^E) be a Hermitian vector bundle on X. Let ∇^E, ∇^L be Hermitian connections on (E, h^E), (L, h^L). Let $R^L = (\nabla^L)^2$, $R^E = (\nabla^E)^2$ be the curvatures of ∇^L, ∇^E. Let g^{TX} be any Riemannian metric on TX compatible with the almost complex structure J. We use the notation from (1.5.14)–(1.5.19) now.

Assumption: (1.5.21) holds for R^L.

Set

$$\mu_0 = \inf_{u \in T_x^{(1,0)}X, \, x \in X} R_x^L(u, \overline{u})/|u|_{g^{TX}}^2 = \inf_{x \in X, \, j} a_j(x) > 0. \tag{1.5.26}$$

Let $\nabla^{\det}$ be a Hermitian connection on $\det(T^{(1,0)}X)$ with curvature $R^{\det}$. We denote by D_p^c, $D_{\pm,p}^c$ the spin^c Dirac operator defined in (1.3.15) associated to $L^p \otimes E$ and $\nabla^{\det}$. For $A \in \Lambda^3(T^*X)$, by (1.3.34), set

$$D_p^{c,A} = D_p^c + {}^cA, \quad D_{\pm,p}^{c,A} = D_{\pm,p}^c + {}^cA. \tag{1.5.27}$$

Theorem 1.5.7. *There exists $C_L > 0$ such that for any $p \in \mathbb{N}$ and any $s \in \Omega^{0,>0}(X, L^p \otimes E) = \bigoplus_{q \geqslant 1} \Omega^{0,q}(X, L^p \otimes E)$,*

$$\|D_p^{c,A}s\|_{L^2}^2 \geqslant (2\mu_0 p - C_L)\|s\|_{L^2}^2. \tag{1.5.28}$$

Especially, for p large enough,

$$\mathrm{Ker}(D_{-,p}^{c,A}) = 0. \tag{1.5.29}$$

Proof. At first, we claim that there exists a constant $C > 0$ such that for any $p \in \mathbb{N}$, $s \in \mathscr{C}^\infty(X, L^p \otimes E)$, we have

$$\|\nabla^{L^p \otimes E}s\|_{L^2}^2 - p\langle \tau s, s\rangle \geqslant -C\|s\|_{L^2}^2. \tag{1.5.30}$$

For $s \in \mathscr{C}^\infty(X, L^p \otimes E)$, by Lemma 1.3.4, Theorem 1.3.5, (1.3.20) and (1.5.17), we get

$$\begin{aligned}
\|D_p^c s\|_{L^2}^2 &= \langle (D_p^c)^2 s, s\rangle = \|\nabla^{\mathrm{Cl}}s\|_{L^2}^2 - p\langle \tau s, s\rangle \\
&+ \left\langle \left(\frac{r^X}{4} + \frac{1}{2}\left(R^F + \frac{1}{2}R^{\det}\right)(e_i, e_j)c(e_i)c(e_j)\right)s, s\right\rangle.
\end{aligned} \tag{1.5.31}$$

From (1.3.8), for $s \in \mathscr{C}^\infty(X, L^p \otimes E)$, the following identity holds:

$$\nabla^{\mathrm{Cl}}s = \nabla^{L^p \otimes E}s + \frac{1}{2}(\nabla^{\det} - \nabla^{\det_1})s - \frac{1}{2}\langle S\overline{w}_l, \overline{w}_m\rangle \, \overline{w}^l \wedge \overline{w}^m \wedge s. \tag{1.5.32}$$

From (1.5.31) and (1.5.32), we know there exists $C > 0$, which does not depend on p, such that

$$0 \leqslant \left\|\left(\nabla^{L^p \otimes E} + \frac{1}{2}(\nabla^{\det} - \nabla^{\det_1})\right)s\right\|_{L^2}^2 - p\langle \tau s, s\rangle + C\|s\|_{L^2}^2. \tag{1.5.33}$$

But $(\nabla^{\det} - \nabla^{\det_1})$ is a purely imaginary 1-form, thus $\nabla_1^E = \nabla^E - \frac{1}{2}(\nabla^{\det} - \nabla^{\det_1})$ is a Hermitian connection on E. Applying ∇_1^E on E for (1.5.33), we get (1.5.30).

Relations (1.3.35), (1.5.17) imply that for $s \in \Omega^{0,\bullet}(X, L^p \otimes E)$,

$$\begin{aligned}
\|D_p^{c,A}s\|_{L^2}^2 &= \|\nabla^A s\|_{L^2}^2 - p\langle \tau s, s\rangle - 2p\langle \omega_d s, s\rangle \\
&+ \left\langle \left(\frac{r^X}{4} + {}^c(R^E + \frac{1}{2}R^{\det}) + {}^c(dA) - 2|A|^2\right)s, s\right\rangle.
\end{aligned} \tag{1.5.34}$$

Now we apply (1.5.30) for E replaced by $\Lambda(T^{*(0,1)}X) \otimes E$ with the Hermitian connection ∇^A in (1.3.33). Then we know that the sum of the first two terms of

(1.5.34) is bounded below by $-C\|s\|_{L^2}^2$. For $s \in \Omega^{0,>0}(X, L^p \otimes E)$ the third term of (1.5.34), $-2p(\omega_d s, s)$ is bounded below by $2\mu_0 p\|s\|_{L^2}^2$, by (1.5.19) and (1.5.26), while the norm of the remaining terms of (1.5.34) is bounded by $C\|s\|_{L^2}^2$. Hence we obtain (1.5.28). The proof of Theorem 1.5.7 is completed. □

Theorem 1.5.8. *There exists $C_L > 0$ such that for $p \in \mathbb{N}$, the spectrum of $(D_p^{c,A})^2$ verifies*

$$\mathrm{Spec}((D_p^{c,A})^2) \subset \{0\} \cup]2p\mu_0 - C_L, +\infty[.$$

Proof. The operator $D_p^{c,A}$ changes the parity of $\Omega^{0,\bullet}(X, L^p \otimes E)$, so Theorem 1.5.7 shows that $(D_p^{c,A})^2$ is invertible on $\Omega^{0,\mathrm{odd}}(X, L^p \otimes E)$ for p large enough and its spectrum is in $]2\mu_0 p - C_L, +\infty[$.

Now, if $s \in \Omega^{0,\mathrm{even}}(X, L^p \otimes E)$ is an eigensection of $(D_p^{c,A})^2$ with $(D_p^{c,A})^2 s = \lambda s$ and $\lambda \neq 0$, then $D_p^{c,A} s \neq 0$ and

$$(D_p^{c,A})^2 D_p^{c,A} s = \lambda D_p^{c,A} s. \tag{1.5.35}$$

As $D_p^{c,A} s \in \Omega^{0,\mathrm{odd}}(X, L^p \otimes E)$, Theorem 1.5.7 yields $\lambda > 2\mu_0 p - C_L$. The proof of Theorem 1.5.8 is complete. □

Remark 1.5.9. From Theorems 1.4.5, 1.5.7, 1.5.8, we get another proof of Theorem 1.5.5.

1.6 Asymptotic of the heat kernel

This section is organized as follows. In Section 1.6.1, we explain the main result, Theorem 1.6.1, the asymptotic of the heat kernel. In the rest of this section, we prove Theorem 1.6.1. In Section 1.6.2, we explain that our problem is local. In Section 1.6.3, we do the rescaling operation on coordinates and compute the limit operators. In Section 1.6.4, we obtain the uniform estimate of the heat kernel. Finally, in Section 1.6.5, we prove Theorem 1.6.1.

1.6.1 Statement of the result

Let (X, J) be a compact complex manifold with complex structure J and $\dim_{\mathbb{C}} X = n$. Let (L, h^L) be a holomorphic Hermitian line bundle on X, and (E, h^E) be a holomorphic Hermitian vector bundle on X. Let ∇^E, ∇^L be the holomorphic Hermitian (i.e., Chern) connections on (E, h^E), (L, h^L). Let R^L, R^E be the curvatures of ∇^L, ∇^E. Let g^{TX} be any Riemannian metric on TX compatible with J. We use the notation in Section 1.5.1, especially D_p was defined in (1.5.20).

For $p \in \mathbb{N}$, we write

$$E_p^j := \Lambda^j(T^{*(0,1)}X) \otimes L^p \otimes E, \quad E_p = \oplus_j E_p^j. \tag{1.6.1}$$

We will denote by ∇^{B,E_p} the connection on E_p defined by (1.4.27).

By (1.4.29), $D_p^2 = 2\Box_p$ is a second order elliptic differential operator with principal symbol $\sigma(D_p^2)(\xi) = |\xi|^2$ for $\xi \in T_x^*X$, $x \in X$. The heat operator $e^{-uD_p^2}$ is well defined for $u > 0$. Let $\exp(-uD_p^2)(x, x')$, $(x, x' \in X)$ be the smooth kernel of the heat operator $\exp(-uD_p^2)$ with respect to the Riemannian volume form $dv_X(x')$. Then

$$\exp(-uD_p^2)(x, x') \in (E_p)_x \otimes (E_p)_{x'}^*. \tag{1.6.2}$$

Especially

$$\exp(-uD_p^2)(x, x) \in \mathrm{End}(E_p)_x = \mathrm{End}(\Lambda(T^{*(0,1)}X) \otimes E)_x, \tag{1.6.3}$$

where we use the canonical identification $\mathrm{End}(L^p) = \mathbb{C}$ for any line bundle L on X. Since D_p^2 preserves the $\mathbb{Z}$-grading of the Dolbeault complex $\Omega^{0,\bullet}(X, L^p \otimes E)$, we get from (D.1.7), that $\exp(-uD_p^2)(x, x') \in \bigoplus_j ((E_p^j)_x \otimes (E_p^j)_{x'}^*)$, especially $\exp(-uD_p^2)(x, x) \in \bigoplus_j \mathrm{End}(\Lambda^j(T^{*(0,1)}X) \otimes E)_x$.

We will denote by det the determinant on $T^{(1,0)}X$. The following result is the main result of this section, and the rest of the section is devoted to its proof.

Theorem 1.6.1. *For each $u > 0$ fixed and any $k \in \mathbb{N}$ we have as $p \to \infty$*

$$\begin{aligned}
\exp(-\frac{u}{p}D_p^2)(x, x) &= (2\pi)^{-n} \frac{\det(\dot{R}^L)\exp(2u\omega_d)}{\det(1 - \exp(-2u\dot{R}^L))} \otimes \mathrm{Id}_E \, p^n + o(p^n) \\
&= \prod_{j=1}^n \frac{a_j(x)\left(1 + (e^{-2ua_j(x)} - 1)\overline{w}^j \wedge i_{\overline{w}_j}\right)}{2\pi(1 - e^{-2ua_j(x)})} \otimes \mathrm{Id}_E \, p^n + o(p^n),
\end{aligned} \tag{1.6.4}$$

in the $\mathscr{C}^k$-norm on $\mathscr{C}^\infty(X, \mathrm{End}(\Lambda(T^{*(0,1)}X) \otimes E))$. Here we use the convention that if an eigenvalue $a_j(x)$ (cf. (1.5.18)) of $\dot{R}_x^L$ is zero, then its contribution for $\det(\dot{R}_x^L)/\det(1 - \exp(-2u\dot{R}_x^L))$ is $1/(2u)$. Finally, the convergence in (1.6.4) is uniform as u varies in any compact subset of $\mathbb{R}_+^*$.

1.6.2 Localization of the problem

Let inj^X be the injectivity radius of (X, g^{TX}), and $\varepsilon \in\,]0, \mathrm{inj}^X /4[$.

As X is compact, there exist $\{x_i\}_{i=1}^{N_0}$ such that $\{U_{x_i} = B^X(x_i, \varepsilon)\}_{i=1}^{N_0}$ is a covering of X. Now we use the normal coordinates as in Section 1.2.1. On U_{x_i}, we identify E_Z, L_Z, $\Lambda(T_Z^{*(0,1)}X)$ to E_{x_i}, L_{x_i}, $\Lambda(T_{x_i}^{*(0,1)}X)$ by parallel transport with respect to the connections ∇^E, ∇^L, $\nabla^{B,\Lambda^{0,\bullet}}$ along the curve $[0,1] \ni u \to uZ$. This induces a trivialization of E_p on U_{x_i}. Let $\{e_i\}_i$ be an orthonormal basis of $T_{x_i}X$. Denote by ∇_U the ordinary differentiation operator on $T_{x_i}X$ in the direction U.

Let $\{\varphi_i\}$ be a partition of unity subordinate to $\{U_{x_i}\}$. For $l \in \mathbb{N}$, we define a Sobolev norm on the lth Sobolev space $\boldsymbol{H}^l(X, E_p)$ by

$$\|s\|_{\boldsymbol{H}^l(p)}^2 = \sum_i \sum_{k=0}^l \sum_{i_1 \dots i_k = 1}^{2n} \|\nabla_{e_{i_1}} \dots \nabla_{e_{i_k}}(\varphi_i s)\|_{L^2}^2. \tag{1.6.5}$$

Lemma 1.6.2. *For any* $m \in \mathbb{N}$, *there exists* $C'_m > 0$ *such that for any* $s \in$ $\mathbf{H}^{2m+2}(X, E_p)$, $p \in \mathbb{N}^*$,

$$\|s\|_{\mathbf{H}^{2m+2}(p)} \leqslant C'_m p^{4m+4} \sum_{j=0}^{m+1} p^{-4j} \|D_p^{2j} s\|_{L^2}. \tag{1.6.6}$$

Proof. Let $\widetilde{e}_i(Z)$ be the parallel transport of e_i with respect to ∇^{TX} along the curve $[0, 1] \ni u \to uZ$. Then $\{\widetilde{e}_i\}_i$ is an orthonormal frame on TX. Let Γ^E, Γ^L, $\Gamma^{B,\Lambda^{0,\bullet}}$ be the corresponding connection forms of ∇^E, ∇^L and $\nabla^{B,\Lambda^{0,\bullet}}$ with respect to any fixed frame for $E, L, \Lambda(T^{*(0,1)}X)$ which is parallel along the curve $[0, 1] \ni u \to uZ$ under the trivialization on U_{x_i}. On U_{x_i}, we have

$$D_p = c(\widetilde{e}_j)\Big(\nabla_{\widetilde{e}_j} + p\Gamma^L(\widetilde{e}_j) + \Gamma^{B,\Lambda^{0,\bullet}}(\widetilde{e}_j) + \Gamma^E(\widetilde{e}_j)\Big). \tag{1.6.7}$$

By Theorem A.1.7, (1.6.7), there exists $C > 0$ (independent on p) such that for any $p \geqslant 1$, $s \in \mathbf{H}^2(X, E_p)$, we have $\|s\|_{\mathbf{H}^1(p)}^2 \leqslant C(\|s\|_{\mathbf{H}^2(p)} + \|s\|_{L^2})\|s\|_{L^2}$, and

$$\|s\|_{\mathbf{H}^2(p)} \leqslant C(\|D_p^2 s\|_{L^2} + p^2\|s\|_{L^1}). \tag{1.6.8}$$

Let Q be a differential operator of order $m \in \mathbb{N}$ with scalar principal symbol and with compact support in U_{x_i}. Then

$$[D_p, Q] = p[c(\widetilde{e}_j)\Gamma^L(\widetilde{e}_j), Q] + \Big[c(\widetilde{e}_j)\Big(\nabla_{\widetilde{e}_j} + \Gamma^{B,\Lambda^{0,\bullet}}(\widetilde{e}_j) + \Gamma^E(\widetilde{e}_j)\Big), Q\Big] \tag{1.6.9}$$

which are differential operators of order $m - 1$, m respectively. By (1.6.8), (1.6.9),

$$\begin{aligned}\|Qs\|_{\mathbf{H}^2(p)} &\leqslant C(\|D_p^2 Qs\|_{L^2} + p^2\|Qs\|_{L^2}) \\ &\leqslant C(\|QD_p^2 s\|_{L^2} + p^2\|Qs\|_{L^2} + p^2\|s\|_{\mathbf{H}^{2m+1}(p)}).\end{aligned} \tag{1.6.10}$$

Using (1.6.10), for $m \in \mathbb{N}$, there exists $C'_m > 0$ such that for $p \geqslant 1$,

$$\|s\|_{\mathbf{H}^{2m+2}(p)} \leqslant C'_m(\|D_p^2 s\|_{\mathbf{H}^{2m}(p)} + p^2\|s\|_{\mathbf{H}^{2m+1}(p)}). \tag{1.6.11}$$

From (1.6.11), we get (1.6.6). $\qquad\square$

Let $f : \mathbb{R} \to [0, 1]$ be a smooth even function such that

$$f(v) = \begin{cases} 1 & \text{for } |v| \leqslant \varepsilon/2, \\ 0 & \text{for } |v| \geqslant \varepsilon. \end{cases} \tag{1.6.12}$$

Definition 1.6.3. For $u > 0$, $\varsigma \geqslant 1$, $a \in \mathbb{C}$, set

$$\begin{aligned}\mathbf{F}_u(a) &= \int_{-\infty}^{+\infty} e^{iva} \exp(-\frac{v^2}{2}) f(\sqrt{u}v) \frac{dv}{\sqrt{2\pi}}, \\ \mathbf{G}_u(a) &= \int_{-\infty}^{+\infty} e^{iva} \exp(-\frac{v^2}{2})(1 - f(\sqrt{u}v)) \frac{dv}{\sqrt{2\pi}}, \\ \mathbf{H}_{u,\varsigma}(a) &= \int_{-\infty}^{+\infty} e^{iva} \exp(-\frac{v^2}{2u})(1 - f(\sqrt{\varsigma}v)) \frac{dv}{\sqrt{2\pi u}}.\end{aligned} \tag{1.6.13}$$

The functions $\mathbf{F}_u(a), \mathbf{G}_u(a)$ are even holomorphic functions. The restrictions of $\mathbf{F}_u, \mathbf{G}_u$ to $\mathbb{R}$ lie in the Schwartz space $\mathcal{S}(\mathbb{R})$. Clearly,

$$\mathbf{G}_u(va) = \mathbf{H}_{v^2, \frac{u}{v^2}}(a), \quad \mathbf{F}_u(vD_p) + \mathbf{G}_u(vD_p) = \exp\left(-\frac{v^2}{2}D_p^2\right). \quad (1.6.14)$$

Let $\mathbf{F}_u(vD_p)(x, x'), \mathbf{G}_u(vD_p)(x, x')$ $(x, x' \in X)$ be the smooth kernels associated to $\mathbf{F}_u(vD_p), \mathbf{G}_u(vD_p)$, calculated with respect to the volume form $dv_X(x')$.

Proposition 1.6.4. *For any $m \in \mathbb{N}$, $u_0 > 0, \varepsilon > 0$, there exists $C > 0$ such that for any $x, x' \in X$, $p \in \mathbb{N}^*$, $u > u_0$,*

$$\left|\mathbf{G}_{\frac{u}{p}}(\sqrt{u/p}D_p)(x, x')\right|_{\mathscr{C}^m} \leqslant Cp^{3m+8n+8}\exp(-\frac{\varepsilon^2 p}{16u}). \quad (1.6.15)$$

Here the $\mathscr{C}^m$ norm is induced by $\nabla^L, \nabla^E, \nabla^{B, \Lambda^{0, \bullet}}$ and h^L, h^E, g^{TX}.

Proof. Due to the obvious relation $i^m a^m e^{iva} = \frac{\partial^m}{\partial v^m}(e^{iva})$, we can integrate by parts in the expression of $a^m \mathbf{H}_{u, \varsigma}(a)$ given by (1.6.13) and obtain that for any $m \in \mathbb{N}$ there exists $C_m > 0$ (which depends on ε) such that for $u > 0, \varsigma \geqslant 1$,

$$\sup_{a \in \mathbb{R}} |a|^m |\mathbf{H}_{u, \varsigma}(a)| \leqslant C_m \varsigma^{\frac{m}{2}} \exp(-\frac{\varepsilon^2}{16u\varsigma}). \quad (1.6.16)$$

Here we use that $z^k \exp(-z^2)$ is bounded on $\mathbb{R}_+$.

Let Q be a differential operator of order $m \in \mathbb{N}$ with scalar principal symbol and with compact support in U_{x_i}. From

$$\langle D_p^{m'}\mathbf{H}_{\frac{u}{p}, 1}(D_p)Qs, s'\rangle = \langle s, Q^*\mathbf{H}_{\frac{u}{p}, 1}(D_p)D_p^{m'}s'\rangle,$$

(C.2.5) (or Theorem D.1.3, or using the Fourier transform as in (1.6.16) and the boundedness of the wave operator e^{iuD_p} in L^2-norm implied by (D.2.16)), (1.6.6) and (1.6.16), we know that for $m, m' \in \mathbb{N}$, there exists $C_{m, m'} > 0$ such that for $p \geqslant 1, u > u_0 > 0$,

$$\|D_p^{m'}\mathbf{H}_{\frac{u}{p}, 1}(D_p)Qs\|_{L^2} \leqslant C_{m, m'}p^{2m+2}\exp(-\frac{\varepsilon^2 p}{16u})\|s\|_{L^2}. \quad (1.6.17)$$

We deduce from (1.6.17) that if P, Q are differential operators of order m, m' with compact support in U_{x_i}, U_{x_j} respectively, then there exists $C > 0$ such that for $p \geqslant 1, u \geqslant u_0$,

$$\|P\mathbf{H}_{\frac{u}{p}, 1}(D_p)Qs\|_{L^2} \leqslant Cp^{2m+2m'+4}\exp(-\frac{\varepsilon^2 p}{16u})\|s\|_{L^2}. \quad (1.6.18)$$

By using the Sobolev inequality and (1.6.14) on $U_{x_i} \times U_{x_j}$, we conclude Proposition 1.6.4. $\quad\square$

Using (1.6.13) and the finite propagation speed, Theorem D.2.1 and (D.2.17), it is clear that for $x, x' \in X$, $\mathbf{F}_{\frac{u}{p}}(\sqrt{\frac{u}{p}}D_p)(x, x')$ only depends on the restriction of D_p to $B^X(x, \varepsilon)$, and is zero if $d(x, x') \geqslant \varepsilon$.

1.6.3 Rescaling of the operator D_p^2

Let $\rho : \mathbb{R} \to [0, 1]$ be a smooth even function such that

$$\rho(v) = 1 \text{ if } |v| < 2; \quad \rho(v) = 0 \text{ if } |v| > 4. \tag{1.6.19}$$

Let Φ_E be the smooth self-adjoint section of $\operatorname{End}(\Lambda(T^{*(0,1)}X) \otimes E)$ on X defined by

$$\Phi_E = \frac{r^X}{4} + {}^c(R^E + \frac{1}{2}R^{\det}) + \frac{\sqrt{-1}}{2}{}^c(\bar{\partial}\partial\Theta) - \frac{1}{8}|(\partial - \bar{\partial})\Theta|^2, \tag{1.6.20}$$

(compare (1.4.29)).

We fix $x_0 \in X$. From now on, we identify $B^{T_{x_0}X}(0, 4\varepsilon)$ with $B^X(x_0, 4\varepsilon)$ as in Section 1.2.1. For $Z \in B^{T_{x_0}X}(0, 4\varepsilon)$, we identify E_Z, L_Z, $\Lambda(T_Z^{*(0,1)}X)$ to E_{x_0}, L_{x_0}, $\Lambda(T_{x_0}^{*(0,1)}X)$ by parallel transport with respect to the connections ∇^E, ∇^L, $\nabla^{B,\Lambda^{0,\bullet}}$ along the curve $[0, 1] \ni u \to uZ$. Thus on $B^X(x_0, 4\varepsilon)$, (E, h^E), (L, h^L), $(\Lambda(T^{*(0,1)}X), h^{\Lambda^{0,\bullet}})$, E_p are identified to the trivial Hermitian bundles $(E_{x_0}, h^{E_{x_0}})$, $(L_{x_0}, h^{L_{x_0}})$, $(\Lambda(T_{x_0}^{*(0,1)}X), h^{\Lambda_{x_0}^{0,\bullet}})$, $(E_{p,x_0}, h^{E_{p,x_0}})$. Let $\Gamma^E, \Gamma^L, \Gamma^{B,\Lambda^{0,\bullet}}$ be the corresponding connection forms of ∇^E, ∇^L and $\nabla^{B,\Lambda^{0,\bullet}}$ on $B^X(x_0, 4\varepsilon)$. Then $\Gamma^E, \Gamma^L, \Gamma^{B,\Lambda^{0,\bullet}}$ are skew-adjoint with respect to $h^{E_{x_0}}$, $h^{L_{x_0}}$, $h^{\Lambda_{x_0}^{0,\bullet}}$.

Denote by ∇_U the ordinary differentiation operator on $T_{x_0}X$ in the direction U. From the above discussion,

$$\nabla^{E_{p,x_0}} = \nabla + \rho(|Z|/\varepsilon)\Big(p\Gamma^L + \Gamma^E + \Gamma^{B,\Lambda^{0,\bullet}}\Big)(Z), \tag{1.6.21}$$

defines a Hermitian connection on $(E_{p,x_0}, h^{E_{p,x_0}})$ on $\mathbb{R}^{2n} \simeq T_{x_0}X$ where the identification is given by

$$\mathbb{R}^{2n} \ni (Z_1, \ldots, Z_{2n}) \longrightarrow \sum_i Z_i e_i \in T_{x_0}X. \tag{1.6.22}$$

Here $\{e_i\}_i$ is an orthonormal basis of $T_{x_0}X$.

Let g^{TX_0} be a metric on $X_0 := \mathbb{R}^{2n}$ which coincides with g^{TX} on $B^{T_{x_0}X}(0, 2\varepsilon)$, and $g^{T_{x_0}X}$ outside $B^{T_{x_0}X}(0, 4\varepsilon)$. Let dv_{X_0} be the Riemannian volume form of (X_0, g^{TX_0}). Let $\Delta^{E_{p,x_0}}$ be the Bochner Laplacian associated to $\nabla^{E_{p,x_0}}$ and dv_{X_0} on X_0. Set

$$L_{p,x_0} = \Delta^{E_{p,x_0}} - p\,\rho(|Z|/\varepsilon)(2\omega_{d,Z} + \tau_Z) - \rho(|Z|/\varepsilon)\Phi_{E,Z}. \tag{1.6.23}$$

Then L_p is a self-adjoint operator with respect to the scalar product (1.3.14) induced by $h^{E_{p,x_0}}$, g^{TX_0}. Moreover, L_{p,x_0} coincides with D_p^2 on $B^{TX}(0, 2\varepsilon)$.

Let dv_{TX} be the Riemannian volume form on $(T_{x_0}X, g^{T_{x_0}X})$. Let $\kappa(Z)$ be the smooth positive function defined by the equation

$$dv_{X_0}(Z) = \kappa(Z)dv_{TX}(Z), \tag{1.6.24}$$

with $k(0) = 1$.

Let $\exp(-uL_{p,x_0})(Z, Z')$, $(Z, Z' \in \mathbb{R}^{2n})$ be the smooth kernel of the heat operator $\exp(-uL_{p,x_0})$ on X with respect to $dv_{X_0}(Z')$.

Lemma 1.6.5. *Under the notation in Proposition 1.6.4, the following estimate holds uniformly on $x_0 \in X$:*

$$\left| \exp\left(-\frac{u}{2p} D_p^2 \right)(x_0, x_0) - \exp\left(-\frac{u}{2p} L_{p,x_0} \right)(0,0) \right| \leqslant C p^{8n+8} \exp\left(-\frac{\varepsilon^2 p}{16u} \right).$$
$$(1.6.25)$$

Proof. Let $\widetilde{\mathbf{F}}_u, \widetilde{\mathbf{G}}_u, \widetilde{\mathbf{H}}_{u,\varsigma}$ be the holomorphic functions on $\mathbb{C}$ such that

$$\widetilde{\mathbf{F}}_u(a^2) = \mathbf{F}_u(a), \quad \widetilde{\mathbf{G}}_u(a^2) = \mathbf{G}_u(a), \quad \widetilde{\mathbf{H}}_{u,\varsigma}(a^2) = \mathbf{H}_{u,\varsigma}(a). \qquad (1.6.26)$$

Then $\widetilde{\mathbf{G}}_u(ua) = \widetilde{\mathbf{H}}_{u,1}(a)$ still verifies (1.6.16). And on $\mathbb{R}^{2n}$, Lemma 1.6.2 still holds uniformly on $x_0 \in X$, if we replace D_p^2 therein by L_{p,x_0}. Thus from the proof of Proposition 1.6.4, we still have (1.6.15) for $\widetilde{\mathbf{G}}_u(uL_{p,x_0})$.

Now by the finite propagation speed (Theorem D.2.1), we know that

$$\mathbf{F}_{\frac{u}{p}}\left(\sqrt{\frac{u}{p}} D_p \right) \quad (x_0, \cdot) = \widetilde{\mathbf{F}}_{\frac{u}{p}}\left(\frac{u}{p} L_{p,x_0} \right)(0, \cdot).$$

Thus, we get (1.6.25) by (1.6.14). $\square$

Let S_L be a unit vector of L_{x_0}. Using S_L, we get an isometry $E_{p,x_0} \simeq (\Lambda(T^{*(0,1)}X) \otimes E)_{x_0} =: \mathbf{E}_{x_0}$. As the operator L_{p,x_0} takes values in $\mathrm{End}(E_{p,x_0}) = \mathrm{End}(\mathbf{E})_{x_0}$ (using the natural identification $\mathrm{End}(L^p) \simeq \mathbb{C}$, which does not depend on S_L), thus our formulas do not depend on the choice of S_L. Now, under this identification, we will consider L_{p,x_0} acting on $\mathscr{C}^\infty(X_0, \mathbf{E}_{x_0})$. For $s \in \mathscr{C}^\infty(\mathbb{R}^{2n}, \mathbf{E}_{x_0})$, $Z \in \mathbb{R}^{2n}$ and $t = \frac{1}{\sqrt{p}}$, set

$$\begin{aligned}
(S_t s)(Z) &= s(Z/t), \\
\nabla_t &= S_t^{-1} t \kappa^{1/2} \nabla^{E_{p,x_0}} \kappa^{-1/2} S_t, \\
L_2^t &= S_t^{-1} \kappa^{1/2} t^2 L_{p,x_0} \kappa^{-1/2} S_t.
\end{aligned} \qquad (1.6.27)$$

Put

$$\begin{aligned}
\nabla_{0,\cdot} &= \nabla \cdot + \frac{1}{2} R_{x_0}^L(Z, \cdot), \\
L_2^0 &= -\sum_i (\nabla_{0,e_i})^2 - 2\omega_{d,x_0} - \tau_{x_0}.
\end{aligned} \qquad (1.6.28)$$

Lemma 1.6.6. *When $t \to 0$, we have*

$$\nabla_{t,\cdot} = \nabla_{0,\cdot} + \mathscr{O}(t), \qquad L_2^t = L_2^0 + \mathscr{O}(t). \qquad (1.6.29)$$

Proof. Let $g_{ij}(Z) = g_Z^{TX_0}(e_i, e_j)$, and let $(g^{ij}(Z))_{ij}$ be the inverse of the matrix $(g_{ij}(Z))_{ij}$. Let $\nabla_{e_i}^{TX_0} e_j = \Gamma_{ij}^k(Z) e_k$. By (1.3.19), we know that on $B(0, 4\varepsilon)$,

$$\Delta^{B,E_p} = -g^{ij}(tZ)\left(\nabla_{e_i}^{B,E_p} \nabla_{e_j}^{B,E_p} - \nabla_{\nabla_{e_i}^{TX} e_j}^{B,E_p} \right). \qquad (1.6.30)$$

From (1.5.17), (1.6.21), (1.6.23), (1.6.27) and (1.6.30), we get

$$
\begin{aligned}
\nabla_{t,\cdot} =& \kappa^{1/2}(tZ)\Big(\nabla_\cdot + \rho(|tZ|/\varepsilon)(t\Gamma_{tZ}^{B,\Lambda^{0,\bullet}} + \frac{1}{t}\Gamma_{tZ}^L + t\Gamma_{tZ}^E)\Big)\kappa^{-1/2}(tZ), \\
L_2^t =& -g^{ij}(tZ)\Big(\nabla_{t,e_i}\nabla_{t,e_j} - t\Gamma_{ij}^k(tZ)\nabla_{t,e_k}\Big) \\
& + \rho(|tZ|/\varepsilon)(-2\omega_{d,tZ} - \tau_{tZ} + t^2\Phi_{E,tZ}).
\end{aligned}
\tag{1.6.31}
$$

Since $g^{ij}(0) = \delta_{ij}$, (1.2.31) and (1.6.31) imply (1.6.29). $\qquad\square$

1.6.4 Uniform estimate on the heat kernel

Let $h^{\mathbf{E}_{x_0}}$ be the metric on $\mathbf{E}_{x_0}$ induced by $h_{x_0}^{\Lambda^{0,\bullet}}, h_{x_0}^E$. We also denote by $\langle\cdot,\cdot\rangle_{0,L^2}$ and $\|\cdot\|_{0,L^2}$ the scalar product and the L^2 norm on $\mathscr{C}^\infty(X_0, \mathbf{E}_{x_0})$ induced by $g^{TX_0}, h^{\mathbf{E}_{x_0}}$ as in (1.3.14). For $s \in \mathscr{C}^\infty(T_{x_0}X, \mathbf{E}_{x_0})$, set

$$
\begin{aligned}
\|s\|_{t,0}^2 &:= \|s\|_0^2 = \int_{\mathbb{R}^{2n}} |s(Z)|_{h^{\mathbf{E}_{x_0}}}^2 \, dv_{TX}(Z), \\
\|s\|_{t,m}^2 &= \sum_{l=0}^m \sum_{i_1,\ldots,i_l=1}^{2n} \|\nabla_{t,e_{i_1}}\cdots\nabla_{t,e_{i_l}}s\|_{t,0}^2.
\end{aligned}
\tag{1.6.32}
$$

We denote by $\langle s', s\rangle_{t,0}$ the inner product on $\mathscr{C}^\infty(X_0, \mathbf{E}_{x_0})$ corresponding to $\|\cdot\|_{t,0}^2$. Let $\boldsymbol{H}_t^m$ be the Sobolev space of order m with norm $\|\cdot\|_{t,m}$. Let $\boldsymbol{H}_t^{-1}$ be the Sobolev space of order -1 and let $\|\cdot\|_{t,-1}$ be the norm on $\boldsymbol{H}_t^{-1}$ defined by $\|s\|_{t,-1} = \sup_{0\neq s'\in\boldsymbol{H}_t^1} |\langle s, s'\rangle_{t,0}|/\|s'\|_{t,1}$. If $A \in \mathscr{L}(\boldsymbol{H}_t^m, \boldsymbol{H}_t^{m'})$ $(m, m' \in \mathbb{Z})$, we denote by $\|A\|_t^{m,m'}$ the norm of A with respect to the norms $\|\cdot\|_{t,m}$ and $\|\cdot\|_{t,m'}$.

Since L_{p,x_0} is formally self-adjoint with respect to $\|\cdot\|_{0,L^2}$, L_2^t is also a formally self-adjoint elliptic operator with respect to $\|\cdot\|_{t,0}^2$, and is a smooth family of operators with parameter $x_0 \in X$.

Theorem 1.6.7. *There exist constants* $C_1, C_2, C_3 > 0$ *such that for* $t \in]0,1]$ *and any* $s, s' \in \mathscr{C}_0^\infty(\mathbb{R}^{2n}, \mathbf{E}_{x_0})$,

$$
\begin{aligned}
\langle L_2^t s, s\rangle_{t,0} &\geqslant C_1\|s\|_{t,1}^2 - C_2\|s\|_{t,0}^2, \\
|\langle L_2^t s, s'\rangle_{t,0}| &\leqslant C_3\|s\|_{t,1}\|s'\|_{t,1}.
\end{aligned}
\tag{1.6.33}
$$

Proof. Now from (1.4.29) and (1.5.17),

$$
\langle L_{p,x_0}s, s\rangle_{0,L^2} = \|\nabla^{E_{p,x_0}}s\|_{0,L^2}^2 + \Big\langle\rho(\tfrac{|Z|}{\varepsilon})(-2p\omega_d - p\tau + \Phi_E)s, s\Big\rangle_{0,L^2}.
\tag{1.6.34}
$$

From (1.6.24), (1.6.27), (1.6.32) and (1.6.34),

$$
\langle L_2^t s, s\rangle_{t,0} = \|\nabla_t s\|_{t,0}^2 - \langle\rho(|tZ|/\varepsilon)(-2\omega_{d,tZ} - \tau_{tZ} + t^2\Phi_{E,tZ})s, s\rangle_{t,0}.
\tag{1.6.35}
$$

From (1.6.35), we get (1.6.33). $\qquad\square$

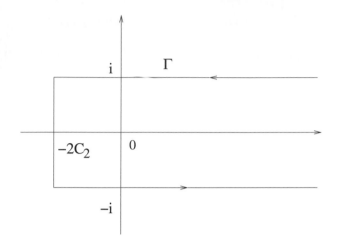

Figure 1.1.

Let Γ be the oriented path in $\mathbb{C}$ defined by Figure 1.1.

Theorem 1.6.8. *There exists $C > 0$ such that for $t \in{]}0,1]$, $\lambda \in \Gamma$, and $x_0 \in X$,*

$$\|(\lambda - L_2^t)^{-1}\|_t^{0,0} \leqslant C,$$
$$\|(\lambda - L_2^t)^{-1}\|_t^{-1,1} \leqslant C(1 + |\lambda|^2). \tag{1.6.36}$$

Proof. As L_2^t is a self-adjoint differential operator, by (1.6.33), $(\lambda - L_2^t)^{-1}$ exists for $\lambda \in \Gamma$. The first inequality of (1.6.36) comes from our choice of Γ. Now, by (1.6.33), for $\lambda_0 \in \mathbb{R}$, $\lambda_0 \leqslant -2C_2$, $(\lambda_0 - L_2^t)^{-1}$ exists, and we have $\|(\lambda_0 - L_2^t)^{-1}\|_t^{-1,1} \leqslant \frac{1}{C_1}$. Then,

$$(\lambda - L_2^t)^{-1} = (\lambda_0 - L_2^t)^{-1} - (\lambda - \lambda_0)(\lambda - L_2^t)^{-1}(\lambda_0 - L_2^t)^{-1}. \tag{1.6.37}$$

Thus (1.6.37) imply for $\lambda \in \Gamma$

$$\|(\lambda - L_2^t)^{-1}\|_t^{-1,0} \leqslant \frac{1}{C_1}\Big(1 + \frac{1}{C}|\lambda - \lambda_0|\Big). \tag{1.6.38}$$

Now we interchange the last two factors in (1.6.37), apply (1.6.38) and obtain

$$\|(\lambda - L_2^t)^{-1}\|_t^{-1,1} \leqslant \frac{1}{C_1} + \frac{|\lambda - \lambda_0|}{C_1{}^2}\Big(1 + \frac{1}{C}|\lambda - \lambda_0|\Big) \tag{1.6.39}$$
$$\leqslant C(1 + |\lambda|^2).$$

The proof of our theorem is complete. $\qquad\square$

Proposition 1.6.9. *Take $m \in \mathbb{N}^*$. There exists $C_m > 0$ such that for $t \in{]}0,1]$, $Q_1, \ldots, Q_m \in \{\nabla_{t,e_i}, Z_i\}_{i=1}^{2n}$ and $s, s' \in \mathscr{C}_0^\infty(\mathbb{R}^{2n}, \mathbf{E}_{x_0})$,*

$$\Big|\langle [Q_1, [Q_2, \ldots, [Q_m, L_2^t] \ldots]]s, s'\rangle_{t,0}\Big| \leqslant C_m \|s\|_{t,1} \|s'\|_{t,1}. \tag{1.6.40}$$

Proof. Note that $[\nabla_{t,e_i}, Z_j] = \delta_{ij}$. Thus by (1.6.31), we know that $[Z_j, L_2^t]$ verifies (1.6.40).

Let $R_\rho^{\Lambda^{0,\bullet}}$, R_ρ^L and R_ρ^E be the curvatures of the connections $\nabla + \rho(|Z|/\varepsilon)\Gamma^{B,\Lambda^{0,\bullet}}$, $\nabla + \rho(|Z|/\varepsilon)\Gamma^L$ and $\nabla + \rho(|Z|/\varepsilon)\Gamma^E$. Then by (1.6.21), (1.6.27),

$$[\nabla_{t,e_i}, \nabla_{t,e_j}] = \left(R_\rho^L + t^2 R_\rho^{\Lambda^{0,\bullet}} + t^2 R_\rho^E\right)(tZ)(e_i, e_j). \tag{1.6.41}$$

Thus from (1.6.31) and (1.6.41), we know that $[\nabla_{t,e_k}, L_2^t]$ has the same structure as L_2^t for $t \in]0, 1]$, i.e., $[\nabla_{t,e_k}, L_2^t]$ is of the type

$$\sum_{ij} a_{ij}(t, tZ)\nabla_{t,e_i}\nabla_{t,e_j} + \sum_i d_i(t, tZ)\nabla_{t,e_i} + c(t, tZ), \tag{1.6.42}$$

and $a_{ij}(t, Z), d_i(t, Z), c(t, Z)$ and their derivatives on Z are uniformly bounded for $Z \in \mathbb{R}^{2n}, t \in [0, 1]$; moreover, they are polynomials in t.

Let $(\nabla_{t,e_i})^*$ be the adjoint of ∇_{t,e_i} with respect to $\langle \cdot, \cdot \rangle_{t,0}$ (see (1.6.32)). Then

$$(\nabla_{t,e_i})^* = -\nabla_{t,e_i} - t(\kappa^{-1}\nabla_{e_i}\kappa)(tZ), \tag{1.6.43}$$

and the last term of (1.6.43) and its derivatives in Z are uniformly bounded in $Z \in \mathbb{R}^{2n}, t \in [0, 1]$.

By (1.6.42) and (1.6.43), (1.6.40) is verified for $m = 1$.

By iteration, we know that $[Q_1, [Q_2, \ldots, [Q_m, L_2^t] \ldots]]$ has the same structure (1.6.42) as L_2^t. By (1.6.43), we get Proposition 1.6.9. $\qquad\square$

Theorem 1.6.10. *For any $t \in]0, 1]$, $\lambda \in \Gamma$, $m \in \mathbb{N}$, the resolvent $(\lambda - L_2^t)^{-1}$ maps $\mathbf{H}_t^m$ into $\mathbf{H}_t^{m+1}$. Moreover for any $\alpha \in \mathbb{N}^{2n}$, there exist $N \in \mathbb{N}$, $C_{\alpha,m} > 0$ such that for $t \in]0, 1]$, $\lambda \in \Gamma$, $s \in \mathscr{C}_0^\infty(X_0, \mathbf{E}_{x_0})$,*

$$\|Z^\alpha(\lambda - L_2^t)^{-1}s\|_{t,m+1} \leqslant C_{\alpha,m}(1 + |\lambda|^2)^N \sum_{\alpha' \leqslant \alpha} \|Z^{\alpha'}s\|_{t,m}. \tag{1.6.44}$$

Proof. For $Q_1, \ldots, Q_m \in \{\nabla_{t,e_i}\}_{i=1}^{2n}$, $Q_{m+1}, \ldots, Q_{m+|\alpha|} \in \{Z_i\}_{i=1}^{2n}$, we can express $Q_1 \ldots Q_{m+|\alpha|}(\lambda - L_2^t)^{-1}$ as a linear combination of operators of the type

$$[Q_1, [Q_2, \ldots [Q_{m'}, (\lambda - L_2^t)^{-1}] \ldots]]Q_{m'+1} \ldots Q_{m+|\alpha|} \quad m' \leqslant m + |\alpha|. \tag{1.6.45}$$

Let $\mathscr{R}_t$ be the family of operators $\mathscr{R}_t = \{[Q_{j_1}, [Q_{j_2}, \ldots [Q_{j_l}, L_2^t] \ldots]]\}$. Clearly, any commutator $[Q_1, [Q_2, \ldots [Q_{m'}, (\lambda - L_2^t)^{-1}] \ldots]]$ is a linear combination of operators of the form

$$(\lambda - L_2^t)^{-1}R_1(\lambda - L_2^t)^{-1}R_2 \ldots R_{m'}(\lambda - L_2^t)^{-1} \tag{1.6.46}$$

with $R_1, \ldots, R_{m'} \in \mathscr{R}_t$.

By Proposition 1.6.9, the norm $\| \quad \|_t^{1,-1}$ of the operators $R_j \in \mathscr{R}_t$ is uniformly bounded by C. By Theorem 1.6.8, we find that there exist $C > 0$ and $N \in \mathbb{N}$ such that the norm $\| \cdot \|_t^{0,1}$ of operators (1.6.46) is dominated by $C(1+|\lambda|^2)^N$. $\square$

Let $e^{-uL_2^t}(Z, Z')$ be the smooth kernels of the operators $e^{-uL_2^t}$ with respect to $dv_{TX}(Z')$. Note that L_2^t are families of differential operators with coefficients in $\mathrm{End}(\mathbf{E}_{x_0}) = \mathrm{End}(\Lambda(T^{*(0,1)}X) \otimes E)_{x_0}$. Let $\pi : TX \times_X TX \to X$ be the natural projection from the fiberwise product of TX on X. Then we can view $e^{-uL_2^t}(Z, Z')$ as smooth sections of $\pi^*(\mathrm{End}(\Lambda(T^{*(0,1)}X) \otimes E))$ on $TX \times_X TX$. Let $\nabla^{\mathrm{End}(\mathbf{E})}$ be the connection on $\mathrm{End}(\Lambda(T^{*(0,1)}X) \otimes E)$ induced by $\nabla^{B,\Lambda^{0,\bullet}}$ and ∇^E. Then $\nabla^{\mathrm{End}(\mathbf{E})}$ induces naturally a $\mathscr{C}^m$-norm for the parameter $x_0 \in X$.

Theorem 1.6.11. *Set $u > 0$ fixed; then for any $m, m' \in \mathbb{N}$, there exists $C > 0$, such that for $t \in]0,1]$, $Z, Z' \in T_{x_0}X$, $|Z|, |Z'| \leqslant 1$,*

$$\sup_{|\alpha|,|\alpha'|\leqslant m} \left| \frac{\partial^{|\alpha|+|\alpha'|}}{\partial Z^\alpha \partial Z'^{\alpha'}} e^{-uL_2^t}(Z, Z') \right|_{\mathscr{C}^{m'}(X)} \leqslant C. \tag{1.6.47}$$

Here $\mathscr{C}^{m'}(X)$ is the $\mathscr{C}^{m'}$ norm for the parameter $x_0 \in X$.

Proof. By (1.6.33) and (1.6.36), (cf. also (C.2.5)), for any $k \in \mathbb{N}^*$,

$$e^{-uL_2^t} = \frac{(-1)^{k-1}(k-1)!}{2\pi i u^{k-1}} \int_\Gamma e^{-u\lambda}(\lambda - L_2^t)^{-k} d\lambda. \tag{1.6.48}$$

For $m \in \mathbb{N}$, let $\mathcal{Q}^m$ be the set of operators $\{\nabla_{t,e_{i_1}} \ldots \nabla_{t,e_{i_j}}\}_{j \leqslant m}$. From Theorem 1.6.10, we deduce that if $Q \in \mathcal{Q}^m$, there are $M \in \mathbb{N}$, $C_m > 0$ such that for any $\lambda \in \Gamma$,

$$\|Q(\lambda - L_2^t)^{-m}\|_t^{0,0} \leqslant C_m(1 + |\lambda|^2)^M. \tag{1.6.49}$$

Observe that if an operator Q_t has the structure and properties after (1.6.42) and $\{a_{ij}(t, Z)\}$ is uniformly positive, then all the above arguments apply for Q_t. Next we study L_2^{t*}, the formal adjoint of L_2^t with respect to (1.6.32). Then L_2^{t*} has the same structure (1.6.31) as the operator L_2^t (in fact, $L_2^{t*} = L_2^t$), especially,

$$\|Q(\lambda - L_2^{t*})^{-m}\|_t^{0,0} \leqslant C_m(1 + |\lambda|^2)^M. \tag{1.6.50}$$

After taking the adjoint of (1.6.50), we get

$$\|(\lambda - L_2^t)^{-m} Q\|_t^{0,0} \leqslant C_m(1 + |\lambda|^2)^M. \tag{1.6.51}$$

From (1.6.48), (1.6.49) and (1.6.51), we have, for $Q, Q' \in \mathcal{Q}^m$,

$$\|Q e^{-uL_2^t} Q'\|_t^{0,0} \leqslant C_m. \tag{1.6.52}$$

Let $\| \quad \|_m$ be the usual Sobolev norm on $\mathscr{C}^\infty(\mathbb{R}^{2n}, \mathbf{E}_{x_0})$ induced by $h^{\mathbf{E}_{x_0}} = h^{\Lambda(T_{x_0}^{*(0,1)}X) \otimes E_{x_0}}$ and the volume form $dv_{TX}(Z)$ as in (1.6.32).

Observe that by (1.6.31), (1.6.32), for any $m \geqslant 0$, there exists $C_m > 0$ such that for $s \in \mathscr{C}^\infty(X_0, \mathbf{E}_{x_0})$, $\mathrm{supp}(s) \subset B^{T_{x_0}X}(0,1)$,

$$\frac{1}{C_m}\|s\|_{t,m} \leqslant \|s\|_m \leqslant C_m\|s\|_{t,m}. \qquad (1.6.53)$$

Now (1.6.52), (1.6.53) together with Sobolev's inequalities implies that if $Q, Q' \in \mathcal{Q}^m$,

$$\sup_{|Z|,|Z'|\leqslant 1} |Q_Z Q'_{Z'} e^{-uL_2^t}(Z, Z')| \leqslant C. \qquad (1.6.54)$$

Thus by (1.6.31), (1.6.54), we derive (1.6.47) for the case when $m' = 0$.

Finally, for U a vector on X,

$$\nabla_U^{\pi^* \, \mathrm{End}(\mathbf{E})} e^{-uL_2^t} = \frac{(-1)^{k-1}(k-1)!}{2\pi i u^{k-1}} \int_\Gamma e^{-u\lambda} \nabla_U^{\pi^* \, \mathrm{End}(\mathbf{E})} (\lambda - L_2^t)^{-k} d\lambda. \qquad (1.6.55)$$

We use a similar formula as (1.6.46) for $\nabla_U^{\pi^* \, \mathrm{End}(\mathbf{E})}(\lambda - L_2^t)^{-k}$, where we replace $\mathscr{R}_t$ by $\{\nabla_U^{\pi^* \, \mathrm{End}(\mathbf{E})} L_2^t\}$. Moreover, we remark that $\nabla_U^{\pi^* \, \mathrm{End}(\mathbf{E})} L_2^t$ is a differential operator on $T_{x_0}X$ with the same structure as L_2^t. Then the above argument yields (1.6.47) for $m' \geqslant 1$. $\qquad \square$

Theorem 1.6.12. *There exists $C > 0$ such that for $t \in [0,1]$,*

$$\left\|\left((\lambda - L_2^t)^{-1} - (\lambda - L_2^0)^{-1}\right) s\right\|_{0,0} \leqslant Ct(1 + |\lambda|^4) \sum_{|\alpha|\leqslant 3} \|Z^\alpha s\|_{0,0}. \qquad (1.6.56)$$

Proof. Remark that by (1.6.31), (1.6.32), for $t \in [0,1]$, $k \geqslant 1$,

$$\|s\|_{t,k} \leqslant C \sum_{|\alpha|\leqslant k} \|Z^\alpha s\|_{0,k}. \qquad (1.6.57)$$

An application of Taylor expansion for (1.6.31) leads to the following equation, if s, s' have compact support:

$$\left|\langle(L_2^t - L_2^0)s, s'\rangle_{0,0}\right| \leqslant Ct\|s'\|_{t,1} \sum_{|\alpha|\leqslant 3} \|Z^\alpha s\|_{0,1}. \qquad (1.6.58)$$

Thus we get

$$\left\|(L_2^t - L_2^0)s\right\|_{t,-1} \leqslant Ct \sum_{|\alpha|\leqslant 3} \|Z^\alpha s\|_{0,1}. \qquad (1.6.59)$$

Note that

$$(\lambda - L_2^t)^{-1} - (\lambda - L_2^0)^{-1} = (\lambda - L_2^t)^{-1}(L_2^t - L_2^0)(\lambda - L_2^0)^{-1}. \qquad (1.6.60)$$

After taking the limit, we know that Theorems 1.6.8–1.6.10 still hold for $t = 0$. From Theorem 1.6.10, (1.6.59) and (1.6.60), we infer (1.6.56). $\qquad \square$

Theorem 1.6.13. *For $u > 0$ fixed, there exists $C > 0$, such that for $t \in]0,1]$, $Z, Z' \in T_{x_0}X$, $|Z|, |Z'| \leqslant 1$,*

$$\left| (e^{-uL_2^t} - e^{-uL_2^0})(Z, Z') \right| \leqslant Ct^{1/(2n+1)}. \tag{1.6.61}$$

Proof. Let $J_{x_0}^0$ be the vector space of square integrable sections of $\mathbf{E}_{x_0}$ over $\{Z \in T_{x_0}X, |Z| \leqslant 2\}$. If $s \in J_{x_0}^0$, put $\|s\|_{(1)}^2 = \int_{|Z| \leqslant 2} |s|_{h^{\mathbf{E}_{x_0}}}^2 dv_{TX}(Z)$. Let $\|A\|_{(1)}$ be the operator norm of $A \in \mathscr{L}(J_{x_0}^0)$ with respect to $\| \quad \|_{(1)}$. Let $u > 0$ fixed. By (1.6.48) and (1.6.56), we get: There exists $C > 0$, such that for $t \in]0,1]$,

$$\begin{aligned}
\|(e^{-uL_2^t} - e^{-uL_2^0})\|_{(1)} &\leqslant \frac{1}{2\pi} \int_\Gamma |e^{-u\lambda}| \, \|(\lambda - L_2^t)^{-1} - (\lambda - L_2^0)^{-1}\|_{(1)} d\lambda \\
&\leqslant C't \int_\Gamma e^{-u\mathrm{Re}(\lambda)}(1 + |\lambda|^4) d\lambda \leqslant Ct.
\end{aligned} \tag{1.6.62}$$

Let $\phi : \mathbb{R}^{2n} \to [0,1]$ be a smooth function with compact support, equal 1 near 0, such that $\int_{T_{x_0}X} \phi(Z) dv_{TX}(Z) = 1$. Take $\nu \in]0,1]$. By the proof of Theorem 1.6.11, $e^{-uL_2^0}$ verifies the similar inequality as in (1.6.47). Thus by Theorem 1.6.11, there exists $C > 0$ such that if $|Z|, |Z'| \leqslant 1$, $U, U' \in \mathbf{E}_{x_0}$,

$$\begin{aligned}
\Bigg| &\left\langle (e^{-uL_2^t} - e^{-uL_2^0})(Z, Z')U, U' \right\rangle \\
&- \int_{T_{x_0}X \times T_{x_0}X} \left\langle (e^{-uL_2^t} - e^{-uL_2^0})(Z - W, Z' - W')U, U' \right\rangle \\
&\qquad \times \frac{1}{\nu^{4n}} \phi(W/\nu)\phi(W'/\nu) dv_{TX}(W) dv_{TX}(W') \Bigg| \leqslant C\nu |U||U'|. \tag{1.6.63}
\end{aligned}$$

On the other hand, by (1.6.62),

$$\begin{aligned}
\Bigg| \int_{T_{x_0}X \times T_{x_0}X} &\left\langle (e^{-uL_2^t} - e^{-uL_2^0})(Z - W, Z' - W')U, U' \right\rangle \\
&\times \frac{1}{\nu^{4n}} \phi(W/\nu)\phi(W'/\nu) dv_{TX}(W) dv_{TX}(W') \Bigg| \leqslant Ct\frac{1}{\nu^{2n}} |U||U'|. \tag{1.6.64}
\end{aligned}$$

By taking $\nu = t^{1/(2n+1)}$, we get (1.6.61). $\qquad \square$

1.6.5 Proof of Theorem 1.6.1

Note that in (1.6.24), $\kappa(0) = 1$. Recall also that $t = 1/\sqrt{p}$. By (1.6.27), for $s \in \mathscr{C}_0^\infty(X_0, E_{x_0})$,

$$\begin{aligned}
(e^{-uL_2^t}s)(Z) &= (S_t^{-1}\kappa^{\frac{1}{2}}e^{-\frac{u}{p}L_p}\kappa^{-\frac{1}{2}}S_t s)(Z) \\
&= \kappa^{\frac{1}{2}}(tZ) \int_{\mathbb{R}^{2n}} \exp(-\frac{u}{p}L_p)(tZ, Z')(S_t s)(Z')\kappa^{\frac{1}{2}}(Z') \, dv_{TX}(Z'). \tag{1.6.65}
\end{aligned}$$

Thus, for $Z, Z' \in T_{x_0}X$,

$$\exp(-\frac{u}{p}L_{p,x_0})(Z, Z') = p^n e^{-uL_2^t}(Z/t, Z'/t)\kappa^{-1/2}(Z')\kappa^{-1/2}(Z). \qquad (1.6.66)$$

By Theorem 1.6.13, (1.6.25), (1.6.66), we get that uniformly on $x_0 \in X$, we have

$$\exp(-\frac{u}{p}D_p^2)(x_0, x_0) - p^n \exp(-uL_{2,x_0}^0)(0, 0) = o(p^n). \qquad (1.6.67)$$

By (1.5.19), (1.6.28), (E.2.4) and (E.2.5), we get with the convention in Theorem 1.6.1,

$$\exp(-uL_2^0)(0, 0) = \frac{1}{(2\pi)^n} \frac{\det(\dot{R}_{x_0}^L)\exp(2u\omega_{d,x_0})}{\det(1 - \exp(-2u\dot{R}_{x_0}^L))}. \qquad (1.6.68)$$

Moreover, for any j fixed,

$$\exp(-2ua_j(x_0)\overline{w}^j \wedge i_{\overline{w}_j}) = 1 + (\exp(-2ua_j(x_0)) - 1)\overline{w}^j \wedge i_{\overline{w}_j}. \qquad (1.6.69)$$

From (1.6.67)–(1.6.69), we get (1.6.4).

If u varies in a compact set of $\mathbb{R}_+^*$, the constant C in (1.6.47) and (1.6.61) is uniformly bounded, so the convergence of (1.6.4) is uniform. The proof of Theorem 1.6.1 is complete.

1.7 Demailly's holomorphic Morse inequalities

We will use the notation of Section 1.6.1 and (1.5.14)–(1.5.19).

Let $X(q)$ be the set of points x of X such that $\sqrt{-1}R_x^L$ is non-degenerate and $\dot{R}_x^L \in \mathrm{End}(T_x^{(1,0)}X)$ has exactly q negative eigenvalues. Set $X(\leqslant q) = \cup_{i=0}^q X(i)$, $X(\geqslant q) = \cup_{i=q}^n X(i)$.

Theorem 1.7.1. *Let X be a compact complex manifold with $\dim X = n$, and let L, E be holomorphic vector bundles on X, $\mathrm{rk}(L) = 1$. As $p \to \infty$, the following strong Morse inequalities hold for every $q = 0, 1, \ldots, n$:*

$$\sum_{j=0}^q (-1)^{q-j} \dim H^j(X, L^p \otimes E) \leqslant \mathrm{rk}(E)\frac{p^n}{n!} \int_{X(\leqslant q)} (-1)^q \left(\frac{\sqrt{-1}}{2\pi}R^L\right)^n + o(p^n),$$

$$(1.7.1)$$

with equality for $q = n$ (asymptotic Riemann–Roch–Hirzebruch formula).

In particular, we get the weak Morse inequalities

$$\dim H^q(X, L^p \otimes E) \leqslant \mathrm{rk}(E)\frac{p^n}{n!} \int_{X(q)} (-1)^q \left(\frac{\sqrt{-1}}{2\pi}R^L\right)^n + o(p^n). \qquad (1.7.2)$$

Proof. For $0 \leqslant q \leqslant n$, set

$$B_q^p = \dim H^q(X, L^p \otimes E). \tag{1.7.3}$$

Remark that the operator D_p^2 preserves the $\mathbb{Z}$-grading of the Dolbeault complex $\Omega^{0,\bullet}(X, L^p \otimes E)$. We will denote by $\mathrm{Tr}_q[e^{-\frac{u}{p}D_p^2}]$ the trace of $e^{-\frac{u}{p}D_p^2}$ acting on $\Omega^{0,q}(X, L^p \otimes E)$, then we have

$$\mathrm{Tr}_q[e^{-\frac{u}{p}D_p^2}] = \int_X \mathrm{Tr}_q\left[e^{-\frac{u}{p}D_p^2}(x,x)\right] dv_X(x). \tag{1.7.4}$$

Lemma 1.7.2. *For any $u > 0$, $p \in \mathbb{N}^*$, $0 \leqslant q \leqslant n$, we have*

$$\sum_{j=0}^q (-1)^{q-j} B_j^p \leqslant \sum_{j=0}^q (-1)^{q-j} \mathrm{Tr}_j\left[\exp(-\frac{u}{p}D_p^2)\right], \tag{1.7.5}$$

with equality for $q = n$.

Proof. If λ is an eigenvalue of D_p^2, set F_j^λ be the corresponding finite-dimensional eigenspace in $\Omega^{0,j}(X, L^p \otimes E)$. We claim that

$$\overline{\partial}^{L^p \otimes E}(F_j^\lambda) \subset F_{j+1}^\lambda, \quad \text{and} \quad \overline{\partial}^{L^p \otimes E,*}(F_{j+1}^\lambda) \subset F_j^\lambda. \tag{1.7.6}$$

In fact, if $s \in F_j^\lambda$, then $D_p^2 s = \lambda s$. By (1.5.20), $\overline{\partial}^{L^p \otimes E}$ commutes with D_p^2, thus $D_p^2 \overline{\partial}^{L^p \otimes E} s = \overline{\partial}^{L^p \otimes E} D_p^2 s = \lambda \overline{\partial}^{L^p \otimes E} s$. In the same way, we get the second equation of (1.7.6).

Thus we have the complex

$$0 \longrightarrow F_0^\lambda \xrightarrow{\overline{\partial}^{L^p \otimes E}} F_1^\lambda \xrightarrow{\overline{\partial}^{L^p \otimes E}} \cdots \xrightarrow{\overline{\partial}^{L^p \otimes E}} F_n^\lambda \longrightarrow 0. \tag{1.7.7}$$

If $\lambda = 0$, then $F_j^0 \simeq H^j(X, L^p \otimes E)$. If $\lambda > 0$, we claim that the complex (1.7.7) is exact. In fact, if $\overline{\partial}^{L^p \otimes E} s = 0$ and $s \in F_j^\lambda$, then by (1.5.20),

$$s = \lambda^{-1} D_p^2 s = \lambda^{-1} \overline{\partial}^{L^p \otimes E}(\overline{\partial}^{L^p \otimes E,*} s). \tag{1.7.8}$$

From (1.7.8), we know $s \in \overline{\partial}^{L^p \otimes E}(F_{j-1}^\lambda)$. Thus for $\lambda > 0$, the complex (1.7.7) is exact and

$$\sum_{j=0}^q (-1)^{q-j} \dim F_j^\lambda = \dim(\overline{\partial}^{L^p \otimes E}(F_q^\lambda)) \geqslant 0 \tag{1.7.9}$$

with equality when $q = n$. Now

$$\mathrm{Tr}_j[\exp(-\frac{u}{p}D_p^2)] = B_j^p + \sum_{\lambda > 0} e^{-\frac{u}{p}\lambda} \dim F_j^\lambda. \tag{1.7.10}$$

(1.7.9) and (1.7.10) yield (1.7.5). $\square$

We denote by $\text{Tr}_{\Lambda^{0,q}}$ the trace on $\Lambda^q(T^{*(0,1)}X)$. By (1.6.69), in the notation of (1.5.19),

$$\text{Tr}_{\Lambda^{0,q}}[\exp(2u\omega_d)] = \sum_{j_1<j_2<\cdots<j_q} \exp\left(-2u\sum_{i=1}^q a_{j_i}(x)\right). \qquad (1.7.11)$$

Thus by the second equality in (1.6.4), $\frac{(\det(\dot{R}^L/(2\pi)))}{\det(1-\exp(-2u\dot{R}^L))}\text{Tr}_{\Lambda^{0,q}}[\exp(2u\omega_d)]$ is uniformly bounded for $x \in X, u > 1, 0 \leqslant q \leqslant n$, and for any $x_0 \in X, 0 \leqslant q \leqslant n$,

$$\lim_{u\to\infty} \frac{\det(\dot{R}^L/(2\pi))\,\text{Tr}_{\Lambda^{0,q}}[\exp(2u\omega_d)]}{\det(1-\exp(-2u\dot{R}^L))}(x_0) = 1_{X(q)}(-1)^q \det\left(\frac{\dot{R}^L}{2\pi}\right)(x_0). \qquad (1.7.12)$$

The function $1_{X(q)}$ is defined by 1 on $X(q)$, 0 otherwise. From Theorem 1.6.1, (1.7.4) and (1.7.5), we have

$$\varlimsup_{p\to\infty} p^{-n} \sum_{j=0}^q (-1)^{q-j} B_j^p$$
$$\leqslant \text{rk}(E)\int_X \frac{\det(\dot{R}^L/(2\pi))\sum_{j=0}^q(-1)^{q-j}\,\text{Tr}_{\Lambda^{0,j}}[\exp(2u\omega_d)]}{\det(1-\exp(-2u\dot{R}^L))}dv_X(x), \qquad (1.7.13)$$

for any q with $0 \leqslant q \leqslant n$ and any $u > 0$. Using (1.7.12), (1.7.13) and dominate convergence, we get

$$\varlimsup_{p\to\infty} p^{-n} \sum_{j=0}^q (-1)^{q-j} B_j^p \leqslant (-1)^q \,\text{rk}(E) \int_{\cup_{i=0}^q X(i)} \det\left(\frac{\dot{R}^L}{2\pi}\right)(x)dv_X(x). \qquad (1.7.14)$$

But (1.5.18) entails

$$\det\left(\frac{\dot{R}^L}{2\pi}\right)(x)dv_X(x) = \prod_j \frac{a_j(x)}{2\pi}dv_X(x) = \left(\frac{\sqrt{-1}}{2\pi}R^L\right)^n/n!. \qquad (1.7.15)$$

Relations (1.7.14), (1.7.15) imply (1.7.1). For $q = n$, we apply (1.7.5) with equality, so we get (1.7.1) with equality. (1.7.2) follows by subtracting inequalities (1.7.1) for q and $q+1$ (or directly from Theorem 1.6.1, (1.7.10) and (1.7.12)).

The proof of Theorem 1.7.1 is complete. □

Problems

Problem 1.1. Verify (1.3.2), (1.3.31) and (1.3.41). With the notation from (1.3.44) verify that

$$\text{Ker}(D^{c,A}) = \text{Ker}((D^{c,A})^2) = \text{Ker}(D_+^{c,A}) \oplus \text{Ker}(D_-^{c,A}).$$

Problem 1.2. In Section 1.3.2, we can always assume that ∇^{Cl} on $\Lambda(T^{*(0,1)}X)\otimes E$ is induced by $\nabla^{TX},\nabla^{\det_1}$ and a Hermitian connection ∇^E_1 on (E,h^E). (Hint: $\frac{1}{2}(\nabla^{\det}-\nabla^{\det_1})$ is a purely imaginary 1-form.)

Problem 1.3. In local coordinates $(x_1,\ldots,x_n)$ of a Riemannian manifold (X,g^{TX}), we set $f_j=\frac{\partial}{\partial x_j}$, $g_{ij}(x)=\langle f_i,f_j\rangle_{g^{TX}}(x)$. Let $(g^{ij}(x))$ be the inverse of the matrix $(g_{ij}(x))$. Verify that in (1.3.19),

$$\Delta^F=-\sum_{ij}g^{ij}(x)\left(\nabla^F_{f_i}\nabla^F_{f_j}-\nabla^F_{\nabla^{TX}_{f_i}f_j}\right).$$

Problem 1.4. In the context of (1.4.5) show that

$$\mathrm{Ker}(D)=\mathrm{Ker}(\overline{\partial})\cap\mathrm{Ker}(\overline{\partial}^*),\quad \mathrm{Im}(\overline{\partial})\cap\mathrm{Im}(\overline{\partial}^*)=0.$$

Thus $\mathrm{Ker}(D)$, $\mathrm{Im}(\overline{\partial})$ and $\mathrm{Im}(\overline{\partial}^*)$ are pairwise orthogonal.

Problem 1.5. Verify Remark 1.4.3 (cf. also [9, §2]). By Theorem A.3.2 for $k\in\mathbb{Z}$, and D^2 is elliptic, for s a distribution with values in E, $D^2s=0$ in the sense of distributions implies that $s\in\Omega^{0,\bullet}(X,E)$ (cf. also [148, Chap. 3], [238, §7.4]). Using this fact, show that $\mathrm{Ker}(D)\subset\Omega^{0,\bullet}(X,E)\cap L^2_{0,\bullet}(X,E)$ is closed in $L^2_{0,\bullet}(X,E)$.

By the Schwartz kernel theorem, $P(x,y)$ is a distribution on $X\times X$ with values in $(\Lambda(T^{*(0,1)}X)\otimes E)_x\otimes(\Lambda(T^{*(0,1)}X)\otimes E)^*_y$. Prove first

$$D^2_xP(x,y)=0,\quad D^2_yP(x,y)=0,$$

in the sense of distributions. Here we identify $(\Lambda(T^{*(0,1)}X)\otimes E)^*_y$ to $(\Lambda(T^{*(0,1)}X)\otimes E)_y$ by the Hermitian product $\langle\cdot,\cdot\rangle_{\Lambda^{0,\bullet}\otimes E}$. Now as $D^2_x+D^2_y$ is an elliptic operator on $X\times X$, $(D^2_x+D^2_y)P(x,y)=0$ in the sense of distributions implies $P(x,y)$ is $\mathscr{C}^\infty$ for $x,y\in X$.

Problem 1.6. Let X be a Kähler manifold.

a) Show that $[\partial,\overline{\partial}^*]=0$, $[\overline{\partial},\partial^*]=0$.

b) Show that Δ commutes with all operators $\partial,\overline{\partial},\partial^*,\overline{\partial}^*,L,\Lambda$.

(Hint: Use Theorem 1.4.11 and (1.3.31).)

Problem 1.7. Verify first (1.5.8). Now let (X,ω,J) be a Kähler manifold. Let $K_X:=\det(T^{*(1,0)}X)$ be the canonical line bundle on X. Set

$$\mathrm{Ric}_\omega=\mathrm{Ric}(J\cdot,\cdot).$$

Using (1.2.55), verify that R^{TX} is a (1,1)-form with values in $\mathrm{End}(TX)$. Using (1.2.5), verify that $\mathrm{Ric}_\omega=\sqrt{-1}R^{K^*_X}=\sqrt{-1}\,\mathrm{Tr}[R^{T^{(1,0)}X}]$.

Problem 1.8. We will use the homogeneous coordinate $(z_0,\ldots,z_n)\in\mathbb{C}^{n+1}$ for $\mathbb{CP}^n\simeq(\mathbb{C}^{n+1}\setminus\{0\})/\mathbb{C}^*$. Denote by $U_i=\{[z_0,\ldots,z_n]\in\mathbb{CP}^n;z_i\neq0\}$, $(i=0,\ldots,n)$, the open subsets of $\mathbb{CP}^n$, and the coordinate charts are defined by $\phi_i:U_i\simeq\mathbb{C}^n$, $\phi_i([z_0,\ldots,z_n])=(\frac{z_0}{z_i},\ldots,\widehat{\frac{z_i}{z_i}},\ldots,\frac{z_n}{z_i})$. (A hat over a variable means that this variable is skipped.)

Let $\mathscr{O}(-1)$ be the tautological line bundle of $\mathbb{C}\mathbb{P}^n$, i.e., $\mathscr{O}(-1) = \{([z], \lambda z) \in \mathbb{C}\mathbb{P}^n \times \mathbb{C}^{n+1}, \lambda \in \mathbb{C}\}$. For any $\alpha = (\alpha_0, \ldots, \alpha_n) \in \mathbb{N}^{n+1}$, the map $\mathbb{C}^{n+1} \ni z \to \prod_{j=0}^{n} z_j^{\alpha_j}$ is naturally identified with a holomorphic section of $\mathscr{O}(-|\alpha|)^* = \mathscr{O}(|\alpha|)$ on $\mathbb{C}\mathbb{P}^n$; we denote it by s_α.

Let $h^{\mathscr{O}(-1)}$ be the Hermitian metric on $\mathscr{O}(-1)$, as a subbundle of the trivial bundle $\mathbb{C}^{n+1}$ on $\mathbb{C}\mathbb{P}^n$, induced by the standard metric on $\mathbb{C}^{n+1}$. Let $h^{\mathscr{O}(1)}$ be the Hermitian metric on $\mathscr{O}(1)$ induced by $h^{\mathscr{O}(-1)}$. Let $\omega_{FS} = \frac{\sqrt{-1}}{2\pi} R^{\mathscr{O}(1)}$ be the $(1,1)$-form associated to $(\mathscr{O}(1), h^{\mathscr{O}(1)})$ defined by (1.5.14).

On U_i, the trivialization of the line bundle $\mathscr{O}(1)$ is defined by $\mathscr{O}(1) \ni s \to s/z_i$, here z_i is considered as a holomorphic section of $\mathscr{O}(1)$.

We work now on $\mathbb{C}^n$ by using $\phi_0 : U_0 \to \mathbb{C}^n$. Verify that for $z \in \mathbb{C}^n$,

$$|s_{(1,0,\ldots,0)}|^2_{h^{\mathscr{O}(1)}}(z) = \left(1 + \sum_{j=1}^{n} |z_j|^2\right)^{-1}.$$

From (1.5.8), verify that for $z \in \mathbb{C}^n$,

$$\omega_{FS}(z) = \frac{\sqrt{-1}}{2\pi} \left(\frac{\sum_{j=1}^{n} dz_j d\overline{z}_j}{1 + \sum_{j=1}^{n} |z_j|^2} - \frac{\sum_{j=1}^{n} \overline{z}_j dz_j \wedge \sum_{k=1}^{n} z_k d\overline{z}_k}{(1 + \sum_{j=1}^{n} |z_j|^2)^2} \right).$$

Thus ω_{FS} is a Kähler form on $\mathbb{C}\mathbb{P}^n$. ω_{FS} is called the Fubini–Study form, and the associated Riemannian metric $g_{FS}^{T\mathbb{C}\mathbb{P}^n}$ is the *Fubini–Study metric* on $\mathbb{C}\mathbb{P}^n$.

Problem 1.9. Let f be a harmonic function on a connected compact manifold X, i.e., $\Delta f = 0$. Show that f is constant on X. (Hint: $\int_X |df|^2 dv_X = \int_X \overline{f}(\Delta f) dv_X$.)

Problem 1.10. Consider a real $(1,1)$-form $\alpha \in \Omega^{1,1}(X)$. Let us choose the local orthonormal frame $\{w_j\}_{j=1}^n$ such that $\alpha = \sqrt{-1} \sum_{j=1}^{n} c_j(x) w^j \wedge \overline{w}^j$ at a given point $x \in X$. For any $u = \sum_{I,J} u_{IJ} w^I \wedge \overline{w}^J \in \Omega^{\bullet,\bullet}(X)$, from (1.4.37), (cf. (1.4.61)), we have

$$[\alpha, \Lambda] u = \sum_{I,J} \left(\sum_{j \in I} c_j(x) + \sum_{j \in J} c_j(x) - \sum_{j=1}^{n} c_j(x) \right) u_{IJ} w^I \wedge \overline{w}^J.$$

Problem 1.11. (a) (*Nakano vanishing theorem*) Let X be a compact Kähler manifold and (E, h^E) be a Nakano-positive vector bundle over X (cf. Definition 1.1.6). Show that $H^q(X, E \otimes K_X) = 0$ for any $q \geqslant 1$.

(b) On $T^{(1,0)}\mathbb{C}\mathbb{P}^n$ we consider the Fubini–Study metric. Show that $T^{(1,0)}\mathbb{C}\mathbb{P}^n \otimes \mathscr{O}(p)$ is Nakano-positive for $p \geqslant 1$. Deduce that $T^{(1,0)}\mathbb{C}\mathbb{P}^n \otimes K^*_{\mathbb{C}\mathbb{P}^n}$ is Nakano-positive and that $H^q(\mathbb{C}\mathbb{P}^n, T^{(1,0)}\mathbb{C}\mathbb{P}^n) = 0$ for $q \geqslant 1$.

Note: The case $q = 1$ in (b) implies that the complex structure of $\mathbb{C}\mathbb{P}^n$ cannot be deformed (cf. [179, Ch.1, Th. γ]).

Problem 1.12. (a) Let (E, h^E) be a holomorphic Hermitian vector bundle. Show that if (E, h^E) is Nakano-positive, then (E, h^E) is Griffiths-positive.

(b) Define the vector bundle $E = \mathbb{C}\mathbb{P}^n \times \mathbb{C}^{n+1}/\mathscr{O}(-1)$ over $\mathbb{C}\mathbb{P}^n$. Show that E is Griffiths-positive but not Nakano-positive.

Note: The notion of Griffiths-positivity is more suitable for the study of ampleness than that of Nakano positivity. For more details see [79, Ch. VI], [217].

Problem 1.13. Verify that if M is a weakly pseudoconvex domain (i.e., the Levi form is positive semi-definite), and L is a positive line bundle on $\overline{M}$, then the spectral gap property for Kodaira Laplacian similar to Theorem 1.5.5 still holds.

Problem 1.14. For $q = n$, prove directly (1.7.1) with equality (use Theorem 1.4.6).

1.8   Bibliographic notes

In Section 1.2.1 we basically follow [15, §1.2]. For basic material concerning manifolds, vector bundles and Riemannian geometry we refer to [85], [252], [140] and [179]. The proof of Lemmas 1.2.3 and 1.2.4 appeared in [10, Appendix II].

A good references for Section 1.3 is [148, Appendix D]. Instead of referring to [148, Appendix D], [160, §2] for a construction of the Clifford connection on $\Lambda(T^{*(0,1)}X)$, we define it here directly and verify its properties. The Atiyah–Singer index theorem was established in [11]. The Riemann–Roch–Hirzebruch theorem appears in Hirzebruch's Habilitation thesis [130] for an algebraic variety X. In [15, Chap. 4], the readers can find a heat kernel approach to the Atiyah–Singer index theorem.

Section 1.3.3 and Theorems 1.4.5, 1.4.7 are taken from [26], where Bismut used them to prove a local index theorem for modified Dirac operators.

The Kähler identities for Kähler manifolds were proved by A. Weil [251] using the primitive decomposition theorem. Ohsawa [187] used the approach of Weil for non-Kähler metrics and showed the existence of the Hermitian torsion operator satisfying the generalized Kähler identities. Theorem 1.4.11 and the Bochner–Kodaira–Nakano formula (1.4.44) were proved in this precise form by Demailly [73]. For (1.4.63) see also Kodaira–Morrow [179, Ch. 3, Th. 6.2].

Bochner–Kodaira–Nakano formulas with boundary term similar to (1.4.72) were proved by Andreotti–Vesentini [7, p. 113] and Griffiths [119, (7.14)]. Estimate (1.4.84) is a more geometric version of the famous Morrey–Kohn–Hörmander estimate [143, 131, 108], which is essential in the solution of the $\overline{\partial}$-Neumann problem (cf. also Section 3.5).

Section 1.5. Theorems 1.5.7 and 1.5.8 are [160, Th. 1.1 and 2.5] if $A = 0$. If $A = 0$, Borthwick–Uribe [43] and Braverman [54] observed also (1.5.29). (1.5.23) was first proved by Bismut and Vasserot [35, Th. 1.1] by using the Bochner–Kodaira–Nakano formula [73, Th. 0.3].

Theorem 1.6.1 was first proved by Bismut in [25] by using probability theory. Demailly [74] and Bouche [48] gave a different approach. Our proof is new and is inspired by the analytic localization techniques of Bismut–Lebeau [33, §11]. Certainly, the argument here works well for the modified Dirac operator.

Theorem 1.7.1 represents Demailly's holomorphic Morse inequalities [72]. The proof in Section 1.7 is Bismut's heat kernel proof of Theorem 1.7.1.

Demailly's work [72] was influenced by Witten's seminal analytic proof of Morse inequalities [253] for a Morse function f with isolated critical points on a compact manifold. In [24], Bismut gave a heat kernel proof of Morse inequalities and of the degenerate Morse inequalities. Subsequently, in [25], he adapted his heat kernel proof of Morse inequalities for Demailly's holomorphic Morse inequalities. Milnor's book [176] is the standard reference for the classical Morse theory. For the analytic proof of classical Morse inequalities, we refer our readers to the interesting recent book [263]. In the literature, there exists another type of holomorphic Morse inequalities [254, 175, 256], which relate the Dolbeault cohomology groups of the fixed point set X^G of a compact connected Lie group G acting on a compact Kähler manifold X to the Dolbeault cohomology groups of X itself.

Chapter 2

Characterization of Moishezon Manifolds

In this chapter we start some basic facts on analytic and complex geometry (divisors, blowing-up, big line bundles), we prove the theorem of Siegel–Remmert–Thimm, that the field of meromorphic functions on a connected compact complex manifold is an algebraic field of transcendence degree less than the dimension of the manifold. Then we study in more detail Moishezon manifolds and their relation to projective manifolds. In particular we prove that a Moishezon manifold is projective if and only if it carries a Kähler metric. We end the section 2.2 by giving the solution of the Grauert–Riemenschneider conjecture as application of the holomorphic Morse inequalities from Theorem 1.7.1.

Section 2.3 is devoted to the Shiffman–Ji–Bonavero–Takayama criterion, which states that Moishezon manifolds can be characterized in terms of integral Kähler currents. We will present a proof using the singular holomorphic Morse inequalities of Bonavero in Section 2.3.2. As application of the singular holomorphic Morse inequalities, we present the computation of the volume of a big line bundle following Boucksom in Section 2.3.3. Moreover, we give some examples of non-projective Moishezon manifolds in Section 2.3.4, showing that the Shiffman–Ji–Bonavero–Takayama criterion is sharp.

Section 2.4 provides an algebraic reformulation of the holomorphic Morse inequalities.

2.1 Line bundles, divisors and blowing-up

In this section we review the basic facts on divisors and blow-ups. Let X be a complex manifold and $\dim X = n$. Let $\mathscr{O}_X$ be the sheaf of holomorphic functions on X. We denote by $\mathscr{O}_X^*$ the sheaf whose sections over an open set U are the units of the ring $\mathscr{O}_X(U)$, and endow $\mathscr{O}_X^*(U)$ with multiplication as a group operation.

Let $\mathscr{S}_X \subset \mathscr{O}_X$ be the subsheaf defined by $\mathscr{S}_{X,x} = \mathscr{O}_{X,x} \setminus \{0_x\}$; for an open set U, $\mathscr{S}_X(U)$ consists of $f \in \mathscr{O}_X(U)$ which do not vanish identically on any connected component of U. Let $\mathscr{M}_X$ be the sheaf associated to the pre-sheaf of rings of quotients $U \mapsto \mathscr{S}_X^{-1}(U)\mathscr{O}_X(U)$, $U \subset X$ open. The sections of $\mathscr{M}_X$ over an open set U are called *meromorphic functions* on U. By definition, $f \in \mathscr{M}_X(U)$ can be written in the neighborhood V of any point as $f = g/h$, where $g \in \mathscr{O}_X(V)$, $h \in \mathscr{S}_X(V)$. We denote by $\mathscr{M}_X^*$ the multiplicative sheaf of germs of non-zero meromorphic functions.

If $P, Q \in \mathbb{C}[z_0, \ldots, z_n]$ are two homogeneous polynomials of same degree, then P/Q defines a meromorphic function on $\mathbb{C}\mathbb{P}^n$, called a *rational function*.

We denote by $\mathscr{L}(X)$ the set of holomorphic line bundles on X, which becomes a group, multiplication being given by tensor product and inverses by dual bundles. Let $L \in \mathscr{L}(X)$ and $\mathscr{U} = (U_\alpha)_\alpha$ be an open cover of X such that $L|_{U_\alpha}$ is trivial and choose holomorphic frames $e_\alpha \in \Gamma(U_\alpha, L)$; then the trivialization $\varphi_\alpha : L|_{U_\alpha} \to \mathbb{C}$ is defined by $\varphi_\alpha(s) = s/e_\alpha$, and the transition functions

$$g_{\alpha\beta} := \varphi_\alpha \circ \varphi_\beta^{-1} = e_\beta/e_\alpha \in \mathscr{O}_X^*(U_\alpha \cap U_\beta) \quad \text{on } U_\alpha \cap U_\beta \neq \emptyset. \quad (2.1.1)$$

It is easy to see that $g_{\alpha\beta} = g_{\beta\alpha}^{-1}$ and $g_{\alpha\beta}g_{\beta\gamma}g_{\gamma\alpha} = 1$ on $U_\alpha \cap U_\beta \cap U_\gamma \neq \emptyset$, so that $(g_{\alpha\beta})$ represents a Čech 1-cocycle with values in $\mathscr{O}_X^*$, called the *associated cocycle* to L and the covering $\{U_\alpha\}$ and the set of frames $\{e_\alpha\}$. This defines a cohomology class $L_{\mathscr{U}} := \{(g_{\alpha\beta})\} \in H^1(\mathscr{U}, \mathscr{O}_X^*)$. Observe that by choosing another set of holomorphic frames $e_\alpha' \in \Gamma(U_\alpha, L)$, we get a cocycle $(g_{\alpha\beta}')$ which differs from $(g_{\alpha\beta})$ by a Čech 1-coboundary. Thus the class $L_{\mathscr{U}} := \{(g_{\alpha\beta})\}$ does not depend on the choice of frames. The same argument shows that two isomorphic line bundles L and L' produce the same cohomology class in $H^1(\mathscr{U}, \mathscr{O}_X^*)$.

Moreover, if we consider a refinement $\mathscr{V}$ of $\mathscr{U}$, $L_{\mathscr{V}} \in H^1(\mathscr{V}, \mathscr{O}_X^*)$ is the image of $L_{\mathscr{U}} \in H^1(\mathscr{U}, \mathscr{O}_X^*)$ by the canonical map $H^1(\mathscr{U}, \mathscr{O}_X^*) \to H^1(\mathscr{V}, \mathscr{O}_X^*)$. Thus (the isomorphism class of) L defines an element in $H^1(X, \mathscr{O}_X^*) = \varinjlim_{\mathscr{W}} H^1(\mathscr{W}, \mathscr{O}_X^*)$ by

taking the direct limit.

Definition 2.1.1. The *Picard group* $\mathrm{Pic}(X)$ of X is the group of isomorphism classes of holomorphic line bundles over X.

By the discussion above, we have a well-defined map

$$\mathrm{Pic}(X) \longrightarrow H^1(X, \mathscr{O}_X^*). \quad (2.1.2)$$

Actually this map is a *group isomorphism*. Indeed, it is well known that we can reconstruct a line bundle isomorphic to L using the cocycle $(g_{\alpha\beta})$. On the disjoint union $\bigsqcup_\alpha(U_\alpha \times \mathbb{C})$, we introduce the equivalence relation $U_\beta \times \mathbb{C} \ni (x, \xi) \sim (y, \eta) \in U_\alpha \times \mathbb{C}$ if $x = y$ and $\eta = g_{\alpha\beta}(x)\xi$. Then $\bigsqcup_\alpha(U_\alpha \times \mathbb{C})/\sim$ can be naturally given the structure of a holomorphic line bundle isomorphic to L.

The isomorphism (2.1.2) allows us to describe L and related objects in terms of the cocycle $(g_{\alpha\beta})$. For example, to any section $s \in H^0(X, L)$, we associate a collection

$$(s_\alpha), \quad s_\alpha \in \mathscr{O}_X(U_\alpha), \quad s_\alpha = s_\beta g_{\alpha\beta} \quad \text{on} \quad U_\alpha \cap U_\beta, \qquad (2.1.3)$$

where s_α is defined by $s = s_\alpha e_\alpha$ on U_α. Conversely, any collection (s_α) as in (2.1.3) defines a holomorphic section $s \in H^0(X, L)$ by setting $s = s_\alpha e_\alpha$ on U_α.

If L and L' are two line bundles, we choose a trivializing covering $\mathscr{U}$ for both. If L (resp. L') is given by the cocycle $(g_{\alpha\beta})$ (resp. $(g'_{\alpha\beta})$), then $L \otimes L'$ is given by the cocycle $(g_{\alpha\beta}g'_{\alpha\beta})$ and L^*, the dual bundle of L, is given by $(g_{\alpha\beta}^{-1})$.

Let L be a holomorphic line bundle and $s \in H^0(X, L \otimes \mathscr{M}_X^*)$ be a meromorphic section of L on X. We define the *divisor* of s by

$$\mathrm{Div}(s) = \textstyle\sum_V \mathrm{ord}_V(s) \cdot V, \qquad (2.1.4)$$

where the sum runs formally over all irreducible analytic hypersurfaces of X and $\mathrm{ord}_V(s) \in \mathbb{Z}$ is the order of s along V, as in (B.1.2). Locally there exist only a finite number of V's such that $\mathrm{ord}_V(s) \neq 0$. Conversely, a Weil divisor defines canonically a holomorphic line bundle.

Definition 2.1.2. A (Weil) *divisor* on X is a locally finite linear combination $D = \sum_i c_i V_i$, where V_i are irreducible analytic hypersurfaces of X and $c_i \in \mathbb{Z}$. The set of all divisors on X is denoted by $\mathrm{Div}(X)$. D is called *effective* if $c_i \geqslant 0$ for all i. The support of D is defined by $\mathrm{supp}(D) = \cup\{V_i : c_i \neq 0\}$. A *divisor with normal crossings* is a divisor of the form $D = \sum_i V_i$ where V_i are distinct irreducible smooth hypersurfaces which intersect transversely; i.e., for any $x \in \mathrm{supp}(D)$, there is a local coordinate system $(U, z_1, \ldots, z_n)$ centered at x such that $\mathrm{supp}(D) \cap U = \{z_1 z_2 \ldots z_r = 0\}$ for some $1 \leqslant r \leqslant n$.

Let $D = \sum_i c_i V_i$ be a divisor on X. Choose an open covering $\mathscr{U} = (U_\alpha)$ such that every V_i has a local defining function $f_{i\alpha} \in \mathscr{O}_X(U_\alpha)$ (i.e., which vanishes at order 1 along V_i). Set

$$f_\alpha := \prod_i f_{i\alpha}^{c_i} \in \mathscr{M}_X^*(U_\alpha), \quad g_{\alpha\beta} := f_\alpha/f_\beta \in \mathscr{O}_X^*(U_\alpha \cap U_\beta). \qquad (2.1.5)$$

Definition 2.1.3. The holomorphic line bundle defined by the cocycle $(g_{\alpha\beta})$ in (2.1.5) is called the line bundle associated to D and is denoted by $\mathscr{O}_X(D)$.

Note that f_α in (2.1.5) defines a meromorphic section s_D of $\mathscr{O}_X(D)$ and $\mathrm{Div}(s_D) = D$, s_D is called the *canonical section* of $\mathscr{O}_X(D)$. Let $\mathscr{M}(D)$ be the space of meromorphic functions f on X which are holomorphic on $X \smallsetminus \cup V_i$ and $\mathrm{ord}_{V_i}(f) \geqslant -c_i$. By definition, we have

Proposition 2.1.4. *The map* $\mathscr{M}(D) \longrightarrow H^0(X, \mathscr{O}_X(D))$, $f \longmapsto f \cdot s_D$ *is bijective.*

For V a complex vector space, the projective space $\mathbb{P}(V)$ is the set of complex lines through the original in V. Let $\mathscr{O}_{\mathbb{P}(V)}(-1)$ be the universal line bundle on $\mathbb{P}(V)$, then $\mathscr{O}_{\mathbb{P}(V)}(-1) = \{([z], \lambda z) \in \mathbb{P}(V) \times V; \lambda \in \mathbb{C}\}$. Especially $\mathbb{C}\mathbb{P}^{n-1} = \mathbb{P}(\mathbb{C}^n)$.

A fundamental construction in complex and algebraic geometry is the blow-up of X with center a point $x \in X$. The idea is to reduce the study of the ideal of holomorphic functions vanishing at x to the study of sections of a divisor, and hence of a line bundle.

Let U be a coordinate neighborhood of the point x, with coordinates $z = (z_1, \ldots, z_n)$, where $z = 0$ corresponds to the point x. Consider the product $U \times \mathbb{C}\mathbb{P}^{n-1}$, where we assume that $[t_1, \ldots, t_n]$ are homogeneous coordinates for $\mathbb{C}\mathbb{P}^{n-1}$. Then let

$$W = \{(z, [t]) \in U \times \mathbb{C}\mathbb{P}^{n-1} : t_j z_k - t_k z_j = 0, \, j, k = 1, \ldots, n\}, \qquad (2.1.6)$$

which is a submanifold of $U \times \mathbb{C}\mathbb{P}^{n-1}$. Then there is a holomorphic projection $\pi : W \longrightarrow U$ given by $\pi(z, [t]) = z$. Moreover, π has the following properties:

$$\begin{aligned} \pi^{-1}(0) &= E = \{0\} \times \mathbb{C}\mathbb{P}^{n-1} \simeq \mathbb{C}\mathbb{P}^{n-1}, \\ \pi|_{W \smallsetminus E} &: W \smallsetminus E \longrightarrow U \smallsetminus \{0\} \quad \text{is a biholomorphism}. \end{aligned} \qquad (2.1.7)$$

We define $\widetilde{X} = \widetilde{X}_x := (X \smallsetminus \{x\}) \sqcup_\pi W$ by pasting $X \smallsetminus \{x\}$ and W via the biholomorphism $\pi : W \smallsetminus E \to U \smallsetminus \{x\}$. Let $N_{E/\widetilde{X}}$ be the holomorphic normal bundle of E in $\widetilde{X}$.

Definition 2.1.5. The map $\pi : \widetilde{X} \to X$ is called the *blow-up of X with center the point $x \in X$*. E is called the *exceptional divisor* of π.

Let us describe the local coordinates near E on $\widetilde{X}_x$. We set $W_i = \{(z, [t]) \in U \times \mathbb{C}\mathbb{P}^{n-1} : t_i \neq 0\}$. Then $\{W_i\}_{i=1}^n$ is an open cover of W and in W_i, we have local coordinates $w_i^j = \frac{t_j}{t_i} = \frac{z_j}{z_i}$, for $j \neq i$, and $w_i^i = z_i$. In these local coordinates, the map π is given by

$$(w_i^1, \ldots, w_i^n) \to (w_i^i \cdot w_i^1, \ldots, w_i^i, \ldots, w_i^i \cdot w_i^n) \qquad (2.1.8)$$

and E is given by $E = \{w_i^i = 0\} = \{z_i = 0\}$. It follows that the transition functions of $\mathscr{O}_{\widetilde{X}}(E)$ over W are $g_{ji} = w_i^j$ on $W_i \cap W_j$.

We consider the holomorphic map $\varrho : E \longrightarrow \mathbb{C}\mathbb{P}^{n-1}$, $\varrho(0, [t]) = [t]$ and the line bundle $\varrho^* \mathscr{O}_{\mathbb{C}\mathbb{P}^{n-1}}(-1)$. Let us denote by $N_{\{x\}/X} \simeq \{x\} \times \mathbb{C}^n$ the normal bundle to $\{x\}$. It is useful to regard $\mathbb{C}\mathbb{P}^{n-1}$ as the projectivized space

$$\mathbb{P}(N_{\{x\}/X}) = \{(x, [t]) : [t] \in \mathbb{C}\mathbb{P}^{n-1}\}.$$

We can describe the bundle $\varrho^* \mathscr{O}_{\mathbb{C}\mathbb{P}^{n-1}}(-1)$ in the following way; if we denote by $\mathscr{O}_{\mathbb{C}\mathbb{P}^{n-1}}(-1) \subset \pi^*(N_{\{x\}/X})$ the tautological subbundle over $E = \mathbb{P}(N_{\{x\}/X})$ such that the fiber above the point $(x, [t]) \in E$ is $\mathbb{C}t \subset N_{\{x\}/X}$, then $\varrho^* \mathscr{O}_{\mathbb{C}\mathbb{P}^{n-1}}(-1) \simeq \mathscr{O}_{\mathbb{P}(N_{\{x\}/X})}(-1)$. Since the transition functions of $\mathscr{O}_{\mathbb{C}\mathbb{P}^n}(-1)$ are $g_{ji} = t_j/t_i$ on $\varrho(W_i \cap W_j \cap E)$, we obtain:

Proposition 2.1.6. *We have an isomorphism of line bundles*

$$N_{E/\widetilde{X}} \simeq \mathscr{O}_{\widetilde{X}}(E)|_E \simeq \varrho^* \mathscr{O}_{\mathbb{C}\,\mathbb{P}^{n-1}}(-1) \simeq \mathscr{O}_{\mathbb{P}(N_{\{x\}/X})}(-1) \,. \qquad (2.1.9)$$

Let $K_{\widetilde{X}} := \det(T^{*(1,0)}\widetilde{X})$, $K_X := \det(T^{*(1,0)}X)$ be the canonical line bundles on $\widetilde{X}, X$. It is also important to calculate the canonical bundle of the blow-up.

Proposition 2.1.7. $K_{\widetilde{X}} = \pi^*(K_X) \otimes \mathscr{O}_{\widetilde{X}}((n-1)E)$.

Proof. The bundle K_X is generated on U by $dz_1 \wedge \cdots \wedge dz_n$ and $\pi^* K_X$ is generated on W_j by

$$\pi^*(dz_1 \wedge \cdots \wedge dz_n) = (w_j^j)^{n-1} dw_j^1 \wedge \cdots \wedge dw_j^n \,. \qquad (2.1.10)$$

If s is the canonical section of $\mathscr{O}_{\widetilde{X}}(E)$, defined by $w_j^j = 0$ in W_j, we have a line bundle isomorphism

$$\pi^*(K_X) \longrightarrow \mathscr{O}_{\widetilde{X}}((1-n)E) \otimes K_{\widetilde{X}}, \quad \xi \to s^{1-n} \otimes \pi^*(\xi) \,. \qquad (2.1.11)$$

The proof of Proposition 2.1.7 is completed. $\qquad\square$

In replacing the point with a divisor, we have to deal with the following situation: if a line bundle L is positive near x, its pull-back $\pi^*(L)$ is only semi-positive near E. We show now how to regain positivity.

Proposition 2.1.8. *Let L be a positive line bundle over a compact complex manifold X. Then for any $x \in X$, there exists $p_0(x)$ such that $\pi^* L^p \otimes \mathscr{O}_{\widetilde{X}}(-E)$ is positive for $p \geqslant p_0(x)$. Moreover, there exists a neighborhood U_x of x such that $p_0(y) = p_0(x)$, for any $y \in U_x$.*

Proof. We have $\mathscr{O}_{\widetilde{X}}(-E)|_E \simeq \varrho^* \mathscr{O}_{\mathbb{C}\,\mathbb{P}^{n-1}}(1)$. We endow $\mathscr{O}_{\widetilde{X}}(-E)|_E$ with the pull-back of a Hermitian metric on $\mathscr{O}_{\mathbb{C}\,\mathbb{P}^{n-1}}(1)$ with positive curvature (cf. Problem 1.8) and then extend this metric (by a partition of unity) to a Hermitian metric on $\mathscr{O}_{\widetilde{X}}(-E)$ over $\widetilde{X}$.

Let h^L be a metric on L with associated curvature R^L such that $\sqrt{-1}R^L$ is a positive $(1,1)$-form on X. Denote $F_p = \pi^* L^p \otimes \mathscr{O}_{\widetilde{X}}(-E)$ with metric induced by the metrics on L and $\mathscr{O}_{\widetilde{X}}(-E)$. Then

$$R^{F_p} = pR^{\pi^* L} + R^{\mathscr{O}_{\widetilde{X}}(-E)} \,. \qquad (2.1.12)$$

Let $g^{T\widetilde{X}}$ be any Riemannian metric on $T\widetilde{X}$ compatible with the complex structure J. Let $N_{E/\widetilde{X}}$ be the holomorphic normal bundle of E in $\widetilde{X}$, we will identify it as the orthogonal complement of $T^{(1,0)}E$ in $T^{(1,0)}\widetilde{X}$ as $\mathscr{C}^\infty$-Hermitian vector bundles.

We have to show that $R^{F_p}(w, \overline{w}) > 0$ for any $w \in ST^{(1,0)}\widetilde{X}$, the sphere bundle of $\widetilde{X}$. The unit normal bundle $N_{E/\widetilde{X}} \cap ST^{(1,0)}\widetilde{X}$ is compact so $R^{\mathscr{O}_{\widetilde{X}}(-E)}$ has a lower bound on it. Moreover, let us consider the differential $\pi_* : T\widetilde{X}|_E \longrightarrow TX$. Since

$\pi(E) = \{x\}$, we have $TE \subset \mathrm{Ker}(\pi_*)$. Thus π_* induces a map $\pi_* : N_{E/\widetilde{X}} \longrightarrow T_x X$, which is actually injective. Indeed, in local coordinates $\{w_i^j\}_{j=1}^n$, the vector fields $\frac{\partial}{\partial w_i^i}$ provide a basis for $N_{E/\widetilde{X}}$ and using (2.1.8), we see that $\pi_*(\frac{\partial}{\partial w_i^i}) \neq 0$. Thus

$$R^{\pi^* L}(w, \overline{w}) = R^L(\pi_*(w), \pi_*(\overline{w})) > 0 \quad \text{for } w \in N_{E/\widetilde{X}} \cap ST^{(1,0)}\widetilde{X}. \quad (2.1.13)$$

Thus, for p large enough, $R^{F_p}(w, \overline{w}) > 0$, for $w \in N_{E/\widetilde{X}} \cap ST^{(1,0)}\widetilde{X}$. On the other hand,

$$R^{F_p}(w, \overline{w}) = R^{\mathscr{O}_{\widetilde{X}}(-E)}(w, \overline{w}) > 0 \quad \text{for } w \in T^{(1,0)}E \subset \mathrm{Ker}(\pi_*). \quad (2.1.14)$$

Summing up, $R^{F_p}(w, \overline{w}) > 0$ for $w \in T^{(1,0)}\widetilde{X}|_E$, and thus in a neighborhood V_E of E, for $p \gg 1$. Using the compacity of $\widetilde{X} \smallsetminus V_E$, we finally find p_0 such that $R^{F_p}(w, \overline{w}) > 0$ for $w \in ST^{(1,0)}\widetilde{X}$ and $p \geqslant p_0$. To see that we can choose p_0 uniformly in a neighborhood of x, we express the equations of the blow-up at $y \in U$ in local coordinates centered at x, namely

$$W_y = \{(z, [t]) \in U \times \mathbb{C}\mathbb{P}^{n-1} : (z_i - y_i)t_j = (z_j - y_j)t_i\}. \quad (2.1.15)$$

By denoting $\psi_{yx} : W_y \longrightarrow W_x$ the isomorphism induced by the translation with y, we have $\psi_{yx}(E_y) = E_x$, $F_p(y) = \psi_{yx}^* F_p(x)$, where $F_p(y)$ is the associated line bundle when we blow up at y. We can perform the construction of metrics on W_x, pull back to W_y and get $R^{F_p(y)} = \psi_{yx}^* R^{F_p(x)}$. $\qquad \square$

We consider now the blow-up with center a compact submanifold $Y \subset X$ of codimension $l \geqslant 2$. The holomorphic normal bundle of Y in X is defined by $N_{Y/X} := (T^{(1,0)}X)|_Y / T^{(1,0)}Y$. The *projectivized normal bundle* $\mathbb{P}(N_{Y/X}) \longrightarrow Y$ is the bundle with fibers the projective spaces $\mathbb{P}(N_{Y/X,y})$ associated to the fibers of $N_{Y/X}$ at $y \in Y$.

Theorem 2.1.9. *There exists a manifold $\widetilde{X}$ together with a holomorphic map $\pi : \widetilde{X} \longrightarrow X$ such that $E = \pi^{-1}(Y)$ is a smooth divisor in $\widetilde{X}$ and $\pi : E \longrightarrow Y$ is a holomorphic fiber bundle which is isomorphic to $\mathbb{P}(N_{Y/X})$. Moreover $\pi : \widetilde{X} \smallsetminus E \longrightarrow X \smallsetminus Y$ is biholomorphic.*

Proof. We generalize the construction given for $Y = \{x\}$, $x \in X$. Let us denote by $E = \mathbb{P}(N_{Y/X})$, $\pi' : \mathbb{P}(N_{Y/X}) \longrightarrow Y$ the canonical projection. Set

$$\widetilde{X} = (X \smallsetminus Y) \sqcup E, \quad \pi = \mathrm{Id}_{X \smallsetminus Y} \sqcup \pi'. \quad (2.1.16)$$

This means that $\widetilde{X}$ is obtained by replacing each point $y \in Y$ by the projective space of the directions normal to Y.

We define an atlas on $\widetilde{X}$ as follows. First every chart on $X \smallsetminus Y$ is taken to be a chart of $\widetilde{X}$, too. If $y \in Y$, let $(U, z_1, \ldots, z_n)$ be local coordinates on X such that $Y \cap U = \{z \in U : z_1 = \cdots = z_l = 0\}$. Hence $z_{l+1}, \ldots, z_n$ are coordinates on

$Y \cap U$ and $\left(\frac{\partial}{\partial z_1}, \ldots, \frac{\partial}{\partial z_l}\right)$ is a holomorphic frame of $N_{Y/X}$ over $U \cap Y$. We denote by $[\xi_1, \ldots, \xi_l]$ the corresponding fiber homogeneous coordinates on $\mathbb{P}(N_{Y/X})$; We set

$$W_i = \{z \in U \smallsetminus Y : z_i \neq 0\} \cup \{(z, [\xi]) \in \mathbb{P}(N_{Y/X}) : \xi_i \neq 0\}. \tag{2.1.17}$$

Then $\{W_i\}_{i=1}^l$ is a covering of $\pi^{-1}(U)$ and we define coordinates on W_i: $w_i^j = \frac{z_j}{z_i}$ for $j \neq i$, $w_i^i = z_i$ and $w_i^k = z_k$ for $k > l$. We can check that, with this atlas, $\widetilde{X}$ becomes a complex manifold. With respect to these coordinates, π can be written

$$\pi : W_i \longrightarrow U , \ (w_i^1, \ldots, w_i^n) \longrightarrow (w_i^i w_i^1, \ldots, w_i^i, \ldots, w_i^i w_i^l, w_i^{l+1}, \ldots, w_i^n).$$

The proof of Theorem 2.1.9 is completed.   □

Definition 2.1.10. The map $\pi : \widetilde{X} \longrightarrow X$ is called the *blow-up of X with center Y* and $E = \pi^{-1}(Y)$ is called the *exceptional divisor*.

As before, let us introduce the subbundle $\mathscr{O}_{\mathbb{P}(N_{Y/X})}(-1) \subset \pi^*(N_{Y/X})$ over $E = \mathbb{P}(N_{Y/X})$ such that the fiber over $(y, [\xi])$ is $\mathbb{C}\xi \subset N_{Y/X,y}$.

Proposition 2.1.11. *The blow-up $\pi : \widetilde{X} \longrightarrow X$ of X with center Y and $l = \operatorname{codim} Y \geqslant 2$, has the following properties:*

(a) $N_{E/\widetilde{X}} \simeq \mathscr{O}_{\widetilde{X}}(E)|_E \simeq \mathscr{O}_{\mathbb{P}(N_{Y/X})}(-1)$, $K_{\widetilde{X}} \simeq \pi^* K_X \otimes \mathscr{O}_{\widetilde{X}}((l-1)E)$.

(b) *If X is compact and L is a positive line bundle over X, then $\pi^*(L^p) \otimes \mathscr{O}_{\widetilde{X}}(-E)$ is positive on $\widetilde{X}$ for $p \gg 0$.*

(c) $H^2(\widetilde{X}, \mathbb{Z}) = H^2(X, \mathbb{Z}) \oplus \mathbb{Z}c_1(\mathscr{O}_{\widetilde{X}}(E))$ *and* $\operatorname{Pic}(\widetilde{X}) = \pi^* \operatorname{Pic}(X) \oplus \mathbb{Z}\mathscr{O}_{\widetilde{X}}(E)$.

Proof. The same proof of Propositions 2.1.6, 2.1.7, 2.1.8 gives (a) and (b). Assertion (c) follows from the computation of Aeppli of the cohomology of modifications.   □

If $\pi : \widetilde{X} \longrightarrow X$ is the blow-up with center Y and Z is a submanifold of X, $Z \not\subseteq Y$, then the closure of $\pi^{-1}(Z \cap (X \smallsetminus Y))$ in $\widetilde{X}$ is called the *strict transform* of Z.

We remind the following important notion.

Definition 2.1.12. Let $A \subset \mathbb{C}\mathbb{P}^N$. A is called *projective manifold* if it is a complex submanifold of $\mathbb{C}\mathbb{P}^N$. A is called an *projective algebraic variety* of $\mathbb{C}\mathbb{P}^N$ if there exist homogeneous polynomials $f_1, \cdots, f_r \in \mathbb{C}[z_0, \ldots, z_N]$ such that $A = \{[z] \in \mathbb{C}\mathbb{P}^N : f_1(z) = \cdots = f_r(z) = 0\}$. If A is also smooth, it is called *projective algebraic manifold*.

We shall use the Hironaka embedded resolution of singularities theorem. Let Z be a closed complex space of a complex manifold X. Consider a sequence of transformations

$$\longrightarrow X_{j+1} \overset{\sigma_{j+1}}{\longrightarrow} X_j \longrightarrow \cdots \longrightarrow X_1 \overset{\sigma_1}{\longrightarrow} X_0 = X$$
$$Z_{j+1} \qquad Z_j \qquad \qquad Z_1 \qquad Z_0 = Z \qquad (2.1.18)$$
$$E_{j+1} \qquad E_j \qquad \qquad E_1 \qquad E_0 = \emptyset$$

where, for each j, $\sigma_{j+1} : X_{j+1} \longrightarrow X_j$ denotes a blowing-up with smooth center $C_j \subset X_j$, Z_{j+1} is the strict transform of Z_j by σ_{j+1} and E_{j+1} is the set of exceptional hypersurfaces in X_{j+1}; i.e., $E_{j+1} = E'_j \cup \sigma_{j+1}^{-1}(C_j)$, where E'_j denotes the set of strict transforms by σ_{j+1} of all hypersurfaces in E_j.

Theorem 2.1.13 (Hironaka). *Let Z be a compact complex subspace of a complex manifold X. Then there exists a finite sequence of blowing-ups* (2.1.18) *with smooth centers C_j such that:*

1. *For each j, either $C_j \subset (Z_j)_{\mathrm{sing}}$, the singular set of Z_j, or Z_j is smooth and $C_j \subset Z_j \cap E_j$.*
2. *Let Z' and E' denote the final strict transform of Z and exceptional set, respectively. Then Z' is smooth and E' has only normal crossings.*

We will need an even more general notion of blow up.

Proposition 2.1.14. *Let X be a compact complex space and $\mathscr{I} \subset \mathscr{O}_X$ be a coherent sheaf of ideals. Then there exists a canonical compact complex space $\widetilde{X}$ and a surjective holomorphic map $\pi : \widetilde{X} \longrightarrow X$ with the following properties:*

(a) *the inverse image sheaf $\pi^{-1}\mathscr{I} \cdot \mathscr{O}_{\widetilde{X}}$ on $\widetilde{X}$ is invertible,*

(b) *if $Y = \mathrm{supp}(\mathscr{O}_X/\mathscr{I})$, then $\pi : \widetilde{X} \smallsetminus \pi^{-1}(Y) \longrightarrow X \smallsetminus Y$ is biholomorphic.*

(c) *If X is projective, then $\widetilde{X}$ is projective too.*

The construction of $\widetilde{X}$ is accomplished in the following way. We set $\mathscr{I}^0 = \mathscr{O}_X$ and consider the graded $\mathscr{O}_X$-algebra $\mathscr{A} := \oplus_{q \geqslant 0} \mathscr{I}^q$. Then $\mathscr{A}$ is of finite presentation, that is, for every point of X there exist a neighborhood U, an integer k and a finitely generated homogeneous ideal $\mathscr{E} \subset \mathscr{O}_U[w_0, w_1, \ldots, w_k]$ such that $\mathscr{A}|_U \sim \mathscr{O}_U[w_0, w_1, \ldots, w_k]/\mathscr{E}$. We consider the subspace $\mathrm{Proj}(\mathscr{A})_U \subset U \times \mathbb{C}\mathbb{P}^k$ defined by $\mathscr{E}$. (For example, if $\mathscr{I}$ is generated on U by independent functions $f_0, \ldots, f_k$, then $\mathrm{Proj}(\mathscr{A})_U$ is given in $U \times \mathbb{C}\mathbb{P}^k$ by the equations $f_i w_j - f_j w_i = 0$, $1 \leqslant i, j \leqslant k$.) By gluing together the subspaces $\mathrm{Proj}(\mathscr{A})_U$ we obtain a complex space $\widetilde{X} = \mathrm{Proj}(\mathscr{A})$. The space $\mathrm{Proj}(\mathscr{A})$ comes with a natural projection $\pi : \mathrm{Proj}(\mathscr{A}) \to X$ and a line bundle $\mathscr{O}_{\mathrm{Proj}(\mathscr{A})}(1)$ whose tensor powers $\mathscr{O}_{\mathrm{Proj}(\mathscr{A})}(q)$ satisfy $\pi_* \mathscr{O}_{\mathrm{Proj}(\mathscr{A})}(q) = \mathscr{I}^q$ for $q \geqslant 0$. If X is projective and $\mathscr{L}$ is an ample line bundle over X then $\pi^* \mathscr{L}^m \otimes \mathscr{O}_{\mathrm{Proj}(\mathscr{A})}(1)$ is very ample for m sufficiently large, hence $\widetilde{X}$ is also projective.

The map $\pi : \widetilde{X} \to X$ is called the *blow up of X along the ideal $\mathscr{I}$* or *with center $(Y, (\mathscr{O}_X/\mathscr{I})|_Y)$*. If Y is smooth, we will talk about a *blow-up with smooth center*. The blow-up of Definition 2.1.10 are blow-ups with center $(Y, (\mathscr{O}_X/\mathscr{I})|_Y)$, where Y is a submanifold and $\mathscr{I}$ the ideal of holomorphic functions vanishing on Y.

Blow-ups are used to resolve the singularities of complex spaces:

Theorem 2.1.15 (Hironaka). *Let X be a compact complex space. Then there exists a finite sequence of blow-ups with smooth centers*

$$X_N \xrightarrow{\sigma_N} \cdots \longrightarrow X_j \xrightarrow{\sigma_j} X_{j-1} \longrightarrow \cdots \longrightarrow X_1 \xrightarrow{\sigma_1} X_0 = X \,, \qquad (2.1.19)$$

such that X_N is non-singular (i.e. a complex manifold) and the center C_{j-1} of the blow-up σ_j is contained in $(X_{j-1})_{\mathrm{sing}}$. If X is projective, X_N is also projective.

Combining Proposition 2.1.14 and Theorem 2.1.15 we obtain:

Proposition 2.1.16. *Let X be a compact complex space and $\mathscr{I} \subset \mathscr{O}_X$ be a coherent sheaf of ideals. Then there exists a compact complex manifold $\widetilde{X}$ and a surjective holomorphic map $\pi : \widetilde{X} \longrightarrow X$ with the properties (a)–(c) from Proposition 2.1.14.*

Since the composition of two blow-ups with smooth center is not in general a blowing-up, we introduce the notion of proper modification. Note, however, that the resulting complex space can be obtained from the original one by blowing up one ideal, by a classical theorem of Hironaka–Rossi.

In the rest of the section we consider only *reduced complex spaces*, if not otherwise stated.

Definition 2.1.17. Let $\varphi : X \longrightarrow Y$ be a holomorphic map between complex irreducible spaces; φ is called a *proper modification* if it is proper, surjective and if there is an analytic set $A \subset Y$ with the following properties:

1. $A \subset Y$ and $\varphi^{-1}(A) \subset X$ are nowhere dense analytic sets,
2. the restriction $\varphi : X \setminus \varphi^{-1}(A) \longrightarrow Y \setminus A$ is biholomorphic.

Theorem 2.1.18. *If $\varphi : X' \longrightarrow X$ is a proper modification, then the canonical morphism $\widetilde{\varphi} : \mathscr{O}_X \to \varphi_* \mathscr{O}_{X'}$ is injective and extends to an isomorphism $\widehat{\varphi} : \mathscr{M}_X \to \varphi_* \mathscr{M}_{X'}$. In particular, the map $\widehat{\varphi}(X) : \mathscr{M}_X(X) \to \mathscr{M}_{X'}(X')$ is an isomorphism.*

More precisely, if $h \in \mathscr{M}_X(X)$ has the local form $h = f/g$ with $f, g \in \mathscr{O}_X(U)$, U open set, then $\widehat{\varphi}(X)h$ has the local form $(f \circ \varphi)/(g \circ \varphi)$ on $\varphi^{-1}(U)$.

Definition 2.1.19. Let X and Y be irreducible complex spaces. A map φ of X to the set of subsets of Y is called a *meromorphic map* of X into Y, written $\varphi : X \dashrightarrow Y$, if the following conditions are fulfilled:

(a) The graph $\mathrm{Graph}(\varphi) = \{(x,y) \in X \times Y : y \in \varphi(x)\}$ is an irreducible analytic subset of $X \times Y$.

(b) The projection on the first factor $\mathrm{pr}_1 : \mathrm{Graph}(\varphi) \to X$ is a proper modification.

To give a meromorphic map is equivalent to giving an analytic subset $G \subset X \times Y$ such that $\mathrm{pr}_1 : G \to X$ is a proper modification. Indeed, then we define φ through $\varphi(x) = \mathrm{pr}_2(\mathrm{pr}_1^{-1}(x))$, where $\mathrm{pr}_2 : G \to Y$ is the projection on the second factor.

Example 2.1.20. (a) Assume that X and Y are projective algebraic varieties. Any rational map of X into Y, i.e., a map whose components are rational functions, is a meromorphic map. Conversely, by the GAGA principle, any meromorphic map of X into Y is actually a rational map.

(b) Let $f_1, \ldots, f_k$ be meromorphic functions on a complex space X, not all identically zero. Let A be a nowhere dense analytic set such that $f_1, \ldots, f_k$ are holomorphic on $U := X \smallsetminus A$. We consider the holomorphic map $\psi : U \to \mathbb{C}^k$, $\psi(x) = (f_1(x), \ldots, f_k(x))$. One can show that the closure G of $\mathrm{Graph}(\psi) \subset U \times \mathbb{C}^k \subset X \times \mathbb{C}\mathbb{P}^k$ in $X \times \mathbb{C}\mathbb{P}^k$ satisfies the condition that $\mathrm{pr}_1 : G \to X$ is a proper modification. Thus $f_1, \ldots, f_k$ define a meromorphic map $X \dashrightarrow \mathbb{C}\mathbb{P}^k$ denoted $x \mapsto [1, f_1(x), \ldots, f_k(x)]$.

(c) In particular, the Kodaira map (cf. Definition 2.2.3 and (2.2.12)) is a meromorphic map: if $S_1, \ldots, S_d$ is a basis of $H^0(X, L)$, the Kodaira map is defined by the meromorphic functions $S_2/S_1, \ldots, S_d/S_1$.

Definition 2.1.21. A meromorphic map $\varphi : X \dashrightarrow Y$ is called *bimeromorphic map* if the projection on the second factor $\mathrm{pr}_2 : \mathrm{Graph}(\varphi) \to Y$ is also a proper modification. Two complex spaces X and Y are called *bimeromorphically equivalent* if there exists a bimeromorphic map $\varphi : X \dashrightarrow Y$.

In this case the analytic set $G = \{(y, x) \in Y \times X : (x, y) \in \mathrm{Graph}(\varphi)\} \subset Y \times X$ defines a meromorphic map $\varphi^{-1} : Y \dashrightarrow X$.

   A holomorphic map between arbitrary complex spaces $f : X \longrightarrow Y$ is called *flat* at a point $x \in X$ if $\mathscr{O}_{X,x}$ is flat when viewed as an $\mathscr{O}_{Y,f(x)}$–module via the canonical morphism $\mathscr{O}_{Y,f(x)} \to \mathscr{O}_{X,x}$. f is called *flat* if it is flat at every $x \in X$.

   If $f : X \to Y$ is flat and X, Y are irreducible, then the fibres $f^{-1}(y)$ with $y \in Y$ have the same dimension.

Theorem 2.1.22 (Hironaka's flattening theorem). *Let $f : X \to Y$ be a holomorphic map of compact complex spaces, where X is not necessarily reduced. Then there exists a commutative diagram*

$$\begin{array}{ccc} \widetilde{X} & \xrightarrow{\ \widetilde{f}\ } & \widetilde{Y} \\ {\scriptstyle \pi_X}\downarrow & & \downarrow{\scriptstyle \pi_Y} \\ X & \xrightarrow{\ f\ } & Y \end{array} \qquad\qquad (2.1.20)$$

where π_Y is the blow-up along a coherent sheaf of ideals $\mathscr{I} \subset \mathscr{O}_Y$, π_X is the blow-up of along the inverse image of $\mathscr{I}$ by f and $\widetilde{f}$ is a flat holomorphic map. Moreover, we can assume that π_Y is a composition of a finite succession of blow-ups with smooth centers.

   The flattening theorem has two important consequences: elimination of indeterminacies and the Chow lemma.

Definition 2.1.23. Let $\varphi : X \dashrightarrow Y$ be a meromorphic map. The smallest nowhere dense analytic set $I(\varphi) \subset X$ such that $\varphi : X \smallsetminus I(\varphi) \to Y$ is holomorphic is called the *set of points of indeterminacy* of φ.

Theorem 2.1.24 (elimination of points of indeterminacy). *Let X, Y be irreducible compact complex spaces and $\varphi : X \dashrightarrow Y$ be a bimeromorphic map. There exists a proper modification $\psi : \widetilde{X} \to X$ obtained as a composition of a finite succession of blow-ups with smooth centers and a holomorphic map $\widetilde{\varphi} : \widetilde{X} \to Y$ such that we have the commutative diagram:*

$$\begin{array}{ccc} \widetilde{X} & & \\ \psi \downarrow & \searrow^{\widetilde{\varphi}} & \\ X & \dashrightarrow & Y \\ & \varphi & \end{array} \qquad (2.1.21)$$

Theorem 2.1.25 (Chow Lemma). *Let X, Y be irreducible compact complex spaces and $\varphi : X \dashrightarrow Y$ be a bimeromorphic map. Then*

(i) *there exists a proper modification $\pi_Y : \widetilde{Y} \to Y$ obtained as a composition of a finite succession of blow-ups with smooth centers and a holomorphic map $\alpha : \widetilde{Y} \to X$ such that $\varphi \circ \alpha = \pi_Y$, and*

(ii) *there exists a proper modification $\pi_X : \widetilde{X} \to X$ obtained as a composition of a finite succession of blow-ups with smooth centers and a holomorphic map $\widetilde{\varphi} : \widetilde{X} \to \widetilde{Y}$ such that the following diagram is commutative:*

$$\begin{array}{ccc} \widetilde{X} & \xrightarrow{\widetilde{\varphi}} & \widetilde{Y} \\ \pi_X \downarrow & & \downarrow \pi_Y \\ X & \dashrightarrow & Y \\ & \varphi & \end{array} \qquad (2.1.22)$$

Proof. By Theorem 2.1.24 (for φ^{-1}) there exists a composition of finitely many blow-ups with smooth centers $\pi_Y : \widetilde{Y} \to Y$ such that the bimeromorphic map $\alpha := \varphi^{-1} \circ \pi_Y$ is holomorphic:

$$\begin{array}{ccc} & \widetilde{Y} & \\ \alpha \swarrow & \downarrow \pi_Y & \\ X & \dashrightarrow & Y \\ & \varphi & \end{array} \qquad (2.1.23)$$

Applying again the same theorem, there exists a composition of finitely many blow-ups with smooth centers $\pi_X : \widetilde{X} \to X$ such that $\widetilde{\varphi} := \alpha^{-1} \circ \pi_X$ is holomorphic. $\qquad \square$

2.2   The Siu–Demailly criterion

Our goal is to prove the Siu–Demailly criterion, which gives a positive answer to the Grauert–Riemenschneider conjecture. The conjecture is the generalization of the Kodaira embedding theorem in connection to the characterization of Moishezon varieties, that is, compact complex spaces with transcendence degree of the meromorphic function field equal to the complex dimension. They are important in algebraic geometry because most of the natural modifications of algebraic varieties can be performed in the category of Moishezon varieties but sometimes not in the category of algebraic varieties.

This section is organized as follows. In Section 2.2.1, we introduce big line bundles and give a characterization in term of the growth of $\dim H^0(X, L^p)$ as $p \to \infty$. In Section 2.2.2, we prove two fundamental results of Moishezon. First, that a Moishezon manifold can be transformed into a projective one by a finite succession of blowing-up along smooth centers. Using this theorem, we show that the obstruction for a Moishezon manifold to be projective is the existence of Kähler metric. Finally, as an application of the holomorphic Morse inequalities, we establish Siu–Demailly criterion which gives a characterization of Moishezon manifolds.

2.2.1   Big line bundles

Let X be a compact connected complex manifold of dimension n and let L be a holomorphic line bundle over X.

It is well known that the space of holomorphic sections of L on X,

$$H^0(X, L) = \{s \in \mathscr{C}^\infty(X, L) : \overline{\partial}^L s = 0\}$$

is finite-dimensional. This follows from the Hodge theory, Theorem 1.4.1. But we start with an elementary proof of this fact which has two advantages: it can be refined to prove Siegel's lemma and generalizes to Andreotti pseudoconcave manifolds.

For $x \in X$, we denote by $\mathscr{I}(x)$ the sheaf of holomorphic functions vanishing at x and by $\mathbf{m}_x \subset \mathscr{O}_{X,x}$ the maximal ideal of the ring of germs of holomorphic functions at x. For a positive integer k we have a canonical residue map $L \to L \otimes (\mathscr{O}_X / \mathscr{I}(x)^{k+1})$ which induces in cohomology a map which associates to each global holomorphic section of L its k-jet at x:

$$J_x^k : H^0(X, L) \longrightarrow H^0(X, L \otimes (\mathscr{O}_X / \mathscr{I}(x)^{k+1})) = L_x \otimes (\mathscr{O}_{X,x} / \mathbf{m}_x^{k+1}). \quad (2.2.1)$$

The right-hand side of (2.2.1) is called the space of k-jets of holomorphic sections of L at x.

For a point $x \in X$, we denote by $P(x, r)$ the polydisc $\{z \in U : |z_i| < r, 1 \leq i \leq n\}$ where $(U, z_1, \ldots, z_n)$ is a holomorphic coordinate system centered at x. The Shilov boundary of $P(x, r)$ is defined by $S(P(x, r)) = \{z \in U : |z_i| = r, 1 \leq$

$i \leq n\}$. It is well known that any holomorphic function in the neighborhood of $\overline{P}(x,r)$, attains its maximum on $\overline{P}(x,r)$ on $S(P(x,r))$.

Lemma 2.2.1. *Let X be a compact complex manifold of dimension n and L be a holomorphic line bundle on X. Then for any points $x_1, \ldots, x_m$ in X and $r_1, \ldots, r_m$ $\in \mathbb{R}_+^*$ such that L is trivial over each $P(x_i, 2r_i)$ and $X \subset \cup_{i=1}^m P(x_i, r_i e^{-1})$, there exists an integer $k = k(L)$ such that if $s \in H^0(X, L)$ vanishes at each point x_i up to order k, then s vanishes identically. Hence $\dim H^0(X, L) \leqslant m\binom{n+k}{k}$.*

Proof. We fix a trivialization of L over each $P(x_i, 2r_i)$. Assume that the line bundle L is given by the holomorphic transition functions

$$c_{ij} : \overline{P}(x_i, r_i) \cap \overline{P}(x_j, r_j) \longrightarrow \mathbb{C}^*. \tag{2.2.2}$$

Set

$$\varphi(L) = \sup\{|c_{ij}(x)| : x \in \overline{P}(x_i, r_i) \cap \overline{P}(x_j, r_j), \, i, j = 1, \ldots, m\}. \tag{2.2.3}$$

Since $c_{ij} = c_{ji}^{-1}$, we have $\varphi(L) \geqslant 1$. Consider a section $s \in H^0(X, L)$ which vanishes up to order $k = \lfloor \log \varphi(L) \rfloor + 1$ at each x_j ($\lfloor u \rfloor$ is the integer part of u). Assume that s is given in the given trivialization of L over $P(x_j, 2r_j)$ by s_j. Set

$$\|s\| = \sup\{|s_j(x)| : x \in \overline{P}(x_j, r_j), \, j = 1, \ldots, m\}. \tag{2.2.4}$$

There exists $q \in \{1, 2, \ldots, m\}$ such that for some $y \in S(P(x_q, r_q))$, $|s_q(y)| = \|s\|$. We can find $j \neq q$ such that $y \in P(x_j, r_j e^{-1})$. Hence $s_q(y) = c_{qj}(y) s_j(y)$ so that

$$\|s\| = |s_q(y)| = |c_{qj}(y) s_j(y)| \leqslant \varphi(L) |s_j(y)|. \tag{2.2.5}$$

By applying the Schwarz inequality (cf. Problem 2.3) to s_j in $P(x_j, r_j)$, we get

$$|s_j(y)| \leqslant \|s\| \, |y|_0^k r_j^{-k} \quad \text{where } |y|_0 = \sup\{|z_j(y)|; 1 \leqslant j \leqslant n\} \leqslant r_j e^{-1}. \tag{2.2.6}$$

Consequently,

$$\|s\| \leqslant \|s\| \, \varphi(L) e^{-k}. \tag{2.2.7}$$

If s is not identically zero, this leads to a contradiction, by our choice of k.

Consider the map

$$\prod_{1 \leqslant j \leqslant m} J_{x_j}^k : H^0(X, L) \longrightarrow \prod_{1 \leqslant j \leqslant m} L_{x_j} \otimes (\mathscr{O}_{X,x_j}/\mathbf{m}_{x_j}^{k+1}), \tag{2.2.8}$$

which sends every section in its k-jets at $x_1, \ldots, x_m$. By the preceding argument, (2.2.8) is injective. Since the dimension of the target space satisfies the desired estimate, we are done. $\qquad\square$

There exists an important link between existence of meromorphic functions and the growth of the dimension of the space of holomorphic sections of a line bundle. We shall exhibit this link by means of Siegel's lemma. For this purpose let us define the Kodaira map.

We give first a geometric description. The linear system associated to $V = H^0(X, L)$ is $|V| = \{\mathrm{Div}(s) : s \in V\}$. Since X is compact, $\mathrm{Div}(s) = \mathrm{Div}(s')$ if and only if $s = \lambda s'$ for some $\lambda \in \mathbb{C}^*$, so that $|V|$ is parameterized by the projective space $\mathbb{P}(V)$. The *base point locus* of the linear system $|V|$ is given by

$$\mathrm{Bl}_{|V|} = \cap_{s \in V} \mathrm{Div}(s) = \{x \in X : s(x) = 0 \quad \text{for all } s \in V\}. \tag{2.2.9}$$

Theorem 2.2.2 (Bertini). *The generic element of the linear system $|H^0(X, L)|$ is smooth away from $\mathrm{Bl}_{|H^0(X,L)|}$, i.e., $\mathrm{Div}(s) \smallsetminus \mathrm{Bl}_{|H^0(X,L)|}$ is a smooth submanifold of X for a generic element s of $H^0(X, L)$.*

For $x \notin \mathrm{Bl}_{|V|}$, the space of divisors $D \in |V|$ containing x forms a hyperplane in $|V|$ or equivalently the space of sections vanishing at x is a hyperplane in V.

Set $d = \dim H^0(X, L)$ and let $\mathbb{G}(d-1, V) = \mathbb{G}(d-1, H^0(X, L))$ be the Grassmannian of hyperplanes of $H^0(X, L)$.

Definition 2.2.3. The *Kodaira map* Φ_V associated to L is defined by

$$\Phi_V : X \smallsetminus \mathrm{Bl}_{|V|} \longrightarrow \mathbb{G}(d-1, H^0(X, L)),$$
$$\Phi_V(x) = \{s \in H^0(X, L) : s(x) = 0\}. \tag{2.2.10}$$

We give now an algebraic definition. Let us identify $\mathbb{P}(V^*)$ with $\mathbb{G}(d-1, V)$ by sending an equivalence class of non-zero forms on $H^0(X, L)$ to their (common) kernel. By composing Φ_V with this identification, we obtain a map

$$\Phi_V : X \smallsetminus \mathrm{Bl}_{|V|} \longrightarrow \mathbb{P}(H^0(X, L)^*) \tag{2.2.11}$$

described as follows. For $x \notin \mathrm{Bl}_{|V|}$, we choose a vector $e_x \in L_x \smallsetminus \{0\}$, then we obtain an element of V^* given by $V \ni s \to s(x)/e_x \in \mathbb{C}$. The class of this element in $\mathbb{P}(V^*)$ does not depend on the choice of e_x and is exactly $\Phi_V(x)$.

To get an analytic description of Φ_V, let us choose a basis $S_1, \ldots, S_d$ of V which gives identifications $V \simeq \mathbb{C}^d$ and $\mathbb{P}(V^*) \simeq \mathbb{C}\mathbb{P}^{d-1}$. We define $\tilde{\Phi}_V$ through the commutative diagram

$$
\begin{array}{ccc}
X \smallsetminus \mathrm{Bl}_{|V|} & \xrightarrow{\Phi_V} & \mathbb{P}(V^*) \\
\downarrow{\scriptstyle \mathrm{Id}} & & \downarrow{\scriptstyle \simeq} \\
X \smallsetminus \mathrm{Bl}_{|V|} & \xrightarrow{\tilde{\Phi}_V} & \mathbb{C}\mathbb{P}^{d-1}.
\end{array}
$$

Let e_L be a local holomorphic frame of L on U and set $S_i = f_i e_L$ with $f_i \in \mathscr{O}_X(U)$. Then

$$\tilde{\Phi}_V(x) = [f_1(x), \ldots, f_d(x)], \quad x \in U. \tag{2.2.12}$$

This form immediately shows that $\widetilde{\Phi}_V$ and Φ_V are holomorphic. If $S_i(x) \neq 0$ then $\widetilde{\Phi}_V(x) = [S_1(x)/S_i(x), \ldots, S_d(x)/S_i(x)] \in \mathbb{C}\mathbb{P}^{d-1}$ which justifies the slightly abusive notation $\widetilde{\Phi}_V(x) = [S_1(x), \ldots, S_d(x)]$.

Definition 2.2.4. We say that

1. $V = H^0(X, L)$ *separates* two points $x \neq y$, if $x, y \notin \mathrm{Bl}_{|V|}$ and $\Phi_V(x) \neq \Phi_V(y)$ (or equivalently, the restriction map $H^0(X, L) \to L_x \oplus L_y$, $s \mapsto s(x) \oplus s(y)$ is surjective).

2. $V = H^0(X, L)$ *gives local coordinates* at a point x if $x \notin \mathrm{Bl}_{|V|}$ and Φ_V is an immersion at x (i.e., $J_x^1 : H^0(X, L) \to L_x \otimes (\mathscr{O}_{X,x}/\mathbf{m}_x^2)$ is surjective).

We will be interested in the Kodaira map associated to high tensor powers of L. We write $V_p := H^0(X, L^p)$ and $\Phi_p := \Phi_{V_p}$.

Definition 2.2.5.

1. The graded ring $\oplus_{p \geqslant 0} H^0(X, L^p)$ separates two points $x \neq y$, if $H^0(X, L^p)$ separates x, y for some p.

2. The graded ring $\oplus_{p \geqslant 0} H^0(X, L^p)$ gives local coordinates at a point x, if $H^0(X, L^p)$ gives local coordinates at x for some p.

3. Set $\varrho_p = \max\{\mathrm{rk}_x \Phi_p : x \in X \smallsetminus \mathrm{Bl}_{|V_p|}\}$ if $V_p \neq \{0\}$, and $\varrho_p = -\infty$ otherwise. The *Kodaira–Iitaka dimension* of L is $\kappa(L) = \max\{\varrho_p : p \in \mathbb{N}^*\}$. A line bundle L is called *big* if $\kappa(L) = \dim X$.

Big line bundles can be characterized in terms of the growth of $\dim H^0(X, L^p)$ as $p \to \infty$. We introduce for this purpose a very useful general estimate.

Lemma 2.2.6 (Siegel's lemma). *Let X be a compact connected complex manifold and L be a holomorphic line bundle on X. Then there exists $C > 0$ such that*

$$\dim H^0(X, L^p) \leqslant C p^{\varrho_p}, \quad \text{for any } p \geqslant 1. \tag{2.2.13}$$

Proof. We modify the proof of Lemma 2.2.1 and use the same notation as there. The set of points where Φ_p has rank less than ϱ_p is a proper analytic set of X, so $\{x \in X : \mathrm{rk}_x \Phi_p = \varrho_p \text{ for any } p \geqslant 1\}$ is dense in X. Let $x_1, \ldots, x_m \in X$ as in Lemma 2.2.1 such that Φ_p has rank ϱ_p at each x_j for any $p \geqslant 1$.

Since Φ_p is of constant rank in a neighborhood of x_j, there exists a submanifold $M_{p,j}$ in the neighborhood of x_j which is transversal in x_j to the fiber $\Phi_p^{-1}(\Phi_p(x_j))$ and $\dim M_{p,j} = \varrho_p$. Consider a section $s \in H^0(X, L^p)$ which vanishes up to order $k = p(\lfloor \log \varphi(L) \rfloor + 1)$ at each x_j along $M_{p,j}$. But s vanishes on the fiber $\Phi_p^{-1}(\Phi_p(x_j))$ which passes through x_j by definition of the Kodaira map, hence s vanishes up to order k at x_j on X. Repeating the argument from Lemma 2.2.1, we obtain

$$\|s\| \leqslant \|s\| \, \varphi(L^p) e^{-k}. \tag{2.2.14}$$

By noting that $\varphi(L^p) = \varphi(L)^p$ we infer that $\|s\| = 0$, so s vanish identically.

Let $\mathbf{m}_{M_{p,j},x_j}$ be the maximal ideal of the ring $\mathcal{O}_{M_{p,j},x_j}$. Consider the map

$$H^0(X, L^p) \longrightarrow \prod_{1 \leqslant j \leqslant m} L^p_{x_j} \otimes (\mathcal{O}_{M_{p,j},x_j}/\mathbf{m}^{k+1}_{M_{p,j},x_j}), \qquad (2.2.15)$$

which sends every section to its jet of order k at x_j along $M_{p,j}$, $j = 1, \ldots, m$. The preceding argument shows that this map is injective. Since the dimension of the target space is less than or equal to $m\binom{\varrho_p+k}{k}$ the desired estimate follows. $\square$

Theorem 2.2.7. *Let X be a compact connected complex manifold of dimension n, and let L be a holomorphic line bundle on X. Then L is big if and only if*

$$\limsup_{p \to \infty} p^{-n} \dim H^0(X, L^p) > 0. \qquad (2.2.16)$$

Proof. We get immediately the "if" part of Theorem 2.2.7 from Lemma 2.2.6.

Conversely, assume that L is big. Let us choose x_0 and m with $\mathrm{rk}_{x_0} \Phi_m = n$. There exist $s_0, \ldots, s_n \in H^0(X, L^m)$ with $s_0(x_0) \neq 0$ and $d(\frac{s_1}{s_0}) \wedge \cdots \wedge d(\frac{s_n}{s_0})|_{x_0} \neq 0$. Let P be a polynomial of degree r in n variables. Then $P(\frac{s_1}{s_0}, \ldots, \frac{s_n}{s_0}) = 0$ implies that $P = 0$, as $(\frac{s_1}{s_0}(x), \ldots, \frac{s_n}{s_0}(x))$ are local coordinates near x_0. Therefore each non-zero homogeneous polynomial Q of degree r in $n+1$ variables produces a non-zero section $Q(s_0, \ldots, s_n) \in H^0(X, L^{mr})$. In fact if $Q(s_0, \ldots, s_r) = 0$, then

$$P\Big(\frac{s_1}{s_0}, \ldots, \frac{s_n}{s_0}\Big) = \frac{1}{s_0^r} Q(s_0, \ldots, s_n) = 0,$$

so $P = 0$ and $Q = 0$. Since the space of homogeneous polynomials of degree r in $n+1$ variables has dimension $\binom{n+r}{r} \geqslant r^n/n!$, we deduce that $\dim H^0(X, L^{mr}) \geqslant r^n/n!$, thus we get (2.2.16). $\square$

2.2.2   Moishezon manifolds

Let X be a compact connected complex manifold of dimension n.

The ring $\mathcal{M}_X(X)$ of global sections of $\mathcal{M}_X$ over X is called the ring of meromorphic functions on X. Certainly, the ring $\mathcal{M}_X(X)$ forms a field.

Let us describe concretely $\mathcal{M}_X(X)$ for a connected projective manifold $X \subset \mathbb{C}\,\mathbb{P}^N$. If $P, Q \in \mathbb{C}[z_0, \ldots, z_N]$ are two homogeneous polynomials of same degree, such that Q does not vanish identically on X, then $P/Q \in \mathcal{M}_X(X)$ is called a *rational function* on X. The set of rational functions forms a subfield $\mathcal{K}_X \subset \mathcal{M}_X(X)$. Actually, by a theorem of Hurwitz, any meromorphic function on a projective manifold is rational, i.e $\mathcal{K}_X = \mathcal{M}_X(X)$.

Definition 2.2.8. We say that $f_1, \cdots, f_k \in \mathcal{M}_X(X)$ are *algebraically dependent*, if there exists a non–trivial polynomial $P \in \mathbb{C}[z_1, \cdots, z_k]$ such that $P(f_1, \cdots, f_k) = 0$ wherever it is defined. The *transcendence degree* of $\mathcal{M}_X(X)$ over $\mathbb{C}$ is the maximal number of algebraically independent meromorphic functions on X. We denote by $a(X)$ the transcendence degree of $\mathcal{M}_X(X)$ over $\mathbb{C}$ and call it the *algebraic dimension* of X. By an *algebraic field of transcendence degree d*, we mean a finite algebraic extension of the field $\mathbb{C}(t_1, \ldots, t_d)$ of all rational functions in d variables. We say that $f_1, \ldots, f_k \in \mathcal{M}_X(X)$ are *analytically dependent* if $df_1(x) \wedge \cdots \wedge df_k(x) = 0$ for any point $x \in X$ where all $f_1, \ldots, f_k$ are holomorphic.

Theorem 2.2.9. *The functions $f_1, \ldots, f_{k+1} \in \mathcal{M}_X(X)$ are analytically dependent if and only if they are algebraically dependent.*

Proof. We prove first that algebraic dependence implies analytic dependence. If $k + 1 > n = \dim X$ there is nothing to prove, so we assume $k + 1 \leqslant n$. Let U be the open set of X where all $f_1, \ldots, f_{k+1}$ are holomorphic. We may also assume without loss of generality that $f_1, \ldots, f_k$ are algebraically independent. Let $Q \in \mathbb{C}[z_1, \ldots, z_{k+1}] \setminus \{0\}$ be a polynomial of minimal degree > 0 in z_{k+1} such that $Q(f_1, \ldots, f_{k+1}) = 0$. By differentiation we obtain

$$\sum_{j=1}^{k+1} \frac{\partial Q}{\partial z_j}(f_1, \ldots, f_{k+1}) df_j = 0, \quad \text{on } U. \tag{2.2.17}$$

Since Q is of minimal degree in z_{k+1}, $\frac{\partial Q}{\partial z_{k+1}}(f_1, \ldots, f_{k+1})$ is not identically zero, so we get from (2.2.17) a non-trivial linear relation between the differentials df_j. This implies $df_1(x) \wedge \cdots \wedge df_{k+1}(x) = 0$ for $x \in U$.

Let us prove now the converse. Assume that $f_1, \ldots, f_{k+1}$ are analytically dependent. Without loss of generality we can assume that $f_1, \ldots, f_k$ are analytically independent. By Problem 2.2 we find a holomorphic line bundle L on X and holomorphic sections $s_0, s_1, \ldots, s_k$ with s_0 not identically zero, such that $f_j = s_j/s_0$, $j = 1, \ldots, k$. Let us choose a covering of X with polydiscs $P(x_i, 2r_i)$, $i = 1, \ldots, m$ as in Lemma 2.2.1 such that:

(a) $L|_{P(x_i, 2r_i)}$ is trivial, for $i = 1, \ldots, m$;

(b) $X = \cup_{i=1}^{m} \overline{P}(x_i, r_i e^{-1})$;

(c) $f_1, \ldots, f_k$ are holomorphic at x_i and $f_1 - f_1(x_i) = z_1^{(i)}, \ldots, f_k - f_k(x_i) = z_k^{(i)}$ can be taken among a set of local holomorphic coordinates at x_i, for any $i = 1, \ldots, m$.

These conditions can be realized by translating the points x_i in the coordinates of $P(x_i, 2r_i)$ and taking into account that the set of points where condition (c) is satisfied is an open dense set of X.

Choose also a holomorphic line bundle F on X and holomorphic sections σ_0, σ_1 with σ_0 not identically zero, such that $f_{k+1} = \sigma_1/\sigma_0$. We can also assume that $F|_{P(x_i, 2r_i)}$ is trivial and f_{k+1} is holomorphic at each x_i, for $i = 1, \ldots, m$.

For $p, q \in \mathbb{N}$ we consider the set of all polynomials $Q \in \mathbb{C}[z_1, \ldots, z_{k+1}] \setminus \{0\}$ of degree $\leqslant p$ in $z_1, \ldots, z_k$ and degree $\leqslant q$ in z_{k+1}. To each such Q we associate the homogeneous polynomial

$$\widetilde{Q}(z_0, z_1, \ldots, z_k, w_0, w_1) = z_0^p w_0^q Q(z_1/z_0, \ldots, z_k/z_0, w_1/w_0). \tag{2.2.18}$$

The set of polynomials $\widetilde{Q}$ forms a vector space $\mathscr{V}(p,q)$ of dimension $\binom{k+p}{k}(q+1)$. We define the linear map

$$\alpha : \mathscr{V}(p,q) \to H^0(X, L^p \otimes F^q), \quad \widetilde{Q} \mapsto \widetilde{Q}(s_0, s_1, \dots, s_k, \sigma_0, \sigma_1). \tag{2.2.19}$$

Set $h = \lfloor p \log \varphi(L) + q \log \varphi(F) \rfloor + 1$, with $\varphi(L)$, $\varphi(F)$ introduced in (2.2.3).

We consider the map ψ which associates to each section $\widetilde{Q}(s_0, s_1, \dots, s_k, \sigma_0, \sigma_1)$ the h-jets of $Q(f_1, \dots, f_{k+1})$ at x_i, $i = 1, \dots, m$. As $f_1, \dots, f_{k+1}$ are analytically dependent, we know $dz_1^{(i)} \wedge \cdots \wedge dz_k^{(i)} \wedge df_{k+1} = 0$ near x_i, thus these jets involve only the coordinates $z_1^{(i)}, \dots, z_k^{(i)}$, $i = 1, \dots, m$. Denote by $\mathbb{C}\{z_1^{(i)}, \dots, z_k^{(i)}\}$ the local ring of convergent power series in the variables $z_1^{(i)}, \dots, z_k^{(i)}$ and by $\mathbf{m}_{x_i}^{(i)}$ the maximal ideal of this ring. Consider the map

$$\psi : \operatorname{Im}(\alpha) \to \prod_{1 \leqslant i \leqslant m} L_{x_i}^p \otimes F_{x_i}^q \otimes \left(\mathbb{C}\{z_1^{(i)}, \dots, z_k^{(i)}\} / (\mathbf{m}_{x_i}^{(i)})^{h+1} \right). \tag{2.2.20}$$

As in Lemma 2.2.1, we show that ψ is injective due to our choice of h. The dimension of the target space of ψ satisfies the estimate

$$d = m \binom{k+h}{k} \leqslant \frac{m}{k!} (h+k)^k = \frac{m}{k!} p^k (\log \varphi(L))^k + o(p^k). \tag{2.2.21}$$

If we choose q such that $q + 1 > m (\log \varphi(L))^k$ and p sufficiently large, we obtain $\dim \mathscr{V}(p,q) > \dim \operatorname{Im}(\alpha)$ and consequently $\operatorname{Ker}(\alpha) \neq 0$, which implies that $f_1, \dots, f_{k+1}$ are algebraically dependent. $\qquad\square$

Proposition 2.2.10. *Let $f_1, \dots, f_k \in \mathscr{M}_X(X)$ be algebraically independent. Then there exists an integer $q = q(f_1, \dots, f_k)$ such that any $f \in \mathscr{M}_X(X)$ which is algebraically dependent on $f_1, \dots, f_k$ satisfies a non-trivial algebraic equation over $\mathbb{C}(f_1, \dots, f_k)$ of degree q.*

Proof. We modify the proof of Theorem 2.2.9. Let us choose a holomorphic line bundle L on X, holomorphic sections $s_0, s_1, \dots, s_k$ and polydiscs $P(x_i, 2r_i)$, $i = 1, \dots, m$ satisfying conditions (a)–(c) as in Theorem 2.2.9. Note that the conditions (a)–(c) remain valid for small translations of $P(x_i, 2r_i)$ within their coordinate patch. Thus we can determine for each x_i a closed neighborhood V_i such that (a)–(c) still hold for all $P(y_i, 2r_i)$ with $y_i \in V_i$. Let us set $P_i = P(y_i, 2r_i)$, with $y_i \in V_i$. We will calculate $\varphi(L)$ with respect to the all possible covering $\{P_i\}_{i=1}^m$.

We find moreover a holomorphic line bundle F on X and holomorphic sections σ_0, σ_1 with σ_0 not identically zero, such that $f = \sigma_1/\sigma_0$. We can assume that $F|_{P_i}$ is trivial. Let us choose $q \in \mathbb{N}$ such that $q + 1 > m(\log \varphi(L))^k$. We can also choose the centers x_i of $P(x_i, 2r_i)$ in V_i such that f is holomorphic at each x_i, for $i = 1, \dots, m$. Proceeding as in the proof of Theorem 2.2.9 (with $f_{k+1} = f$), we can pick p sufficiently large such that f satisfies a non-trivial algebraic equation over $\mathbb{C}(f_1, \dots, f_k)$ of degree at most q. $\qquad\square$

Theorem 2.2.11 (Siegel-Remmert-Thimm). *Let X be a compact connected complex manifold. The field of meromorphic functions $\mathcal{M}_X(X)$ is an algebraic field of transcendence degree $a(X) \leqslant n = \dim X$. Therefore, there exists a projective algebraic variety V such that $\dim V \leqslant n$ and $\mathcal{M}_X(X)$ is isomorphic to the rational function field $\mathcal{M}_V(V)$.*

Proof. Let $f_1, \ldots, f_k$ be a maximal set of algebraically independent meromorphic functions. We choose f_{k+1} such that its degree d over $\mathbb{C}(f_1, \ldots, f_k)$ is maximal, which is possible by Proposition 2.2.10.

We show that $\mathcal{M}_X(X) = \mathbb{C}(f_1, \ldots, f_k, f_{k+1})$. It is obvious that $\mathbb{C}(f_1, \ldots, f_k, f_{k+1}) \subset \mathcal{M}_X(X)$. Let $g \in \mathcal{M}_X(X)$, then the field $\mathbb{C}(f_1, \ldots, f_k, f_{k+1}, g)$ is a finite algebraic extension of $\mathbb{C}(f_1, \ldots, f_k)$, so there exists a primitive element $h \in \mathcal{M}_X(X)$ such that $\mathbb{C}(f_1, \ldots, f_k, f_{k+1}, g) = \mathbb{C}(f_1, \ldots, f_k, h)$. For a field extension $K_1 \subset K$, we set $[K : K_1] = \dim_{K_1} K$. We have

$$d \geqslant [\mathbb{C}(f_1, \ldots, f_k, h) : \mathbb{C}(f_1, \ldots, f_k)]$$
$$= [\mathbb{C}(f_1, \ldots, f_k, h) : \mathbb{C}(f_1, \ldots, f_k, f_{k+1})] \cdot [\mathbb{C}(f_1, \ldots, f_k, f_{k+1}) : \mathbb{C}(f_1, \ldots, f_k)].$$

Since the last factor equals d, the first factor equals one, so $\mathbb{C}(f_1, \ldots, f_k, h) = \mathbb{C}(f_1, \ldots, f_k, f_{k+1})$ and $g \in \mathbb{C}(f_1, \ldots, f_k, f_{k+1})$. Hence $\mathcal{M}_X(X)$ is a finite algebraic extension of $\mathbb{C}(f_1, \ldots, f_k)$. Since algebraic and analytic dependence coincide, we see immediately $a(X) \leqslant n$.

Let $Q \in \mathbb{C}(f_1, \ldots, f_k)[t]$ be the minimal polynomial of f_{k+1} over $\mathbb{C}(f_1, \ldots, f_k)$. We chase denominators of $f_1, \ldots, f_k$ and divide off factors involving only in the variables $f_1, \ldots, f_k$ in Q, so we may assume that $Q \in \mathbb{C}[f_1, \ldots, f_k, t]$ and is irreducible. Let $A = \{z \in \mathbb{C}^{k+1} : Q(z) = 0\}$ be the affine algebraic variety defined by Q and let $V \subset \mathbb{C}\mathbb{P}^{k+1}$ be its projective closure. Then V is an irreducible projective algebraic variety with $\dim V = k$ and its field of rational functions is isomorphic to $\mathbb{C}(f_1, \ldots, f_k, f_{k+1})$. $\square$

Definition 2.2.12. A compact connected complex manifold X is called a *Moishezon manifold* if it possesses $\dim X$ algebraically independent meromorphic functions.

Observe that $z_1/z_0, \cdots, z_n/z_0$ are algebraically independent rational functions on $\mathbb{C}\mathbb{P}^n$, where $[z_0, \cdots, z_n]$ are homogeneous coordinates. Hence $\mathbb{C}\mathbb{P}^n$ is Moishezon. Moreover, if X' is a connected compact complex manifold, and $\varphi : X' \longrightarrow X$ is a proper modification, then by Theorem 2.1.18, X is Moishezon if and only if X' is Moishezon.

Recall that a holomorphic map $\pi : V \to Y$ between complex spaces is called a *ramified covering* if there exists a nowhere dense analytic set $A \subset Y$ such that $\pi : V \setminus \pi^{-1}(A) \to Y \setminus A$ is locally biholomorphic.

Theorem 2.2.13. *If $V \subset \mathbb{C}\mathbb{P}^N$ is a connected projective manifold of dimension n, then there exists a projection π of $\mathbb{C}\mathbb{P}^N$ on a projective n-plane $\mathbb{C}\mathbb{P}^n \subset \mathbb{C}\mathbb{P}^N$ such that $\pi|_V$ is finite ramified cover. Therefore, any connected projective manifold is Moishezon.*

Theorem 2.2.14. *A compact connected complex manifold X is Moishezon if and only if there exists a projective algebraic variety Y with $\dim Y = \dim X$ and a bimeromorphic map $\varphi : X \dashrightarrow Y$.*

Proof. Assume that X is Moishezon. Consider the projective algebraic variety $Y \subset \mathbb{C}\mathbb{P}^N$ with $\dim Y = \dim X = n$ constructed at the end of Theorem 2.2.11. By resolution of singularities (Theorem 2.1.15) and since the blow-up of a projective manifold is projective (Proposition 2.1.14), we can assume that Y is smooth. Let $\mathbb{C}[\xi_0, \ldots, \xi_N]$ be a homogeneous coordinate ring of Y (i.e., the quotient of the polynomial ring by the ideal of Y). Then $\xi_1/\xi_0, \ldots, \xi_N/\xi_0 \in \mathscr{M}_Y(Y)$ and we denote by $\varphi_1, \ldots, \varphi_N \in \mathscr{M}_X(X)$ the corresponding elements through the isomorphism $\mathscr{M}_X(X) \cong \mathscr{M}_Y(Y)$.

Consider the meromorphic mapping $\psi : X \dashrightarrow \mathbb{C}\mathbb{P}^N$, $\psi = [1, \varphi_1, \ldots, \varphi_N]$ (cf. Example 2.1.20). Let $G = \mathrm{Graph}(\psi)$ and denote $\mathrm{pr}_1 : G \to X$. Consider a resolution of singularities $\sigma : X' \to G$ and set $\rho := \mathrm{pr}_2 \circ \sigma : X' \to Y$. The holomorphic map ρ is called the *algebraic reduction* of X. It is clear that X' is smooth and bimeromorphically equivalent to X. Moreover, $\rho^* : \mathscr{M}_Y(Y) \to \mathscr{M}_{X'}(X')$ is an isomorphism. Consider the Stein factorization (Theorem B.1.10) of $\rho : X' \to Y$:

$$
\begin{array}{ccc}
X' & \xrightarrow{\ \rho\ } & Y \\
 & {\scriptstyle \alpha}\searrow & \ \uparrow{\scriptstyle \beta} \\
 & & Z
\end{array}
\tag{2.2.22}
$$

Then β is a finite holomorphic map which is bimeromorphic, since $\mathscr{M}_Z(Z)$ and $\mathscr{M}_Y(Y)$ are isomorphic via β. By Zariski's Main Theorem B.1.11 we infer that β is biholomorphic. By the properties of the Stein factorization, the fibers of α are connected, hence the fibers of ρ are connected. Now, the map $\psi : X \dashrightarrow Y$ is generically locally biholomorphic so this holds for ρ also. We deduce that ρ is a proper modification, hence X', Y are bimeromorphically equivalent and thus also X and Y. $\qquad\square$

We can also give a simple characterization of Moishezon manifolds in terms of order of growth of spaces of sections of line bundles.

Theorem 2.2.15. *A compact connected complex manifold X is Moishezon if and only if it carries a big line bundle.*

Proof. If X is Moishezon, then there exist $n = \dim X$ algebraically independent meromorphic functions. Using Problem 2.2, we can find a line bundle L such that these functions have the form $s_1/s_0, \ldots, s_n/s_0$ where $s_0, \ldots, s_n \in H^0(X, L)$. By Theorem 2.2.9 these functions are also analytically independent, i.e., $d(s_1/s_0) \wedge \cdots \wedge d(s_n/s_0) \neq 0$ on the set where the left–hand side is defined. By completing $\{s_0, \cdots, s_n\}$ to a basis of $H^0(X, L)$, we see that the Kodaira map $\Phi_1 : X \setminus \mathrm{Bl}_{|H^0(H,L)|} \longrightarrow \mathbb{P}(H^0(H, L)^*)$ has maximal rank i.e. $\rho_1 = n$ and hence $\kappa(L) = n$.

Conversely, if L is big, there exists $p > 0$ such that $\varrho_p = n$. Thus there exist $s_0, \ldots, s_n \in H^0(X, L^p)$ such that $d(s_1/s_0) \wedge \cdots \wedge d(s_n/s_0) \neq 0$ outside a nowhere dense analytic set. This means that $s_1/s_0, \ldots, s_n/s_0$ are n analytically independent and so algebraically independent meromorphic functions. $\qquad\square$

Moishezon showed that even more than Theorem 2.2.14 holds:

Theorem 2.2.16 (Moishezon). *If X is a Moishezon manifold, then there exists a proper modification $\pi : \widehat{X} \longrightarrow X$, obtained by a finite number of blow-ups with smooth centers, such that $\widehat{X}$ is a projective algebraic manifold.*

Proof. By Theorem 2.2.14 there exists a projective algebraic variety Y, $\dim Y = \dim X$, and a bimeromorphic map $\varphi : Y \dashrightarrow X$. Applying Theorem 2.1.25 (ii) (Chow Lemma) we find holomorphic maps $\pi_X : \widetilde{X} \to X$, $\pi_Y : \widetilde{Y} \to Y$ and $\widetilde{\varphi} : \widetilde{Y} \to \widetilde{X}$ such that π_X, π_Y are compositions of finitely many blow-ups with smooth centers, and the diagram

$$
\begin{array}{ccc}
\widetilde{Y} & \xrightarrow{\ \widetilde{\varphi}\ } & \widetilde{X} \\
{\scriptstyle \pi_Y}\big\downarrow & & \big\downarrow{\scriptstyle \pi_X} \\
Y & \xrightarrow[\ \varphi\ ]{} & X
\end{array}
\tag{2.2.23}
$$

commutes. Using Hironaka's flattening theorem 2.1.22 for the map $\dot{\varphi} : \widetilde{Y} \to \dot{X}$ we may assume that $\dot{\varphi}$ is flat and $\pi_X : \widetilde{X} \to X$ is the composition of finitely many blow-ups with smooth centers. Now $\widetilde{\varphi}$ is flat and bimeromorphic, therefore a finite holomorphic map which is a proper modification. By Zariski's main theorem B.1.11 we deduce that $\widetilde{\varphi}$ is a biholomorphic map. On the other hand, $\widetilde{Y}$ is the successive blow-up of Y along coherent ideal sheaves hence also projective (Proposition 2.1.14). This completes the proof. $\qquad\square$

Remark 2.2.17. Artin introduced the notion of algebraic space and showed that Moishezon manifolds carry a structure of proper algebraic space over $\mathbb{C}$. Thus, it is possible to apply the algebraic versions of the elimination of indeterminacy points and Chow lemma in order to prove Theorem 2.2.16.

Since Moishezon manifolds are not so far from projective ones, they share a lot of features of projective manifolds. We prove now that the Hodge decomposition holds on Moishezon manifolds. Let's recall first the definition of the spectral sequence associated to a filtered complex.

Let $(E = \oplus_{j=0}^k E^j, d : E^\bullet \to E^{\bullet+1})$ be a complex of vector spaces, i.e., $d^2 = 0$. The cohomology of the complex is

$$
H^j(E) = \frac{\mathrm{Ker}(d) \cap E^j}{\mathrm{Im}(d) \cap E^j}, \quad H^\bullet(E) = \bigoplus_j H^j(E).
\tag{2.2.24}
$$

A subcomplex (V, d) of (E, d) is given by subspaces $V^j \subset E^j$ with $dV^j \subset V^{j+1}$. A filtered complex $(F^p E^\bullet, d)$ is a decreasing sequence of subcomplexes

$$E^\bullet = F^0 E^\bullet \supset F^1 E^\bullet \supset \cdots \supset F^m E^\bullet \supset F^{m+1} E^\bullet = 0. \qquad (2.2.25)$$

Certainly, the filtration $F^p E^\bullet$ on $E^\bullet$ also induces a filtration $F^p II^\bullet(E)$ on the cohomology by $F^p H^q(E) = \frac{\text{Ker}(d) \cap F^p E^q}{\text{Im}(d) \cap F^p E^q}$. The associated graded cohomology is

$$\text{Gr}H^\bullet(E) = \bigoplus_{p,q} \text{Gr}^p H^q(E), \quad \text{with } \text{Gr}^p H^q(E) = \frac{F^p H^q(E)}{F^{p+1} H^q(E)}. \qquad (2.2.26)$$

Let $(F^p E^\bullet, d)$ be a filtered complex, for $r \in \mathbb{N}$, we define

$$E_r^{p,q} = \frac{\{u \in F^p E^{p+q} : du \in F^{p+r} E^{p+q+1}\}}{d(F^{p-r+1} E^{p+q-1}) + F^{p+1} E^{p+q}}, \quad E_r = \bigoplus_{p,q} E_r^{p,q}, \qquad (2.2.27)$$

and for $[u] \in E_r^{p,q}$,

$$d_r[u] = [du] \in \frac{\{v \in F^{p+r} E^{p+q+1} : dv \in F^{p+2r} E^{p+q+2}\}}{d(F^{p+1} E^{p+q}) + F^{p+r+1} E^{p+q+1}} = E_r^{p+r, q-r+1}. \qquad (2.2.28)$$

Then we verify that

$$d_r : E_r^{p,q} \to E_r^{p+r, q-r+1}, \quad d_r^2 = 0,$$
$$H^\bullet(E_r) = E_{r+1}. \qquad (2.2.29)$$

Moreover, from (2.2.27), we have

$$E_0^{p,q} = \frac{F^p E^{p+q}}{F^{p+1} E^{p+q}}, \quad E_r^{p,q} = \text{Gr}^p H^{p+q}(E) \text{ for } r \geqslant m+1. \qquad (2.2.30)$$

The above sequence (E_r, d_r) $(r \geqslant 0)$ is called the *spectral sequence* associated to the filtered complex $(F^p E^\bullet, d)$. The last statement in (2.2.30) is usually written $E_r \Rightarrow H^\bullet(E)$, and we say that the spectral sequence abuts to $H^\bullet(E)$.

For a compact connected complex manifold X, we denote

$$H^{p,q}(X) = H^{p,q}(X, \mathbb{C}).$$

We define a filtration on the de Rham complex $(\Omega(X), d)$ by

$$\Omega^j(X) = \bigoplus_{p+q=j} \Omega^{p,q}(X), \quad F^p \Omega^j(X) = \bigoplus_{p' \geqslant p} \Omega^{p', j-p'}(X). \qquad (2.2.31)$$

The corresponding spectral sequence $(E_r(X), d_r)$ $(r \geqslant 0)$ is called the *Fröhlicher spectral sequence* which abuts to $H^\bullet(X, \mathbb{C})$, the de Rham cohomology of X.

As $d = \partial + \bar{\partial}$, by (2.2.27)-(2.2.30), we get

$$E_0^{p,q}(X) = \Omega^{p,q}(X), \quad d_0 = \bar{\partial}, \quad E_1^{p,q}(X) = H^{p,q}(X). \qquad (2.2.32)$$

Theorem 2.2.18. *If X is a Moishezon manifold, then the Hodge decomposition holds:*

    (a) $H^j(X, \mathbb{C}) \cong \oplus_{p+q=j} H^q(X, \mathcal{O}_X^p) \cong \oplus_{p+q=j} H^{p,q}(X)$,

    (b) $H^{p,q}(X) \cong \overline{H^{q,p}(X)}$.

Proof. Let $\pi : \widehat{X} \longrightarrow X$ be a modification as in Theorem 2.2.16. Since $\widehat{X}$ is Kähler, the Hodge decomposition holds for $\widehat{X}$.

Let us remark first that $\pi^* : H^{p,q}(X) \to H^{p,q}(\widehat{X})$ induced by the pull-back of forms is an injective morphism . To see this, consider a $\bar{\partial}$-closed form $\alpha \in \Omega^{p,q}(X)$ such that $\pi^*(\alpha) = \bar{\partial}\beta$ on $\widehat{X}$. Take an arbitrary $\bar{\partial}$-closed form $\gamma \in \Omega^{n-p,n-q}(X)$, where $n = \dim X$. By Stokes formula we have $\int_{\widehat{X}} \pi^*(\alpha) \wedge \pi^*(\gamma) = 0$. Since π is a modification, we have $\int_X \alpha \wedge \gamma = \int_{\widehat{X}} \pi^*(\alpha) \wedge \pi^*(\gamma) = 0$. By Serre duality we know that the linear map

$$H^{p,q}(X) \to (H^{n-p,n-q}(X))^*, \quad [\alpha] \to \int_X \alpha \wedge [\cdot]$$

is an isomorphism. We deduce that $[\alpha] = 0$, so π^* is injective. We show similarly that $\pi^* : H^j(X, \mathbb{C}) \to H^j(\widehat{X}, \mathbb{C})$ is injective.

If $\alpha \in H^{p,q}(X)$, then $\pi^*\alpha \in H^{p,q}(\widehat{X})$. By Corollary 1.4.13 for $\widehat{X}$, we know that $\partial \pi^* \alpha = 0$. Thus $\partial\alpha = 0$ on X, as $\pi^*\partial\alpha = \partial\pi^*\alpha = 0$. This means $d\alpha = 0$ for any $\alpha \in H^{p,q}(X)$. But by (2.2.28)-(2.2.29), this implies that $E_1^{p,q}(X) = E_r^{p,q}(X)$ for any $r \geq 1$, together with (2.2.30), we get (a).

To prove (b), we observe that $H^j(X, \mathbb{C})$ and $\overline{H^j(X, \mathbb{C})}$ are $\mathbb{C}$-anti-isomorphic, so $H^j(X, \mathbb{C}) \cong \oplus_{p+q=j} \overline{H^{p,q}(X)}$. Now, $\pi^*(\bar{\alpha}) = \overline{\pi^*(\alpha)}$ for any $\alpha \in \Omega^{p,q}(X)$. Therefore $\pi^*\big(\overline{H^{q,p}(X)}\big) = \overline{\pi^*(H^{q,p}(X))} \subset H^{p,q}(\widehat{X})$ and $\pi^*\big(\overline{H^{q,p}(X)} \cap H^{p_1,q_1}(X)\big) = \{0\}$ if $(q,p) \neq (q_1, p_1)$. Since π^* is injective, we must have $\overline{H^{q,p}(X)} \cong H^{p,q}(X)$. $\square$

As an application of the Hodge decomposition, let us study the analytic cohomology classes.

Definition 2.2.19. Given an n-dimensional compact complex manifold X, we call a $2k$-cycle $\sum_i c_i Y_i$ in X ($c_i \in \mathbb{Z}$) *analytic* if Y_i are compact k-dimensional analytic sets. We call analytic 2-cycles *curves*. A homology class in $H_{2k}(X, \mathbb{Z})$ ($k \in \mathbb{N}$) is called *analytic*, if it represented by an analytic $2k$-cycle. We denote by $A_{2k}(X, \mathbb{Z}) \subset H_{2k}(X, \mathbb{Z})$ and the subgroup generated by analytic homology classes and set $A_{2k}(X, \mathbb{Q}) = A_{2k}(X, \mathbb{Z}) \otimes_{\mathbb{Z}} \mathbb{Q}$ and $A_{2k}(X, \mathbb{R}) = A_{2k}(X, \mathbb{Q}) \otimes_{\mathbb{Q}} \mathbb{R}$.

Two curves C, C' are called *numerically equivalent*, written $C \equiv C'$, if $D \cdot C = D \cdot C'$ for any divisor D on X. Here $D \cdot C = \int_C c_1(\mathcal{O}_X(D))$ is the intersection number of D with C. Two divisors D, D' are called *numerically equivalent*, written $D \equiv D'$ if $D \cdot C = D' \cdot C$ for any irreducible curve C in X.

Recall that there exists an isomorphism $\Psi : H_{2k}(X, \mathbb{Z}) \to H^{2n-2k}(X, \mathbb{Z})$ (by Poincaré duality), which associates to each homology class $[Y] \in H_{2k}(X, \mathbb{Z})$ of a

$2k$-cycle Y a cohomolgy class $[\eta_Y] \in H^{2n-2k}(X, \mathbb{Z}) \subset H^{2n-2k}(X, \mathbb{R})$ of a closed $(2n - 2k)$-form η_Y such that

$$\int_Y \alpha = \int_X \alpha \wedge \eta_Y, \quad \text{for any closed } 2k\text{-form } \alpha. \tag{2.2.33}$$

η_Y is called the *fundamental class* of Y. A cohomology class of $H^{2n-2k}(X, \mathbb{Z})$ is called analytic if it is the fundamental class of an analytic $2k$-cycle.

For a form $\alpha \in \Omega^{2k}(X)$ we denote by $\alpha^{p,q} \in \Omega^{p,q}(X)$ the component of α in the decomposition $\Omega^{2k}(X) = \oplus_{p+q=2k}\Omega^{p,q}(X)$.

Assume that Y is an analytic $2k$-cycle. By the Hodge decomposition theorem 2.2.18 (i), we can choose a representative η_Y in the fundamental class of Y such that $\eta_Y^{p,q}$ are d- and $\bar{\partial}$-closed for any p, q. Y being analytic also entails

$$\int_Y \alpha = \int_Y \alpha^{k,k}. \tag{2.2.34}$$

If $(p, q) \neq (n - k, n - k)$ we deduce from (2.2.33) and (2.2.34) that $\int_X \alpha \wedge \eta_Y^{p,q} = 0$ for any d-closed $\alpha \in \Omega^\bullet(X)$. By another form of the Poincaré duality, the pairing $\psi : H^\bullet(X, \mathbb{C}) \times H^\bullet(X, \mathbb{C}) \to \mathbb{C}$, given by $\psi(\alpha, \beta) = \int_X \alpha \wedge \beta$, is non-degenerate (which can also be seen from the Hodge theory for de Rham cohomology). Hence $[\eta_Y^{p,q}] = 0 \in H^\bullet(X, \mathbb{C})$. (Using the spectral sequence we see that actually $[\eta_Y^{p,q}] = 0 \in H^{p,q}(X)$.) Therefore, $[\eta_Y] = [\eta_Y^{n-k,n-k}] \in H^{2n-2k}(X, \mathbb{C})$ and the de Rham cohomology class of η_Y in $H^{2n-2k}(X, \mathbb{C})$ (i.e., the fundamental class of Y) lies in $H^{n-k,n-k}(X)$.

For $(1, 1)$-classes on Moishezon manifolds the converse is also true, namely the Lefschetz theorem on $(1, 1)$-classes holds.

Theorem 2.2.20. *Let X be a Moishezon manifold. Then for any $\alpha \in H^{1,1}(X) \cap H^2(X, \mathbb{Z})$, there exists a divisor $D \in \mathrm{Div}(X)$ with $\alpha = c_1(\mathscr{O}_X(D))$. Hence, α is the fundamental class of D, i.e. $\alpha = \eta_D$.*

Proof. Remark that the last assertion follows from the equality

$$c_1(\mathscr{O}_X(D)) = \eta_D, \tag{2.2.35}$$

valid for any divisor D, so it suffices to find $D \in \mathrm{Div}(X)$ with $\alpha = c_1(\mathscr{O}_X(D))$. Let (F, h^F) be a holomorphic Hermitian line bundle on X such that $c_1(F, h^F) = \alpha$ (cf. Lemma 2.3.5).

We assume first that X is projective. Consider a positive line bundle (L, h^L) on X. By the Kodaira-Serre vanishing theorem 1.5.6 and the Riemann-Roch-Hirzebruch theorem 1.4.6,

$$\dim H^0(X, L^p \otimes F) = \frac{1}{n!} \int_X \left(\frac{\sqrt{-1}}{2\pi} R^L\right)^n p^n + \mathscr{O}(p^{n-1}),$$

for $p \in \mathbb{N}$ sufficiently large. Since the coefficient of p^n is positive, $H^0(X, L^p \otimes F) \neq 0$ for some fixed large p (cf. also (1.7.1) with $q = 1$). Hence $L^p \otimes F \cong \mathcal{O}_X(D)$ where D is the divisor defined by a non-trivial $s \in H^0(X, L^p \otimes F)$. Applying this argument for F trivial we see that $L^p \cong \mathcal{O}_X(D_1)$ for $D_1 \in \mathrm{Div}(X)$ and finally $F \cong \mathcal{O}_X(D - D_1)$.

For general X, consider a modification $\pi : \widehat{X} \longrightarrow X$ as in Theorem 2.2.16. Since $\widehat{X}$ is projective, there exists $\widehat{D} \in \mathrm{Div}(\widehat{X})$ such that $\pi^*(F) \cong \mathcal{O}_{\widehat{X}}(\widehat{D})$. Let $A \subset X$ be an analytic set of codimension $\geqslant 2$ such that $\pi : \widehat{X} \smallsetminus \pi^{-1}(A) \to X \smallsetminus A$ is biholomorphic (note that the centers of blow-up have codimension $\geqslant 2$).

By the Remmert-Stein extension theorem B.1.7, the image of $\widehat{D}$ in $X \smallsetminus A$ extends over A to a divisor D on X. Then $F \cong \mathcal{O}_X(D)$ on $X \smallsetminus A$, which means that there exists a holomorphic section s of $F \otimes \mathcal{O}_X(-D)$ over $X \smallsetminus A$ which has no zeros. In trivialization patches s is given by a holomorphic function which extends holomorphically over A (by Riemann's second extension theorem B.1.6). Hence s extends to a holomorphic section $\widetilde{s}$ of $F \otimes \mathcal{O}_X(-D)$ over X. But $\widetilde{s}$ cannot have zeros (they would form an analytic set of codimension 1), therefore $F \otimes \mathcal{O}_X(-D)$ is trivial over X, i.e., $F \cong \mathcal{O}_X(D)$ on X. $\qquad\square$

Our next goal is to prove that a Moishezon Kähler manifold is projective and prepare for this purpose some lemmas.

Lemma 2.2.21. *Let X be a Moishezon manifold. Then for any $\beta \in H^{n-1,n-1}(X) \cap H^{2n-2}(X, \mathbb{Z})$, there is a curve $\sum_i n_i C_i$, $n_i \in \mathbb{Z}$ whose fundamental class is β. The assertion still hold if we consider cohomology with values in $\mathbb{Q}$ and curves with rational coefficients.*

Proof. Let us first assume that X is projective and we fix an embeding in a projective space $\mathbb{C}\mathbb{P}^m$. By the hard Lefschetz theorem we have a commutative diagram

$$
\begin{array}{ccc}
H^{1,1}(X) \cap H^2(X, \mathbb{Z}) & \xrightarrow{\ \Phi\ } & H^{n-1,n-1}(X) \cap H^{2n-2}(X, \mathbb{Z}) \\
\Psi_1 \downarrow & & \downarrow \Psi_2 \\
A_{2n-2}(X, \mathbb{Z}) & \xrightarrow{\ \Xi\ } & H_2(X, \mathbb{Z})
\end{array}
\tag{2.2.36}
$$

where Φ is the isomorphism given by the multiplication with ω_{FS}^{n-1} (ω_{FS} being the restriction of the Fubini-Study form to X) and Ψ_1, Ψ_2 are induced by the Poincaré duality. Finally, Ξ is the intersection with $n - 1$ hyperplanes in $\mathbb{C}\mathbb{P}^m$, which means that the image of Ψ_2 is exactly $A_2(X, \mathbb{Z})$. By Theorem 2.2.20, Ψ_1 is an isomorphism. Since Φ is an isomorphism and Ξ injective, we have an isomorphism $\Psi_2 : H^{n-1,n-1}(X) \cap H^{2n-2}(X, \mathbb{Z}) \to A_2(X, \mathbb{Z})$.

Assume now that X is just Moishezon and consider a modification $\pi : \widehat{X} \longrightarrow X$ as in Theorem 2.2.16. The class $\pi^*(\beta)$ is represented by a curve $\sum_{i \in I} n_i C_i$, where $C_i \subset \widehat{X}$. Then β is represented by $\sum_{j \in J} n_j \pi(C_j)$, where the sum is restricted to the indices $j \in I$ such that $\pi(C_j)$ is a curve in X (note that by Remmert's proper image theorem B.1.12, the image of analytic sets through proper holomorphic maps are analytic sets). Indeed, for any $\bar{\partial}$-closed form $\eta \in \Omega^2(X)$ we have

$$
\int_X \beta \wedge \eta = \int_{\widehat{X}} \pi^*(\beta) \wedge \pi^*(\eta) = \sum_{i \in I} n_i \int_{C_i} \pi^*(\eta)
$$
$$
= \sum_{i \in J} n_i \int_{C_i} \pi^*(\eta) = \sum_{i \in J} n_i \int_{\pi(C_i)} \eta .
$$

(2.2.37)

Since $H^\bullet(X, \mathbb{Q}) = H^\bullet(X, \mathbb{Z}) \otimes_{\mathbb{Z}} \mathbb{Q}$, we also get the $\mathbb{Q}$-coefficient version. $\square$

Lemma 2.2.22. *The numerical and homological equivalence for curves and divisors in a Moishezon manifold coincide.*

Proof. Let $D \in \mathrm{Div}(D)$. We have to show that $c_1(\mathscr{O}_X(D)) = 0$ in $H^2(X, \mathbb{Q})$ if and only if $D \cdot C = 0$ for all curves $C \subset X$. The "only if" part is trivial. We prove the "if" part. Assume that $D \cdot C = 0$ for all curves $C \subset X$. We have to show that

$$
\int_X \alpha \wedge \psi = 0
$$

(2.2.38)

for any d-closed form $\psi \in \Omega^{2n-2}(X)$, where α is a $(1,1)$-form representing $c_1(\mathscr{O}_X(D))$. By the Hodge decomposition

$$
H^{2n-2}(X, \mathbb{C}) = H^{n-1,n-1}(X) \oplus H^{n,n-2}(X) \oplus H^{n-2,n}(X)
$$

and since α is of bidegree $(1,1)$ it suffices to show (2.2.38) for $\psi \in \Omega^{n-1,n-1}(X)$. Moreover, since $c_1(\mathscr{O}_X(D)) \in H^{1,1}(X) \cap H^2(X, \mathbb{Q})$, it's enough to consider forms $\psi \in \Omega^{n-1,n-1}(X)$ with $[\psi] \in H^{n-1,n-1}(X) \cap H^{2n-2}(X, \mathbb{Q})$. Using Lemma 2.2.21 we know that $[\psi]$ is represented by a curve with rational coefficients. But then (2.2.38) follows from the hypothesis.

The proof that for a curve C, $[C] = 0$ if and only if $D \cdot C = 0$ for any $D \in \mathrm{Div}(X)$, is analogous. $\square$

Lemma 2.2.23. *Let X be a Moishezon manifold. Then the canonical morphism $\Phi : A_{2n-2}(X, \mathbb{Q}) \to A_2(X, \mathbb{Q})^*$ given by $\Phi([D])([C]) = D \cdot C$ is an isomorphism.*

Proof. Consider a linear map $\varphi : A_2(X, \mathbb{Q}) \to \mathbb{Q}$. We identify now $A_2(X, \mathbb{Q})$ to $H^{n-1,n-1}(X) \cap H^{2n-2}(X, \mathbb{Q})$ by Lemma 2.2.21. The extension $\varphi_{\mathbb{R}} : A_2(X, \mathbb{R}) = H^{n-1,n-1}(X) \to \mathbb{R}$ is given by a $\bar{\partial}$-closed form ψ: $\varphi_{\mathbb{R}}(\gamma) = \int_X \gamma \wedge \psi$, where γ is a $\bar{\partial}$-closed $(n-1, n-1)$-form. The trivial extension $\widehat{\varphi} : H^{2n-2}(X, \mathbb{Q}) \to \mathbb{Q}$ of φ is given by an element of $H^2(X, \mathbb{Q})$. Therefore $[\psi] \in H^{1,1}(X) \cap H^2(X, \mathbb{Q})$.

Theorem 2.2.20 provides $D \in \mathrm{Div}(X)$ and $k \in \mathbb{N}^*$ such that

$$
[\psi] = \frac{1}{k} c_1(\mathscr{O}_X(D)), \quad \text{so} \quad \varphi([C]) = \frac{1}{k} D \cdot C.
$$

By Lemma 2.2.22, the class of $\frac{1}{k} D$ in $A_{2n-2}(X, \mathbb{Q})$ is uniquely determined by this property. $\square$

For a compact curve $C \subset X$ we define its *multiplicity* as

$$m(C) = \sup_{x \in C} m_x(C),$$

where $m_x(C)$ is the multiplicity at x (see Theorem and Definition B.1.3).

Lemma 2.2.24 (Seshadri's ampleness criterion). *Let X be a Moishezon manifold and $D \in \mathrm{Div}(X)$. If there exists $\varepsilon > 0$ such that $D \cdot C > \varepsilon m(C)$ for all irreducible curves $C \subset X$, then $\mathscr{O}_X(D)$ is positive.*

Lemma 2.2.25. *Let X be a Moishezon manifold. Assume that there exists $\varepsilon > 0$ and $\varphi \in A_2(X, \mathbb{Q})^*$ such that $\varphi([C]) \geqslant \varepsilon m(C)$ for any irreducible curve $C \subset X$. Then X is projective.*

Proof. By Lemma 2.2.23 there exist $k \in \mathbb{N}^*$, $D \in A_{2n-2}(X, \mathbb{Z})$ such that $\varphi([C]) = \frac{1}{k} D \cdot C$ for any curve $C \subset X$. By Seshadri's criterion we deduce that $\mathscr{O}_X(D)$ is positive and X is projective by the Kodaira embedding theorem 5.1.12. $\qquad\square$

Theorem 2.2.26 (Moishezon). *If X is a Moishezon Kähler manifold, then X is projective.*

Proof. Let ω be a Kähler form on X. We consider the linear map $\varphi : A_2(X; \mathbb{R}) \to \mathbb{R}$, $\varphi([C]) = \int_C \omega$. Lemma 2.2.23 entails that there exists $D \in A_{2n-2}(X, \mathbb{R})$, with $\varphi([C]) = D \cdot C$ for all curves $C \subset X$. Set $D = \sum_{i=1}^k m_i D_i$, where $m_i \in \mathbb{R}$, $D_i \in \mathrm{Div}(X)$. We consider sequences of rational numbers $m_{i\nu} \to m_i$ and set $D^\nu = \sum_{i=1}^k m_{i\nu} D_i$. We define

$$\alpha_\nu = \sum_{i=1}^k (m_i - m_{i\nu}) \left(\frac{\sqrt{-1}}{2\pi} R^{\mathscr{O}_X(D_i)} \right), \qquad (2.2.39)$$

where $R^{\mathscr{O}_X(D_i)}$ is the curvature of a Chern connection on $\mathscr{O}_X(D_i)$. Then $\alpha_\nu \to 0$ uniformly, as $\nu \to \infty$. Thus, there exists ν_0 such that $\omega - \alpha_\nu$ is strictly positive for $\nu \geqslant \nu_0$. Let $\nu \geqslant \nu_0$ be fixed and set $\widetilde{\omega} = \omega - \alpha_\nu$. Then $\widetilde{\omega}$ is Kähler and $\int_C \widetilde{\omega} = D^\nu \cdot C$. This shows that by substituting ω by $\widetilde{\omega}$ and D by D^ν we can assume $D \in \mathrm{Div}(X)$.

We denote by $\pi_x : \widetilde{X}_x \to X$ the blow-up of X with center x and by E_x the exceptional divisor. Due to the compactness of X we can find $\varepsilon > 0$ and a Hermitian metric on $\mathscr{O}_{\widetilde{X}_x}(E_x)$ such that $\widehat{\omega}_x = \pi_x^*(\omega) - \varepsilon \frac{\sqrt{-1}}{2\pi} R^{\mathscr{O}_{\widetilde{X}_x}(E_x)}$ is positive for any $x \in X$ (cf. the proof of Lemma 2.1.8). Consider a curve $C \subset X$ and a point $x \in C$. Let $\widehat{C}$ be the strict transform of C in $\widetilde{X}_x$. Then $\varphi([C]) = D \cdot C = \int_C \omega = \int_{\widehat{C}} \pi_x^*(\omega)$. It is easy to see that $m_x(C) = E_x \cdot \widehat{C}$. Thus

$$\begin{aligned}
\int_C \omega - \varepsilon m_x(C) &= \int_C \omega - \varepsilon E_x \cdot \widehat{C} \\
&= \int_{\widehat{C}} \pi_x^*(\omega) - \varepsilon \int_{\widehat{C}} \frac{\sqrt{-1}}{2\pi} R^{\mathscr{O}_{\widetilde{X}_x}(E_x)} = \int_{\widehat{C}} \widehat{\omega}_x > 0.
\end{aligned} \qquad (2.2.40)$$

Since ε does not depend on C and x, we can apply Lemma 2.2.25 and obtain that X is projective. □

We come now to a characterization à la Kodaira of the Moishezon manifolds. In view of Theorem 2.2.16, there exists a projective proper modification $\pi : X' \longrightarrow X$ if X is Moishezon. By taking the push-forward of the positive line bundle on X', we obtain in general a sheaf on X, which outside a proper analytic set A is locally free and has a smooth metric of positive curvature. Such sheaves are called *quasi-positive*. The question is, whether this property characterizes Moishezon manifolds. Since the Moishezon property is bimeromorphically invariant, we can blow up X in order to obtain a manifold X'' possessing a line bundle with semi-positive curvature everywhere and positive outside a proper analytic set. If we show that X'' is Moishezon, it follows that X is Moishezon too.

Grauert-Riemenschneider conjecture. If X possesses a smooth Hermitian line bundle which is semi-positive everywhere and positive on an open dense set, then X is Moishezon. Therefore, X is Moishezon if and only if X carries a quasi-positive sheaf.

The following theorem gives a positive answer to the above conjecture.

Theorem 2.2.27 (Siu–Demailly). *Let X be a compact connected complex manifold of dimension n and (L, h^L) be a holomorphic Hermitian line bundle over X. Then X is Moishezon if one of the following conditions is verified*

(i) *(Siu's criterion)* (L, h^L) *is semi-positive on X and positive at one point.*
(ii) *(Demailly's criterion)* (L, h^L) *satisfies*

$$\int_{X(\leqslant 1)} \left(\tfrac{\sqrt{-1}}{2\pi} R^L \right)^n > 0. \tag{2.2.41}$$

Proof. We apply Theorem 1.7.1 for $q = 1$ and obtain

$$\dim H^0(X, L^p) \geqslant \frac{p^n}{n!} \int_{X(\leqslant 1)} \left(\tfrac{\sqrt{-1}}{2\pi} R^L \right)^n + o(p^n). \tag{2.2.42}$$

First observe that if (L, h^L) is semi-positive, then $\sqrt{-1}R^L$ has no negative eigenvalue, i.e., $X(1) = \emptyset$. If (L, h^L) is moreover positive at at least one point, then $X(0) \neq \emptyset$ and of course $\int_{X(0)} \left(\tfrac{\sqrt{-1}}{2\pi} R^L \right)^n > 0$. Thus condition (i) implies (ii), so it suffices to prove that (ii) implies that X is Moishezon.

Siegel's lemma 2.2.6, (2.2.41) and (2.2.42) show that there exist $C_1, C_2 > 0$ and $p_0 \in \mathbb{N}$ such that for $p > p_0$,

$$C_2 p^{\varrho_p} \geqslant \dim H^0(X, L^p) \geqslant C_1 p^n. \tag{2.2.43}$$

Therefore $\varrho_p = n$ for $p \gg p_0$, so $\kappa(L) = n$ and L is big. By Theorem 2.2.15, we conclude that X is Moishezon. □

2.3 The Shiffman–Ji–Bonavero–Takayama criterion

This section is dedicated to the proof of Shiffman's conjecture that Moishezon manifolds can be characterized in terms of integral Kähler currents. The conjecture was first solved by Ji–Shiffman. We will present a proof using the singular holomorphic Morse inequalities of Bonavero in Section 2.3.2. Later, in Section 6.2, we prove the criterion using the holomorphic Morse inequalities on non-compact manifolds derived from the expansion of the Bergman kernel.

This section is organized as follows. In Section 2.3.1, we introduce singular Hermitian metrics on line bundles. In Section 2.3.2, we establish the singular holomorphic Morse inequalities of Bonavero from the holomorphic Morse inequalities, then prove Shiffman's conjecture. In Section 2.3.3, we study the volume of big line bundles. In Section 2.3.4, we present an example of a non-projective Moishezon manifold.

2.3.1 Singular Hermitian metrics on line bundles

The proof of the Grauert–Riemenschneider criterion from Section 2.2 deals with smooth Hermitian line bundles. However, smooth Hermitian metrics with semi-positive curvature do not *characterize* Moishezon manifolds, since there exist examples of Moishezon manifolds which do not possess a line bundle satisfying (2.2.41) for a *smooth* metric, cf. Section 2.3.4. Actually, the solution of the Grauert–Riemenschneider conjecture just shows that Moishezon manifolds can be characterized in terms of *quasi-positive analytic sheaves*. Nevertheless, one can conjecture another characterization, in terms of currents. Since the pioneering work of P. Lelong, currents have had a deep impact on complex analysis. While intrinsically linked to analytic objects, they are much more flexible.

Let X be a complex manifold. Let L be a holomorphic line bundle over X.

Definition 2.3.1. A *singular Hermitian metric* h^L on L is a choice of a sesquilinear, Hermitian-symmetric form $h_x^L : L_x \times L_x \to \mathbb{C} \cup \{\infty\}$ on each fiber L_x, such that, for any local holomorphic frame e_L of L on U, with *local weights* $\varphi \in L^1_{\mathrm{loc}}(U)$, we have

$$|e_L|_{h^L}^2 = h^L(e_L, e_L) = e^{-2\varphi} \in [0, \infty]. \qquad (2.3.1)$$

Here $L^1_{\mathrm{loc}}(U)$ is the space of local integrable functions on U. If the local weights $\varphi \in \mathscr{C}^\infty(U)$, we obtain the usual definition of a Hermitian metric.

If s is an arbitrary holomorphic section of L on U, $s = f\, e_L$ with $f \in \mathscr{O}_X(U)$ we cannot set $|s|_{h^L}^2 = |f|^2 |e_L|_{h^L}^2$ everywhere (e.g., if $s(x) = 0$ and $|e_L|_{h^L}(x) = \infty$). Of course, the equality holds a.e. (almost everywhere). Moreover

$$\log |s|_{h^L}^2 = \log |f|^2 - 2\varphi \quad \text{in } L^1_{\mathrm{loc}}(U). \qquad (2.3.2)$$

In terms of an open covering $\{U_\alpha\}$ and holomorphic frames e_α of L on U_α, the metric is given by a family $\{h_\alpha\}$ of functions $h_\alpha = |e_\alpha|_{h^L}^2 : U_\alpha \to [0, \infty]$ with local weights $\varphi_\alpha = -(1/2) \log h_\alpha \in L^1_{\mathrm{loc}}(U_\alpha)$. If $(g_{\alpha\beta} \in \mathscr{O}^*_X(U_\alpha \cap U_\beta))$ is the cocycle of L, then $h_\beta = |g_{\alpha\beta}|^2 h_\alpha$ on $U_\alpha \cap U_\beta$, so

$$\log h_\beta = \log |g_{\alpha\beta}|^2 + \log h_\alpha \quad \text{in } L^1_{\mathrm{loc}}(U_\alpha \cap U_\beta). \tag{2.3.3}$$

Since $\partial\bar{\partial} \log |g_{\alpha\beta}|^2 = 0$, (2.3.3) allows us to define the generalized curvature.

Definition 2.3.2. The *curvature current* and the Chern–Weil representative of the first Chern class of a singular Hermitian line bundle (L, h^L) are the global $(1,1)$-currents $R^L = R^{(L,h^L)}$, $c_1(L, h^L)$ defined locally by

$$R^{(L,h^L)} = -\partial\bar{\partial} \log h_\alpha = 2\partial\bar{\partial}\varphi_\alpha, \quad c_1(L, h^L) = \frac{\sqrt{-1}}{2\pi} R^{(L,h^L)}. \tag{2.3.4}$$

$c_1(L, h^L)$ is obviously closed and integral, representing the Chern class of L in the de Rham cohomology group $H^2(X, \mathbb{R})$ (which as we saw in Appendix B.2, can be also calculated using currents).

Another way to define a singular Hermitian metric h^L on L is to give a smooth Hermitian metric h^L_0 and a function $\varphi \in L^1_{\mathrm{loc}}(X)$ and then set $h^L = h^L_0 e^{-2\varphi}$. From (1.5.9), the curvature current is then

$$R^{(L,h^L)} = R^{(L,h^L_0)} + 2\partial\bar{\partial}\varphi, \quad c_1(L, h^L) = c_1(L, h^L_0) + \frac{\sqrt{-1}}{\pi} \partial\bar{\partial}\varphi. \tag{2.3.5}$$

Theorem 2.3.3 (Poincaré–Lelong formula). *Let s be a meromorphic section of L on X and h^L be a singular Hermitian metric on L. Then*

$$\frac{\sqrt{-1}}{2\pi} \partial\bar{\partial} \log |s|^2_{h^L} = [\mathrm{Div}(s)] - c_1(L, h^L). \tag{2.3.6}$$

Proof. Indeed, writing locally $s = f_\alpha e_\alpha$ and plugging in (2.3.2), (2.3.6) follows from the Poincaré–Lelong formula (B.2.18).   □

Example 2.3.4. Consider a non-trivial meromorphic section s of L on X. We define a singular Hermitian metric h^L on L (associated to s) by

$$|s(x)|_{h^L} = 1 \quad \text{for } x \in X \smallsetminus \mathrm{Div}(s). \tag{2.3.7}$$

We conclude from (2.3.6) that the curvature of the singular Hermitian metric (2.3.7) is given by

$$c_1(L, h^L) = [\mathrm{Div}(s)]. \tag{2.3.8}$$

Under the notation in (2.1.5), the canonical section s_D of $\mathscr{O}_X(D)$ is defined by $\{f_\alpha\}$ on U_α. The singular Hermitian metric $h^{\mathscr{O}_X(D)}$ associated to s_D is defined by (2.3.7). (2.3.8) reads

$$c_1(\mathscr{O}_X(D), h^{\mathscr{O}_X(D)}) = [D]. \tag{2.3.9}$$

The following result is quite useful. For another proof in the case ω is smooth, see Proposition 1.5.3.

Lemma 2.3.5. *Let ω be a real, closed, $(1,1)$-current of order 0. If the cohomology class of ω in $H^2(X, \mathbb{R})$ is integral, there exists a holomorphic line bundle L, endowed with a singular Hermitian metric h^L, such that $c_1(L, h^L) = \omega$ in the sense of currents.*

Proof. Let $\mathscr{U} = \{U_\alpha\}$ be a covering of X with geodesically convex open sets; hence all intersections of sets in $\mathscr{U}$ are contractible. We can also assume that U_α are biholomorphic to open balls in the Euclidean space. For a sheaf $\mathscr{F}$ on X, we denote by $C^q(\mathscr{U}, \mathscr{F})$, $Z^q(\mathscr{U}, \mathscr{F})$ and $B^q(\mathscr{U}, \mathscr{F})$ the spaces of q-cochains, q-cocycles and q-coboundaries of $\mathscr{F}$ with respect to $\mathscr{U}$. Let $\delta : C^q(\mathscr{U}, \mathscr{F}) \to C^{q+1}(\mathscr{U}, \mathscr{F})$ be the coboundary operator in Čech cohomology.

We start by solving $\omega = 2\sqrt{-1}\,\partial\bar{\partial}u_\alpha$ on U_α with $u_\alpha \in L^1_{\mathrm{loc}}(U, \mathbb{R})$, using Lemma B.2.20. Define $h_\alpha := \exp(-4\pi u_\alpha)$. We will construct a line bundle L such that h_α is a singular metric on L. Writing $d^c := \sqrt{-1}(\bar{\partial} - \partial)$, we have $\omega = dd^c u_\alpha$. Let

$$u_{\alpha\beta} = u_\beta - u_\alpha \in L^1_{\mathrm{loc}}(U_\alpha \cap U_\beta). \tag{2.3.10}$$

Then $dd^c u_{\alpha\beta} = 0$, so $u_{\alpha\beta}$ is pluriharmonic and there exists $v_{\alpha\beta} \in \mathscr{C}^\infty(U_\alpha \cap U_\beta, \mathbb{R})$ with

$$dv_{\alpha\beta} = d^c u_{\alpha\beta}. \tag{2.3.11}$$

Set

$$c_{\alpha\beta\gamma} := v_{\alpha\beta} + v_{\beta\gamma} + v_{\gamma\alpha}. \tag{2.3.12}$$

Since $dc_{\alpha\beta\gamma} = d^c(u_{\alpha\beta} + u_{\beta\gamma} + u_{\gamma\alpha}) = d^c 0 = 0$, $c_{\alpha\beta\gamma}$ are constant, so $(c_{\alpha\beta\gamma}) \in Z^2(\mathscr{U}, \mathbb{R})$. The de Rham isomorphism and Leray Theorem for acyclic coverings (note that $\mathscr{U}$ is acyclic for the constant sheaf $\mathbb{R}$) induce a canonical isomorphism between the de Rham $H^2(X, \mathbb{R})$ and the Čech $H^2(\mathscr{U}, \mathbb{R})$ cohomology groups. Let us describe the image of the class $[\omega]$ through this isomorphism. If $\varphi_\alpha := d^c u_\alpha$, we have $d\varphi_\alpha = \omega$ and $(\delta\varphi)_{\alpha\beta} = \varphi_\beta - \varphi_\alpha = d^c u_{\alpha\beta} = dv_{\alpha\beta}$. Then $\delta(v_{\alpha\beta}) = (c_{\alpha\beta\gamma})$ is the image of $[\omega]$ under the de Rham isomorphism. Hence $(c_{\alpha\beta\gamma})$ is integral; i.e., there exists a 1-cochain $\{b_{\alpha\beta}\} \in C^1(\mathscr{U}, \mathbb{R})$ such that for all α, β, γ,

$$m_{\alpha\beta\gamma} := c_{\alpha\beta\gamma} + b_{\alpha\beta} + b_{\beta\gamma} + b_{\gamma\alpha} \in \mathbb{Z}. \tag{2.3.13}$$

Let

$$f_{\alpha\beta} = u_{\alpha\beta} + \sqrt{-1}(v_{\alpha\beta} + b_{\alpha\beta}), \quad \rho_{\alpha\beta} = \exp(-2\pi f_{\alpha\beta}). \tag{2.3.14}$$

By (2.3.11), the $f_{\alpha\beta}$ are holomorphic functions, and by (2.3.10), (2.3.12) and (2.3.13),

$$f_{\alpha\beta} + f_{\beta\gamma} + f_{\gamma\alpha} = \sqrt{-1}\,m_{\alpha\beta\gamma}. \tag{2.3.15}$$

By (2.3.15), $\rho_{\alpha\beta}\rho_{\beta\gamma}\rho_{\gamma\alpha} = \exp(-2\pi\sqrt{-1}\,m_{\alpha\beta\gamma}) = 1$, and thus $(\rho_{\alpha\beta}) \in Z^1(\mathscr{U}, \mathscr{O}_X^*)$ is the cocycle of a holomorphic line bundle L. Moreover, (h_α) defines indeed a

Hermitian metric h^L on L:

$$h_\beta h_\alpha^{-1} = \exp(-4\pi u_{\alpha\beta}) = |\exp(-4\pi f_{\alpha\beta})| = |\rho_{\alpha\beta}|^2 \,. \qquad (2.3.16)$$

Finally,

$$\omega = dd^c u_\alpha = -\tfrac{1}{4\pi} dd^c \log h_\alpha = -\tfrac{\sqrt{-1}}{2\pi} \partial\bar\partial \log h_\alpha = c_1(L, h^L) \,. \qquad (2.3.17)$$

This achieves the proof. $\qquad\qquad\qquad\qquad\qquad\qquad\qquad\qquad\qquad\qquad\square$

Recall that a *Kähler current* ω on X is a closed, strictly positive $(1,1)$-current ω. Back to Moishezon manifolds we have the following.

Lemma 2.3.6. *Any compact Moishezon manifold X of dimension n carries an integral Kähler current ω whose singular support is contained in a proper analytic set. Therefore, there exists a holomorphic line bundle L, endowed with a singular Hermitian metric h^L, smooth outside an analytic set, such that $c_1(L, h^L) = \omega$ in the sense of currents.*

Proof. By Moishezon's theorem 2.2.16, there exists a projective manifold $\widehat{X}$ and a proper modification $\pi : \widehat{X} \to X$. Let $(\widehat{L}, h^{\widehat{L}})$ be a positive line bundle with smooth metric $h^{\widehat{L}}$ on $\widehat{X}$, set $\widehat\omega = c_1(\widehat{L}, h^{\widehat{L}})$. The push-forward $\omega := \pi_*\widehat\omega$ is a $(1,1)$-current defined by

$$(\pi_*\widehat\omega, \varphi) := (\widehat\omega, \pi^*\varphi) = \int_{\widehat{X}} \widehat\omega \wedge \pi^*\varphi, \qquad (2.3.18)$$

for any $(n-1, n-1)$-form φ on X. We show that $\widehat\omega$ forms an integral Kähler current on X. It is clear that the current $\pi_*\widehat\omega$ is smooth outside the analytic set where π is not biholomorphic. Let θ be an arbitrary positive smooth $(1,1)$-form on X. Then $\pi^*\theta$ is a semi-positive smooth $(1,1)$-form on $\widehat{X}$. Hence there exists $C > 0$ such that $\widehat\omega \geqslant C\pi^*\theta$. We deduce that for any positive $(n-1, n-1)$-form φ on X

$$(\pi_*\widehat\omega, \varphi) = \int_{\widehat{X}} \widehat\omega \wedge \pi^*\varphi \geqslant \int_{\widehat{X}} C\pi^*\theta \wedge \pi^*\varphi = \int_X C\theta \wedge \varphi = (C\theta, \varphi) > 0,$$

thus $\pi_*\widehat\omega$ is positive. It is also clear that $d\pi_*\widehat\omega = \pi_*d\widehat\omega = 0$. Let $A \subset X$ be a nowhere dense analytic set such that $\pi : \widehat{X} \smallsetminus \pi^{-1}(A) \to X \smallsetminus A$ is biholomorphic. Since π is a composition of blow-ups in codimension $\geqslant 2$, A is of codimension $\geqslant 2$. We consider a divisor $\widehat{D}$ such that $\widehat\omega$ represents the first Chern class of $\mathscr{O}_{\widehat{X}}(\widehat{D})$. As shown in the proof of Theorem 2.2.20, $\pi(\widehat{D} \smallsetminus \pi^{-1}(A))$ extends by the Remmert-Stein theorem B.1.7 to a divisor D on X. Then ω represents the first Chern class of $\mathscr{O}_X(D)$ and thus $[\omega] \in H^2(X, \mathbb{Z})$. $\qquad\qquad\square$

The converse is also true:

Theorem 2.3.7. *A compact connected complex manifold is Moishezon if and only if it admits an integral Kähler current with singular support contained in a proper analytic set.*

There are two strategies to prove Theorem 2.3.7, both based on holomorphic Morse inequalities for dim $H^0(X, L^p)$ for a suitable line bundle L. One can work directly with the given current and apply the Morse inequalities on the complement of the analytic set which contains the singular support. We develop this method in Section 6.2.

Another possibility is to apply Demailly's approximation Theorem 2.3.10 and suppose that the current has analytic singularities (cf. Definition 2.3.9) which allows us to apply Bonavero's holomorphic Morse inequalities (see Section 2.3.2). With the help of Theorem 2.3.10, Theorem 2.3.7 has its equivalent version, Theorem 2.3.8 which was conjectured by Shiffman. Note that in the proof of Theorem 2.3.7 in Section 6.2, we do not use Demailly's approximation theorem 2.3.10. We will prove Theorems 2.3.7 and 2.3.8 in the next section.

Theorem 2.3.8 (Ji–Shiffman). *A compact connected complex manifold is Moishezon if and only if it admits a strictly positive singular polarization* (L, h^L).

2.3.2 Bonavero's singular holomorphic Morse inequalities

Bonavero gave a proof of Theorem 2.3.8 in terms of his singular holomorphic Morse inequalities. One crucial ingredient of the proof is the approximation result of Demailly which permits us to apply the singular holomorphic Morse inequalities, Theorem 2.3.18 (or Theorem 2.3.7). We start by recalling the notion of singular Hermitian metric with analytic singularities and state the approximation Theorem 2.3.10. After introducing the notion of Nadel multiplier sheaf, we prove the singular holomorphic Morse inequalities in Theorem 2.3.18. This permits us to show the characterization of Moishezon manifolds in Theorem 2.3.28; it contains also Theorems 2.3.7 and 2.3.8.

Let X be a complex manifold of dimension n, and let L be a holomorphic line bundle on X.

Definition 2.3.9. A real *function with analytic singularities* is a locally integrable function φ on X which has locally the form

$$\varphi = \frac{c}{2} \log \Big(\sum_{j \in J} |f_j|^2 \Big) + \psi, \qquad (2.3.19)$$

where J is at most countable, f_j are non-vanishing holomorphic functions and ψ is smooth and c is a $\mathbb{Q}_+$-valued, locally constant function on X. In particular the singular support of φ and $\partial \bar{\partial} \varphi$ is an *analytic set*. By a *Hermitian metric with analytic singularities* we understand a singular Hermitian metric h^L on L with $h^L = h_0^L e^{-2\varphi}$, where h_0^L is a smooth Hermitian metric on L and $\varphi : X \to \mathbb{R}$ has analytic singularities. A $(1,1)$-current ω is called *current with analytic singularities* if $\omega = 2\sqrt{-1}\partial\bar{\partial}\varphi + \omega_0$ where ω_0 is a smooth representative of the de Rham class of ω and $\varphi : X \to \mathbb{R}$ has analytic singularities.

Note that a function with analytic singularities is quasi-plurisubharmonic (cf. Definition B.2.16). With this notion we can state the approximation theorem.

Theorem 2.3.10 (Demailly). *Let X be a compact complex manifold, and consider a closed $(1,1)$-current $\omega = 2\sqrt{-1}\partial\overline{\partial}\varphi + \omega_0$ on X, where ω_0 is a smooth $(1,1)$-form in the same de Rham cohomology class as ω and φ is a quasi-plurisubharmonic function. Let θ be a smooth $(1,1)$-form such that $\omega \geqslant \theta$, and let Θ be a Hermitian form on X as in $(1.2.49)$. Then there exists a sequence $\{\varphi_m\}_{m=1}^{\infty} \subset L_{\mathrm{loc}}^1(X)$ of real functions with analytic singularities such that for $m \to \infty$, the following assertions hold:*

(i) *φ_m is decreasing and converges pointwise and in $L_{\mathrm{loc}}^1(X)$ to φ;*
(ii) *$\omega_m := 2\sqrt{-1}\partial\overline{\partial}\varphi_m + \omega_0$ is in the same de Rham cohomology class as ω, and converges weakly to ω;*
(iii) *$\omega_m \geqslant \theta - \varepsilon_m\Theta$, where $\varepsilon_m > 0$ converges to 0.*

Remark 2.3.11. It was shown by Boucksom that we can choose ω_m such that $\omega_{m,\mathrm{ac}} \to \omega_{\mathrm{ac}}$ as $m \to \infty$.

Remark 2.3.12. To explain the form $(2.3.19)$ of the approximating functions, let us consider the local part of the proof of Theorem 2.3.10. Let ψ be a plurisubharmonic function defined in the unit ball $\mathbb{B}^n \subset \mathbb{C}^n$. Let $\{f_i^m\}_{i=1}^{\infty}$ be an orthonormal basis of the Hilbert space $\{f \in \mathscr{O}_{\mathbb{C}^n}(\mathbb{B}^n) : \int_{\mathbb{B}^n} |f|^2 e^{-2m\psi} dZ < \infty\}$, where dZ is the canonical Euclidean volume form on $\mathbb{C}^n$. Then $\frac{1}{2m}\log(\sum_i |f_i^m|^2) \to \psi$ pointwise and in L_{loc}^1 on $\mathbb{B}^n$ as $m \to \infty$. By carefully using a partition of unity, Demailly shows that one can glue the local approximations and obtain a global approximation φ_m having analytic singularities.

The new element in Bonavero's approach is the introduction of the Nadel multiplier sheaf in the holomorphic Morse inequalities.

Definition 2.3.13. Let $\varphi \in L_{\mathrm{loc}}^1(X, \mathbb{R})$. The *Nadel multiplier ideal sheaf* $\mathscr{I}(\varphi)$ is the ideal subsheaf of germs of holomorphic functions $f \in \mathscr{O}_{X,x}$ such that $|f|^2 e^{-2\varphi}$ is integrable with respect to the Lebesgue measure in local coordinates near x. Let $h^L = h_0^L e^{-2\varphi}$ be a singular Hermitian metric on L where h_0^L is smooth and $\varphi \in L_{\mathrm{loc}}^1(X, \mathbb{R})$. The Nadel multiplier ideal sheaf of h^L is defined by $\mathscr{I}(h^L) = \mathscr{I}(\varphi)$, which does not depend on the choice of φ.

Remark 2.3.14. Let E be a holomorphic vector bundle on X. Since $\mathscr{I}(\varphi)$ is a subsheaf of $\mathscr{O}_X$, we have

$$H^0(X, E \otimes \mathscr{I}(\varphi)) \subset H^0(X, E). \qquad (2.3.20)$$

An important property of the multiplier ideal sheaves is the coherence.

Theorem 2.3.15 (Nadel). *Let h^L be a singular Hermitian metric on L. Assume that $c_1(L, h^L) \geqslant C\Theta$ for some real constant C and a smooth Hermitian form Θ on X as in $(1.2.49)$. Then $\mathscr{I}(h^L)$ is coherent.*

The proof is an application of Theorem B.4.6 for a small coordinate ball $B^X(x, \varepsilon)$ with the trivial bundle endowed with the singular metric

$$h^L \cdot |z - x|^{-(2n+2j)} \quad \text{for} \quad j \in \mathbb{N}.$$

Example 2.3.16. (1) Let φ be bounded from below near $x \in X$. Then $\mathscr{I}(\varphi)_x = \mathscr{O}_{X,x}$.

(2) Let $X = \mathbb{C}^n$ and for $c_1, \ldots, c_k \in \mathbb{R}^*_+$, $p \in \mathbb{N}^*$, set

$$\varphi_p(z) = \frac{p}{2} \log(|z_1|^{2c_1} + \cdots + |z_k|^{2c_k}), \quad z \in \mathbb{C}^n.$$

Set $P(0, r) = \{z \in \mathbb{C}^n : |z_j| < r \text{ for any } 1 \leqslant j \leqslant n\}$, $r > 0$, and let dZ be the Euclidean volume form on $\mathbb{C}^n$. Then the holomorphic functions z^β ($\beta \in \mathbb{N}^n$) are orthogonal with respect to the norm $\|f\|^2_{\varphi_p} = \int_{P(0,r)} |f|^2 e^{-2\varphi_p} dZ$, and $z^\beta \in \mathscr{O}_{\mathbb{C}^n, 0}$ is an element of $\mathscr{I}(\varphi_p)_0$ if and only if $\|z^\beta\|^2_{\varphi_p} < \infty$.

Passing to polar coordinates, we see that $\|z^\beta\|^2_{\varphi_p} < \infty$ if and only if

$$\int_{[0,r]^k} \frac{u_1^{((2\beta_1+2)/c_1)-1} \cdots u_k^{((2\beta_k+2)/c_k)-1}}{(u_1^2 + \cdots + u_k^2)^p} \, du_1 \cdots du_k < \infty.$$

Using the homogeneity, this is the case if and only if

$$\int_0^r \frac{t^{2\sum_{j=1}^k ((\beta_j+1)/c_j)-k}}{t^{2p}} \, t^{k-1} \, dt < \infty,$$

that is, if and only if $2\sum_{j=1}^k \frac{\beta_j+1}{c_j} - k - 2p + (k-1) > -1$. Hence $\mathscr{I}(\varphi_p)_0$ is generated by $\prod_{j=1}^k z_j^{\beta_j}$ over $\mathscr{O}_{\mathbb{C}^n,0}$, with $\beta_j \in \mathbb{N}$ and $\sum_{j=1}^k (\beta_j + 1)/c_j > p$. If $c_1 = \cdots = c_k$, the condition on $\beta_1, \ldots, \beta_k$ is $\sum_{j=1}^k \beta_j \geqslant \lfloor pc_1 \rfloor - k + 1$.

(3) If $D_j = g_j^{-1}(0)$ are smooth divisors with transversal intersections with $g_j \in \mathscr{O}_X(X)$, and $c_1, \ldots, c_k \in \mathbb{R}^*_+$. Set $\varphi = \sum_{j=1}^k c_j \log |g_j|$. Then

$$\mathscr{I}(\varphi) = \mathscr{O}_X(-\sum_j \lfloor c_j \rfloor D_j). \tag{2.3.21}$$

Indeed $f \in \mathscr{I}(\varphi)_x$ if and only if $|f|^2 |g_1|^{-2c_1} \cdots |g_k|^{-2c_k} \in L^1_{loc}$. Since g_j are functionally independent, this means that $2d_j - 2c_j > -2$ where d_j is the order of vanishing of f along D_j. Thus $d_j \geqslant \lfloor c_j \rfloor$ and we know that the sections of the line bundle $\mathscr{O}_X(-\sum_j \lfloor c_j \rfloor D_j)$ can be identified with holomorphic functions vanishing along D_j to order $\geqslant \lfloor c_j \rfloor$.

Definition 2.3.17. Let h^L be a Hermitian metric with analytic singularities on L with curvature R^L. We define the set $X(q) = X(q, c_1(L, h^L))$ *of points of index* q to be the open set of points $x \in X$ such that h^L is smooth at x, $\dot{R}_x^L \in \mathrm{End}(T_x^{(1,0)}X)$ is invertible and has exactly q negative eigenvalues. We set as usual $X(\leqslant q) = X(0) \cup \cdots \cup X(q)$.

Theorem 2.3.18 (Bonavero). *Let X be a compact complex manifold of dimension n and let L be a holomorphic line bundle on X equipped with a Hermitian metric h^L with analytic singularities. Let (E, h^E) be a holomorphic Hermitian vector bundle on X. Then for $0 \leqslant q \leqslant n$, we have*

$$\dim H^q(X, L^p \otimes E \otimes \mathscr{I}(h^{L^p}))$$
$$\leqslant \operatorname{rk}(E) \cdot \frac{p^n}{n!} \int_{X(q)} (-1)^q c_1(L, h^L)^n + o(p^n), \qquad (2.3.22)$$

$$\sum_{j=0}^{q} (-1)^{q-j} \dim H^j(X, L^p \otimes E \otimes \mathscr{I}(h^{L^p}))$$
$$\qquad\qquad (2.3.23)$$
$$\leqslant \operatorname{rk}(E) \cdot \frac{p^n}{n!} \int_{X(\leqslant q)} (-1)^q c_1(L, h^L)^n + o(p^n)\,.$$

Proof. We will suppose in the sequel that X is connected. Thus c from (2.3.19) takes a constant positive rational value on X.

The idea is to construct a proper modification $\widetilde{\pi} : \widetilde{X} \longrightarrow X$ such that $(\widetilde{L}, h^{\widetilde{L}}) = (\widetilde{\pi}^* L, \widetilde{\pi}^* h^L)$ has singularities along a codimension 1 analytic set, which means that $\mathscr{I}(h^{\widetilde{L}})$ is invertible. Then we apply the holomorphic Morse inequalities from Theorem 1.7.1 on $\widetilde{X}$ and relate the cohomology spaces on $\widetilde{X}$ with those on X.

First step: blowing up the singularities of h^L. Let us consider φ, c and f_j as in Definition 2.3.9. We first make sense of the "ideal generated by the functions f_j". The functions f_j are given just locally and we wish to have a globally defined ideal. We introduce the ideal $\mathscr{I} \subset \mathscr{O}_X$ by

$$\mathscr{I}_x = \{ f \in \mathscr{O}_{X,x} : \text{there exists } C > 0 \text{ with } |f| \leqslant C e^{\varphi/c}$$
$$\text{in a neighborhood of } x \}, \quad (2.3.24)$$

for any $x \in X$. This is a globally defined sheaf which coincides with the integral closure of the ideal generated by the f_j's on each open set where φ has the form (2.3.19) (Briançon–Skoda theorem).

We now blow X up along the ideal $\mathscr{I}$.

Lemma 2.3.19. *There exists a proper modification $\widetilde{\pi} : \widetilde{X} \longrightarrow X$ (a composition of finitely many blow-ups with smooth centers) such that the local weight $\widetilde{\varphi}$ of the metric $h^{\widetilde{L}} = \widetilde{\pi}^* h^L$ on $\widetilde{L} = \widetilde{\pi}^* L$ has the form $\widetilde{\varphi} = c \sum_j c_j \log |g_j| + \widetilde{\psi}$ in appropriate local holomorphic coordinates centered in a given point $\widetilde{x}$, where $\widetilde{\psi}$ is smooth, $c_j \in \mathbb{N}^*$ and g_j are irreducible in $\mathscr{O}_{\widetilde{X}, \widetilde{x}}$ and define a divisor with normal crossings.*

Proof. By Proposition 2.1.16, there exists a blow-up $\sigma : X' \longrightarrow X$ along $\mathscr{I}$ such that the pull-back $\sigma^{-1} \mathscr{I} \cdot \mathscr{O}_{X'}$ is an invertible sheaf. Theorem 2.1.25 shows that there exists a composition $\widetilde{\pi} : \widetilde{X} \longrightarrow X$ of finitely many blow-ups with smooth centers which dominates $\sigma : X' \longrightarrow X$:

$$
\begin{array}{ccc}
 & & \widetilde{X} \\
 & {}^{\rho}\swarrow & \downarrow{}^{\widetilde{\pi}} \\
X' & \xrightarrow{\ \sigma\ } & X
\end{array}
$$

The pull-back $\widetilde{\pi}^{-1} \mathscr{I} \cdot \mathscr{O}_{\widetilde{X}}$ is also invertible. Let us denote by g the local generator of the ideal generated by $f_j \circ \widetilde{\pi}$. Then $f_j \circ \widetilde{\pi} = g \cdot h_j$, where h_j have no common zeros. In view of (2.3.19), the local weight of $\widetilde{\pi}^* h^L$ has the form

$$
\begin{aligned}
\widetilde{\varphi} &= \tfrac{c}{2} \log \Big(\sum_j |f_j \circ \widetilde{\pi}|^2 \Big) + \psi \circ \widetilde{\pi} \\
&= \tfrac{c}{2} \log |g|^2 + \tfrac{c}{2} \log \Big(\sum_j |h_j|^2 \Big) + \psi \circ \widetilde{\pi} \qquad (2.3.25) \\
&= \tfrac{c}{2} \log |g|^2 + \widetilde{\psi} \, ,
\end{aligned}
$$

where $\widetilde{\psi} := \tfrac{c}{2} \log \Big(\sum_j |h_j|^2 \Big) + \psi \circ \widetilde{\pi}$ is smooth.

We consider the decomposition $g = \prod g_j^{c_j}$ of g in irreducible factors in $\mathscr{O}_{\widetilde{X}, \widetilde{x}}$. We introduce global divisors $\widetilde{D}_j$ given locally by $\widetilde{D}_j = \{g_j = 0\}$. By Theorem 2.1.13, we can further blow up to make the divisor $\sum D_j$ defined by g_j with normal crossings. $\qquad\square$

Let $\widetilde{\pi} : \widetilde{X} \longrightarrow X$, c, c_j and $\widetilde{D}_j$ as in Lemma 2.3.19. For any Hermitian vector bundle (F, h^F) on X, we set $(\widetilde{F}, h^{\widetilde{F}}) := (\widetilde{\pi}^* F, \widetilde{\pi}^* h^{\widetilde{F}})$. By Example 2.3.16 we have

$$
\mathscr{I}(h^{\widetilde{L}^p}) = \mathscr{O}_{\widetilde{X}} \Big(- \sum_j \lfloor c\, c_j p \rfloor \widetilde{D}_j \Big). \qquad (2.3.26)
$$

Second step: holomorphic Morse inequalities on $\widetilde{X}$. We take advantage of the fact that $\mathscr{I}(h^{\widetilde{L}^p})$ is invertible and we write $\widetilde{L}^p \otimes \mathscr{I}(h^{\widetilde{L}^p})$ as a tensor power of a fixed line bundle, in order to apply the holomorphic Morse inequalities in the smooth case.

We fix $r \in \mathbb{N}, m \in \mathbb{N}^*$ such that $c = r/m$. Set

$$
\widetilde{D} = r\sum_j c_j \widetilde{D}_j, \quad \widehat{L} = \widetilde{L}^m \otimes \mathscr{O}_{\widetilde{X}}(-\widetilde{D}). \qquad (2.3.27)
$$

Let $h^{\mathscr{O}_{\widetilde{X}}(-\widetilde{D})}$ be the singular Hermitian metric on $\mathscr{O}_{\widetilde{X}}(-\widetilde{D})$ defined in Example 2.3.4. Let $h^{\widehat{L}} = h^{\widetilde{L}^m} \otimes h^{\mathscr{O}_{\widetilde{X}}(-\widetilde{D})}$ be the metric on $\widehat{L}$ induced by $h^{\widetilde{L}^m}$ and $h^{\mathscr{O}_{\widetilde{X}}(-\widetilde{D})}$.

Lemma 2.3.20. *The metric $h^{\widehat{L}}$ is smooth on $\widetilde{X}$, and*

$$
R^{(\widehat{L}, h^{\widehat{L}})} = m\, R^{(\widetilde{L}, h^{\widetilde{L}})} = m\, \widetilde{\pi}^* R^{(L, h^L)} \quad on \ \widetilde{X} \setminus \mathrm{supp}(\widetilde{D}). \qquad (2.3.28)
$$

Proof. The local weight of $h^{\widetilde{L}}$ is $m\,\widetilde{\varphi} - r\sum_j c_j \log|g_j| = m\,\widetilde{\psi}$ which is smooth. Since $\partial\overline{\partial}\log|g_j| = 0$ outside $\widetilde{D}$, we have $R^{(\widetilde{L},h^{\widetilde{L}})} = 2m\partial\overline{\partial}\widetilde{\psi} = 2m\,\partial\overline{\partial}\widetilde{\varphi}$ on $\widetilde{X}\setminus\operatorname{supp}(\widetilde{D})$. The proof of Lemma 2.3.20 is completed. $\qquad\square$

Lemma 2.3.21. *As $p\to\infty$, the following inequality holds:*

$$\sum_{j=0}^{q}(-1)^{q-j}\dim H^j(\widetilde{X},\widetilde{L}^p\otimes\widetilde{E}\otimes\mathscr{I}(h^{\widetilde{L}^p}))$$

$$\leqslant \operatorname{rk}(E)\frac{p^n}{n!}\int_{X(\leqslant q)}(-1)^q c_1(L,h^L)^n + o(p^n). \quad (2.3.29)$$

Proof. Observe first that by (2.3.26) and (2.3.27), we have for $p'\in\mathbb{N}$:

$$\mathscr{I}(h^{\widetilde{L}^{mp'}}) = \mathscr{O}_{\widetilde{X}}(-p'\widetilde{D}), \quad \widehat{L}^{p'} = \widetilde{L}^{mp'}\otimes\mathscr{I}(h^{\widetilde{L}^{mp'}}). \quad (2.3.30)$$

By Theorem 1.7.1, Lemma 2.3.20 and (2.3.30) with $p = mp'$, we obtain

$$\sum_{j=0}^{q}(-1)^{q-j}\dim H^j(\widetilde{X},\widetilde{L}^p\otimes\widetilde{E}\otimes\mathscr{I}(h^{\widetilde{L}^p}))$$

$$\begin{aligned}
&\leqslant \operatorname{rk}(E)\frac{p'^n}{n!}\int_{\widetilde{X}(\leqslant q)}(-1)^q c_1(\widehat{L},h^{\widehat{L}})^n + o(p'^n) \\
&= \operatorname{rk}(E)\frac{p^n}{n!}\int_{\widetilde{X}(\leqslant q)\setminus\operatorname{supp}(\widetilde{D})}(-1)^q c_1(\widetilde{L},h^{\widetilde{L}})^n + o(p^n) \\
&= \operatorname{rk}(E)\frac{p^n}{n!}\int_{X(\leqslant q)}(-1)^q c_1(L,h^L)^n + o(p^n).
\end{aligned} \quad (2.3.31)$$

Now we show that the inequality holds for any p. We write $p = p'm + m'$ (where $c = r/m$ as in (2.3.27), $0\leqslant m' < m$, $p',m'\in\mathbb{N}$); then $\lfloor c\,c_j p\rfloor = rc_j p' + \lfloor c\,c_j m'\rfloor$. Set $\widetilde{E}_{m'} = \widetilde{L}^{m'}\otimes\widetilde{E}\otimes\mathscr{O}_{\widetilde{X}}(-\sum_j\lfloor c\,c_j m'\rfloor\widetilde{D}_j)$. We infer from (2.3.27)

$$\widetilde{L}^p\otimes\widetilde{E}\otimes\mathscr{O}_{\widetilde{X}}(-\sum_j\lfloor c\,c_j p\rfloor\widetilde{D}_j) = \widehat{L}^{p'}\otimes\widetilde{E}_{m'}. \quad (2.3.32)$$

We can now argue as in the first part, by taking a smooth Hermitian metric on $\widetilde{E}_{m'}$ and the smooth metric $h^{\widehat{L}}$ on $\widehat{L}$ as in Lemma 2.3.20. $\qquad\square$

Third step: relation to the cohomology on X.

Lemma 2.3.22 (Skoda). *Let $U\subset\mathbb{C}^n$ be open, and let $A\subset U$ be an analytic subset. If $f\in L^2_{\mathrm{loc}}(U)$ is holomorphic on $U\setminus A$, then f is holomorphic on U.*

Proof. As in Definition B.1.1, we decompose the analytic set $A = A_{\mathrm{reg}}\cup A_{\mathrm{sing}}$ where A_{reg} is the set of regular points of A. By Lemma B.1.2, A_{reg}, A_{sing} are analytic subsets of X and $\dim A_{\mathrm{sing}} < \dim A_{\mathrm{reg}}$.

Arguing by induction on the dimension of A, we see that it suffices to prove the assertion for U a neighborhood of a regular point $x \in A$. We can thus assume that $x = 0$, $A \cap U \subset \{z_1 = 0\}$. We will show that $\bar{\partial} f = 0$ in the sense of distributions on U. Then $\bar{\partial}^* \bar{\partial} f = 0$ and the regularity theorem A.3.4 implies that f is smooth and hence holomorphic on U. We have to show that

$$\int_U f \bar{\partial} s = 0 \quad \text{for any } s \in \Omega_0^{n, n-1}(U). \tag{2.3.33}$$

Consider $\chi \in \mathscr{C}^\infty(\mathbb{R})$ such that $\chi(t) = 0$ for $t \leqslant 1/2$, $\chi(t) = 1$ for $t \geqslant 1$. Set $\chi_\varepsilon : \mathbb{C}^n \longrightarrow \mathbb{R}$, $\chi_\varepsilon(z) = \chi(|z_1|/\varepsilon)$. Since $\bar{\partial} f = 0$ on $U \smallsetminus A$ and $\mathrm{supp}(\chi_\varepsilon s) \subset U \smallsetminus A$, we have $\int_U f \bar{\partial}(\chi_\varepsilon s) = 0$ and thus

$$\int_U \chi_\varepsilon f \bar{\partial} s = - \int_U f \bar{\partial} \chi_\varepsilon \wedge s. \tag{2.3.34}$$

By the Cauchy Schwarz inequality

$$\left| \int_U f \bar{\partial} \chi_\varepsilon \wedge s \right| \leqslant \int_{|z_1| \leqslant \varepsilon} |f s|^2 \, dZ \int_{\mathrm{supp}(s)} |\bar{\partial} \chi_\varepsilon|^2 \, dZ \,,$$

where dZ is the Euclidean volume form on $\mathbb{C}^n$.

Since $f \in L^2_{\mathrm{loc}}(U)$, $\int_{|z_1| \leqslant \varepsilon} |f s|^2 \, dZ \longrightarrow 0$, for $\varepsilon \to 0$. On the other hand, there exist constants $C, C' > 0$ such that

$$\int_{\mathrm{supp}(s)} |\bar{\partial} \chi_\varepsilon|^2 \, dZ \leqslant \frac{C}{\varepsilon^2} \mathrm{vol}\left(\mathrm{supp}(s) \cap \{|z_1| \leqslant \varepsilon\}\right) \leqslant C'.$$

Thus $\int_U f \bar{\partial} \chi_\varepsilon \wedge s \longrightarrow 0$, for $\varepsilon \to 0$. Since $f \in L^2_{\mathrm{loc}}(U)$, $\int_U \chi_\varepsilon f \bar{\partial} s \longrightarrow \int_U f \bar{\partial} s$ as $\varepsilon \to 0$. From (2.3.34), we get (2.3.33). The proof of Lemma 2.3.22 is completed. $\square$

Lemma 2.3.23. Let $\pi : X' \longrightarrow X$ be a proper modification. Assume φ is a quasi-plurisubharmonic function on X. Then

$$\pi_*(K_{X'} \otimes \mathscr{I}(\varphi \circ \pi)) = K_X \otimes \mathscr{I}(\varphi). \tag{2.3.35}$$

Proof. Let $A \subset X$ be a proper analytic set such that $\pi : X' \smallsetminus \pi^{-1}(A) \longrightarrow X \smallsetminus A$ is biholomorphic. If $V \subset X$ is open and f is a section of $K_X \otimes \mathscr{I}(\varphi)$ over V, then f is a holomorphic $(n, 0)$-form with $(\sqrt{-1})^{n^2} f \wedge \bar{f} e^{-2\varphi} \in L^1_{\mathrm{loc}}(V)$ with $n = \dim X$. Since φ is bounded from above, it follows that $f \in L^2_{\mathrm{loc}}(V)$. By the change of variable formula,

$$\int_{V \smallsetminus A} (\sqrt{-1})^{n^2} f \wedge \bar{f} e^{-2\varphi} = \int_{\pi^{-1}(V) \smallsetminus \pi^{-1}(A)} (\sqrt{-1})^{n^2} \pi^* f \wedge \overline{\pi^* f} e^{-2\varphi \circ \pi},$$

thus f is a section of $K_X \otimes \mathscr{I}(\varphi)$ over V if and only if $\pi^* f$ is a section of $K_{X'} \otimes \mathscr{I}(\varphi \circ \pi)$ over $\pi^{-1}(V)$ from Lemma 2.3.22. $\square$

Let $\pi : X' \longrightarrow X$ be the blow-up of X with smooth center $Y \subset X$. Let $\mathscr{A}$ by a sheaf of abelian groups on X' and $q \in \mathbb{N}$. Recall that the qth direct image of $\mathscr{A}$ by π is the sheaf on X associated to the pre-sheaf $U \longrightarrow H^q(\pi^{-1}(U), \mathscr{A})$, $U \subset X$ open set, and is denoted by $R^q \pi_* \mathscr{A}$.

Theorem 2.3.24 (Leray). *If $R^q \pi_* \mathscr{A} = 0$ for $q \geqslant 1$, then there is a canonical isomorphism*

$$H^l(X, \pi_* \mathscr{A}) \simeq H^l(X', \mathscr{A}), \quad \text{for any } l \geqslant 0. \tag{2.3.36}$$

Proposition 2.3.25. *Assume that in the neighborhood of any point of the exceptional divisor D of π, a local weight of the metric h^L satisfies*

$$\varphi \circ \pi = c \log |f| + \psi, \tag{2.3.37}$$

for some $c > 0$, f is a local definition function of D and ψ is quasi-plurisubharmonic. Then, for any $p > 1/c$, $q \geqslant 0$,

$$H^q\big(X', K_{X'} \otimes \pi^*(L^p \otimes E) \otimes \mathscr{I}\big(\pi^* h^{L^p}\big)\big) \simeq H^q\big(X, K_X \otimes L^p \otimes E \otimes \mathscr{I}\big(h^{L^p}\big)\big). \tag{2.3.38}$$

Proof. Recall that a locally integrable function with analytic singularities is quasi-plurisubharmonic. For any holomorphic line bundle F, endowed with a singular Hermitian metric h^F with analytic singularities on X, by (2.3.35),

$$\pi_*(K_{X'} \otimes \pi^* F \otimes \mathscr{I}(\pi^* h^F)) = K_X \otimes F \otimes \mathscr{I}(h^F). \tag{2.3.39}$$

In order to show (2.3.38) by applying Leray's theorem 2.3.24, we have to verify that the higher direct image sheaves vanish:

$$R^q \pi_*(K_{X'} \otimes \pi^*(L^p \otimes E) \otimes \mathscr{I}(\pi^* h^{L^p})) = 0, \quad q \geqslant 1. \tag{2.3.40}$$

The sheaf $R^q \pi_*(K_{X'} \otimes \pi^*(L^p \otimes E)) \otimes \mathscr{I}(\pi^* h^{L^p}))$ is supported on Y and its fiber over $y \in Y$ is

$$F_{p,y} = \varinjlim_{U \ni y} H^q(\pi^{-1}(U), K_{X'} \otimes \pi^*(L^p \otimes E)) \otimes \mathscr{I}(\pi^* h^{L^p})), \tag{2.3.41}$$

where U runs over the neighborhoods of y in X. Since the question is local, we can assume that L and E are trivial, when proving (2.3.40).

Let U be a Stein neighborhood (cf. Definition B.3.3) of y (on which L and E are trivial) and let φ be a local weight of h^L over U. From (2.3.37), we infer

$$\mathscr{I}(p \varphi \circ \pi) = \mathscr{O}_{X'}(-\lfloor pc \rfloor D) \otimes \mathscr{I}\big(p\psi + (pc - \lfloor pc \rfloor) \log |f|\big), \tag{2.3.42}$$

hence

$$\begin{aligned} &H^q(\pi^{-1}(U), K_{X'} \otimes \mathscr{I}(\pi^* h^{L^p})) \\ &= H^q\big(\pi^{-1}(U), K_{X'} \otimes \mathscr{O}_{X'}(-\lfloor pc \rfloor D) \otimes \mathscr{I}(p\psi + (pc - \lfloor pc \rfloor) \log |f|)\big). \end{aligned} \tag{2.3.43}$$

Note that we can assume ψ to be plurisubharmonic; by hypothesis, $\psi = \psi_1 + \psi_2$, with ψ_1 plurisubharmonic and ψ_2 smooth, so ψ_2 has no influence on the Nadel multiplier ideal sheaves from (2.3.43), i.e., $\mathscr{I}(p\psi + (pc - \lfloor pc \rfloor) \log |f|) = \mathscr{I}(p\psi_1 + (pc - \lfloor pc \rfloor) \log |f|)$.

We endow $F = \mathscr{O}_{X'}(-\lfloor pc \rfloor D)$ with a smooth Hermitian metric h^F with positive curvature on $\pi^{-1}(U)$. For this we argue as in Propositions 2.1.8 and 2.1.11; the role of the positive line bundle therein is played now by $(U \times \mathbb{C}, e^{-\eta})$, where η is a strictly plurisubharmonic function in the neighborhood of $\overline{U}$. Then $h^F e^{-p\psi - (pc - \lfloor pc \rfloor) \log |f|}$ has positive curvature on $\pi^{-1}(U)$ (since ψ and $\log |f|$ are plurisubharmonic).

We remark further that $\pi^{-1}(U)$ is a weakly 1-complete manifold. Indeed, if $\rho : U \to \mathbb{R}$ is a strictly plurisubharmonic exhaustion function, then $\rho \circ \pi : \pi^{-1}(U) \to \mathbb{R}$ is a plurisubharmonic exhaustion function which is strictly plurisubharmonic outside the exceptional divisor D. Then, for $\lambda : \mathbb{R} \to \mathbb{R}$ increasing fast enough and strictly convex (i.e., $\lambda'' > 0$),

$$\Theta = \sqrt{-1}(R^F + \partial\overline{\partial}\lambda(\rho \circ \pi)), \tag{2.3.44}$$

delivers the associated $(1, 1)$-form of a complete Kähler metric on $\pi^{-1}(U)$. Moreover, the curvature of $h^F e^{-p\psi - (pc - \lfloor pc \rfloor) \log |f|} e^{-\lambda(\rho \circ \pi)}$ on $\pi^{-1}(U)$ dominates this metric on $\pi^{-1}(U)$. As a consequence of Theorem B.4.7, we obtain that the right-hand side of (2.3.43) vanishes for $q \geqslant 1$. Thus (2.3.38) follows from (2.3.43) together with the Leray theorem and (2.3.39). The proof of Proposition 2.3.25 is completed. $\square$

We apply now Proposition 2.3.25 at each step of the blowing-up of $\mathscr{I}$ performed in Lemma 2.3.19. In doing so, we replace at the first step E by $E \otimes K_X^*$. The hypothesis (2.3.37) is satisfied, since in Lemma 2.3.19 the local weight φ has analytic singularities and the centers of the blow-ups are included in the singular locus of the metric. Thus at each step (2.3.38) holds and for all $q \geqslant 0$ and p large enough:

$$H^q(X, L^p \otimes E \otimes \mathscr{I}(h^{L^p})) \simeq H^q(\widetilde{X}, \widetilde{L}^p \otimes \widetilde{E} \otimes K_{\widetilde{X}} \otimes \widetilde{K_X^*} \otimes \mathscr{I}(h^{\widetilde{L}^p})). \tag{2.3.45}$$

Applying Lemma 2.3.21 to the right-hand side term of (2.3.45), we obtain (2.3.23). Finally, (2.3.22) follows by subtracting inequalities (2.3.23) for q and $q + 1$.

We conclude thus the proof of Theorem 2.3.18. $\square$

Corollary 2.3.26. *In the same conditions as in Theorem* 2.3.18,

$$\dim H^0(X, L^p \otimes E \otimes \mathscr{I}(h^{L^p})) \geqslant \mathrm{rk}(E) \cdot \frac{p^n}{n!} \int_{X(\leqslant 1)} c_1(L, h^L)^n + o(p^n). \tag{2.3.46}$$

Proof. By (2.3.23) for $q = 1$, we get immediately (2.3.46). $\square$

Remark 2.3.27. We can reformulate Theorem 2.3.18 using the absolute continuous component of the curvature current (cf. Definition B.2.12). Namely let $\omega \in c_1(L)$

be a current with analytic singularities. Then $\omega = \omega_0 + \frac{\sqrt{-1}}{\pi}\partial\bar{\partial}\varphi$, where $\omega_0 \in c_1(L)$ is smooth, and φ has analytic singularities. By Lemma 2.3.5, there exists a smooth Hermitian metric h_0^L on L with $c_1(L, h_0^L) = \omega_0$. Setting $h^L = h_0^L e^{-2\varphi}$, we obtain a Hermitian metric with analytic singularities on L such that $c_1(L, h^L) = \omega$.

In view of Definition 2.3.17, we define

$$\mathscr{I}(p\omega) := \mathscr{I}(h^{L^p}), \quad X(q, \omega) = X(q, c_1(L, h^L)), \quad X(\le q, \omega) = \bigcup_{i \le q} X(i, \omega).$$

$$(2.3.47)$$

Note that $X(q, \omega) \subset X \smallsetminus \operatorname{sing supp}(\omega)$. Outside the singular support $\operatorname{sing supp}(\omega)$ of ω, we have $\omega_{\mathrm{ac}} = \omega$, so (2.3.23) takes the form:

$$\dim H^0(X, L^p \otimes \mathscr{I}(p\omega)) \ge \frac{p^n}{n!} \int_{X(\le 1, \omega)} \omega_{\mathrm{ac}}^n + o(p^n), \quad p \longrightarrow \infty. \quad (2.3.48)$$

We summarize now the different characterizations of Moishezon manifolds (including Theorems 2.3.7 and 2.3.8) in terms of singular Hermitian metrics.

Theorem 2.3.28. *Let X be a compact connected complex manifold of dimension n. Then the following are equivalent:*

(1) *X is Moishezon.*

(2) *X admits an integral Kähler current with singular support contained in a proper analytic set.*

(3) *(Ji–Shiffman) X admits an integral Kähler current.*

(4) *X admits an integral Kähler current with analytic singularities.*

(5) *(Bonavero) X admits a closed integral current ω with analytic singularities such that $\int_{X(\le 1, \omega)} \omega_{\mathrm{ac}}^n > 0$.*

(6) *(Takayama) X admits a closed integral current ω with singular support contained in a proper analytic set Σ such that ω is positive in a neighborhood of Σ and $\int_{X(\le 1, \omega)} \omega_{\mathrm{ac}}^n > 0$.*

Proof. (1) $\Rightarrow$ (2) follows from Lemma 2.3.6. (2) $\Rightarrow$ (3) is trivial.

(3) $\Rightarrow$ (4). Let ω be an integral Kähler current. Let Θ be a smooth positive $(1, 1)$-form on X such that $\omega \ge \Theta$. Theorem 2.3.10 shows the existence of currents ω_ε, $1 > \varepsilon > 0$, in the same de Rham cohomology class as ω and having analytic singularities, such that $\omega_\varepsilon \ge (1 - \varepsilon)\Theta$.

(4) $\Rightarrow$ (5). If ω is a Kähler current with analytic singularities with $\Sigma :=$ $\operatorname{singsupp}(\omega)$, then $X(1, \omega) = \emptyset$, $X(0, \omega) = X \smallsetminus \Sigma$, thus $\int_{X(\le 1, \omega)} \omega_{\mathrm{ac}}^n = \int_{X \smallsetminus \Sigma} \omega^n > 0$.

(5) $\Rightarrow$ (1). Corollary 2.3.26 (or rather (2.3.48)) and (2.3.20) show that there exists $C > 0$ such that for $p \in \mathbb{N}^*$,

$$\dim H^0(X, L^p) \ge \dim H^0(X, L^p \otimes \mathscr{I}(p\omega)) \ge C\, p^n, \quad (2.3.49)$$

and therefore L is big and X is Moishezon.

(2) $\Rightarrow$ (6) is trivial. (6) $\Rightarrow$ (5) follows from Theorem 2.3.10. This achieves the proof. We remark that (6) $\Rightarrow$ (1) is proved independently by Problem 6.2. $\qquad \square$

2.3.3 Volume of big line bundles

Let X be a compact connected complex manifold of dimension n. Let L be a holomorphic line bundle on X.

We characterize now the big line bundles in terms of singular Hermitian metrics, first on projective and then on general manifolds.

Theorem 2.3.29. *If X is projective, then the following conditions are equivalent:*

(1) *L is big.*

(2) *For $p \in \mathbb{N}$ large enough, there exists a decomposition of divisors $L^p = \mathscr{O}_X(A + D)$ with D effective and $\mathscr{O}_X(A)$ positive on X.*

(3) *L has a singular Hermitian metric h^L with strictly positive curvature current and with analytic singularities.*

Proof. (1) $\Rightarrow$ (2). As X is a submanifold of $\mathbb{C}\mathbb{P}^N$, we denote it by $\phi : X \hookrightarrow \mathbb{C}\mathbb{P}^N$, then the base locus of the linear system $|H^0(X, \phi^* \mathscr{O}_{\mathbb{C}\mathbb{P}^N}(1))|$ is the empty set. By Bertini's theorem 2.2.2, there exists $s \in H^0(X, \phi^* \mathscr{O}_{\mathbb{C}\mathbb{P}^N}(1))$ such that $A = \mathrm{Div}(s)$ is a smooth submanifold of X. Then we have the exact sequence of sheaves $0 \to L^p \otimes \mathscr{O}_X(-A) \longrightarrow L^p \longrightarrow L^p|_A \to 0$ which induces the exact sequence

$$0 \longrightarrow H^0(X, L^p \otimes \mathscr{O}_X(-A)) \longrightarrow H^0(X, L^p) \longrightarrow H^0(A, L^p|_A). \qquad (2.3.50)$$

Since $\dim A = n - 1$, by (1.7.2) or (2.2.13), we have $\dim H^0(A, L^p|_A) \leqslant c_1 p^{n-1}$, and by assumption $\dim H^0(X, L^p) \geqslant c_2 p^n$ for some $c_1, c_2 > 0$ and p large enough. Therefore $\dim H^0(X, L^p \otimes \mathscr{O}_X(-A)) > 0$ for p large enough, so there exists an effective divisor D with $L^p \otimes \mathscr{O}_X(-A) = \mathscr{O}_X(D)$.

(2) $\Rightarrow$ (3). Let $h^{\mathscr{O}_X(A)}$ be a smooth metric on $\mathscr{O}_X(A)$ with $\sqrt{-1}R^{\mathscr{O}_X(A)} > 0$. We endow $\mathscr{O}_X(D)$ with the metric $h^{\mathscr{O}_X(D)}$ constructed in Example 2.3.4 and by (2.3.9), we have $c_1(\mathscr{O}_X(D), h^{\mathscr{O}_X(D)}) = [D] \geqslant 0$. The induced metric $h^L = (h^{\mathscr{O}_X(A)} \otimes h^{\mathscr{O}_X(D)})^{1/p}$ has curvature

$$\sqrt{-1}R^L = \frac{\sqrt{-1}}{p}\left[R^{\mathscr{O}_X(A)} + R^{\mathscr{O}_X(D)}\right] > 0.$$

(3) $\Rightarrow$ (1). This follows from Corollary 2.3.26 as in (2.3.49). $\qquad \square$

About the characterization of big line bundles, we have a parallel result to Theorem 2.3.28.

Theorem 2.3.30. *The following are equivalent:*

(1) L *is big.*

(2) L *admits a singular Hermitian metric h^L, smooth outside a proper analytic set and whose curvature is a strictly positive current.*

(3) *(Ji–Shiffman) L admits a singular Hermitian metric h^L, whose curvature is a strictly positive current.*

(4) L *admits a singular Hermitian metric h^L with analytic singularities, whose curvature is a strictly positive current.*

(5) *(Bonavero) L admits a singular Hermitian metric h^L with analytic singularities such that $\int_{X(\leqslant 1)} c_1(L, h^L)^n > 0$.*

(6) *(Takayama) L admits a singular Hermitian metric h^L, whose curvature is smooth outside a proper analytic set Σ, positive in a neighborhood of Σ, and $\int_{X(\leqslant 1)} c_1(L, h^L)^n > 0$.*

Proof. The proof is parallel to the proof of Theorem 2.3.28. As an example, we prove $(1) \Rightarrow (3)$.

If L is big, by Theorem 2.2.15, X is Moishezon. Let $\pi : \widehat{X} \longrightarrow X$ be a proper modification with $\widehat{X}$ projective (Theorem 2.2.16). Set $\widehat{L} = \pi^* L$. Since $\pi^* : H^0(X, L^p) \longrightarrow H^0(\widehat{X}, \widehat{L}^p)$ is injective, thus $\limsup_{p \to \infty} p^{-n} \dim H^0(\widehat{X}, \widehat{L}^p) > 0$, so $\widehat{L}$ is big (by Theorem 2.2.7). By $(1) \Rightarrow (3)$ in Theorem 2.3.29, there exists a singular Hermitian metric $h^{\widehat{L}}$ on $\widehat{L}$ with $c_1(\widehat{L}, h^{\widehat{L}}) > 0$. We consider $h^L = \pi_* h^{\widehat{L}}$. Then $c_1(L, \pi_* h^{\widehat{L}}) = \pi_* c_1(\widehat{L}, h^{\widehat{L}})$ is also strictly positive (see the argument in Lemma 2.3.6). $\qquad\square$

After characterizing big line bundles, we wish to find a measure of the bigness.

Definition 2.3.31. The *volume of the holomorphic line bundle L* is defined by

$$\mathrm{vol}(L) := \limsup_{p \to \infty} \frac{n!}{p^n} \dim H^0(X, L^p). \tag{2.3.51}$$

Thus, L is big if and only if $\mathrm{vol}(L) > 0$. If L is positive on X, by the Kodaira–Serre vanishing theorem 1.5.6 and by the asymptotic Riemann–Roch–Hirzebruch formula (1.7.1) (for $q = n$), we have

$$\mathrm{vol}(L) = \int_X c_1(L)^n \tag{2.3.52}$$

which gives a nice description of the volume in terms of the first Chern class.

If L is semi-positive, then for any $q \geqslant 1$, since $X(q) = \emptyset$, by (1.7.2), for $p \to \infty$, $\dim H^q(X, L^p) = o(p^n)$. Thus (1.7.1) for $q = n$ shows again that (2.3.52) holds.

We consider next numerically effective line bundles, which are the counterpart of semi-positive line bundles in the algebraic geometry.

Definition 2.3.32. A holomorphic line bundle F over a projective manifold M is said to be *numerically effective, nef* for short, if $F \cdot C = \int_C c_1(F) \geq 0$ for every curve $C \subset M$.

It is easily seen that a semi-positive line bundle is nef, but the converse is not true.

Proposition 2.3.33 (Kleiman). *If X is projective and L is nef, then for any q-dimensional subvariety $Y \subset X$, with the integral $\int_Y$ taken over Y_{reg} as in (B.2.16),*

$$L^q \cdot Y := \int_Y c_1(L)^q \geq 0. \tag{2.3.53}$$

Let us recall the Nakai–Moishezon ampleness criterion:

Theorem 2.3.34 (Nakai–Moishezon). *If X is projective, then L is positive if and only if $L^q \cdot Y > 0$ for every q-dimensional subvariety $Y \subset X$.*

From this, we easily infer:

Proposition 2.3.35. *If X is a projective manifold on which a positive line bundle F and a positive $(1,1)$-form Θ are given. The following properties are equivalent:*

(a) *L is nef;*
(b) *For any integer $p \geq 1$, the line bundle $L^p \otimes F$ is positive;*
(c) *For every $\varepsilon > 0$, there is a smooth metric h_ε^L on L such that $c_1(L, h_\varepsilon^L) \geq -\varepsilon\Theta$.*

Proof. (a) $\Rightarrow$ (b). If L is nef and F is positive, then clearly $L^p \otimes F$ satisfies the Nakai–Moishezon criterion, hence $L^p \otimes F$ is positive.

(b) $\Rightarrow$ (c). Let h^F be a Hermitian metric on F such that $c_1(F, h^F) > 0$. Condition (c) is independent of the choice of the positive $(1,1)$-form Θ; thus we set $\Theta = c_1(F, h^F)$. If $L^p \otimes F$ is positive, there exists a metric $h^{L^p \otimes F}$ of positive curvature. Then for the curvature R^L of $(L, h^L = (h^{L^p \otimes F} \otimes h^{F^*})^{1/p})$,

$$\sqrt{-1}R^L = \frac{1}{p}(\sqrt{-1}R^{L^p \otimes F} - \sqrt{-1}R^F) \geq -\frac{1}{p}\sqrt{-1}R^F = -\frac{2\pi}{p}\Theta. \tag{2.3.54}$$

When p is large enough in (2.3.54), we get (c).

(c) $\Rightarrow$ (a). Under hypothesis (c), we get $L \cdot C = \int_C c_1(L, h_\varepsilon^L) \geq -\varepsilon \int_C \Theta$ for every curve C and $\varepsilon > 0$, hence $L \cdot C \geq 0$ and L is nef. □

Since there need not exist any curve in an arbitrary compact complex manifold X, Proposition 2.2.35 (c) will be taken as a definition of nefness:

Definition 2.3.36. L is said to be *nef* if for every $\varepsilon > 0$, there is a smooth Hermitian metric h_ε^L on L such that $c_1(L, h_\varepsilon^L) \geq -\varepsilon\Theta$.

Proposition 2.3.37. *If (X, ω) is a compact Kähler manifold and L is a nef line bundle. Then for any $q \geq 1$,*

$$\dim H^q(X, L^p) = o(p^n), \quad \text{as } p \to \infty. \tag{2.3.55}$$

Proof. We set $\omega_\varepsilon = c_1(L, h_\varepsilon^L)$ for (L, h_ε^L) with h_ε^L as in Proposition 2.3.35 (c). We apply the weak holomorphic Morse inequalities (1.7.2) as $p \to \infty$, we get

$$\dim H^q(X, L^p) \leqslant \frac{p^n}{n!} \int_{X(q,\omega_\varepsilon)} (-1)^q \omega_\varepsilon^n + o(p^n). \qquad (2.3.56)$$

The characteristic function $1_{X(q,\omega_\varepsilon)}$ is defined by 1 on $X(q, \omega_\varepsilon)$, 0 otherwise. Let us remark that

$$0 \leqslant \frac{(-1)^q}{n!} \omega_\varepsilon^n 1_{X(q,\omega_\varepsilon)} \leqslant \frac{1}{q!} (\varepsilon\omega)^q \wedge \frac{1}{(n-q)!} (\omega_\varepsilon + \varepsilon\omega)^{n-q} \quad \text{on } X. \qquad (2.3.57)$$

For a closed form ϑ, we will denote by $[\vartheta]$ its cohomology class. Then $[\omega_\varepsilon] = c_1(L)$. As ω is closed, from (2.3.56) and (2.3.57), we get

$$\dim H^q(X, L^p) \leqslant \frac{p^n \varepsilon^q}{q!(n-q)!} \int_X [\omega]^q \cdot (c_1(L) + \varepsilon[\omega])^{n-q} + o(p^n). \qquad (2.3.58)$$

When we take $\varepsilon \to 0$, we get (2.3.55). $\qquad \square$

Corollary 2.3.38. *If (X, ω) is a compact Kähler manifold of dimension n and L is a nef line bundle on X. Then L is big if and only if $\int_X c_1(L)^n > 0$. If this is the case, we have $\mathrm{vol}(L) = \int_X c_1(L)^n$.*

We continue our train of thought and wish to compute the volume of big line bundles which are not necessarily nef. For this we need to introduce additional notions.

Definition 2.3.39. A $\mathbb{Q}$-*divisor* on a compact connected complex manifold X, $\dim X = n$, is an element D of the vector space $\mathrm{Div}_{\mathbb{Q}}(X) := \mathrm{Div}(X) \otimes_{\mathbb{Z}} \mathbb{Q}$. Then D is a finite linear combination $D = \sum_i c_i V_i$, where V_i are irreducible analytic hypersurfaces of X and $c_i \in \mathbb{Q}$. Obviously $\mathrm{Div}(X) \subset \mathrm{Div}_{\mathbb{Q}}(X)$ and elements of $\mathrm{Div}(X)$ are called integral divisors. $D \in \mathrm{Div}_{\mathbb{Q}}(X)$ is called *effective* if $c_i \geqslant 0$ for all i. A *ample* (*resp. big*) $\mathbb{Q}$-*divisor* is a $\mathbb{Q}$-divisor D such that there exists $m \in \mathbb{Z}$ with $mD \in \mathrm{Div}(X)$ and $\mathscr{O}_X(mD)$ is positive (resp. big). The *volume of a divisor* $D \in \mathrm{Div}(X)$ is defined by $\mathrm{vol}(D) := \mathrm{vol}(\mathscr{O}_X(D))$. The *volume of a $\mathbb{Q}$-divisor* is defined by $\mathrm{vol}(D) := \mathrm{vol}(mD)/m^n$ where $m \in \mathbb{Z}$ satisfies $mD \in \mathrm{Div}(X)$. The definition is independent of the choice of m with this property.

A divisor $D \in \mathrm{Div}(X)$ is said to have a *Zariski decomposition* if there exist a nef $\mathbb{Q}$-divisor N and an effective $\mathbb{Q}$-divisor E on X such that
   (i) $D = N + E$ as $\mathbb{Q}$-divisors.
   (ii) The canonical inclusion of $H^0(X, \mathscr{O}_X(mN))$ in $H^0(X, \mathscr{O}_X(mD))$ is surjective for every multiple m of the denominator of N.

If a Zariski decomposition exists, then $\mathrm{vol}(D) = \mathrm{vol}(N)$, so the computation of the volume of L reduces to that of the nef part N. The following result, due to Fujita, states that we can recover most of the volume of D from the volume of a positive $\mathbb{Q}$-divisor on a modification:

Theorem 2.3.40 (Fujita). *If X is a projective manifold and $D \in \mathrm{Div}(X)$ is a big divisor on X, then, for every $\varepsilon > 0$, there exists a proper modification $\mu_\varepsilon : X_\varepsilon \longrightarrow X$, an ample $\mathbb{Q}$-divisor F_ε and an effective $\mathbb{Q}$-divisor E_ε on X_ε such that:*

(i) *$\mu_\varepsilon^* D = F_\varepsilon + E_\varepsilon$ as $\mathbb{Q}$-divisors,*

(ii) *$|\mathrm{vol}(F_\varepsilon) - \mathrm{vol}(D)| < \varepsilon$.*

A decomposition as in (i) is called an *approximate Zariski decomposition*. Approximate Zariski decompositions enable us to compute the volume of big line bundles on Kähler manifolds. We can actually give a formula for a larger class of bundles. Recall from Theorem 2.3.30 that the first Chern class of a big line bundle contains a strictly positive current.

Definition 2.3.41. A holomorphic line bundle F on X is called *pseudo-effective* if there exists a positive $(1,1)$-current ω such that $[\omega] = c_1(F)$. A divisor D is called pseudo-effective if $\mathscr{O}_X(D)$ is pseudo-effective.

From Lemma 2.3.5, we deduce that F is pseudo-effective exactly when F has a singular Hermitian metric whose curvature is a positive current. If D is an effective divisor and we endow $\mathscr{O}_X(D)$ with the singular metric from Example 2.3.4, we see by (2.3.9) that $\mathscr{O}_X(D)$ has a singular metric whose curvature current is positive. Thus every effective divisor is pseudo-effective.

To approximate $\mathrm{vol}(D)$, the following two results are useful.

Theorem 2.3.42 (Demailly). *Under the same hypotheses as in Theorem 2.3.10, there exist closed currents ω_m, $m \geqslant 1$, such that*

(i) *$\omega_m = 2\sqrt{-1}\partial\bar{\partial}\varphi_m + \omega_0$, where φ_m are smooth real functions, and ω_0 is a $\mathscr{C}^\infty$ representative of ω,*

(ii) *ω_m converge weakly to ω, for $m \to \infty$, and pointwise on the set where the Lelong number of ω vanishes.*

(iii) *$\omega_m \geqslant \theta - C\lambda_m\Theta$, where $C > 0$ depends only on (X, Θ), λ_m are continuous and $\lambda_m \to \nu(\omega, x)$ for all $x \in X$ as $m \to \infty$, where $\nu(\omega, x)$ is the Lelong number of ω at x, cf. (B.2.19).*

Lemma 2.3.43. *If (X, Θ) is a compact Kähler manifold and if ω is a closed positive $(1,1)$-current on X, then $\int_X \Theta^{n-k} \wedge \omega_{\mathrm{ac}}^k$ are finite and can be bounded in terms of Θ and the cohomology class of ω only.*

Proof. We first observe that the Lelong numbers $\nu(\omega, x)$ in (B.2.19) are bounded. In fact, taking $r > 0$ small enough, we have

$$\nu(\omega, x) < \nu(\omega, x, r) < C \int_X \Theta^{n-1} \wedge \omega, \tag{2.3.59}$$

and the last integral depends only on the cohomology class of ω.

Using Theorem 2.3.42, we find smooth forms ω_ε approximating ω and such that $\omega_\varepsilon \geqslant -C\Theta$, where $C > 0$ depends only on $[\omega]$ and Θ. Hence

$$\int_X (\omega_\varepsilon + C\Theta)^k \wedge \Theta^{n-k} = \int_X [\omega + C\Theta]^k \cdot [\Theta]^{n-k}, \qquad (2.3.60)$$

does not depend on ε. Since $\omega_\varepsilon + C\Theta$ converges to $\omega_{\mathrm{ac}} + C\Theta$ almost everywhere, the result follows from Fatou's lemma. $\qquad \square$

We need the following application of the holomorphic Morse inequalities.

Lemma 2.3.44. *If (X, Θ) is a compact Kähler manifold and L is a pseudo-effective line bundle on X and $\omega \in c_1(L)$ is a closed positive current, one has:*

$$\mathrm{vol}(L) \geqslant \int_X \omega_{\mathrm{ac}}^n . \qquad (2.3.61)$$

Proof. We choose a sequence ω_k of currents with analytic singularities as in Remark 2.3.11, i.e., in such a way that $\omega_{k,\mathrm{ac}} \to \omega_{\mathrm{ac}}$. We denote by $\lambda_1 \leqslant \cdots \leqslant \lambda_n$ (resp. $\lambda_1^{(k)} \leqslant \cdots \leqslant \lambda_n^{(k)}$) the eigenvalues of ω_{ac} (resp. $\omega_{k,\mathrm{ac}}$) with respect to Θ. We have by assumption and Remark B.2.13 that $\lambda_1 \geqslant 0$, $\lambda_1^{(k)} \geqslant -\varepsilon_k, \varepsilon_k \to 0$ and $\lambda_j^{(k)}(x) \to \lambda_j(x)$ almost everywhere.

We may assume that $\int_X \omega_{\mathrm{ac}}^n > 0$, which means that the set $A := \{\lambda_1 > 0\}$ has positive measure. For each small $\delta > 0$, Egoroff's lemma delivers some $B_\delta \subset A$ such that $\lambda_1^{(k)} \to \lambda_1$ uniformly on B_δ and also $A - B_\delta$ has measure less than δ. Thus $B_\delta \subset X(0, \omega_k)$ for k big enough, and consequently $\limsup \int_{X(0,\omega_k)} \omega_{k,\mathrm{ac}}^n \geqslant \int_{B_\delta} \liminf \omega_{k,\mathrm{ac}}^n = \int_{B_\delta} \omega_{\mathrm{ac}}^n$, using Fatou's lemma. Letting now $\delta \to 0$, we get

$$\limsup_{k \to \infty} \int_{X(0,\omega_k)} \omega_{k,\mathrm{ac}}^n \geqslant \int_A \omega_{\mathrm{ac}}^n = \int_X \omega_{\mathrm{ac}}^n .$$

Since by (2.3.48) we have $\int_{X(0,\omega_k)} \omega_{k,\mathrm{ac}}^n \leqslant \mathrm{vol}(L) - \int_{X(1,\omega_k)} \omega_{k,\mathrm{ac}}^n$ for every k, we can conclude our argument if we can show that $-\int_{X(1,\omega_k)} \omega_{k,\mathrm{ac}}^n \to 0$. But

$$0 \leqslant -\omega_{k,\mathrm{ac}}^n 1_{X(1,\omega_k)} \leqslant n\varepsilon_k \Theta \wedge (\omega_{k,\mathrm{ac}} + \varepsilon_k\Theta)^{n-1},$$

hence

$$0 \leqslant -\int_{X(1,\omega_k)} \omega_{k,\mathrm{ac}}^n \leqslant n\varepsilon_k \int_X \Theta \wedge (\omega_{k,\mathrm{ac}} + \varepsilon_k\Theta)^{n-1} . \qquad (2.3.62)$$

From Lemma 2.3.43, it follows that the integrals in (2.3.62) are bounded, which ends the proof of Lemma 2.3.44. $\qquad \square$

Theorem 2.3.45 (Boucksom). *If (X, Θ) is a compact Kähler manifold and L is pseudo-effective, then*

$$\mathrm{vol}(L) = \sup \left\{ \int_X \omega_{\mathrm{ac}}^n : \omega \in c_1(L) \text{ closed positive current} \right\}. \qquad (2.3.63)$$

Proof. By Lemma 2.3.44, we need just to construct a family of positive currents ω_ε, $\varepsilon > 0$, such that $\int_X \omega_{\varepsilon,\mathrm{ac}}^n \to \mathrm{vol}(L)$, for $\varepsilon \to 0$. If $\mathrm{vol}(L) = 0$, (2.3.63) is clear from (2.3.61). So we can assume $\mathrm{vol}(L) > 0$. Then L is big, hence X is Moishezon. But as X is Kähler, from Theorem 2.2.26, X is projective.

Observe that $H^0(X, L^m) \neq 0$ for some m so $L^m \simeq \mathcal{O}_X(D)$ for the effective divisor $D = \{s = 0\}$, with a non-trivial $s \in H^0(X, L^m)$. Since $\mathrm{vol}(L) = \mathrm{vol}(D)/m^n$ and $c_1(L) = c_1(\mathcal{O}_X(D))/m$, it is enough to prove (2.3.63) for D. We apply thus Fujita's theorem 2.3.40 for D; for any $\varepsilon > 0$, there exists a proper modification $\mu_\varepsilon :$ $X_\varepsilon \longrightarrow X$ and a decomposition $\mu_\varepsilon^* D = F_\varepsilon + E_\varepsilon$, where F_ε is an ample $\mathbb{Q}$-divisor, E_ε is an effective $\mathbb{Q}$-divisor, and $|\mathrm{vol}(D) - \mathrm{vol}(F_\varepsilon)| < \varepsilon$. Let $\vartheta_\varepsilon \in c_1(\mathcal{O}_{X_\varepsilon}(F_\varepsilon))$ be a Kähler form. Consider the push-forward $\omega_\varepsilon = \mu_{\varepsilon,*}(\vartheta_\varepsilon + [E_\varepsilon])$ (where $[E_\varepsilon]$ is the current of integration on E_ε). Then $\omega_\varepsilon \in c_1(\mathcal{O}_X(D))$ is a positive current. Moreover,

$$\int_X \omega_{\varepsilon,\mathrm{ac}}^n = \int_{X_\varepsilon} (\vartheta_\varepsilon + [E_\varepsilon])_{\mathrm{ac}}^n = \int_{X_\varepsilon} \vartheta_\varepsilon^n = \int_{X_\varepsilon} c_1(\mathcal{O}_{X_\varepsilon}(F_\varepsilon))^n = \mathrm{vol}(F_\varepsilon). \quad (2.3.64)$$

Hence $|\mathrm{vol}(D) - \int_X \omega_{\varepsilon,\mathrm{ac}}^n| < \varepsilon$, which achieves the proof of Theorem 2.3.45. $\square$

Therefore, one has a Grauert–Riemenschneider-type criterion:

Corollary 2.3.46 (Boucksom). *If (X, Θ) is Kähler, and L is pseudo-effective and its Chern class $c_1(L)$ contains a current ω with $\int_X \omega_{\mathrm{ac}}^n > 0$, then $\mathrm{vol}(L) > 0$ and L is big. Especially, X is projective.*

Actually, we can make Theorem 2.3.45 precise as follows.

Theorem 2.3.47. *Under the assumption in Theorem 2.3.45, assume that $\int_X \Theta^n = 1$. Then there exists a singular Hermitian metric h^L on L with positive curvature current $\omega := c_1(L, h^L)$ and $\omega_{\mathrm{ac}}^n = \mathrm{vol}(L)\Theta^n$ almost everywhere. In particular, ω realizes the maximum in (2.3.63).*

The proof is based on the well-known

Theorem 2.3.48 (Calabi–Yau). *Let (X, Θ) be a compact Kähler manifold with $\int_X \Theta^n = 1$. Then for any Kähler cohomology class $\theta \in H^{1,1}(X, \mathbb{R})$, there exists a unique Kähler form ω in θ such that $\omega^n = (\int_X \theta^n)\Theta^n$ on X.*

2.3.4 Some examples of Moishezon manifolds

We already observed that connected projective manifolds are Moishezon. Are there other Moishezon manifolds than projective ones? If X is a compact complex manifold of dimension 1, then X is a projective manifold (Problem 5.6). From Theorem 2.2.16, we infer easily that a smooth connected complex surface is Moishezon if and only if it is projective (Problem 2.5). Thus we should look for an example of non-projective Moishezon manifold starting with dimension 3.

In this section, we present an example of Kollár, used by Kollár and Bonavero, to show that there exist Moishezon manifolds without line bundles satisfying the conditions from the Siu–Demailly criterion. Therefore for this class of manifolds only the Shiffman–Ji–Bonavero–Takayama criterion applies.

Let $Q = \{[z] \in \mathbb{C}\,\mathbb{P}^3 : z_0 z_1 = z_2 z_3\}$ be a smooth quadric. The map

$$\varphi : \mathbb{C}\,\mathbb{P}^1 \times \mathbb{C}\,\mathbb{P}^1 \longrightarrow Q, \quad \varphi([u_0, u_1], [v_0, v_1]) = [u_0 v_0, u_1 v_1, u_0 v_1, u_1 v_0] \quad (2.3.65)$$

is an isomorphism and consequently $H_2(Q, \mathbb{Z}) \simeq \mathbb{Z}^2$ is generated by

$$\begin{aligned}
L_1 &= \varphi([0, 1] \times \mathbb{C}\,\mathbb{P}^1) = \{z_0 = z_2 = 0\}, \\
L_2 &= \varphi(\mathbb{C}\,\mathbb{P}^1 \times [0, 1]) = \{z_0 = z_3 = 0\}.
\end{aligned} \quad (2.3.66)$$

We use the notation in (2.3.53) now. The intersection numbers of these curves are

$$L_1 \cdot L_1 = L_2 \cdot L_2 = 0, \quad L_1 \cdot L_2 = 1. \quad (2.3.67)$$

Any divisor $D \subset Q$ is characterized by the pair $(a, b) = (D \cdot L_1, D \cdot L_2) \in \mathbb{Z}^2$ called the type of D. For example the canonical divisor (line bundle) K_Q is of type $(-2, -2)$. We have

$$\begin{aligned}
K_{\mathbb{C}\,\mathbb{P}^3} &= \mathcal{O}_{\mathbb{C}\,\mathbb{P}^3}(-4), \quad \mathcal{O}_{\mathbb{C}\,\mathbb{P}^3}(Q)|_Q = N_{Q/\mathbb{C}\,\mathbb{P}^3}, \\
K_Q &= K_{\mathbb{C}\,\mathbb{P}^3} \otimes \mathcal{O}_{\mathbb{C}\,\mathbb{P}^3}(Q)|_Q = \mathcal{O}_{\mathbb{C}\,\mathbb{P}^3}(-2)|_Q.
\end{aligned} \quad (2.3.68)$$

Let q_1, q_2 be the projections from $\mathbb{C}\,\mathbb{P}^1 \times \mathbb{C}\,\mathbb{P}^1$ to the first and second factor. For any $l, m \in \mathbb{N}^*$, consider a section of the line bundle

$$q_1^* \mathcal{O}_{\mathbb{C}\,\mathbb{P}^1}(l) \times q_2^* \mathcal{O}_{\mathbb{C}\,\mathbb{P}^1}(m) \quad \text{on} \quad \mathbb{C}\,\mathbb{P}^1 \times \mathbb{C}\,\mathbb{P}^1 \simeq Q.$$

Its zero set is $C_{l,m}$. From Theorem 1.4.6, as $K_{C_{l,m}} = \mathcal{O}_Q(C_{l,m}) \otimes K_Q|_{C_{l,m}}$, the genus $g_{l,m} = \dim H^1(C_{l,m}, \mathcal{O}_{C_{l,m}})$ of $C_{l,m}$ is given by

$$\begin{aligned}
2(g_{l,m} - 1) &= C_{l,m} \cdot (C_{l,m} + K_Q), \\
C_{l,m} \cdot C_{l,m} &= 2lm, \quad C_{l,m} \cdot K_Q = -2(l + m).
\end{aligned} \quad (2.3.69)$$

Thus by Bertini's theorem 2.2.2, there exists a smooth curve $C_{l,m} \subset Q$ of type (l, m), genus $(l - 1)(m - 1)$ and degree $l + m$, where the degree of $C_{l,m}$ is

$$\int_{C_{l,m}} c_1(\mathcal{O}_{\mathbb{C}\,\mathbb{P}^3}(1)).$$

Let us blow up $\mathbb{C}\,\mathbb{P}^3$ along $C_{l,m}$ and obtain $\pi_1 : \widetilde{X} \longrightarrow \mathbb{C}\,\mathbb{P}^3$, with exceptional divisor $E_{l,m} \simeq \mathbb{P}(N_{C_{l,m}/\mathbb{C}\,\mathbb{P}^3})$, the projectivization of the normal bundle of $C_{l,m}$ in $\mathbb{C}\,\mathbb{P}^3$. By Proposition 2.1.11 (b),

$$\mathrm{Pic}(\widetilde{X}) = \mathrm{Pic}(\mathbb{C}\,\mathbb{P}^3) \oplus \mathbb{Z} \cdot \mathcal{O}_{\widetilde{X}}(E_{l,m}) = \mathbb{Z} \cdot \mathcal{O}_{\mathbb{C}\,\mathbb{P}^3}(1) \oplus \mathbb{Z} \cdot \mathcal{O}_{\widetilde{X}}(E_{l,m}). \quad (2.3.70)$$

Denote by $\widetilde{Q}, \widetilde{L}_1, \widetilde{L}_2$ the strict transforms of Q, L_1, L_2.

Proposition 2.3.49. *We have* $N_{\widetilde{Q}/\widetilde{X}} \cdot \widetilde{L}_1 = 2 - l$, $N_{\widetilde{Q}/\widetilde{X}} \cdot \widetilde{L}_2 = 2 - m$.

Proof. By Proposition 2.1.11 (a), $K_{\widetilde{X}} = \pi_1^* K_{\mathbb{C}\mathbb{P}^3} \otimes \mathscr{O}_{\widetilde{X}}(E_{l,m})$. From the exact sequence $0 \longrightarrow T\widetilde{Q} \longrightarrow T\widetilde{X}|_{\widetilde{Q}} \longrightarrow N_{\widetilde{Q}/\widetilde{X}} \longrightarrow 0$, we deduce $N_{\widetilde{Q}/\widetilde{X}} = K_{\widetilde{Q}} \otimes K_{\widetilde{X}}^*|_{\widetilde{Q}}$. Hence for $i = 1, 2$,

$$N_{\widetilde{Q}/\widetilde{X}} \cdot \widetilde{L}_i = K_{\widetilde{Q}} \cdot \widetilde{L}_i - \pi_1^* K_{\mathbb{C}\mathbb{P}^3} \cdot \widetilde{L}_i - \mathscr{O}_{\widetilde{X}}(E_{l,m}) \cdot \widetilde{L}_i$$
$$= K_Q \cdot L_i - K_{\mathbb{C}\mathbb{P}^3} \cdot L_i - C_{l,m} \cdot L_i. \tag{2.3.71}$$

Since $K_Q = \mathscr{O}_{\mathbb{C}\mathbb{P}^3}(-2)|_Q$ and $K_{\mathbb{C}\mathbb{P}^3} = \mathscr{O}_{\mathbb{C}\mathbb{P}^3}(-4)$, we conclude our proof. $\square$

We shall take in the sequel $l = 3$. Therefore $N_{\widetilde{Q}/\widetilde{X}} \cdot \widetilde{L}_1 = -1$ and the restriction of $N_{\widetilde{Q}/\widetilde{X}}$ to $\widetilde{L}_1$ is isomorphic to $\mathscr{O}_{\mathbb{C}\mathbb{P}^1}(-1)$. We invoke now the Fujiki–Nakano criterion for exceptional divisors.

Theorem 2.3.50 (Fujiki–Nakano). *Let X be a compact complex manifold and $D \subset X$ a smooth divisor isomorphic to the projectivization $\mathbb{P}(F)$ of a holomorphic vector bundle F over a compact complex manifold Y. Let $\sigma : \mathbb{P}(F) \longrightarrow Y$ be the projection. Assume that $N_{D/X} \simeq \mathscr{O}_{\mathbb{P}(F)}(-1)$. Then there exists a complex manifold X' containing Y as a submanifold and a map $\pi : X \longrightarrow X'$ such that π is the blow-up of X' along Y and $\pi|_D = \sigma$.*

From (2.3.65) and (2.3.66), we know that $\widetilde{Q} \simeq \widetilde{L}_1 \times \widetilde{L}_2$ and $\widetilde{L}_1$, $\widetilde{L}_2$ are biholomorphic to $\mathbb{C}\mathbb{P}^1$. As for $x \in \widetilde{L}_2$, $N_{\widetilde{Q}/\widetilde{X}}|_{\widetilde{L}_1 \times \{x\}} \simeq \mathscr{O}_{\widetilde{L}_1}(-1)$, applying Theorem 2.3.50 to $\sigma : \widetilde{Q} \to \widetilde{L}_2$, there exists a complex manifold X_m and a map $\pi_2 : \widetilde{X} \longrightarrow X_m$ such that π_2 is the blow-up of rational curve C_m with $N_{C_m/X_m} \simeq \mathscr{O}_{\mathbb{C}\mathbb{P}^1}(2 - m) \otimes \mathbb{C}^2$ and such that the exceptional divisor of π_2 is exactly $\widetilde{Q}$.

Since X_m is bimeromorphically equivalent to $\mathbb{C}\mathbb{P}^3$, X_m is Moishezon. Moreover, the Picard group of X_m is isomorphic to $\mathbb{Z}$, by Prop. 2.1.11 and (2.3.70).

Theorem 2.3.51 (Kollár–Bonavero). (a) *If $m > 3$, there exists no nef and big holomorphic line bundle L on X_m.*

(b) *If $m > 5$, there exists no holomorphic Hermitian line bundle (L, h^L) on X_m such that $\int_{X_m(\leqslant 1)} c_1(L, h^L)^3 > 0$.*

Corollary 2.3.52. *X_m is a non-projective Moishezon manifold for $m > 3$.*

Proof of Theorem 2.3.51. (a) Let L be a non-trivial holomorphic line bundle on X_m. Then by (2.3.70), there exist $k, r \in \mathbb{N}$ such that

$$\pi_2^* L = \pi_1^* \mathscr{O}_{\mathbb{C}\mathbb{P}^3}(r) \otimes \mathscr{O}_{\widetilde{X}}(-kE_{3,m}).$$

Since $\widetilde{L}_1$ is a fiber of π_2, we have $\pi_2^* L \cdot \widetilde{L}_1 = 0$, but $E_{3,m} \cdot L_1 = 3$, $\mathscr{O}_{\mathbb{C}\mathbb{P}^3}(1) \cdot L_1 = 1$, and hence $r = 3k$ and then

$$\pi_2^* L = \pi_1^* \mathscr{O}_{\mathbb{C}\mathbb{P}^3}(3k) \otimes \mathscr{O}_{\widetilde{X}}(-kE_{3,m}). \tag{2.3.72}$$

For any fiber $\widetilde{F}$ of π_1, we have $\pi_2^* L \cdot \widetilde{L}_2 = k(3-m)$ and $\pi_2^* L \cdot \widetilde{F} = k$ (cf. Prop. 2.1.11 (a)). Thus, for $m > 3$, $\pi_2^* L$ cannot be nef since its intersection with $\widetilde{L}_2$ is negative.

(b) Denote by $\mathscr{L}$ the generator of the Picard group of X_m which satisfies $\pi_2^* \mathscr{L} = \pi_1^* \mathcal{O}_{\mathbb{C}\mathbb{P}^3}(3) \otimes \mathcal{O}_{\widetilde{X}}(-E_{3,m})$. Let us observe that

$$H^0(X_m, \mathscr{L}^q) = 0 \quad \text{for any} \quad q < 0. \tag{2.3.73}$$

Indeed, by Proposition 2.1.4, $H^0(X_m, \mathscr{L}^q)$ can be identified with the space of sections of $\mathcal{O}_{\mathbb{C}\mathbb{P}^3}(3q)$ having a pole of order $\leqslant |q|$ along $C_{3,m}$. Any such sections extends to a holomorphic section of $\mathcal{O}_{\mathbb{C}\mathbb{P}^3}(3q)$. But for $q < 0$, $\mathcal{O}_{\mathbb{C}\mathbb{P}^3}(3q)$ has only the trivial global section 0. From (2.3.73), the Kodaira–Iitaka dimension of $\mathscr{L}^*$ is $-\infty$. Since X_m is Moishezon with Picard group $\mathbb{Z}$, we deduce that $\mathscr{L}$ is big.

Next, we prove that

$$K_{X_m} = \mathscr{L}^{-2}. \tag{2.3.74}$$

We have $K_{\widetilde{X}} = \pi_2^* K_{X_m} \otimes \mathcal{O}_{\widetilde{X}}(\widetilde{Q})$ by construction. If $\widetilde{F}$ is a fiber of π_1, $\pi_2^* K_{X_m} \cdot \widetilde{F} = -2$, and $\pi_2^* \mathscr{L} \cdot \widetilde{F} = 1$, hence (2.3.74). By Serre duality and (2.3.74), we obtain

$$H^3(X_m, \mathscr{L}^p) \simeq H^0(X_m, \mathscr{L}^{-p} \otimes K_{X_m})^* = 0, \quad \text{for } p > -2. \tag{2.3.75}$$

Assume now that $\mathscr{L}$ has a Hermitian metric $h^{\mathscr{L}}$ as in (b). (1.7.2) implies then that

$$h_p := \dim H^0(X_m, \mathscr{L}^p) - \dim H^1(X_m, \mathscr{L}^p) \geqslant Cp^3 + o(p^3) \tag{2.3.76}$$

for $C = \frac{1}{6} \int_{X_m(\leqslant 1)} \left(\frac{\sqrt{-1}}{2\pi} R^{\mathscr{L}} \right)^3 > 0$. On the other hand, we know from the Riemann–Roch–Hirzebruch theorem 1.4.6 and (2.3.75) that

$$h_p + \dim H^2(X_m, \mathscr{L}^p) = c_1(\mathscr{L})^3 \cdot \frac{p^3}{6} + o(p^3). \tag{2.3.77}$$

We will show that $c_1(\mathscr{L})^3 \leqslant 0$ for $m > 5$, which is a contradiction to the existence of the Hermitian metric $h^{\mathscr{L}}$ as in (b) from (2.3.76) and (2.3.77). We compute now:

$$\begin{aligned}
c_1(\mathscr{L})^3 =& c_1 \left(\pi_1^* \mathcal{O}_{\mathbb{C}\mathbb{P}^3}(3) \otimes \mathcal{O}_{\widetilde{X}}(-E_{3,m}) \right)^3 \\
=& c_1 \left(\mathcal{O}_{\mathbb{C}\mathbb{P}^3}(3) \right)^3 - 3 \int_{E_{3,m}} c_1 \left(\pi_1^* \mathcal{O}_{\mathbb{C}\mathbb{P}^3}(3) \right)^2 \\
& + 3 \int_{\widetilde{X}} c_1 \left(\pi_1^* \mathcal{O}_{\mathbb{C}\mathbb{P}^3}(3) \right) \wedge c_1 \left(\mathcal{O}_{\widetilde{X}}(E_{3,m}) \right)^2 - E_{3,m}^3.
\end{aligned} \tag{2.3.78}$$

We calculate next the terms of the right-hand side of (2.3.78). It is well known that $c_1 \left(\mathcal{O}_{\mathbb{C}\mathbb{P}^3}(3) \right)^3 = 27$ and $\int_{E_{3,m}} c_1 \left(\pi_1^* \mathcal{O}_{\mathbb{C}\mathbb{P}^3}(3) \right)^2 = 0$.

Remark that $c_1\big(\mathscr{O}_{\widetilde{X}}(E_{3,m})\big)|_{E_{3,m}} = -h$, where $h = c_1\big(\mathscr{O}_{\mathbb{P}(N_{C_{3,m}/\mathbb{C}\mathbb{P}^3})}(1)\big)$, by Prop. 2.1.11. Hence

$$
\begin{aligned}
\int_{\widetilde{X}} c_1\big(\pi_1^*\mathscr{O}_{\mathbb{C}\mathbb{P}^3}(3)\big) \wedge c_1\big(\mathscr{O}_{\widetilde{X}}(E_{3,m})\big)^2 &= -\int_{E_{3,m}} \pi_1^* c_1\big(\mathscr{O}_{\mathbb{C}\mathbb{P}^3}(3)\big) \wedge h \\
&= -\int_{C_{3,m}} c_1\big(\mathscr{O}_{\mathbb{C}\mathbb{P}^3}(3)\big) = -3(3+m),
\end{aligned}
\tag{2.3.79}
$$

since the degree of $C_{3,m}$ in $\mathbb{C}\mathbb{P}^3$ is $3+m$. To compute $E_{3,m}^3$, we use the formula

$$
h^2 + \pi_1^* c_1(N_{C_{3,m}/\mathbb{C}\mathbb{P}^3})h + \pi_1^* c_2(N_{C_{3,m}/\mathbb{C}\mathbb{P}^3}) = 0,
\tag{2.3.80}
$$

which in our case becomes $h^2 + \pi_1^* c_1(N_{C_{3,m}/\mathbb{C}\mathbb{P}^3})h = 0$, as $\dim C_{3,m} = 1$, thus $c_2(N_{C_{3,m}/\mathbb{C}\mathbb{P}^3}) = 0$. Thus

$$
E_{3,m}^3 = \int_{E_{3,m}} h^2 = \int_{E_{3,m}} \pi_1^* c_1(N_{C_{3,m}/\mathbb{C}\mathbb{P}^3}^*)h = \int_{C_{3,m}} c_1(N_{C_{3,m}/\mathbb{C}\mathbb{P}^3}^*).
\tag{2.3.81}
$$

For the latter integral, we use the exact sequence

$$
0 \to T^{(1,0)}C_{3,m} \to T^{(1,0)}\mathbb{C}\mathbb{P}^3|_{C_{3,m}} \to N_{C_{3,m}/\mathbb{C}\mathbb{P}^3} \to 0,
$$

which gives

$$
\int_{C_{3,m}} c_1(N_{C_{3,m}/\mathbb{C}\mathbb{P}^3}^*) = \int_{C_{3,m}} c_1(\mathscr{O}_{\mathbb{C}\mathbb{P}^3}(-4)) - 2g_{3,m} + 2 = -6 - 8m.
\tag{2.3.82}
$$

Finally we get

$$
c_1(\mathscr{L})^3 = 27 - 27 - 9m + 6 + 8m = 6 - m \leqslant 0 \quad \text{for } m > 5.
\tag{2.3.83}
$$

This concludes the proof of Theorem 2.3.51. $\qquad\square$

2.4 Algebraic Morse inequalities

The curvature integrals appearing in the holomorphic Morse inequalities (1.7.1) and (1.7.2) are neither topological nor algebraic invariants. That is why it is interesting to have an algebraic reformulation.

Theorem 2.4.1 (Demailly). *Let* $L = F \otimes G^*$ *where* F, G *are holomorphic nef line bundles over a compact connected Kähler manifold* X *of dimension* n. *Then for any* $q = 0, 1, \ldots, n$, *there is a strong holomorphic Morse inequality for* $p \to \infty$:

$$
\sum_{j=0}^{q} (-1)^{q-j} \dim H^j(X, L^p) \leqslant \frac{p^n}{n!} \sum_{j=0}^{q} (-1)^{q-j} \binom{n}{j} \int_X c_1(F)^{n-j} \wedge c_1(G)^j + o(p^n).
\tag{2.4.1}
$$

Proof. Let ω be a Kähler form on X. For $\varepsilon > 0$ consider Hermitian metrics h_ε^F, h_ε^G on F and G such that $\Theta_\varepsilon(F) := c_1(F, h_\varepsilon^F) + \varepsilon\omega$ and $\Theta_\varepsilon(G) := c_1(G, h_\varepsilon^G) + \varepsilon\omega$ are positive forms on X. We denote by $\lambda_1^\varepsilon \geqslant \cdots \geqslant \lambda_n^\varepsilon > 0$ the eigenvalues of $\Theta_\varepsilon(G)$ with respect to $\Theta_\varepsilon(F)$. Then $h_\varepsilon^L = h_\varepsilon^F(h_\varepsilon^G)^{-1}$ is a Hermitian metric on L and $c_1(L, h_\varepsilon^L) = c_1(F, h_\varepsilon^F) - c_1(G, h_\varepsilon^G) = \Theta_\varepsilon(F) - \Theta_\varepsilon(G)$. The eigenvalues of $c_1(L, h_\varepsilon^L)$ with respect to $\Theta_\varepsilon(F)$ are $1 - \lambda_1^\varepsilon \leqslant \cdots \leqslant 1 - \lambda_n^\varepsilon$. Hence $c_1(L, h_\varepsilon^L)^n = \prod_{j=1}^n (1 - \lambda_j^\varepsilon)\Theta_\varepsilon(F)^n$ and $X(\leqslant q, c_1(L, h_\varepsilon^L)) = \{x \in X : \lambda_{q+1}^\varepsilon(x) < 1\}$. By applying (1.7.1), we obtain for $p \to \infty$:

$$\sum_{j=0}^q (-1)^{q-j} \dim H^j(X, L^p) \leqslant \frac{p^n}{n!} \int_{\{\lambda_{q+1}^\varepsilon < 1\}} (-1)^q \prod_{j=1}^n (1 - \lambda_j^\varepsilon)\Theta_\varepsilon(F)^n + o(p^n)\,.$$
(2.4.2)

One calculates that $\binom{n}{j}\Theta_\varepsilon(F)^{n-j} \wedge \Theta_\varepsilon(G)^j = \sigma_n^j(\lambda^\varepsilon)\Theta_\varepsilon(F)^n$ where $\sigma_n^j(\lambda^\varepsilon)$ is the jth elementary function in $\lambda_1^\varepsilon, \ldots, \lambda_n^\varepsilon$. As ω is closed, we have as in (2.3.58)

$$\int_X c_1(F)^{n-j} \wedge c_1(G)^j = \lim_{\varepsilon \to 0} \int_X (c_1(F) + \varepsilon[\omega])^{n-j} \wedge (c_1(G) + \varepsilon[\omega])^j$$

$$= \lim_{\varepsilon \to 0} \int_X \Theta_\varepsilon(F)^{n-j} \wedge \Theta_\varepsilon(G)^j = \frac{j!\,(n-j)!}{n!} \lim_{\varepsilon \to 0} \int_X \sigma_n^j(\lambda^\varepsilon)\Theta_\varepsilon(F)^n\,. \quad (2.4.3)$$

In order to achieve the proof, we need to show that

$$\sum_{j=0}^q (-1)^{q-j} \sigma_n^j(\lambda^\varepsilon) - 1_{\{\lambda_{q+1}^\varepsilon < 1\}}(-1)^q \prod_{j=1}^n (1 - \lambda_j^\varepsilon) \geqslant 0\,, \qquad (2.4.4)$$

which follows by induction on n. □

Corollary 2.4.2 (Trapani, Siu). *In the same hypothesis as in Theorem 2.4.1, we have for* $p \to \infty$:

$$\dim H^0(X, L^p) - \dim H^1(X, L^p) \geqslant \frac{p^n}{n!} \int_X \left(c_1(F)^n - nc_1(F)^{n-1} \wedge c_1(G) \right) - o(p^n)\,.$$
(2.4.5)

Corollary 2.4.3. *Assume that F and G are nef line bundles over a compact connected Kähler manifold X of dimension n, and F is moreover big. Then some tensor power of $F^m \otimes G^*$ has a section if*

$$m > n\left(\int_X c_1(F)^{n-1} \wedge c_1(G) \right)\left(\int_X c_1(F)^n \right)^{-1}\,. \qquad (2.4.6)$$

Finally let us mention the following algebraic version of the singular holomorphic Morse inequalities, which is obtained with the same proof as of Theorem 2.4.1, by using (2.3.22) instead of (1.7.1). Let $D = \sum c_i D_i$ be a rational effective divisor, where $c_i \in \mathbb{Q}_+$ and D_i are irreducible divisors. If g_i are the local generators of D_i we set $\mathscr{I}(D) := \mathscr{I}(\sum c_i \log|g_i|)$. If D_i intersect transversely, then $\mathscr{I}(D) = \mathscr{O}_X(-\lfloor D \rfloor)$, where $\lfloor D \rfloor := \sum \lfloor c_i \rfloor D_i$.

Theorem 2.4.4 (Bonavero). *Let X be a compact connected Kähler manifold and F, G two holomorphic line bundles over X. Assume that G is nef and there exists $m \in \mathbb{N}^*$, a nef line bundle H and an effective divisor D such that $F^m = H \otimes \mathcal{O}_X(D)$. Then for $q = 0, 1, \ldots, n = \dim X$, we have as $p \to \infty$,*

$$\sum_{j=0}^{q} (-1)^{q-j} \dim H^j \left(X, (F \otimes G^*)^p \otimes \mathcal{I}(\tfrac{p}{m} D) \right)$$

$$\leqslant \frac{p^n}{n!} \sum_{j=0}^{q} (-1)^{q-j} \binom{n}{j} m^{-n+j} \int_X c_1(H)^{n-j} \wedge c_1(G)^j + o(p^n). \tag{2.4.7}$$

Problems

Problem 2.1 ([2]). Let X be a complex manifold and $f \in \mathcal{M}_X(X)$.

(1) Show that there exists an open cover $\{U_i\}$ of X and $r_i \in \mathcal{O}_X(U_i)$, $s_i \in \mathcal{S}_X(U_i)$ such that the germs $r_{i,x}, s_{i,x} \in \mathcal{O}_{X,x}$ are coprime for any $x \in U_i$ and $f|_{U_i} = \frac{r_i}{s_i}$. (Hint: $\mathcal{O}_{X,x}$ is a unique factorization domain and if $r_x, s_x \in \mathcal{O}_{X,x}$ are coprime, then $r_y, s_y \in \mathcal{O}_{X,y}$ are coprime for y in a neighborhood of x, cf. [120, p.10])

(2) Show that there exists $g_{ij} \in \mathcal{O}_X^*(U_i \cap U_j)$ with $r_i = g_{ij} r_j$ and $s_i = g_{ij} s_j$ on $U_i \cap U_j$. (Hint: $r_i s_j = r_j s_i$ on $U_i \cap U_j$.)

(3) Show that $g_{ij} g_{jk} g_{ki} = 1$ on $U_i \cap U_j \cap U_k \neq \emptyset$. Deduce that there exists a holomorphic line bundle L on X and $S_0, S_1 \in H^0(X, L)$ with $S_0 \not\equiv 0$ and $f = S_1/S_0$. In fact, $L = \mathcal{O}_X(\mathrm{Pol}(f))$ where $\mathrm{Pol}(f)$ is the divisor on X defined by

$$\mathrm{Pol}(f) = - \sum_{V, \, \mathrm{ord}_V(f) < 0} \mathrm{ord}_V(f) \, V.$$

Problem 2.2 ([2]). Let X be a complex manifold and $f_1 \ldots, f_k \in \mathcal{M}_X(X)$. Show that there exists a holomorphic line bundle L on X and $S_0, S_1, \ldots, S_k \in H^0(X, L)$ with $S_0 \not\equiv 0$ and $f_i = S_i/S_0$ for $i = 1, \ldots, k$. (Hint: Apply Problem 2.1 and find line bundles L_i and holomorphic sections $S_0^{(i)}, S_1^{(i)}$ such that $S_0^{(i)} \not\equiv 0$ and $f_i = S_1^{(i)}/S_0^{(i)}$. Take $L = L_1 \otimes \cdots \otimes L_k$ and $S_0 = S_0^{(1)} \otimes \cdots \otimes S_0^{(k)}$, $S_i = S_0^{(1)} \otimes \cdots \otimes S_1^{(i)} \otimes \cdots \otimes S_0^{(k)}$ for $i = 1, \ldots, k$.)

Problem 2.3 (Schwarz inequality). For $z \in \mathbb{C}^n$, set $|z|_0 = \sup\{|z_j| : 1 \leqslant j \leqslant n\}$. Denote $P(0, r) = \{z \in \mathbb{C}^n : |z|_0 < r\}$ for $r > 0$. Let $f \in \mathscr{C}^0(\overline{P(0, r)}) \cap \mathcal{O}_X(P(0, r))$ which vanishes up to order $m \in \mathbb{N}$ at 0.

(a) Let $z \in \overline{P(0, r)}$, $z \neq 0$ fixed. Show that $g : \{t \in \mathbb{C} : |t| \leqslant r\} \longrightarrow \mathbb{C}$, $g(t) = t^{-m} f(tz/|z|_0)$ is continuous and holomorphic in $\{t \in \mathbb{C} : |t| < r\}$. (Hint: Write $f(z) = \sum_{k=m}^{\infty} P_k(z)$ with homogeneous polynomials P_k of degree k.)

(b) Show that $|g(t)| \leqslant \|f\|_\infty r^{-m}$ for $|t| = r$.

(c) Deduce the Schwarz inequality:

$$|f(z)| \leqslant \|f\|_\infty \frac{|z|_0^m}{r^m}, \quad z \in \overline{P(0, r)}.$$

Problem 2.4. Let X be a projective manifold, $\dim X = n$, let L be a nef line bundle and E be a holomorphic vector bundle over X. Show that for $q \geqslant 0$,

$$\dim H^q(X, L^p \otimes E) = \mathcal{O}(p^{n-q}), \quad p \to \infty.$$

(Hint: Induction on n. Prove the result for $n = 1$ and for $q = 0$ and n arbitrary. For the induction step consider D a smooth positive divisor. Using Proposition 2.3.35 (b) and the Kodaira–Serre vanishing theorem 1.5.6, show that for $k \gg 1$, $H^q(X, \mathcal{O}_X(kD) \otimes L^p \otimes E) = 0$, for all $p \geqslant 0$, $q \geqslant 1$. Fix a large k and conclude by using the exact cohomology sequence of the sheaf exact sequence

$$0 \longrightarrow L^p \longrightarrow L^p \otimes \mathcal{O}_X(kD) \longrightarrow (L^p \otimes \mathcal{O}_X(kD))|_{D'} \longrightarrow 0$$

and the induction hypothesis (note that $\dim D' = n - 1$).) Here D' is a smooth divisor of X which is the zero-set of a section of $\mathcal{O}_X(kD)$.

Problem 2.5.

(a) Let $\pi : \widetilde{X} \to X$ be the blow-up of a complex manifold X, $\dim X = n$, with center $x \in X$. Show that $\widetilde{X}$ is projective if and only if X is projective.
(b) Prove the following theorem of Chow and Kodaira: A complex connected surface is Moishezon if and only if X is projective.

(Hint: For (a), show that $X \smallsetminus \{x\}$ is quasiprojective and apply [139, Cor. 3, p. 328]. For (b), use the fact that the center of a blow-up of a surface can be only zero dimensional; apply Theorem 2.2.16.)

2.5   Bibliographic notes

The material from Section 2.1 is classical; good sources are [120, 79]. For Proposition 2.1.11 (c), one may consult [1, p. 269], [124, Ch. II. Prob. 8.5 (a)]. Bertini's Theorem 2.2.2 can be read at [120, p. 137]. For more details on the blow up with center a submanifold see [79, VII, §12], [107, p. 162]. Hironaka's famous resolution of singularities Theorem 2.1.13 stems from [128, Main Th. I]. An elementary and constructive proof is given by Bierstone and Milman [22]. For the definition and properties of the blow up of a coherent ideal (Prop. 2.1.14) and for the notion of ample line bundle (invertible sheaf) on a complex space, we refer to [124, Ch. II, §7]. The proof of Theorem 2.1.18 is in [107, p. 186]. A constructive proof of the Hironaka–Rossi theorem is given in [114]. The notion of meromorphic map we used here is due to Remmert [204]. For properties of flat morphisms, see [107, Ch. 3]. The flattening theorem, the elimination of indeterminacies of meromorphic maps and the Chow lemma can be found in Hironaka [128, 129].

Section 2.2. Siegel's lemma 2.2.6 stems from Siegel [223]. We have followed here the account of Andreotti [2], where the method of Siegel is extended to pseudoconcave complex spaces. Theorem 2.2.11 of Siegel-Remmert-Thimm saying that $\mathscr{M}_X(X)$ is an algebraic field was proved in [239, 223, 203]. In the case of complex tori, the result was communicated in 1860 by Riemann to Hermite and it was stated by Weierstrass in 1869. A proof of Hurwitz Theorem is contained in [107, §4.7], [120, p. 168].

For Theorem 2.2.13, we refer to [216, I.§5.4]. Moishezon's theorems 2.2.16, 2.2.26 appeared in [177]. We have followed [195, 196] in the proof of Theorem 2.2.26. Another proof follows from the techniques of [81]. The Seshadri criterion can be found in [123]; the proof given there holds for Moishezon manifolds without change. For the hard Lefschetz theorem we refer to [120, p. 122]. Formula (2.2.35) can be found in [120, p. 141]. The Poincaré duality and the spectral sequences are treated in [120].

The Grauert–Riemenschneider [118] conjecture was proved by Siu [224, 225] and Demailly [72]. Another notable application of the strong Morse inequalities is to the proof of the effective Matsusaka theorem by Siu [226, 78].

Section 2.3. The characterization of Moishezon manifolds Theorem 2.3.28 given by Ji–Shiffman [134], Bonavero [41] and Takayama [234], has many applications. An important one is the projectivity criterion for hyperkähler manifolds given by Huybrechts [133].

For the singular holomorphic Morse inequalities of Bonavero, we followed closely [41]. For the Briançon–Skoda theorem, see [79, VIII.10.4]. For the Leray and de Rham isomorphism we refer to [79, IV.6], [120].

The systematic use of singular Hermitian metrics was initiated by Nadel [180] and Demailly [76] (building on previous work of Bombieri [39] and Skoda [229]). For the Nadel coherence theorem 2.3.15 see [180], [78, Prop. 5.7].

Demailly's approximation theorems 2.3.10, 2.3.42 stems from [75, Th. 1.1]. The refinement from Remark 2.3.11 has been shown by Boucksom in [50, Th. 2.4].

Proposition 2.3.33 is proved in [139]. The Nakai-Moishezon criterion (Theorem 2.3.34) can be found in Kleiman [139]. The Definition 2.3.36 of nefness on arbitrary compact manifolds is introduced by Demailly-Peternell-Schneider [82].

On a surface, it is known that the Zariski decomposition exists, but this is no longer true in general by a counter-example of Cutkovsky. The Fujita theorem 2.3.40 stems from [109], cf. also [80]. Formula (2.3.63) for the volume of a pseudo-effective line bundle was proved by Boucksom [50]. The Calabi-Yau theorem 2.3.48 can be found in [257]. Using techniques from the proof of the Nakai-Moishezon criterion of Demailly-Paun [81], Boucksom extends the Grauert-Riemenschneider criterion from Corollary 2.3.46 to non-necessarily rational pseudo-effective classes. For a general reference about the volume of a line bundle, we refer to Lazarsfeld [149, §2.2.C, §11.4].

It seems that the first example of non-projective Moishezon spaces appeared as folklore in Russia during the 50's and was named in Princeton the "algebraic

Sputnik". Hironaka was the first to write down an example, and we refer to the books of Hartshorne [124, Ex. 3.4.2] and Shafarevich [216, VIII. §3.3]. Grauert constructed an example of a two-dimensional non-projective normal Moishezon space with a singular point in [116, §4]. Kollár's example is given in [146]. The material of Section 2.3.4 is taken from [41]. For (2.3.80) see [120, p. 606].

Section 2.4. The algebraic Morse inequalities were obtained by Demailly [78]. The case $q = 1$ (Corollary 2.4.2) was proved before by Siu [226] and Trapani [245]. Theorem 2.4.4 appeared in [40]. Angelini gave an algebraic version in [8].

Using a mean value estimate of eigensections of the Kodaira Laplacian, Berndtsson [20] shows that the assertion of Problem 2.4 holds for a compact complex manifold X and a semi-positive line bundle L (and E trivial). Generalizing this method, Berman [16] reproved the holomorphic Morse inequalities (1.7.1).

Chapter 3

Holomorphic Morse Inequalities on Non-compact Manifolds

We start by the L^2 Hodge theory on non-compact Hermitian manifolds in Section 3.1. In Section 3.2, we prove holomorphic Morse inequalities for the L^2-cohomology in a quite general context, namely, when the fundamental estimate (3.2.2) holds. This gives a fairly general method which may be applied in many situations. The main idea, going back to Witten, is to show that the spectral spaces of the Laplacian, corresponding to small eigenvalues, inject in the spectral spaces of the Laplacian with Dirichlet boundary conditions on a smooth relatively domain, The asymptotic of the latter operator is calculated in Theorem 3.2.9. For a compact manifold we recover of course Theorem 1.7.1.

In Section 3.3, we specialize the abstract Morse inequalities to a geometric situation and we assume that we have a line bundle which is uniformly positive on a complete Hermitian manifold. In Section 3.4, we establish Siu–Demailly's type criterion for complex compact spaces with isolated singularities. Section 3.5 treats the q-convex and weakly 1-complete manifolds, while in Section 3.6 we treat the case of covering manifolds.

We will revisit some of these topics in Chapter 6 and treat them from the point of view of Bergman kernels.

3.1 L^2-cohomology and Hodge theory

In this section, we establish the L^2 Hodge theory on non-compact Hermitian manifolds and introduce the fundamental estimate (Poincaré inequality at infinity), which entails the strong Hodge decomposition.

Let (X, J, Θ) be a complex Hermitian manifold and let (E, h^E) be a holomorphic Hermitian vector bundle on X. Let dv_X be the Riemannian volume form of X. The boundary of X may be non-empty.

We start with a review of Hilbertian extension of differential operators. Let $R : \Omega_0^{0,\bullet}(X, E) \to L_{0,\bullet}^2(X, E)$ be a differential operator (e.g., $\overline{\partial}^E$). Then R is preclosed, so there exists closed extensions of R. In general, at least two possibilities present themselves to extend R. The first choice is to consider the closure $\overline{R}$ whose graph is the closure of the graph of R. This is called the *minimal extension* of R denoted also $R_{\min}$.

We denote by R^* the formal adjoint of R. This is a differential operator satisfying $\langle Rs_1, s_2 \rangle = \langle s_1, R^* s_2 \rangle$ for all $s_1, s_2 \in \Omega_0^{0,\bullet}(X, E)$. For $s_1 \in L_{0,\bullet}^2(X, E)$, we can still calculate Rs_1 in the sense of currents: Rs_1 is the current defined by

$$\langle Rs_1, s_2 \rangle = \langle s_1, R^* s_2 \rangle, \quad \text{for } s_2 \in \Omega_0^{0,\bullet}(X, E). \tag{3.1.1}$$

Lemma 3.1.1. *The operator $R_{\max}$ defined by*

$$\begin{aligned}
&\mathrm{Dom}(R_{\max}) = \{s \in L_{0,\bullet}^2(X, E) : Rs \in L_{0,\bullet}^2(X, E)\}, \\
&R_{\max} s = Rs, \quad \text{for } s \in \mathrm{Dom}(R_{\max}),
\end{aligned} \tag{3.1.2}$$

is a densely defined, closed extension, called the maximal extension *of R.*

We leave the proof as an exercise. In the sequel we will work mainly with the maximal extension of R, so we will simplify the notation and write $R = R_{\max}$. The Hilbert space adjoint of $R_{\max}$ is denoted by R_H^*. Note that $R_H^* \subset (R^*)_{\max}$. The two operators are in general not equal, but $R_{\max} = (R_H^*)_H^*$.

The Kodaira Laplacian $\Box^E = \overline{\partial}^E \overline{\partial}^{E,*} + \overline{\partial}^{E,*} \overline{\partial}^E$ is a densely defined, positive operator, so we can consider its Friedrichs extension (cf. Definition C.1.7). However, we will describe now a self-adjoint extension of $\Box^E$ which is very useful in conjunction with the study of L^2 cohomology. Consider the complex of closed, densely defined operators

$$L_{0,q-1}^2(X, E) \xrightarrow{T = \overline{\partial}^E} L_{0,q}^2(X, E) \xrightarrow{S = \overline{\partial}^E} L_{0,q+1}^2(X, E), \tag{3.1.3}$$

where T and S are the maximal extensions of $\overline{\partial}^E$. Let T^*, S^* be their Hilbert space adjoints, thus $T^* = \overline{\partial}_H^{E,*}$, $S^* = \overline{\partial}_H^{E,*}$. Note that $\mathrm{Im}(T) \subset \mathrm{Ker}(S)$, so $ST = 0$.

Proposition 3.1.2 (Gaffney). *The operator defined by*

$$\begin{aligned}
&\mathrm{Dom}(\Box^E) = \{s \in \mathrm{Dom}(S) \cap \mathrm{Dom}(T^*) : Ss \in \mathrm{Dom}(S^*), T^* s \in \mathrm{Dom}(T)\}, \\
&\Box^E s = S^* S s + T T^* s, \quad \text{for } s \in \mathrm{Dom}(\Box^E),
\end{aligned} \tag{3.1.4}$$

is a positive self-adjoint extension of the Kodaira Laplacian, called the Gaffney *extension. The quadratic form associated to $\Box^E$ is the form Q given by*

$$\begin{aligned}
&\mathrm{Dom}(Q) := \mathrm{Dom}(S) \cap \mathrm{Dom}(T^*), \\
&Q(s_1, s_2) = \langle Ss_1, Ss_2 \rangle + \langle T^* s_1, T^* s_2 \rangle, \quad \text{for } s_1, s_2 \in \mathrm{Dom}(Q).
\end{aligned} \tag{3.1.5}$$

Proof. It is easy to show that $\square^E$ is closed (since T and S are) and positive. We need to show that it is self-adjoint.

Note that $T = (T^*)^*_H$. By Lemma C.1.3, we know that $(1 + TT^*)^{-1}$ and $(1 + S^*S)^{-1}$ are bounded self-adjoint operators on $L^2_{0,q}(X, E)$. We have

$$\mathrm{Im}(1 + TT^*)^{-1} = \mathrm{Dom}(TT^*), \quad \mathrm{Im}(1 + S^*S)^{-1} = \mathrm{Dom}(S^*S). \tag{3.1.6}$$

We define a bounded self-adjoint operator R by

$$R = (1 + TT^*)^{-1} + (1 + S^*S)^{-1} - 1. \tag{3.1.7}$$

Since $R = (1 + S^*S)^{-1} - TT^*(1 + TT^*)^{-1}$, $ST = 0$ and (3.1.6), we get

$$\mathrm{Im}(R) \subset \mathrm{Dom}(S^*S), \quad S^*SR = S^*S(1 + S^*S)^{-1}. \tag{3.1.8}$$

In the same way, from $T^*S^* = 0$, (3.1.6) and (3.1.7), we have

$$\mathrm{Im}(R) \subset \mathrm{Dom}(TT^*), \quad TT^*R = TT^*(1 + TT^*)^{-1}. \tag{3.1.9}$$

Thus, from (3.1.6)–(3.1.9), we have $\mathrm{Im}(R) \subset \mathrm{Dom}(\square^E)$ and

$$(1 + \square^E)R = R + S^*S(1 + S^*S)^{-1} + TT^*(1 + TT^*)^{-1} = 1. \tag{3.1.10}$$

Since $1 + \square^E$ is injective, we get $R = (1 + \square^E)^{-1}$. By Lemma C.1.2, $1 + \square^E$ and hence $\square^E$ are self-adjoint.

We prove now that Q is the associated quadratic form of $\square^E$ in the sense of Proposition C.1.4. First remark that Q is a closed form (Problem 3.2). By Proposition C.1.5, there exists a self-adjoint, positive operator B with $Q = Q_B$. The domain of B is given by (C.1.6). For $s \in \mathrm{Dom}(\square^E)$, it is clear that $s \in \mathrm{Dom}(Q_B)$ and s satisfies (C.1.6) with $v = \square^E s$. Therefore $s \in \mathrm{Dom}(B)$ and $Bs = \square^E s$, i.e., $\square^E \subset B$ and, since both operators are self-adjoint, $B = \square^E$ (for, in general, $B_1 \subset B$ implies $B^* \subset B_1^*$). $\qquad\square$

Friedrichs' lemma about the identity of weak and strong derivatives is very useful.

Lemma 3.1.3 (Friedrichs). *Let* $Rf = \sum g_k \partial f / \partial x_k + gf$ *be a differential operator of order 1 on an open set* $U \subset \mathbb{R}^m$, *with coefficients* $g_k \in \mathscr{C}^1(U)$, $g \in \mathscr{C}^0(U)$. *Then for any* $s \in L^2(\mathbb{R}^m)$ *with compact support in* U, $\rho \in \mathscr{C}^\infty_0(\mathbb{R}^m)$, *we have*

$$\lim_{\varepsilon \longrightarrow 0} \| R(s * \rho_\varepsilon) - (Rs) * \rho_\varepsilon \|_{L^2} = 0. \tag{3.1.11}$$

Here we denote by dx *the Euclidean volume form on* $\mathbb{R}^m$, *and*

$$\rho_\varepsilon(x) = \frac{1}{\varepsilon^m} \rho(x/\varepsilon), \quad (s * \rho_\varepsilon)(y) = \int_{\mathbb{R}^m} s(y - x) \, \rho_\varepsilon(x) \, dx. \tag{3.1.12}$$

The Gaffney and Friedrichs extensions are in general not equal, but they coincide in all the cases we consider (Kodaira Laplacian on complete manifolds or with $\bar\partial$-Neumann boundary conditions). However, the quadratic forms of the two extensions are always equal for elements with compact support (Problem 3.3). This can be expressed also as the following regularity result for elements of the domain of the quadratic Q.

Proposition 3.1.4. *The domain of the quadratic form* (3.1.5) *satisfies* $\mathrm{Dom}(Q) \subset \boldsymbol{H}^1(X, \Lambda^\bullet(T^{*(0,1)}X) \otimes E, \mathrm{loc})$.

Proof. The statement follows from general arguments about the regularity of solutions of the elliptic operator $\bar\partial^E + \bar\partial^{E,*}$ (cf. Theorem A.3.4), but we give here a direct argument. Let $s \in \mathrm{Dom}(Q)$ and $\varphi \in \mathscr{C}_0^\infty(X)$. We have to show that $\varphi s \in \boldsymbol{H}_0^1(K, \Lambda^\bullet(T^{*(0,1)}X) \otimes E)$, for $K = \mathrm{supp}(\varphi)$. It easy to see that $\varphi s \in \mathrm{Dom}(\bar\partial^E)$ and $\varphi s \in \mathrm{Dom}(\bar\partial_H^{E,*})$. Without loss of generality, we can replace s by φs and assume that s has compact support in K.

After using a partition of unity in order to divide the support of s in small pieces contained in coordinate charts, by Lemma 3.1.3, we obtain a sequence $\{s_k\} \subset \Omega_0^{0,\bullet}(X, E)$ such that $s_k \longrightarrow s$, $\bar\partial^E s_k \longrightarrow \bar\partial^E s$, $\bar\partial_H^{E,*} s_k \longrightarrow \bar\partial_H^{E,*} s$ in L^2. By Gårding's inequality (A.3.5) applied to $s_{k_1} - s_{k_2}$, we obtain that $\{s_k\}$ is a Cauchy sequence in $\boldsymbol{H}_0^1(K, \Lambda^\bullet(T^{*(0,1)}X) \otimes E)$ and therefore $s \in \boldsymbol{H}_0^1(K, \Lambda^\bullet(T^{*(0,1)}X) \otimes E)$. $\square$

Definition 3.1.5. The space of *harmonic forms* $\mathscr{H}^{0,\bullet}(X, E)$ is defined by

$$\mathscr{H}^{0,\bullet}(X, E) := \mathrm{Ker}(\Box^E) = \{s \in \mathrm{Dom}(\Box^E) : \Box^E s = 0\}. \tag{3.1.13}$$

The qth *reduced L^2 Dolbeault cohomology* is defined by

$$\overline{H}_{(2)}^{0,q}(X, E) := \mathrm{Ker}(\bar\partial^E) \cap L_{0,q}^2(X, E) \Big/ \left[\mathrm{Im}(\bar\partial^E) \cap L_{0,q}^2(X, E)\right], \tag{3.1.14}$$

where $[V]$ denotes the closure of the space V.

The qth (non-reduced) *L^2 Dolbeault cohomology* is defined by

$$H_{(2)}^{0,q}(X, E) := \mathrm{Ker}(\bar\partial^E) \cap L_{0,q}^2(X, E) \Big/ \mathrm{Im}(\bar\partial^E) \cap L_{0,q}^2(X, E). \tag{3.1.15}$$

For $q, r \geq 0$, we set:

$$H_{(2)}^{r,q}(X, E) := H_{(2)}^{0,q}(X, E \otimes \Lambda^r(T^{*(1,0)}X)), \quad H_{(2)}^q(X, E) := H_{(2)}^{0,q}(X, E). \tag{3.1.16}$$

By (3.1.4), we see that

$$\mathscr{H}^{0,\bullet}(X, E) = \mathrm{Ker}(\bar\partial^E) \cap \mathrm{Ker}(\bar\partial_H^{E,*}). \tag{3.1.17}$$

Using the notation of (3.1.3), by (C.1.2), we have

$$\mathrm{Im}(T)^\perp = \mathrm{Ker}(T^*) = \big(\mathrm{Ker}(T^*) \cap \mathrm{Ker}(S)\big) \oplus \big(\mathrm{Ker}(T^*) \cap \mathrm{Ker}(S)^\perp\big). \tag{3.1.18}$$

But by (C.1.2), $\mathrm{Ker}(S)^{\perp} = [\mathrm{Im}(S^*)]$ and by $S^*T^* = 0$, $[\mathrm{Im}(S^*)] \subset \mathrm{Ker}(T^*)$. Thus

$$\mathrm{Im}(T)^{\perp} = \mathscr{H}^{0,\bullet}(X, E) \oplus [\mathrm{Im}(S^*)]. \tag{3.1.19}$$

Likewise, $\mathrm{Ker}(S) = [\mathrm{Im}(T)] \oplus \big(\mathrm{Ker}(S) \cap \mathrm{Im}(T)^{\perp} \big)$ and by $\mathrm{Im}(T)^{\perp} = \mathrm{Ker}(T^*)$, we get

$$\mathrm{Ker}(S) = \mathscr{H}^{0,\bullet}(X, E) \oplus [\mathrm{Im}(T)]. \tag{3.1.20}$$

Since $L^2_{0,q}(X, E) = \mathrm{Im}(T)^{\perp} \oplus [\mathrm{Im}(T)]$, we deduce the *weak Hodge decomposition*:

$$L^2_{0,\bullet}(X, E) = \mathscr{H}^{0,\bullet}(X, E) \oplus \left[\mathrm{Im}(\overline{\partial}^E) \right] \oplus \left[\mathrm{Im}(\overline{\partial}_H^{E,*}) \right],$$
$$\mathrm{Ker}(\overline{\partial}^E) = \mathscr{H}^{0,\bullet}(X, E) \oplus \left[\mathrm{Im}(\overline{\partial}^E) \right]. \tag{3.1.21}$$

By (3.1.21), we have a canonical isomorphism

$$\overline{H}^{0,\bullet}_{(2)}(X, E) \simeq \mathscr{H}^{0,\bullet}(X, E), \tag{3.1.22}$$

which associates to each cohomology class its unique harmonic representative.

However, from the point of view of solving the $\overline{\partial}^E$ equation, the (non-reduced) L^2 Dolbeault cohomology $H^{0,\bullet}_{(2)}(X, E)$ is more interesting. A general condition for the range of $\overline{\partial}^E$ to be closed and for the finiteness of the L^2 cohomology is as follows.

Proposition 3.1.6.

(i) *A necessary and sufficient condition for* $\mathrm{Im}\,(\overline{\partial}^E) \cap L^2_{0,q}(X, E)$ *and* $\mathrm{Im}(\overline{\partial}_H^{E,*}) \cap L^2_{0,q}(X, E)$ *to be closed is that there exists* $C > 0$ *such that*

$$\|s\|^2_{L^2} \leqslant C(\|\overline{\partial}^E s\|^2_{L^2} + \|\overline{\partial}_H^{E,*} s\|^2_{L^2}),$$
$$s \in \mathrm{Dom}(\overline{\partial}^E) \cap \mathrm{Dom}(\overline{\partial}_H^{E,*}) \cap L^2_{0,q}(X, E), \ s \perp \mathscr{H}^{0,q}(X, E). \tag{3.1.23}$$

(ii) *Assume that from every sequence* $s_k \in \mathrm{Dom}(\overline{\partial}^E) \cap \mathrm{Dom}(\overline{\partial}_H^{E,*}) \cap L^2_{0,q}(X, E)$ *with* $\|s_k\|_{L^2}$ *bounded and*

$$\overline{\partial}^E s_k \longrightarrow 0 \text{ in } L^2_{0,q+1}(X, E), \quad \overline{\partial}_H^{E,*} s_k \longrightarrow 0 \text{ in } L^2_{0,q-1}(X, E), \tag{3.1.24}$$

one can select a strongly convergent subsequence. Then both

$$\mathrm{Im}(\overline{\partial}^E) \cap L^2_{0,q}(X, E), \quad \mathrm{Im}(\overline{\partial}_H^{E,*}) \cap L^2_{0,q}(X, E)$$

are closed. Moreover, $\mathscr{H}^{0,q}(X, E)$ *is finite-dimensional and*

$$H^{0,q}_{(2)}(X, E) \simeq \mathscr{H}^{0,q}(X, E). \tag{3.1.25}$$

Proof. We use the notation from (3.1.3).

(i) Assume that (3.1.23) holds. Then, for any $s \in \mathrm{Dom}(T^*) \cap [\mathrm{Im}(T)]$, we have $Ss = 0$, hence $\|s\|_{L^2}^2 \leqslant C\|T^*s\|_{L^2}^2$. Likewise, for any $s \in \mathrm{Dom}(S) \cap [\mathrm{Im}(S^*)]$, we have $T^*s = 0$, thus $\|s\|_{L^2}^2 \leqslant C\|Ss\|_{L^2}^2$. By Lemma C.1.1, T and S have closed range. (In fact, set $T_1 = T$ on $\mathrm{Dom}(T) \subset L_{0,q-1}(X, E)$, and $T_1 = 0$ on $L_{0,q}(X, E)$, then we apply Lemma C.1.1 to T_1.)

Conversely if T and S have closed range, by (3.1.21), any $s \in \mathrm{Dom}(S) \cap \mathrm{Dom}(T^*)$, $s \perp \mathscr{H}^{0,q}(X, E)$ can be decomposed in orthogonal sum $s = s_1 + s_2$, $s_1 \in \mathrm{Im}(T)$, $s_2 \in \mathrm{Im}(S^*)$, and (3.1.23) follows from Lemma C.1.1.

(ii) The space $\mathscr{H}^{0,q}(X, E)$ is closed, so a Hilbert space. The hypothesis implies that the unit ball

$$B = \{s \in L_{0,q}^2(X, E) : \|s\|_{L^2} \leqslant 1, \, Ss = 0, \, T^*s = 0\} \subset \mathscr{H}^{0,q}(X, E)$$

is compact. Therefore $\mathscr{H}^{0,q}(X, E)$ is finite-dimensional.

Assume that (3.1.23) were false. Then there exists $\{s_k\} \subset \mathrm{Dom}(T^*) \cap \mathrm{Dom}(S)$, $s_k \perp \mathscr{H}^{0,q}(X, E)$ such that

$$\|s_k\|_{L^2}^2 > k(\|T^*s_k\|_{L^2}^2 + \|Ss_k\|_{L^2}^2). \tag{3.1.26}$$

Setting $w_k = s_k / \|s_k\|_{L^2}$, we have $\|T^*w_k\|_{L^2}^2 + \|Sw_k\|_{L^2}^2 < 1/k$. By hypothesis, we can extract a convergent subsequence (still denoted by $\{w_k\}$). Put $w = \lim_k w_k$. Then $\|w\|_{L^2} = 1$ and $w \perp \mathscr{H}^{0,q}(X, E)$. But $w_k \to w$, $Sw_k \to 0$ implies $w \in \mathrm{Ker}(S)$ and likewise we get $w \in \mathrm{Ker}(T^*)$, hence $w \in \mathscr{H}^{0,q}(X, E)$. This means $w = 0$ which contradicts $\|w\|_{L^2} = 1$. Thus (3.1.23) holds true. □

In the geometric situations we shall encounter, the hypotheses of Proposition 3.1.6 are implied by the following estimate.

Definition 3.1.7. We say that the *fundamental estimate* holds in bidegree $(0, q)$ if there exists a compact set K in the interior of X and $C > 0$ such that

$$\|s\|_{L^2}^2 \leqslant C\left(\|\overline{\partial}^E s\|_{L^2}^2 + \|\overline{\partial}_H^{E,*}s\|_{L^2}^2 + \int_K |s|^2 \, dv_X\right), \tag{3.1.27}$$

for $s \in \mathrm{Dom}(\overline{\partial}^E) \cap \mathrm{Dom}(\overline{\partial}_H^{E,*}) \cap L_{0,q}^2(X, E)$.

Theorem 3.1.8. *If the fundamental estimate* (3.1.27) *holds in bidegree* $(0, q)$, *then*

(i) *The operators* $\overline{\partial}^E$ *on* $L_{0,q-1}^2(X, E)$ *and* $\square^E$ *on* $L_{0,q}^2(X, E)$ *have closed range and we have the* strong Hodge decomposition:

$$L_{0,q}^2(X, E) = \mathscr{H}^{0,q}(X, E) \oplus \mathrm{Im}(\overline{\partial}^E\overline{\partial}_H^{E,*}) \oplus \mathrm{Im}(\overline{\partial}_H^{E,*}\overline{\partial}^E),$$
$$\mathrm{Ker}(\overline{\partial}^E) \cap L_{0,q}^2(X, E) = \mathscr{H}^{0,q}(X, E) \oplus \left(\mathrm{Im}(\overline{\partial}^E) \cap L_{0,q}^2(X, E)\right). \tag{3.1.28}$$

Moreover, $\mathscr{H}^{0,q}(X, E)$ *is finite-dimensional. We have a canonical isomorphism*

$$\mathscr{H}^{0,q}(X, E) \to H_{(2)}^{0,q}(X, E), \quad s \mapsto [s]. \tag{3.1.29}$$

(ii) *There exists a bounded operator G on $L^2_{0,q}(X, E)$, called the* Green operator, *such that*
$$\Box^E G = G\Box^E = \mathrm{Id} - P, \quad PG = GP = 0, \qquad (3.1.30)$$
where P is the orthogonal projection from $L^2_{0,q}(X, E)$ onto $\mathscr{H}^{0,q}(X, E)$.

Proof. Consider a sequence $\{s_k\} \subset \mathrm{Dom}(\overline{\partial}^E) \cap \mathrm{Dom}(\overline{\partial}^{E,*}_H) \cap L^2_{0,q}(X, E)$ with $\{\|s_k\|_{L^2}\}$ bounded and $\|\overline{\partial}^{E,*}_H s_k\|_{L^2} + \|\overline{\partial}^E s_k\|_{L^2} \longrightarrow 0$, for $k \to \infty$.

Let $\varphi \in \mathscr{C}^\infty_0(X)$ such that $\varphi = 1$ on K; then
$$Q(\varphi s_k, \varphi s_k) + \|\varphi s_k\|^2_{L^2} = \|\overline{\partial}^E(\varphi s_k)\|^2_{L^2} + \|\overline{\partial}^{E,*}_H(\varphi s_k)\|^2_{L^2} + \|\varphi s_k\|^2_{L^2} \qquad (3.1.31)$$
is also bounded. Let X' be a compact manifold with boundary containing $\mathrm{supp}(\varphi)$ in its interior. By Proposition 3.1.4, $\varphi s_k \in \boldsymbol{H}^1_0(X', \Lambda^q(T^{*(0,1)}X) \otimes E)$. By applying Gårding's inequality (A.3.5) and (3.1.31), we obtain that $(\|\varphi s_k\|_1)$ is bounded. By Rellich's theorem A.3.1,
$$\left(\boldsymbol{H}^1_0(X', \Lambda^q(T^{*(0,1)}X) \otimes E), \|\cdot\|_1\right) \hookrightarrow \left(L^2_{0,q}(X, E), \|\cdot\|_{L^2}\right) \qquad (3.1.32)$$
is compact. We can select therefore a convergent subsequence in $L^2_{0,q}(X, E)$, denoted also $\{\varphi s_k\}$. Since $\varphi = 1$ on K, it follows that $\{s_k|_K\}$ converges in $\|\cdot\|_{L^2}$. By estimate (3.1.27), this entails that $\{s_k\}$ converges in $L^2_{0,q}(X, E)$.

Proposition 3.1.6 implies that (3.1.23) holds and $\mathscr{H}^{0,q}$ is finite-dimensional. From (3.1.23), we infer that
$$\|s\|_{L^2} \leqslant C\|\Box^E s\|_{L^2}, \quad \text{for } s \in \mathrm{Dom}(\Box^E), \ s \perp \mathrm{Ker}(\Box^E). \qquad (3.1.33)$$
Therefore $\Box^E$ has closed range. Since $\Box^E$ is self-adjoint, by (C.1.2), we have
$$L^2_{0,q}(X, E) = \mathrm{Im}(\Box^E) \oplus \mathrm{Ker}(\Box^E) = \mathrm{Im}(\overline{\partial}^E \overline{\partial}^{E,*}_H) \oplus \mathrm{Im}(\overline{\partial}^{E,*}_H \overline{\partial}^E) \oplus \mathscr{H}^{0,q}(X, E).$$
By (3.1.33), there exists a bounded inverse G of $\Box^E$ on $\mathrm{Im}(\Box^E)$. We extend G to $L^2_{0,q}(X, E)$ by setting $G = 0$ on $\mathscr{H}^{0,q}(X, E)$. We obtain thus a bounded operator G on $L^2_{0,q}(X, E)$ (bounded by $1/C$, where C is the constant from (3.1.33)), satisfying $\mathrm{Ker}(G) = \mathscr{H}^{0,q}(X, E)$ and $\mathrm{Im}(G) = \mathrm{Im}(\Box^E)$. The proof of Theorem 3.1.8 is completed. $\qquad\square$

Corollary 3.1.9. *Assume that the fundamental estimate (3.1.27) holds in bidegree $(0, q)$.*

(i) *If $f \in \mathrm{Im}(\overline{\partial}^E) \cap L^2_{0,q}(X, E)$, the unique solution $s \perp \mathrm{Ker}(\overline{\partial}^E) \cap L^2_{0,q-1}(X, E)$ of the equation $\overline{\partial}^E s = f$ is given by $s = \overline{\partial}^{E,*}_H G f$.*
(ii) *The operator G maps $L^2_{0,q}(X, E) \cap \Omega^{0,q}(X, E)$ into itself.*

Assertion (i) is immediate from the preceding proof. Finally, assertion (ii) follows from the interior regularity for the elliptic operator $\Box^E$ (Theorem A.3.4).

Remark 3.1.10. If X is compact, then the fundamental estimate (3.1.27) holds in bidegree $(0, q)$ for any $q \in \mathbb{N}$. Thus from Theorem 3.1.8, we get the Hodge decomposition on compact manifolds, Theorem 1.4.1.

3.2   Abstract Morse inequalities for the L^2-cohomology

We shall examine a general situation which permits us to prove asymptotic Morse inequalities for the L^2-Dolbeault cohomology groups of complex manifolds.

We show in Section 3.2.1 that the fundamental estimate allows us to reduce the study of the spectral spaces of the Kodaira Laplacian on the whole manifold to the study of spectral spaces of the Kodaira Laplacian with Dirichlet boundary conditions on a smooth relatively compact domain, whose asymptotic we determine in Section 3.2.2. In Section 3.2.3, we apply then Theorem 3.2.9 in order to prove the abstract holomorphic Morse inequalities.

3.2.1   The fundamental estimate

Let (X, J, Θ) be a complex Hermitian manifold of dimension n. Let (L, h^L) and (E, h^E) be holomorphic Hermitian vector bundles on X with $\mathrm{rk}(L) = 1$. Let dv_X be the Riemannian volume form on X associated to the metric $g^{TX} = \Theta(\cdot, J\cdot)$ on TX. From (1.5.20) and (1.6.1), we will denote simply that

$$E_p^j = \Lambda^j(T^{*(0,1)}X) \otimes L^p \otimes E, \quad E_p = \oplus_j E_p^j,$$
$$\overline{\partial}_p^E := \overline{\partial}^{L^p \otimes E}, \quad \overline{\partial}_p^{E,*} := \overline{\partial}^{L^p \otimes E,*}, \quad \Box_p := \Box^{L^p \otimes E}. \tag{3.2.1}$$

We postulate a general estimate for the quadratic form of the Kodaira Laplacian $\Box_p$ acting on the bundle $L^p \otimes E$, which implies estimates from above of the spectral function.

Definition 3.2.1 (fundamental estimate). We say that the *fundamental estimate* holds in bidegree $(0, q)$ for forms with values in $L^p \otimes E$ if there exists a compact $K \subset X$ and $C_0 > 0$ such that for sufficiently large p, we have

$$\|s\|_{L^2}^2 \leqslant \frac{C_0}{p}\big(\|\overline{\partial}_p^E s\|_{L^2}^2 + \|\overline{\partial}_p^{E,*} s\|_{L^2}^2\big) + C_0 \int_K |s|^2 \, dv_X \,, \tag{3.2.2}$$
$$\text{for } s \in \mathrm{Dom}(\overline{\partial}_p^E) \cap \mathrm{Dom}(\overline{\partial}_p^{E,*}) \cap L_{0,q}^2(X, L^p \otimes E).$$

K is called the exceptional compact set of the estimate.

Let us consider the Gaffney extension $\Box_p$ of the Kodaira Laplacian acting on $L^p \otimes E$ which we normalize by $\frac{1}{p}\Box_p$. (3.2.2) is of course a variant with parameters of the fundamental estimate (3.1.27). It allows us to compare the spectral spaces of $\frac{1}{p}\Box_p$ on X and of $\frac{1}{p}\Box_p$ with Dirichlet boundary conditions on a relatively compact domain U containing K.

Definition 3.2.2. Let $U \subset X$ be an open set. The Friedrichs extension (cf. Definition C.1.7) of the positive operator $\frac{1}{p}\Box_p : \mathscr{C}_0^\infty(U, E_p) \to L^2(U, E_p)$ is called the operator $\frac{1}{p}\Box_p$ *with Dirichlet boundary conditions*. We denote the Kodaira Laplacian with Dirichlet boundary conditions by $\Box_{p,U}$ and the quadratic form associated to $\frac{1}{p}\Box_{p,U}$ by $\overline{Q}_{p,U}$.

Lemma 3.2.3. *If U is relatively compact, then $\mathrm{Dom}(\Box_{p,U}) = \boldsymbol{H}_0^2(U, E_p)$ (cf. Appendix A.3), and $\Box_{p,U}$ has discrete spectrum.*

Proof. It is easy to see that $\mathrm{Dom}(\overline{Q}_{p,U}) = \boldsymbol{H}_0^1(U, E_p)$. By Gårding's inequality (A.3.5) and Rellich's theorem A.3.1, we deduce that

$$
\begin{aligned}
&(\mathrm{Dom}(\overline{Q}_{p,U}), \|\cdot\|_{\overline{Q}_{p,U}}) \hookrightarrow L^2(U, E_p), \\
&\text{with } \|s\|_{\overline{Q}_{p,U}} := (\overline{Q}_{p,U}(s,s) + \|s\|_{L^2}^2)^{1/2},
\end{aligned}
\tag{3.2.3}
$$

is a compact operator. By Proposition C.2.4, $\Box_{p,U}$ has discrete spectrum. Using Definition C.1.7, and Theorem A.3.4, we know $\mathrm{Dom}(\Box_{p,U}) = \boldsymbol{H}_0^2(U, E_p)$. Note that $\Box_p^* = (\Box_p)_{\max}$. $\qquad\square$

According to Proposition 3.1.2, the quadratic form associated to $\frac{1}{p}\Box_p$ is

$$
\begin{aligned}
&Q_p(s,s) = \frac{1}{p}\left(\|\overline{\partial}_p^E s\|_{L^2}^2 + \|\overline{\partial}_p^{E,*} s\|_{L^2}^2 \right), \\
&s \in \mathrm{Dom}(Q_p) = \mathrm{Dom}(\overline{\partial}_p^E) \cap \mathrm{Dom}(\overline{\partial}_p^{E,*}).
\end{aligned}
\tag{3.2.4}
$$

Let $\{E_\lambda(\frac{1}{p}\Box_p)\}_\lambda$ be the spectral resolution of $\frac{1}{p}\Box_p$ and let $\mathscr{E}(\lambda, \frac{1}{p}\Box_p) = \mathrm{Im}(E_\lambda(\frac{1}{p}\Box_p))$ be the corresponding spectral spaces. All these objects decompose in a direct sum according to the decomposition of forms after bidegree.

Let us fix an open, relatively compact neighborhood U of K with smooth boundary. Let $\{E_\lambda(\frac{1}{p}\Box_{p,U})\}_\lambda$ be the spectral resolution of $\frac{1}{p}\Box_{p,U}$ and for $q \in \mathbb{N}$, set:

$$
\begin{aligned}
\mathscr{E}^q(\lambda, \tfrac{1}{p}\Box_p) &= \mathrm{Im}(E_\lambda(\tfrac{1}{p}\Box_p)) \cap L^2_{0,q}(X, L^p \otimes E), \\
\mathscr{E}^q(\lambda, \tfrac{1}{p}\Box_{p,U}) &= \mathrm{Im}(E_\lambda(\tfrac{1}{p}\Box_{p,U})) \cap L^2_{0,q}(X, L^p \otimes E), \\
N^q(\lambda, \tfrac{1}{p}\Box_p) &= \dim \mathscr{E}^q(\lambda, \tfrac{1}{p}\Box_p), \\
N^q(\lambda, \tfrac{1}{p}\Box_{p,U}) &= \dim \mathscr{E}^q(\lambda, \tfrac{1}{p}\Box_{p,U}).
\end{aligned}
\tag{3.2.5}
$$

One of the tools for the proof of the holomorphic Morse inequalities is to estimate $N^\bullet(\lambda, \frac{1}{p}\Box_p)$ from above and from below.

Proposition 3.2.4 (decomposition principle). *The operators $\frac{1}{p}\Box_p$ and $\frac{1}{p}\Box_{p,X\setminus U}$ have the same essential spectrum.*

Proof. Let $\varphi \in \mathscr{C}_0^\infty(X)$ be a nonnegative function which is equal to 1 on a neighborhood of $\overline{U}$, the closure of U. Let U_1 be a relatively compact neighborhood of the support of φ. Let $\{f_k\}_{k\in\mathbb{N}}$ be an orthonormal characteristic sequence for $(\frac{1}{p}\Box_p, \lambda)$ for some $\lambda \geqslant 0$ as in Theorem C.3.3. Rellich's theorem A.3.1 implies that $\{\varphi f_k\}_{k\in\mathbb{N}}$ is compact in $\boldsymbol{H}_0^2(U_1, E_p)$. Thus, by passing to a subsequence of $\{f_k\}_{k\in\mathbb{N}}$, if necessary, we may assume that $\{\varphi f_k\}_{k\in\mathbb{N}}$ is convergent.

Then we set $g_k = f_{2k} - f_{2k-1}$ $(k \geqslant 1)$. We see that $\{g_k\}_{k\in\mathbb{N}}$ is non-compact, $\lim_{k\to\infty}\|(\frac{1}{p}\Box_p - \lambda\,\mathrm{Id})g_k\|_{L^2} = 0$, and $\varphi g_k \to 0$ in the Sobolev space $\boldsymbol{H}_0^2(U_1, E_p)$. Then $\lim_{k\to\infty}\|(\frac{1}{p}\Box_{p,X\smallsetminus\overline{U}} - \lambda\,\mathrm{Id})(1-\varphi)g_k\|_{L^2} = 0$, and consequently

$$\{\widetilde{g}_k\}_{k\in\mathbb{N}} \quad \text{with} \quad \widetilde{g}_k := \frac{(1-\varphi)g_k}{\|(1-\varphi)g_k\|_{L^2}},$$

is a characteristic sequence for $(\frac{1}{p}\Box_{p,X\smallsetminus\overline{U}}, \lambda)$. So $\sigma_{\mathrm{ess}}(\frac{1}{p}\Box_p) \subset \sigma_{\mathrm{ess}}(\frac{1}{p}\Box_{p,X\smallsetminus\overline{U}})$. We trivially have $\sigma_{\mathrm{ess}}(\frac{1}{p}\Box_{p,X\smallsetminus\overline{U}}) \subset \sigma_{\mathrm{ess}}(\frac{1}{p}\Box_p)$. $\qquad\square$

Proposition 3.2.4 and its consequence, Theorem 3.2.5(1) will not be used in the sequel (e.g., for the proof of Theorem 3.2.5(2)), but they provide a good description of the spectrum of $\frac{1}{p}\Box_p$. Consider a smooth function ρ on X, $0 \leqslant \rho \leqslant 1$ such that $\rho = 1$ on K and $\rho = 0$ on $X \smallsetminus \overline{U}$. Set $C_1 = \sup|d\rho|^2$.

Theorem 3.2.5. *If the fundamental estimate* (3.2.2) *holds in bidegree* $(0,q)$, *then*

(1) *$\frac{1}{p}\Box_p$ on $L_{0,q}^2(X, L^p \otimes E)$ has only discrete spectrum in $[0, 1/C_0]$ for large enough p.*

(2) *There exists a constant C_2 depending only on C_0 and C_1 such that for $\lambda < 1/(2C_0)$, the following maps are injective for p sufficiently large,*

$$\begin{aligned}\mathscr{E}^q(\lambda, \tfrac{1}{p}\Box_p) &\longrightarrow \mathscr{E}^q(3C_0\lambda + C_2 p^{-1}, \tfrac{1}{p}\Box_{p,U}),\\ s &\longrightarrow E_{3C_0\lambda + C_2 p^{-1}}(\tfrac{1}{p}\Box_{p,U})(\rho s).\end{aligned} \tag{3.2.6}$$

In particular for any $\lambda < 1/(2C_0)$, $p \gg 1$,

$$N^q(\lambda, \tfrac{1}{p}\Box_p) \leqslant N^q(3C_0\lambda + C_2 p^{-1}, \tfrac{1}{p}\Box_{p,U}). \tag{3.2.7}$$

Proof. (1) By Proposition 3.2.4, $\frac{1}{p}\Box_p$ has the same essential spectrum as $\frac{1}{p}\Box_{p,X\smallsetminus\overline{U}}$. The fundamental estimate (3.2.2) shows then that $\overline{Q}_{p,X\smallsetminus\overline{U}}(s) \geqslant \frac{1}{C_0}\|s\|^2$, $s \in \mathrm{Dom}(\overline{Q}_{p,X\smallsetminus\overline{U}})$, since $\mathrm{Dom}(\overline{Q}_{p,X\smallsetminus\overline{U}}) \subset \mathrm{Dom}(Q_p)$. It follows that $\frac{1}{p}\Box_{p,X\smallsetminus\overline{U}}$ has no essential spectrum in $[0, \frac{1}{C_0}]$ and $\frac{1}{p}\Box_p$ has the same property.

(2) For $s \in \mathrm{Dom}(\overline{\partial}_p^E) \cap \mathrm{Dom}(\overline{\partial}_p^{E,*})$, we obtain by Leibniz's formula

$$\|\overline{\partial}_p^E(\rho s)\|_{L^2}^2 + \|\overline{\partial}_p^{E,*}(\rho s)\|_{L^2}^2 = \|\rho\overline{\partial}_p^{E,*}s + \overline{\partial}\rho \wedge s\|_{L^2}^2 + \|\rho\overline{\partial}_p^{E,*}s + i_{\overline{\partial}\rho}s\|_{L^2}^2.$$

Using the inequality $(x+y)^2 \leqslant \frac{3}{2}x^2 + 3y^2$ together with the triangle inequality, we obtain with $C_1 = \sup|d\rho|^2$,

$$\|\overline{\partial}_p^E(\rho s)\|_{L^2}^2 + \|\overline{\partial}_p^{E,*}(\rho s)\|_{L^2}^2 \leqslant \frac{3}{2}(\|\overline{\partial}_p^E s\|_{L^2}^2 + \|\overline{\partial}_p^{E,*}s\|_{L^2}^2) + 6C_1\|s\|_{L^2}^2. \tag{3.2.8}$$

Let $s \in \mathcal{E}^q(\lambda, \frac{1}{p}\Box_p)$, $\lambda < 1/(2C_0)$. Then by (C.3.2) and (3.2.2), we get

$$Q_p(s, s) = \frac{1}{p}\left(\|\overline{\partial}_p^E s\|_{L^2}^2 + \|\overline{\partial}_p^{E,*} s\|_{L^2}^2\right) \leqslant \lambda\|s\|_{L^2}^2,$$

$$\|s\|_{L^2}^2 \leqslant 2C_0 \int_K |s|^2\, dv_X. \tag{3.2.9}$$

By (3.2.8) and (3.2.9), we have

$$Q_{p,U}(\rho s, \rho s) \leqslant \frac{3}{2} Q_p(s, s) + \frac{6C_1}{p}\|s\|_{L^2}^2$$

$$\leqslant \left(\frac{3}{2}\lambda + \frac{6C_1}{p}\right) \cdot 2C_0 \int_K |s|^2\, dv_X. \tag{3.2.10}$$

We set $C_2 = 12C_0 C_1$. Inequality (3.2.10) and $E_{3C_0\lambda + C_2 p^{-1}}(\frac{1}{p}\Box_{p,U})(\rho s) = 0$ imply $\rho s = 0$, by Problem 3.4. Since $\rho = 1$ on K, it follows that $s = 0$ on K and by (3.2.9), we infer $s = 0$. Thus (3.2.6) is injective. $\qquad\square$

As for a lower bound of the counting function $N^q(\lambda, \frac{1}{p}\Box_p)$, we have a general result which does not depend on the fundamental estimate.

Lemma 3.2.6. *The following estimate from below holds for any $q \in \mathbb{N}$:*

$$N^q(\lambda, \tfrac{1}{p}\Box_p) \geqslant N^q(\lambda, \tfrac{1}{p}\Box_{p,U}). \tag{3.2.11}$$

Proof. Let us denote by $\lambda_0 \leqslant \lambda_1 \leqslant \cdots$ the spectrum of $\frac{1}{p}\Box_{p,U}$ acting on $(0,q)$-forms. Let $\{s_i\}_i$ be an orthonormal basis which consists of eigenforms corresponding to the eigenvalues $\{\lambda_i\}_i$; if we let $\widetilde{s}_i = 0$ on $X \smallsetminus U$ and $\widetilde{s}_i = s_i$ on U, then $\widetilde{s}_i \in \mathrm{Dom}(Q_p)$ and $Q_p(\widetilde{s}_i, \widetilde{s}_j) = \delta_{i,j}\lambda_i$. Let V_λ^0 be the subspace spanned by $\{s_i : \lambda_i \leqslant \lambda\}$ in $L^2_{0,q}(U, L^p \otimes E)$ and V_λ the closed subspace spanned by $\{\widetilde{s}_i : \lambda_i \leqslant \lambda\}$ in $L^2_{0,q}(X, L^p \otimes E)$. Then $\dim V_\lambda = \dim V_\lambda^0 = N^q(\lambda, \frac{1}{p}\Box_{p,U})$.

If f is a linear combination of $\{\widetilde{s}_i : \lambda_i \leqslant \lambda\}$, then $Q_p(f, f) \leqslant \lambda\|f\|_{L^2}^2$ and, as $\mathrm{Dom}(Q_p)$ is complete in the graph norm, we obtain

$$V_\lambda \subset \mathrm{Dom}(Q_p) \text{ and } Q_p(f, f) \leqslant \lambda\|f\|_{L^2}^2, \text{ for } f \in V_\lambda. \tag{3.2.12}$$

Glazman's lemma C.3.1 implies now Lemma 3.2.6. $\qquad\square$

3.2.2 Asymptotic distribution of eigenvalues

We use the notation and assumption in Section 1.6.1, except that X is a compact complex manifold with boundary ∂X.

As in Section 3.2.1, we denote by $D_{p,X}^2 = 2\Box_{p,X}$ the operator $D_p^2 = 2\Box_p$ with Dirichlet boundary condition (cf. (D.1.25)) on X.

For $j = 0, 1, \ldots$, let $\lambda_j^{p,q}$ be the eigenvalues (counted with multiplicities) of $\frac{1}{p}\Box_{p,X}$ acting on $(0,q)$-forms with values in $L^p \otimes E$. We define the counting function $N^q(\lambda, \frac{1}{p}\Box_{p,X})$ by (3.2.5), and consider the measure μ_p^q on $\mathbb{R}_+$,

$$\mu_p^q = p^{-n} \frac{d}{d\lambda} N^q(\lambda, \tfrac{1}{p}\Box_{p,X}) = p^{-n} \sum_{j=0}^{\infty} \delta(\lambda_j^{p,q}). \qquad (3.2.13)$$

By integrating the function $e^{-u\lambda}$ against this measure, we obtain the trace of the operator $\exp(-\frac{u}{2p}D_{p,X}^2)$, so using (1.7.4), we get

$$\int_0^{\infty} e^{-u\lambda} d\mu_p^q(\lambda) = p^{-n} \sum_{j=0}^{\infty} \exp(-u\,\lambda_j^{p,q}) = p^{-n} \operatorname{Tr}_q \left[\exp(-\frac{u}{2p}D_{p,X}^2)\right]. \qquad (3.2.14)$$

Proposition 3.2.7. *For any $u > 0$ fixed, the following relation holds:*

$$\lim_{p \to \infty} \int_0^{\infty} e^{-u\lambda} d\mu_p^q(\lambda) = \operatorname{rk}(E) \int_X \frac{\det(\dot{R}^L/(2\pi)) \operatorname{Tr}_{\Lambda^{0,q}}[\exp(u\omega_d)]}{\det(1 - \exp(-u\dot{R}^L))}. \qquad (3.2.15)$$

The convergence in (3.2.15) is uniform as u varies in any compact subset of $\mathbb{R}_+^*$.

Proof. Let N be the normal bundle of ∂X in X; we identify it as the orthogonal complement of $T\partial X$ in TX. Let e_n be the inward pointing unit normal at any boundary point of X. For $y_0 \in \partial X, 0 \leqslant x_n \leqslant \epsilon$, by using $N_{y_0} \ni x_n e_n \to \exp_{y_0}^X(x_n e_n)$, we get a neighborhood of ∂X for $\epsilon > 0$ small enough, and we identify $\partial X \times [0, \epsilon[$ as a neighborhood of ∂X in this way.

At first, when we choose $\{x_i\}$ in Section 1.6.2, we will take first $x_i \in \partial X$ for $1 \leqslant i \leqslant k_0$ such that $B^X(x_i, \varepsilon/2)$ covers ∂X, then we choose $\{x_i\}_{i=k_0+1}^{N_0}$ such that $\{B^X(x_i, \varepsilon/2)\}_{i=k_0+1}^{N_0}$ covers $X \smallsetminus \cup_{i=1}^{k_0} B^X(x_i, \varepsilon)$ and $\cup_{i=k_0+1}^{N_0} B^X(x_i, \varepsilon) \cap \partial X = \emptyset$. We still define $\|s\|_{H^l(p)}^2$ as in (1.6.5). By basic elliptic estimate, Theorem A.3.2, the Dirichlet boundary condition is an elliptic boundary problem for D_p^2, thus (1.6.8) still holds for $s \in H_0^2(X, E_p)$ (cf. (3.2.1) for E_p). Thus for $s \in H_0^{2m+2}(X, E_p)$, $p \in \mathbb{N}$, we have the analogue of (1.6.6),

$$\|s\|_{H^{2m+2}(p)} \leqslant C_m' p^{4m+4} \sum_{j=0}^{m+1} p^{-4j} \|D_p^{2j} s\|_{L^2}. \qquad (3.2.16)$$

Recall that $\widetilde{\mathbf{F}}_u, \widetilde{\mathbf{G}}_u, \widetilde{\mathbf{H}}_{u,\varsigma}$ are the holomorphic functions on $\mathbb{C}$ defined by (1.6.26). By (3.2.16) as in (1.6.18), for any differential operators P, Q of order m, m' with compact support in $U_{x_i} := B^X(x_i, \varepsilon)$, U_{x_j} respectively, there exists $C > 0$ such that for $p \geqslant 1, u \geqslant u_0, s \in \mathscr{C}_0^\infty(X, E_p)$,

$$\|P\widetilde{\mathbf{H}}_{\frac{u}{p}, p^\theta}(D_{p,X}^2)Qs\|_{L^2} \leqslant Cp^{2(m+m'+2)(\theta+1)} \exp\left(-\frac{\varepsilon^2 p^{1-\theta}}{16u}\right) \|s\|_{L^2}. \qquad (3.2.17)$$

By (1.6.14), $\widetilde{\mathbf{G}}_{\frac{u}{p^{1-\theta}}}(\frac{u}{p}D_{p,X}^2) = \widetilde{\mathbf{H}}_{\frac{u}{p},p^\theta}(D_{p,X}^2)$. On $U_{x_i} \times U_{x_j}$, by using Sobolev inequality from Theorem A.1.6, we get the following generalization of (1.6.15) : For any $m \in \mathbb{N}$, $0 \leqslant \theta < 1$, $u_0 > 0$, $\varepsilon > 0$, there exists $C > 0$ such that for any $x, x' \in X$, $p \in \mathbb{N}^*$, $u > u_0$,

$$\left| \widetilde{\mathbf{G}}_{\frac{u}{p^{1-\theta}}}(\frac{u}{p}D_{p,X}^2)(x,x') \right|_{\mathscr{C}^m} \leqslant C p^{2(m+4n+4)(1+\theta)} \exp\left(-\frac{\varepsilon^2 p^{1-\theta}}{16u}\right). \quad (3.2.18)$$

Using (1.6.13) and finite propagation speed, Theorem D.2.1 and (D.2.17), it is clear that for $x, x' \in X$, $\widetilde{\mathbf{F}}_{u/p^{1-\theta}}(\frac{u}{p}D_{p,X}^2)(x,x')$ only depends on the restriction of D_p^2 to $B^X(x, \varepsilon p^{-\theta/2})$, and is zero if $d(x,x') \geqslant \varepsilon p^{-\theta/2}$.

Now we embed X into a closed manifold $\widetilde{X}$ (which need not be complex) and we extend the bundles $\Lambda(T^{*(0,1)}X)$, L, E to $\widetilde{X}$ with smooth metrics and smooth Hermitian connections (cf. Problem 3.5), moreover, we extend the smooth section Φ_E in (1.6.20) to $\widetilde{X}$, thus we get a generalized Laplacian $\widetilde{L}_p$ on $\widetilde{X}$ such that the restriction of $\widetilde{L}_p$ on X is D_p^2. Certainly, (3.2.18) also holds for $\widetilde{\mathbf{G}}_{\frac{u}{p^{1-\theta}}}(\frac{u}{p}\widetilde{L}_p)(x,x')$. Thus from (1.6.14) and (3.2.18), we get uniformly for $x_0 \in X \setminus (\partial X \times [0, \varepsilon p^{-\theta/2}])$, $p \geqslant 1$, $u > u_0$,

$$\left| \exp(-\frac{u}{2p}D_{p,X}^2)(x_0,x_0) - \exp(-\frac{u}{2p}\widetilde{L}_p)(x_0,x_0) \right|$$
$$\leqslant C p^{8(n+1)(1+\theta)} \exp\left(-\frac{\varepsilon^2 p^{1-\theta}}{16u}\right). \quad (3.2.19)$$

For $y_0 \in \partial X$, we use the coordinate $\mathcal{V} \times [0, \epsilon[$ of y_0 in X such that $\mathcal{V}$ is a normal coordinate of y_0 in ∂X. We still trivialize the bundles $\Lambda(T^{*(0,1)}X)$, L, E by the parallel transport with respect to the connections $\nabla^{B,\Lambda^{0,\bullet}}$, ∇^L, ∇^E along the curve $[0,1] \ni v \to vZ$. Now we extend the operator $D_{p,X}^2$ to L_{p,D,x_0} on $\mathbb{R}_+^{2n} = \mathbb{R}^{2n-1} \times \mathbb{R}_+^*$ by (1.6.23) with the Dirichlet boundary condition (cf. (D.1.25)). Then by the argument in Lemma 1.6.5 and (3.2.18), for $0 \leqslant x_n \leqslant \epsilon$,

$$\left| \exp(-\frac{u}{2p}D_{p,X}^2)((y_0,x_n),(y_0,x_n)) - \exp(-\frac{u}{2p}L_{p,D})((0,x_n),(0,x_n)) \right|$$
$$\leqslant C p^{8(n+1)(1+\theta)} \exp(-\frac{\varepsilon^2 p^{1-\theta}}{16u}). \quad (3.2.20)$$

We fix the coordinate such that $e_n = e_{2n}$. In our coordinate, for $j \neq 2n$, we have $g_{j,2n}(Z) = 0$ in (1.6.31), thus we have

$$L_2^t = -(\nabla_{t,e_{2n}})^2 - \sum_{i,j=1}^{2n-1} g^{ij}(tZ)\left(\nabla_{t,e_i}\nabla_{t,e_j} - t\Gamma_{ij}^k(tZ)\nabla_{t,e_k}\right)$$
$$+ \rho(|tZ|/\epsilon)(-2\omega_{d,tZ} - \tau_{tZ} + t^2\Phi_{E,tZ}). \quad (3.2.21)$$

We denote by $L_{2,D}^t$ the corresponding operator L_2^t with the Dirichlet boundary condition. Let $\boldsymbol{H}_{0,t}^m$ be the Sobolev space which is the completion of $\mathscr{C}_0^\infty(\mathbb{R}_+^{2n}, \mathbf{E}_{x_0})$ with respect to the norm $\| \ \ \|_{t,m}$. Then Theorems 1.6.7–1.6.9 still hold for $L_{2,D}^t$ and $s, s' \in \mathscr{C}_0^\infty(\mathbb{R}_+^{2n}, \mathbf{E}_{x_0})$ without any change. Especially, $(\lambda - L_{2,D}^t)^{-1}$ exists for $t \in]0,1]$, $\lambda \in \Gamma$.

To prove the analogue of Theorem 1.6.10, we need to prove that for $Q_0, \ldots,$ $Q_m \in \{\nabla_{t,e_i}\}_{i=1}^{2n}$, $Q_{m+1}, \ldots, Q_{m+|\alpha|} \in \{Z_i\}_{i=1}^{2n}$, and $Q_{m+1} \cdots Q_{m+|\alpha|} = Z^\alpha$, there exist $M \in \mathbb{N}$, $C_{\alpha,m} > 0$ such that for $t \in]0,1]$, $\lambda \in \Gamma$, $s \in \mathscr{C}^\infty(\mathbb{R}_+^{2n}, \mathbf{E}_{x_0})$ with compact support,

$$\|Q_0 \cdots Q_{m+|\alpha|}(\lambda - L_{2,D}^t)^{-1}s\|_{t,0} \leqslant C_{\alpha,m}(1 + |\lambda|^2)^M \sum_{\alpha' \leqslant \alpha} \|Z^{\alpha'}s\|_{t,m}. \quad (3.2.22)$$

The main point here is that $\{\nabla_{t,e_i}\}_{i=1}^{2n-1}$ preserves the Dirichlet boundary condition, but $\nabla_{t,e_{2n}}$ does not preserve the Dirichlet boundary condition, thus the following equation only holds for $1 \leqslant i \leqslant 2n-1$,

$$[\nabla_{t,e_i}, (\lambda - L_{2,D}^t)^{-1}] = (\lambda - L_{2,D}^t)^{-1}[\nabla_{t,e_i}, L_2^t](\lambda - L_{2,D}^t)^{-1}. \quad (3.2.23)$$

If $Q_0, \ldots, Q_m \in \{\nabla_{t,e_i}\}_{i=1}^{2n-1}$, then the same proof of Theorem 1.6.10 gives (3.2.22), as $\{\nabla_{t,e_i}\}_{i=1}^{2n-1}$ preserves the Dirichlet boundary condition. If there are only one Q_i such that $Q_i = \nabla_{t,e_{2n}}$, then by using (1.6.41), we can assume that $Q_0 = \nabla_{t,e_{2n}}$, and the same proof of Theorem 1.6.10 still gives (3.2.22). If there are at least two $\nabla_{t,e_{2n}}$ in $Q_0, \ldots, Q_m$, then by using (1.6.41) and $[\nabla_{t,e_i}, Z_j] = \delta_{ij}$, we move all $\nabla_{t,e_{2n}}$ to the right-hand side of $Q_0 \cdots Q_{m+|\alpha|}$, thus we express $Q_0 \cdots Q_{m+|\alpha|}$ as a linear combination of operators of type

$$Q_{j_1} \cdots Q_{j_l}(\nabla_{t,e_{2n}})^k(\lambda - L_{2,D}^t)^{-1}; \quad (3.2.24)$$

with $k \leqslant m + 1, l + k \leqslant m + |\alpha| + 1$ and $Q_{j_i} \in \{\nabla_{t,e_i}\}_{i=1}^{2n-1} \cup \{Z_i\}_{i=1}^{2n}$. If $k < 2$, we apply the previous argument to get the estimate (3.2.22). If $k \geqslant 2$, then we replace $(\nabla_{t,e_{2n}})^2(\lambda - L_{2,D}^t)^{-1}$ in (3.2.24) by

$$1 + (-\lambda + L_2^t + (\nabla_{t,e_{2n}})^2)(\lambda - L_{2,D}^t)^{-1}. \quad (3.2.25)$$

By (3.2.21), the degree of $\nabla_{t,e_{2n}}$ in $L_2^t + (\nabla_{t,e_{2n}})^2$ is at most 1, and we can continue the process to get (3.2.24) with $k \leqslant 1$. Thus we have proved (3.2.22).

By (3.2.22), as in (1.6.49), (1.6.51), for $s \in \mathscr{C}_0^\infty(\mathbb{R}_+^{2n}, \mathbf{E}_{x_0})$, $Q, Q' \in \mathcal{Q}^m$,

$$\|Q(\lambda - L_{2,D}^t)^{-m}s\|_{t,0} \leqslant C_m(1 + |\lambda|^2)^M \|s\|_{t,0},$$
$$\|(\lambda - L_{2,D}^t)^{-m}Qs\|_{t,0} \leqslant C_m(1 + |\lambda|^2)^M \|s\|_{t,0}. \quad (3.2.26)$$

Thus

$$\|Q_Z Q'_{Z'} \exp(-\frac{u}{2}L_{2,D}^t)\|_t^{0,0} \leqslant C_m. \quad (3.2.27)$$

Let $\| \cdot \|_m$ be the usual Sobolev norm on $\mathscr{C}_0^\infty(\overline{\mathbb{R}_+^{2n}}, \mathbf{E}_{x_0})$ induced by $h^{\mathbf{E}_{x_0}} = h^{\Lambda(T_{x_0}^{*(0,1)}X) \otimes E_{x_0}}$ and the volume form $dv_{TX}(Z)$ as in (1.6.32). Observe that by (1.6.31), (1.6.32), for $m \geqslant 0$, there exists $C_m > 0$ such that for $s \in \mathscr{C}_0^\infty(\overline{\mathbb{R}_+^{2n}}, \mathbf{E}_{x_0})$, $\mathrm{supp}(s) \subset B^{T_{x_0}X}(0, q)$,

$$\frac{1}{C_m}(1+q)^{-m}\|s\|_{t,m} \leqslant \|s\|_m \leqslant C_m(1+q)^m\|s\|_{t,m}. \tag{3.2.28}$$

Now (3.2.27), (3.2.28) together with Sobolev's inequalities, Theorem A.2.2, implies that

$$\left|\exp(-\frac{u}{2}L_{2,D}^t)(Z, Z')\right| \leqslant C(1+|Z|+|Z'|)^{2n+2}. \tag{3.2.29}$$

By (1.6.66), (3.2.29), we get for $0 \leqslant x_\mathbf{n} \leqslant \epsilon$,

$$\left|\exp(-\frac{u}{2p}L_{p,D})((0, x_\mathbf{n}), (0, x_\mathbf{n}))\right| \leqslant Cp^n(1+\sqrt{p}|x_\mathbf{n}|)^{2n+2}. \tag{3.2.30}$$

From (3.2.20) and (3.2.30), we get for $0 \leqslant x_\mathbf{n} \leqslant \varepsilon p^{-\theta/2}$,

$$p^{-n}\left|\exp(-\frac{u}{2p}D_{p,X}^2)((y_0, x_\mathbf{n}), (y_0, x_\mathbf{n}))\right|$$
$$\leqslant Cp^{8(n+1)(1+\theta)}\exp(-\frac{\varepsilon^2 p^{1-\theta}}{16u}) + Cp^{(n+1)(1-\theta)}. \tag{3.2.31}$$

Now, we fix $0 < \theta < 1$ such that $(n+1)(1-\theta) < \theta/2$, then by Theorem 1.6.1 and (3.2.14), (3.2.19), (3.2.31) and dominated convergence, we get for $u > 0$ fixed,

$$\lim_{p\to\infty} \int_0^\infty e^{-u\lambda}d\mu_p^q(\lambda) = \lim_{p\to\infty} p^{-n}\int_X \mathrm{Tr}_q\left[e^{-\frac{u}{2p}D_{p,X}^2}(x, x)\right] dv_X(x)$$
$$= \lim_{p\to\infty} \int_{X\setminus(\partial X\times[0,\varepsilon p^{-\theta/2}[)} p^{-n}\mathrm{Tr}_q[\exp(-\frac{u}{2p}D_{p,X}^2)(x, x)]dv_X(x)$$
$$+ \lim_{p\to\infty} \int_{\partial X\times[0,\varepsilon p^{-\theta/2}[} p^{-n}\mathrm{Tr}_q[\exp(-\frac{u}{2p}D_{p,X}^2)(x, x)]dv_X(x) \tag{3.2.32}$$
$$= \mathrm{rk}(E) \int_X \frac{\det(\dot{R}^L/(2\pi))\,\mathrm{Tr}_{\Lambda^{0,q}}[\exp(u\,\omega_d)]}{\det(1 - \exp(-u\dot{R}^L))}.$$

From Theorem 1.6.1 and (3.2.31), the convergence in (3.2.32) is uniform as u varies in any compact subset of $\mathbb{R}_+^*$. The proof of Proposition 3.2.7 is complete. $\qquad\square$

We denote by $\mathscr{D}'(\mathbb{R})$ (resp. $\mathscr{S}'(\mathbb{R})$) the space of distributions (resp. temperate distributions) on $\mathbb{R}$. Consider $g \in \mathscr{D}'(\mathbb{R})$ supported in $[0, \infty[$. Assume that there exists $u_0 \in \mathbb{R}$ such that for any $u > u_0$, $\lambda \mapsto e^{-\lambda u}g(\lambda)$ belongs to $\mathscr{S}'(\mathbb{R})$. The

Laplace transform of g is the holomorphic function $L(g) : \{\mathrm{Re}(z) > u_0\} \to \mathbb{C}$ defined by

$$L(g)(z) = (g(\lambda), e^{-\lambda z}), \tag{3.2.33}$$

where $z = u + iv$. Relation (3.2.14) gives the Laplace transform of the measure μ_p^q and (3.2.15) describes the limit of the Laplace transforms as $p \to \infty$. Since we wish to calculate the limit of the sequence $(\mu_p^q)_{p \geqslant 1}$, we need the following.

Theorem 3.2.8. *A holomorphic function $G : \{\mathrm{Re}(z) > u_0\} \to \mathbb{C}$, $u_0 \in \mathbb{R}$, is the Laplace transform of a distribution $g \in \mathscr{D}'(\mathbb{R})$ with support in $[0, \infty[$ if and only if there exists a polynomial $Q \in \mathbb{R}[t]$ and $u_1 > u_0$ such that $|G(z)| \leqslant Q(|z|)$ for $\mathrm{Re}(z) > u_1$. The inverse Laplace transform $L^{-1}(G)$ of G is given for any $u > u_0$ by*

$$L^{-1}(G)(\lambda) = \frac{1}{2\pi i} \int_{u-i\infty}^{u+i\infty} G(z)e^{\lambda z} \, dz. \tag{3.2.34}$$

As in Section 3.2.1, we use the notation (1.5.14)–(1.5.20). Especially, $a_i(x)$ are eigenvalues of $\dot{R}^L(x) \in \mathrm{End}(T_x^{(1,0)}X)$. Let us denote by $l = l(x)$ the rank of $\dot{R}^L(x)$ and order the eigenvalues so that $|a_1(x)| \geqslant \cdots \geqslant |a_l(x)| > 0 = a_{l+1}(x) = \cdots = a_n(x)$. For $I \subset \{1, \ldots, n\}$, we set

$$a_I(x) = \sum_{i \in I} a_i(x), \quad \complement I = \{1, \ldots, n\} \smallsetminus I. \tag{3.2.35}$$

Let $H_+(t) = 0$ for $t < 0$, and 1 for $t \geqslant 0$ be the Heaviside function. We introduce the function

$$\nu_{R^L(x)}(\lambda) = \frac{(2\pi)^{-n}}{(n-l)!} \, |a_1(x) \cdots a_l(x)|$$

$$\times \sum_{(\alpha_1, \ldots, \alpha_l) \in \mathbb{N}^l} (t^{n-l} H_+(t)) \Big|_{t = \lambda - \sum_j (\alpha_j + \frac{1}{2})|a_j(x)|}, \tag{3.2.36}$$

where we set $t^0 = 1$ for any $t \in \mathbb{R}$. To shorten the notation, we set

$$\nu_{R^L}^q(\lambda, x) = \sum_{|I| = q} \nu_{R^L(x)}\big(\lambda + \tfrac{1}{2}(a_{\complement I}(x) - a_I(x))\big). \tag{3.2.37}$$

Theorem 3.2.9. *There exists an at most countable set $\mathscr{A}^q \subset \mathbb{R}$ such that we have for $\lambda \in \mathbb{R}_+ \smallsetminus \mathscr{A}^q$:*

$$\lim_{p \to \infty} p^{-n} N^q(\lambda, \tfrac{1}{p}\square_{p,X}) = I^q(X, \lambda) := \mathrm{rk}(E) \int_X \nu_{R^L}^q(\lambda, x) \, dv_X. \tag{3.2.38}$$

Moreover,

$$\lim_{\lambda \to +0} I^q(X, \lambda) = I^q(X, 0) = \frac{1}{n!} \mathrm{rk}(E) \int_{X(q)} (-1)^q \Big(\tfrac{\sqrt{-1}}{2\pi} R^L\Big)^n. \tag{3.2.39}$$

Proof. Taking into account (1.7.11), we can reformulate (3.2.15) as follows: for any $u > 0$,

$$\lim_{p \to \infty} \int_0^\infty e^{-u\lambda} \, d\mu_p^q(\lambda) \tag{3.2.40}$$

$$= \mathrm{rk}(E) \int_X \sum_{|I|=q} \frac{\prod_{j=1}^l |a_j|}{(2\pi)^n} \cdot \frac{\exp(\frac{u}{2}(a_{\complement I} - a_I - \sum_{j=1}^l |a_j|))}{u^{n-l} \prod_{j=1}^l (1 - e^{-u|a_j|})} \, dv_X \, .$$

Let us denote the integrand in (3.2.40) by $F(x, u)$. As a function of u, $F(x, u)$ can be extended to a holomorphic function $F(x, z)$ in $\{z \in \mathbb{C} : \mathrm{Re}(z) > 0\}$. There exists $C > 0$ such that for $u \in \mathbb{C}$ and $a \in \mathbb{R}$ we have $|(1 - e^{-u|a|})^{-1}| \leq \max\{\frac{C}{\mathrm{Re}(u)|a|}, 2\}$. Then for any $u_0 > 0$ we have $|F(x, u)| \leq C$ uniformly on $x \in X$, $\mathrm{Re}(u) > u_0$. We see from Theorem 3.2.8, that $F(x, u)$ admits an inverse Laplace transform $L^{-1}(F(x, \cdot))$ for each fixed x. We have (cf. Problem 3.6),

$$\int_0^\lambda L^{-1}(F(x, \cdot))(\lambda_0) \, d\lambda_0 = \nu_{R^L}^q(\lambda, x). \tag{3.2.41}$$

Lemma 3.2.10. *Consider a sequence of measures μ_p on $\mathbb{R}$ and a function $\Omega :$ $]c, \infty[\longrightarrow \mathbb{R}$ such that when $p \to \infty$, $L(\mu_p)(\lambda) \to \Omega(\lambda)$ for $\lambda \in \mathbb{R}$, $\lambda > c$. Then there exists a measure μ on $\mathbb{R}$ with support in $[0, \infty[$ such that $L(\mu)(\lambda) = \Omega(\lambda)$ for $\lambda \in \mathbb{R}$, $\lambda > c$, and $\mu_p \to \mu$ in the sense of distributions.*

By Lemma 3.2.10 the sequence of measures $(\mu_p^q)_{p \geq 1}$ converges in the sense of distributions to the measure μ^q satisfying:

$$\mu^q([0, \lambda]) = \mathrm{rk}(E) \int_X \nu_{R^L}^q(\lambda, x) \, dv_X \, , \tag{3.2.42}$$

especially the measure μ^q is zero on $]-\infty, 0[$. In other words, for any $\varphi \in \mathscr{C}_0^\infty(\mathbb{R})$, we have

$$\lim_{p \to \infty} (\mu_p^q, \varphi) = (\mu^q, \varphi). \tag{3.2.43}$$

Now, take $\varphi_k : \mathbb{R} \to [0, 1]$ smooth, $\mathrm{supp}(\varphi_k) \subset \,]-2, \lambda[$, and $\varphi_k \nearrow 1$ on $[-1, \lambda[$. Then from (3.2.43), we have for any $k \in \mathbb{N}$,

$$\liminf_{p \to \infty} \mu_p^q([0, \lambda]) \geq \lim_{p \to \infty} (\mu_p^q, \varphi_k) = (\mu^q, \varphi_k). \tag{3.2.44}$$

Relations (3.2.13) and (3.2.44) yield by taking $k \to \infty$:

$$\liminf_{p \to \infty} p^{-n} N^q(\lambda, \tfrac{1}{p}\square_{p,X}) \geq \mu^q([0, \lambda[). \tag{3.2.45}$$

In the same way, we choose $\varphi_k \in \mathscr{C}_0^\infty(\mathbb{R})$, such that $\varphi_k = 1$ on $[-1, \lambda]$, and $\varphi_k \searrow 1_{[0, \lambda]}$ on $[0, \infty[$. Then we get as in (3.2.45),

$$\limsup_{p \to \infty} p^{-n} N^q(\lambda, \frac{1}{p}\square_{p,X}) \leq \lim_{\lambda_0 \to +\lambda} \mu^q([0, \lambda_0]) = \mu^q([0, \lambda]). \tag{3.2.46}$$

Let us denote by $\mathscr{A}^q$ the countable set where $\lambda \to \mu^q([0,\lambda])$ is possibly discontinuous. Then by (3.2.45) and (3.2.46), $p^{-n}N^q(\lambda, \frac{1}{p}\square_{p,X}) \to \mu^q([0,\lambda])$ as $p \to \infty$, for all $\lambda \in \mathbb{R}_+ \setminus \mathscr{A}^q$ and formula (3.2.38) holds true.

We compute the behavior of $I^q(X,\lambda)$ for $\lambda \to +0$. First it is clear that

$$\lim_{\lambda \to +0} I^q(X,\lambda) = I^q(X,0) = \text{rk}(E)\int_X \nu^q_{R^L}(0,x)\,dv_X . \tag{3.2.47}$$

Since $\frac{1}{2}(a_{\complement I} - a_I) - \sum_j(\alpha_j + \frac{1}{2})|a_j| \leqslant 0$ for all $\alpha \in \mathbb{N}^l$, it is clear that for a given $\alpha \in \mathbb{N}^l$,

$$\left. (t^{n-l}H_+(t)) \right|_{t = \frac{1}{2}(a_{\complement I} - a_I) - \sum_j(\alpha_j + \frac{1}{2})|a_j|} = 0$$

unless $l = n$ (i.e., $\sqrt{-1}R^L(x)$ is non-degenerate) and

$$\tfrac{1}{2}(a_{\complement I} - a_I) - \sum_j(\alpha_j + \tfrac{1}{2})|a_j| = 0.$$

The last equality holds if and only if $\alpha_1 = \cdots = \alpha_n = 0$, $a_j < 0$ for $j \in I$ and $a_j > 0$ for $j \in \complement I$. In particular, if $\nu_{R^L(x)}(\frac{1}{2}(a_{\complement I} - a_I)) \neq 0$, $\sqrt{-1}R^L(x)$ is non-degenerate and has exactly q negative eigenvalues.

For $x \in X(q)$, set $I(x) = \{j : a_j(x) < 0\}$. For $|I| = q$, it follows that

$$\nu_{R^L(x)}\left(\tfrac{1}{2}(a_{\complement I} - a_I)\right) = \begin{cases} (2\pi)^{-n}|a_1(x)\ldots a_n(x)| & \text{for } I = I(x), \\ 0 & \text{for } I \neq I(x). \end{cases} \tag{3.2.48}$$

By (1.7.15) and (3.2.48), we get (3.2.39). $\qquad\qquad\qquad\qquad\qquad\square$

3.2.3 Morse inequalities for the L^2 cohomology

Let (X, Θ) be a complex Hermitian manifold of dimension n. Let (L, h^L) and (E, h^E) be holomorphic Hermitian vector bundles on X with $\text{rk}(L) = 1$.

Let $\left(\text{Dom}(\overline{\partial}_p^E) \cap L^2_{0,\bullet}(X, L^p \otimes E), \overline{\partial}_p^E\right)$ be the L^2-Dolbeault complex of densely defined closed operators. Since $\overline{\partial}_p^E$ commutes with $\frac{1}{p}\square_p$, it commutes with the spectral projections, i.e.,

$$\overline{\partial}_p^E E_\lambda\big(\tfrac{1}{p}\square_p\big) = E_\lambda\big(\tfrac{1}{p}\square_p\big)\overline{\partial}_p^E. \tag{3.2.49}$$

We obtain therefore a sub-complex

$$\left(\mathscr{E}^\bullet(\lambda, \tfrac{1}{p}\square_p), \overline{\partial}_p^E\right) \longrightarrow \left(\text{Dom}(\overline{\partial}_p^E) \cap L^2_{0,\bullet}(X, L^p \otimes E), \overline{\partial}_p^E\right). \tag{3.2.50}$$

The cohomology of this complex is denoted by $H^\bullet\big(\mathscr{E}^\bullet(\lambda, \tfrac{1}{p}\square_p), \overline{\partial}_p^E\big)$.

Proposition 3.2.11. *If the fundamental estimate (3.2.2) holds for $(0,q)$-forms with $q \leqslant m$, then for $q \leqslant m$ and $\lambda \geqslant 0$,*

$$H^q\big(\mathscr{E}^\bullet(\lambda, \tfrac{1}{p}\square_p), \overline{\partial}_p^E\big) \simeq \mathscr{H}^{0,q}(X, L^p \otimes E) \simeq H^{0,q}_{(2)}(X, L^p \otimes E). \qquad (3.2.51)$$

Proof. By Theorem 3.1.8, the strong Hodge decomposition holds in bidegrees $(0,q)$, for any $q \leqslant m$. Let us restrict (3.1.28) to the complex (3.2.50):

$$\mathscr{E}^q(\lambda, \tfrac{1}{p}\square_p) = \mathscr{H}^{0,q}(X, L^p \otimes E) \oplus \big(\operatorname{Im}(\overline{\partial}_p^E) \cap \mathscr{E}^q(\lambda, \tfrac{1}{p}\square_p)\big)$$

$$\oplus \big(\operatorname{Im}(\overline{\partial}_p^{E,*}) \cap \mathscr{E}^q(\lambda, \tfrac{1}{p}\square_p)\big). \qquad (3.2.52)$$

Due to (3.2.49), we see that

$$\begin{aligned}
\operatorname{Im}(\overline{\partial}_p^E) \cap \mathscr{E}^q(\lambda, \tfrac{1}{p}\square_p) &= \operatorname{Im}(\overline{\partial}_p^E | \mathscr{E}^q(\lambda, \tfrac{1}{p}\square_p)), \\
\operatorname{Im}(\overline{\partial}_p^{E,*}) \cap \mathscr{E}^q(\lambda, \tfrac{1}{p}\square_p) &= \operatorname{Im}(\overline{\partial}_p^{E,*} | \mathscr{E}^q(\lambda, \tfrac{1}{p}\square_p)).
\end{aligned} \qquad (3.2.53)$$

Therefore

$$\operatorname{Ker}(\overline{\partial}_p^E | \mathscr{E}^q(\lambda, \tfrac{1}{p}\square_p)) = \mathscr{H}^{0,q}(X, L^p \otimes E) \oplus \operatorname{Im}\big(\overline{\partial}_p^E | \mathscr{E}^q(\lambda, \tfrac{1}{p}\square_p)\big). \qquad (3.2.54)$$

By (3.1.29) and (3.2.54) , we get (3.2.51). $\qquad \square$

The following *algebraic Morse inequalities* which were essentially established in (1.7.9), are very useful.

Lemma 3.2.12. *Let*

$$0 \longrightarrow V^0 \xrightarrow{d^0} V^1 \xrightarrow{d^1} \cdots \xrightarrow{d^{n-1}} V^n \longrightarrow 0$$

be a complex of vector spaces. Let $H^i(V^\bullet) = \operatorname{Ker}(d^i)/\operatorname{Im}(d^{j-1})$ with $\operatorname{Im}(d^{-1}) = 0$. If $\dim V^q < +\infty$ for any $q \leqslant m$, then

$$\sum_{j=0}^q (-1)^{q-j} \dim H^j(V^\bullet) \leqslant \sum_{j=0}^q (-1)^{q-j} \dim V^j, \quad \text{for } q \leqslant m. \qquad (3.2.55)$$

Proof. Set $n_j = \dim \operatorname{Ker}(d^j)$, $m_j = \dim \operatorname{Im}(d^j)$. Then

$$\dim V^j = n_j + m_j, \quad \dim H^j(V^\bullet) = n_j - m_{j-1} \quad (\text{with } m_{-1} = 0)$$

and for $q \leqslant m$,

$$\sum_{j=0}^q (-1)^{q-j} \dim V^j = m_q + \sum_{j=0}^q (-1)^{q-j} \dim H^j(V^\bullet). \qquad (3.2.56)$$

The inequalities (3.2.55) follow from (3.2.56). $\qquad \square$

We are prepared to state the Morse inequalities for the L^2 cohomology.

Theorem 3.2.13. *Assume that there exists $m \in \mathbb{N}$ such that the fundamental estimate (3.2.2) holds for $(0, q)$-forms with any $q \leqslant m$. Let $U \subset X$ be a relatively compact open set with smooth boundary such that $K \Subset U$. Then as $p \to \infty$, the following strong Morse inequalities hold for any $q \leqslant m$:*

$$\sum_{j=0}^{q}(-1)^{q-j} \dim H_{(2)}^{0,j}(X, L^p \otimes E) \leqslant \mathrm{rk}(E)\frac{p^n}{n!} \int_{U(\leqslant q)} (-1)^q \left(\frac{\sqrt{-1}}{2\pi}R^L\right)^n + o(p^n).$$
(3.2.57)

In particular, we get the weak Morse inequalities for any $q \leqslant m$:

$$\dim H_{(2)}^{0,q}(X, L^p \otimes E) \leqslant \mathrm{rk}(E)\frac{p^n}{n!} \int_{U(q)} (-1)^q \left(\frac{\sqrt{-1}}{2\pi}R^L\right)^n + o(p^n). \quad (3.2.58)$$

Proof. Lemma 3.2.12 applied to the complex (3.2.50) together with Proposition 3.2.11 delivers

$$\sum_{j=0}^{q}(-1)^{q-j} \dim H_{(2)}^{0,j}(X, L^p \otimes E) \leqslant \sum_{j=0}^{q}(-1)^{q-j}N^j\left(\lambda, \tfrac{1}{p}\Box_p\right). \quad (3.2.59)$$

We estimate the right-hand side of (3.2.59). Theorem 3.2.5 and (3.2.11) show that

$$N^j\left(\lambda, \tfrac{1}{p}\Box_{p,U}\right) \leqslant N^j\left(\lambda, \tfrac{1}{p}\Box_p\right) \leqslant N^j\left(3C_0\lambda + C_2 p^{-1}, \tfrac{1}{p}\Box_{p,U}\right). \quad (3.2.60)$$

Thus (3.2.38) and (3.2.60) imply for $\lambda \notin \mathscr{A}^j$,

$$N^j\left(\lambda, \tfrac{1}{p}\Box_p\right) \geqslant p^n I^j(U, \lambda) + o(p^n), \quad \text{for } p \to \infty. \quad (3.2.61)$$

Let us consider $\lambda < 1/(2C_0)$ and $\delta > 0$. For $p > C_2/\delta$, we have

$$N^j\left(3C_0\lambda + C_2 p^{-1}, \tfrac{1}{p}\Box_{p,U}\right) \leqslant N^j\left(3C_0\lambda + \delta, \tfrac{1}{p}\Box_{p,U}\right). \quad (3.2.62)$$

For $3C_0\lambda + \delta \notin \mathscr{A}^j$, we get from (3.2.38) and (3.2.62),

$$\limsup_{p\to\infty} p^{-n} N^j\left(\lambda, \tfrac{1}{p}\Box_p\right) \leqslant \lim_{p\to\infty} p^{-n} N^j\left(3C_0\lambda + \delta, \tfrac{1}{p}\Box_{p,U}\right) = I^j(U, 3C_0\lambda + \delta).$$

Since $\nu_{R^L}^j$ is right-continuous in λ and bounded on U, we can use the Lebesgue dominated convergence theorem to let $\delta \to 0$. By (3.2.38) and the above equation, we get

$$N^j\left(\lambda, \tfrac{1}{p}\Box_p\right) \leqslant p^n I^j(U, 3C_0\lambda) + o(p^n), \quad \text{for } p \to \infty. \quad (3.2.63)$$

By (3.2.59), (3.2.61) and (3.2.63), we deduce

$$\sum_{j=0}^{q}(-1)^{q-j} \dim H_{(2)}^{0,j}(X, L^p \otimes E) \leqslant p^n \sum_{j=0}^{q}(-1)^{q-j} I^j\left(U, \lambda(q - j)\right) + o(p^n),$$
(3.2.64)

where $\lambda(r) = \lambda$ for r odd, $\lambda(r) = 3C_0\lambda$ for all r even. By passing to the limit $\lambda \to +0$ such that $\lambda(r) \notin \cup_{j \leqslant q}\mathscr{A}^j$, we obtain (3.2.57) invoking (3.2.39). $\qquad \square$

Remark 3.2.14. If X is compact, as in Remark 3.1.10, the hypotheses of Theorem 3.2.13 are trivially satisfied for $m = n$, so that Theorem 1.7.1 is a special case of Theorem 3.2.13.

Remark 3.2.15. In a similar manner as in Theorem 3.2.13, if the fundamental estimate (3.2.2) holds in bi-degree $(0, q)$ for any $q \geqslant m$, then for any $r \geqslant m$,

$$\sum_{j=r}^{n} (-1)^{j-r} \dim H_{(2)}^{0,j}(X, L^p \otimes E) \leqslant \mathrm{rk}(E) \frac{p^n}{n!} \int_{X(\geqslant r)} (-1)^r \left(\frac{\sqrt{-1}}{2\pi} R^L \right)^n + o(p^n). \tag{3.2.65}$$

Back to the general situation, observe that as in (2.2.42), we can estimate from below the dimension of the space

$$H_{(2)}^0(X, L^p \otimes E) = \{ s \in L_{0,0}^2(X, L^p \otimes E) \ : \ \bar{\partial}_p^E s = 0 \}, \tag{3.2.66}$$

by applying (3.2.57) for $q - 1$, provided the fundamental estimate holds in bidegrees $(0, 0)$ and $(0, 1)$. We show in the sequel that we can get such an estimate if we only assume that the fundamental estimate (3.2.2) holds in bidegree $(0, 1)$, although (3.2.57) for $q = 1$ might not hold.

Theorem 3.2.16. *Assume that the fundamental estimate (3.2.2) holds for $(0, 1)$-forms. Let U be a relatively compact open set with smooth boundary such that $K \Subset U$. Then, for $p \to \infty$,*

$$\dim H_{(2)}^0(X, L^p \otimes E) \geqslant \mathrm{rk}(E) \frac{p^n}{n!} \int_{U(\leqslant 1)} \left(\frac{\sqrt{-1}}{2\pi} R^L \right)^n + o(p^n). \tag{3.2.67}$$

Note that the left side of (3.2.67) may be infinite.

Proof. For $\lambda > 0$, let $\bar{\partial}_{p,\lambda}^E : \mathscr{E}^0(\lambda, \frac{1}{p}\square_p) \longrightarrow \mathscr{E}^1(\lambda, \frac{1}{p}\square_p)$ be the restriction of $\bar{\partial}_p^E$. By the definition of $\mathscr{E}^0(\lambda, \frac{1}{p}\square_p)$, $\bar{\partial}_{p,\lambda}^E$ is a bounded operator, and by (3.2.5),

$$\begin{aligned} \mathrm{Ker}(\bar{\partial}_{p,\lambda}^E) &= H_{(2)}^0(X, L^p \otimes E), \\ N^0(\lambda, \tfrac{1}{p}\square_p) &= \dim \mathrm{Ker}(\bar{\partial}_{p,\lambda}^E) + \dim \mathrm{Im}(\bar{\partial}_{p,\lambda}^E). \end{aligned} \tag{3.2.68}$$

Obviously $\dim \mathrm{Im}(\bar{\partial}_{p,\lambda}^E) \leqslant N^1(\lambda, \frac{1}{p}\square_p)$. For $\lambda < 1/(2C_0)$ and sufficiently large p, by Theorem 3.2.5, $N^1(\lambda, \frac{1}{p}\square_p)$ is finite-dimensional, and by (3.2.68),

$$\dim H_{(2)}^0(X, L^p \otimes E) \geqslant N^0(\lambda, \tfrac{1}{p}\square_p) - N^1(\lambda, \tfrac{1}{p}\square_p). \tag{3.2.69}$$

We repeat now the proof of Theorem 3.2.13 to estimate $N^1(\lambda, \frac{1}{p}\square_p)$ from above and $N^0(\lambda, \frac{1}{p}\square_p)$ from below, and observe that by the latter estimate, we applied Lemma 3.2.6 which does not require any hypothesis on the spectrum of $\square_p$ on $(0, 0)$-forms. $\qquad \square$

3.3   Uniformly positive line bundles

In this section, we apply the results from the previous one to the study of the L^2 cohomology of complex manifolds satisfying certain curvature conditions. If X is a complete Kähler manifold and L a positive line bundle on X, the L^2 estimates of Andreotti–Vesentini–Hörmander allow us to find a lot of sections of $L^p \otimes K_X$. We prove now a "compact perturbation" of this result. In this case the underlying complete metric is no more assumed to be Kähler, but we assume instead the existence of a uniformly positive line bundle outside a compact set. As an application we prove the Nadel–Tsuji theorem in Corollary 3.3.7 and the holomorphic Morse inequalities for hyperconcave manifolds in Theorem 3.4.9.

Let (X, Θ) be a complex Hermitian manifold of dimension n and (E, h^E) be a holomorphic Hermitian vector bundle on X.

We start by proving the Andreotti–Vesentini density lemma. For a linear operator R between Hilbert spaces, the *graph-norm* is defined by $\mathrm{Dom}(R) \ni s \longmapsto \|s\| + \|Rs\|$.

Lemma 3.3.1 (Andreotti–Vesentini). *Assume that (X, Θ) is complete. Let $\overline{\partial}^E$ and $\overline{\partial}^{E,*}$ be the maximal extensions as in Lemma 3.1.1. Then $\Omega_0^{0,\bullet}(X, E)$ is dense in $\mathrm{Dom}(\overline{\partial}^E)$, $\mathrm{Dom}(\overline{\partial}^{E,*})$, $\mathrm{Dom}(\overline{\partial}^E) \cap \mathrm{Dom}(\overline{\partial}^{E,*})$ in the graph norms of $\overline{\partial}^E$, $\overline{\partial}^{E,*}$ and $\overline{\partial}^E + \overline{\partial}^{E,*}$, respectively.*

Proof. We first reduce the proof to the case of a compactly supported form s. The completeness of the metric implies the existence of a sequence $\{\varphi_k\}_k \subset \mathscr{C}_0^\infty(X)$, such that $0 \leqslant \varphi_k \leqslant 1$, $\varphi_{k+1} = 1$ on $\mathrm{supp}(\varphi_k)$, $|d\varphi_k| \leqslant 1/2^k$ for every $k \geqslant 1$ and $\{\mathrm{supp}(\varphi_k)\}_k$ exhaust X. To construct this sequence, we first construct an exhaustive function $\varphi : X \longrightarrow \mathbb{R}$ with $|d\varphi| < 1$. This is done by smoothing the distance to a point (we can assume that X is connected). Next, consider a smooth function $\rho : \mathbb{R} \longrightarrow [0, 1]$ such that $\rho = 0$ on $]-\infty, -2]$, $\rho = 1$ on $[-1, \infty[$ and $0 \leqslant \rho' \leqslant 2$. Then $\varphi_k = \rho(-\varphi/2^{k+1})$ satisfies the conditions above.

Thus there exists $C > 0$ such that for $s \in \mathrm{Dom}(\overline{\partial}^E) \cap \mathrm{Dom}(\overline{\partial}^{E,*})$, we have $\varphi_k s \in \mathrm{Dom}(\overline{\partial}^E) \cap \mathrm{Dom}(\overline{\partial}^{E,*})$ and

$$
\begin{aligned}
\|\overline{\partial}^E(\varphi_k s) - \varphi_k \overline{\partial}^E s\|_{L^2} &\leqslant C\|s\|_{L^2}/2^k, \\
\|\overline{\partial}^{E,*}(\varphi_k s) - \varphi_k \overline{\partial}^{E,*} s\|_{L^2} &\leqslant C\|s\|_{L^2}/2^k.
\end{aligned}
\tag{3.3.1}
$$

Hence $\{\varphi_k s\}$ converges to s in the graph norm. So to prove the assertion, we can start with a form s having compact support in X. But then the approximation in the graph norm follows from the Friedrichs lemma 3.1.3. $\square$

As a by-product of Friedrichs lemma, we obtain also:

Corollary 3.3.2. *Assume that*

$$s \in L^2_{0,q}(X,E),\ w \in L^2_{0,q+1}(X,E),\ \overline{\partial}^E s \in L^2_{0,q+1}(X,E)\ and\ \overline{\partial}^{E,*} w \in L^2_{0,q}(X,E).$$

Suppose also that s and w have compact support. Then $\langle \overline{\partial}^E s, w\rangle = \langle s, \overline{\partial}^{E,*} w\rangle$.

Proof. We may assume that s, w have support in a trivialization patch diffeomorphic to $\mathbb{R}^{2n}$. We denote $w_\varepsilon = w * \rho_\varepsilon$ as in Lemma 3.1.3. We have:

$$
\begin{aligned}
\langle \overline{\partial}^E s, w\rangle &= \lim_{\varepsilon \to 0} \langle (\overline{\partial}^E s)_\varepsilon, w_\varepsilon\rangle = \lim_{\varepsilon \to 0} \langle \overline{\partial}^E s_\varepsilon, w_\varepsilon\rangle = \lim_{\varepsilon \to 0} \langle s_\varepsilon, \overline{\partial}^{E,*} w_\varepsilon\rangle \\
&= \lim_{\varepsilon \to 0} \langle s_\varepsilon, (\overline{\partial}^{E,*} w)_\varepsilon\rangle = \langle s, \overline{\partial}^{E,*} w\rangle.
\end{aligned}
\tag{3.3.2}
$$

From (3.3.2), we get Lemma 3.3.2. $\qquad\square$

Corollary 3.3.3. *If (X,Θ) is complete, then $\overline{\partial}^{E,*}_H = \overline{\partial}^{E,*}$, that is the Hilbert space adjoint and the maximal extension of the formal adjoint of $\overline{\partial}^E$ coincide.*

Proof. It is clear that $\overline{\partial}^{E,*}_H \subset \overline{\partial}^{E,*}$. Conversely if $s \in \mathrm{Dom}(\overline{\partial}^{E,*})$, from Lemma 3.3.1, there exist $s_k \in \Omega^{0,\bullet}_0(X,E)$ with $s_k \longrightarrow s$, $\overline{\partial}^{E,*} s_k \longrightarrow \overline{\partial}^{E,*} s$. Then by definition of $\overline{\partial}^E$,

$$\langle \overline{\partial}^E w, s_k\rangle = \langle w, \overline{\partial}^{E,*} s_k\rangle, \quad \text{for } w \in \mathrm{Dom}(\overline{\partial}^E).
\tag{3.3.3}$$

The limit of this equality for $k \to \infty$ gives

$$\langle \overline{\partial}^E w, s\rangle = \langle w, \overline{\partial}^{E,*} s\rangle, \quad \text{for } w \in \mathrm{Dom}(\overline{\partial}^E).$$

Thus $s \in \mathrm{Dom}(\overline{\partial}^{E,*}_H)$ and $\overline{\partial}^{E,*}_H s = \overline{\partial}^{E,*} s$. $\qquad\square$

Corollary 3.3.4. *If (X,Θ) is complete, then $\square^E = \square^E|_{\Omega^{0,\bullet}(X,E)}$ is essentially self-adjoint. In particular the Gaffney and Friedrichs extensions coincide.*

Proof. We will show that $\square^E_{\max}$ is self-adjoint. Since $\square^E|_{\Omega^{0,\bullet}(X,E)}$ is symmetric, $\square^E_{\max} = (\square^E)^*_H$. This implies the closure $\overline{\square^E}$ of $\square^E$ satisfies

$$\overline{\square^E} = ((\square^E)^*_H)^*_H = (\square^E_{\max})^*_H = \square^E_{\max} = (\square^E)^*_H.
\tag{3.3.4}$$

It follows that the closure $\overline{\square^E}$ is self-adjoint and $\square^E$ is essentially self-adjoint.

Let $s \in \mathrm{Dom}(\square^E_{\max})$. Since $\square^E$ is elliptic, we have $s \in \boldsymbol{H}^2(X,\Lambda(T^{*(0,1)}X)\otimes E,\mathrm{loc})$ by Theorem A.3.4. Thus $\overline{\partial}^{E,*} s, \overline{\partial}^E s \in L^2_{0,\bullet}(X,E,\mathrm{loc})$ and by Corollary

3.3.2, we can integrate by parts if s is multiplied with a smooth compactly supported function. Let φ_k be the family of functions defined in the proof of Lemma 3.3.1. We obtain:

$$
\begin{aligned}
\|\varphi_k\overline{\partial}^E s\|_{L^2}^2 + \|\varphi_k\overline{\partial}^{E,*} s\|_{L^2}^2 &= \langle \varphi_k^2\overline{\partial}^E s, \overline{\partial}^E s\rangle + \langle s, \overline{\partial}^E(\varphi_k^2\overline{\partial}^{E,*} s)\rangle \\
&= \langle \varphi_k^2 s, \Box^E s\rangle - 2\langle \overline{\partial}\varphi_k \wedge s, \varphi_k\overline{\partial}^E s\rangle + 2\langle s, \overline{\partial}\varphi_k \wedge (\varphi_k\overline{\partial}^{E,*} s)\rangle \\
&\leqslant \langle \varphi_k^2 s, \Box^E_{\max} s\rangle + 2^{-k}(2\|\varphi_k\overline{\partial}^E s\|_{L^2}\|s\|_{L^2} + 2\|\varphi_k\overline{\partial}^{E,*} s\|_{L^2}\|s\|_{L^2}) \\
&\leqslant \langle \varphi_k^2 s, \Box^E_{\max} s\rangle + 2^{-k}(\|\varphi_k\overline{\partial}^E s\|_{L^2}^2 + \|\varphi_k\overline{\partial}^{E,*} s\|_{L^2}^2 + 2\|s\|_{L^2}^2).
\end{aligned}
\tag{3.3.5}
$$

We get therefore the Stampacchia type inequality

$$
\|\varphi_k\overline{\partial}^E s\|_{L^2}^2 + \|\varphi_k\overline{\partial}^{E,*} s\|_{L^2}^2 \leqslant \frac{1}{1-2^{-k}}(\langle \varphi_k^2 s, \Box^E_{\max} s\rangle + 2^{1-k}\|s\|_{L^2}^2).
$$

By letting $k \to \infty$, we obtain $\|\overline{\partial}^E s\|_{L^2}^2 + \|\overline{\partial}^{E,*} s\|_{L^2}^2 \leqslant \langle s, \Box^E_{\max} s\rangle$, in particular $\overline{\partial}^E s, \overline{\partial}^{E,*} s$ are in $L^2_{0,\bullet}(X, E)$. This implies

$$
\langle s, \Box^E_{\max} s_1\rangle = \langle \overline{\partial}^E s, \overline{\partial}^E s_1\rangle + \langle \overline{\partial}^{E,*} s, \overline{\partial}^{E,*} s_1\rangle \quad s, s_1 \in \mathrm{Dom}(\Box^E_{\max}), \tag{3.3.6}
$$

because the equality holds for $\varphi_k s$ and s_1, and since we have $\varphi_k s \to s$, $\overline{\partial}^E(\varphi_k s) \to \overline{\partial}^E s$ and $\overline{\partial}^{E,*}(\varphi_k s) \to \overline{\partial}^{E,*} s$ in L^2 by the proof of Lemma 3.3.1. An analogous calculation shows that the right-hand side of (3.3.6) equals $\langle \Box^E_{\max} s, s_1\rangle$. Thus

$$
\langle s, \Box^E_{\max} s_1\rangle = \langle \Box^E_{\max} s, s_1\rangle, \quad s, s_1 \in \mathrm{Dom}(\Box^E_{\max}) \tag{3.3.7}
$$

which means that $\Box^E_{\max} \subset (\Box^E_{\max})^*$. But $\Box^E_{\max}$ is the maximal extension of $\Box^E$ so that $\Box^E_{\max} = (\Box^E_{\max})^*$. $\square$

Theorem 3.3.5. *Let (L, h^L) be a holomorphic Hermitian line bundle on a complex Hermitian manifold (X, Θ). If (X, Θ) is complete and there exist $K \Subset X$ and $C_0 > 0$ such that $\sqrt{-1}R^L \geqslant C_0\,\Theta$ on $X \smallsetminus K$,*

(i) *then, for $p \to \infty$,*

$$
\dim H^{n,0}_{(2)}(X, L^p) \geqslant \frac{p^n}{n!}\int_{X(\leqslant 1)}\left(\frac{\sqrt{-1}}{2\pi}R^L\right)^n + o(p^n), \tag{3.3.8}
$$

where $H^{n,0}_{(2)}(X, L^p)$ is the space of $(n,0)$-forms with values in L^p which are L^2 with respect to any metric on X and the metric h^L on L.

(ii) *Assume moreover that the Hermitian torsion $\mathcal{T} = [i(\Theta), \partial\Theta]$ is bounded and $R^{\det} = R^{K_X^*}$ is bounded from below with respect to Θ. Then, for $p \to \infty$,*

$$
\dim H^0_{(2)}(X, L^p) \geqslant \frac{p^n}{n!}\int_{X(\leqslant 1)}\left(\frac{\sqrt{-1}}{2\pi}R^L\right)^n + o(p^n), \tag{3.3.9}
$$

where $H^0_{(2)}(X, L^p)$ is the space of holomorphic sections in L^p which are L^2 with respect to the Hermitian form Θ on X and h^L on L.

Remark 3.3.6. We can state Theorem 3.3.5 (i) without reference to the auxiliary metric Θ, by saying that (L, h^L) is positive outside a compact set K and the curvature $\sqrt{-1}R^L$ defines a complete metric on X (by extending it to a metric over K).

Proof. (i) Let us endow X with a Hermitian form ω_0 such that $\omega_0 = \frac{\sqrt{-1}}{2\pi}R^L$ outside K, which is complete, for $\omega_0 \geqslant C_0\Theta$ on $X \smallsetminus K$. Since ω_0 is Kähler outside K, the Bochner–Kodaira–Nakano formula (1.4.44) gives

$$\langle \square^{L^p}s, s \rangle \geqslant p\langle [\sqrt{-1}R^L, i(\omega_0)]s, s \rangle, \quad \text{for } s \in \Omega_0^{n,1}(X \smallsetminus K, L^p). \qquad (3.3.10)$$

By (1.4.61) (cf. also Problem 1.10), we know that $\langle [\sqrt{-1}R^L, i(\omega_0)]s, s \rangle \geqslant a_1(x)|s|^2$, where $a_1 \leqslant \cdots \leqslant a_n$ are the eigenvalues of $\sqrt{-1}R^L$ with respect to ω_0. In our case, $a_1 = \cdots = a_n = 2\pi$ outside K. Thus for $s \in \Omega_0^{n,1}(X \smallsetminus K, L^p) = \Omega_0^{0,1}(X \smallsetminus K, L^p \otimes K_X)$,

$$\|s\|_{L^2}^2 \leqslant \frac{1}{p}\left(\|\overline{\partial}^{L^p}s\|_{L^2}^2 + \|\overline{\partial}^{L^p,*}s\|_{L^2}^2 \right). \qquad (3.3.11)$$

Let U be any open set with smooth boundary such that $K \Subset U \Subset X$. Choose $\rho \in \mathscr{C}^\infty(X)$ such that $\rho = 1$ on a neighborhood of K and $\mathrm{supp}(\rho) \subset U$. Applying (3.3.11) for $(1-\rho)s$ and using (3.2.8), we obtain the fundamental estimate (3.2.2) (with a slightly larger K) in bidegree $(0,1)$ for all $s \in \Omega_0^{0,1}(X, L^p \otimes K_X)$.

By Lemma 3.3.1, we infer that (3.2.2) holds true in bidegree $(0,1)$ with K_X as E therein. We conclude by Theorem 3.2.16 that as $p \to \infty$,

$$\dim H^{n,0}_{(2)}(X, L^p)_0 \geqslant p^n \int_{U(\leqslant 1)} \left(\frac{\sqrt{-1}}{2\pi}R^L \right)^n + o(p^n). \qquad (3.3.12)$$

Here $H^{n,0}_{(2)}(X, L^p)_0$ is the L^2-cohomology group with respect to the metric ω_0 on X. For any $(n,0)$-form s with values in L, and any metrics g^{TX}, g_1^{TX} on X, with Riemannian volume forms dv_X, $dv_{X,1}$, respectively, we have pointwise $|s|^2_{g^{TX}}dv_X = |s|^2_{g_1^{TX}}dv_{X,1}$. Thus the L^2 condition for $(n,0)$-forms does not depend on the metric on X, so

$$H^{n,0}_{(2)}(X, L^p)_0 = H^{n,0}_{(2)}(X, L^p) \qquad (3.3.13)$$

where in the latter group, the L^2 condition is with respect to an arbitrary metric on X.

(ii) Due to the bound $\sqrt{-1}R^L \geqslant C_0\Theta$ on $X \smallsetminus K$ and since $R^{K_X^*}$ is bounded from below, there exists $C_1 > 0$ such that for $s \in \Omega^{0,1}(X \smallsetminus K, L^p)$, we have pointwise on $X \smallsetminus K$,

$$\left\langle R^{L^p \otimes K_X^*}(w_j, \overline{w}_k)\overline{w}^k \wedge i_{\overline{w}_j}s, s \right\rangle \geqslant (pC_0 - C_1)|s|^2. \qquad (3.3.14)$$

As $\mathcal{T} = [i(\Theta), \partial\Theta]$ is bounded, from the Bochner–Kodaira–Nakano formula (1.4.64) and (3.3.14), there exists $C > 0$ such that for $s \in \Omega_0^{0,1}(X \smallsetminus K, L^p)$,

$$3\langle \square^{L^p}s, s \rangle \geqslant (2pC_0 - C)\|s\|_{L^2}^2. \qquad (3.3.15)$$

We can thus proceed just as in the proof of (i). $\qquad\square$

Corollary 3.3.7 (Nadel–Tsuji). *Let* (X, Θ) *be a complete Kähler manifold with* $R^{\det} \leqslant -\Theta$. *Then we have the following estimate:*

$$\dim H^0_{(2)}(X, K_X^p) \geqslant \frac{p^n}{n!} \int_X \left(\frac{\sqrt{-1}}{2\pi} R^{K_X} \right)^n + o(p^n), \qquad (3.3.16)$$

where $R^{K_X} = -R^{\det}$ *is the curvature of the canonical bundle* K_X *equipped with the metric induced from* Θ.

Proof. At first, $T = 0$ as Θ is Kähler. Since (3.3.14) holds for $L = K_X$, (3.3.15) still holds. We get (3.3.16) by the same argument in Theorem 3.3.5 (ii).   □

3.4   Siu–Demailly criterion for isolated singularities

In this section, we will prove Theorems 3.4.10, 3.4.14 (generalizations of the Siu–Demailly criterion for complex compact spaces with isolated singularities) by applying the holomorphic Morse inequalities and Theorems 3.3.5 and 3.4.9.

It is useful to consider a more general class of manifolds which include as a special case the regular locus of compact complex spaces with isolated singularities.

Definition 3.4.1. A complex manifold X is called *hyperconcave* or *hyper 1-concave* if there exists a smooth function $\varphi : X \longrightarrow] - \infty, u]$ where $u \in \mathbb{R}$, such that $X_c := \{\varphi > c\} \Subset X$ for all $c \in] - \infty, u]$ and φ is strictly plurisubharmonic outside a compact set.

Example 3.4.2. (i) Let Y be a compact complex manifold, S a complete pluripolar set. By definition, S is *complete pluripolar*, if there exists a neighborhood W of S and a plurisubharmonic function $\psi : W \longrightarrow [-\infty, \infty[$ such that $S = \psi^{-1}(-\infty)$. If ψ is strictly plurisubharmonic, then $X = Y \setminus S$ is hyperconcave. Conversely, it is shown in Theorem 6.3.9 that any hyperconcave manifold M is biholomorphic to a complement of a pluripolar set in a compact manifold.

(ii) Let X be a compact complex space with isolated singularities. Then the regular locus X_{reg} is hyperconcave. Indeed, let $\{U_\alpha\}$ be pairwise disjoint neighborhoods of the singular points $\{x_\alpha\}$ and let $\iota_\alpha : U_\alpha \hookrightarrow \mathbb{C}^{N_\alpha}$ be holomorphic embeddings. We may assume that the singular points are mapped to the origin, $\iota_\alpha(x_\alpha) = 0$. The function $z \longmapsto \log |z|^2$ is strictly plurisubharmonic on $\mathbb{C}^{N_\alpha}$. By taking its pull-back to each U_α through ι_α, we obtain a strictly plurisubharmonic function on $X_{\mathrm{reg}} \cap (\cup_\alpha U_\alpha)$ which tends to $-\infty$ at the singular points. By extending this function to X_{reg} by means of a partition of unity, we get a function as in the definition.

(iii) If X is a complete Kähler manifold of finite volume and bounded negative sectional curvature, X is hyperconcave as shown by Siu–Yau in Theorem 6.3.8.

Definition 3.4.3. We will call a connected complex manifold X *Andreotti pseudo-concave* if there exists a non-empty open set $M \Subset X$ with smooth boundary ∂M

such that the Levi form of M restricted to the analytic tangent space $T^{(1,0)}\partial M$ has at least one negative eigenvalue at each point of ∂M.

Immediate examples of Andreotti pseudoconcave manifolds are connected q-concave manifolds, $q \leqslant n - 1$. Indeed, let X be a connected q-concave manifold with associated function $\varphi : X \longrightarrow]u, v]$ as in Definition B.3.2. The definition function of $X_c = \{\varphi > c\} \Subset X$ is $c - \varphi$ and for c sufficiently close to u, the Levi form of $c - \varphi$ has at least $n - q + 1$ negative eigenvalues in a neighborhood of ∂X_c. Thus, the restriction of the Levi form on the analytic tangent space $T^{(1,0)}\partial X_c$ has at least $n - q \geqslant 1$ negative eigenvalues.

Lemma 3.4.4. *In Definition 3.4.3, for each point $x \in \overline{M}$, we can choose holomorphic coordinates (U_x, x) and a coordinate polydisc $P(x, r) \subset U_x$ centered at x such that the Silov boundary $S(P(x, r)) = \{y \in U_x : |y_i| = r\} \subset M$.*

Proof. Let ρ be the defining function of M near $x = 0$ as in Definition B.3.7. After a change of coordinates, we may assume that $\mathscr{L}_{\rho,0}(v, \overline{v}) < 0$ for $v = (1, 0, \ldots, 0)$ and $\partial\rho(0) = dz_n$. The Taylor expansion of ρ gives:

$$\rho(z) = \rho(0) + 2\operatorname{Re}\left(\partial\rho(0) \cdot z + Q_\rho(z, z)\right) + \mathscr{L}_{\rho,0}(z, \overline{z}) + o(|z|^2), \qquad (3.4.1)$$

where $Q_\rho(z, z) = \sum_{j,k} \frac{\partial^2 \rho}{\partial z_j \partial z_k}(0) z_j z_k$. Therefore

$$\rho(z_1, 0, \ldots, 0, z_n) = 2\operatorname{Re}(h(z)) + \frac{\partial^2 \rho}{\partial z_1 \partial \overline{z}_1}(0)|z_1|^2$$
$$+ \mathscr{O}(|z_1||z_n|) + \mathscr{O}(|z_n|^2) + o(|z_1|^2 + |z_n|^2), \qquad (3.4.2)$$

where $h(z) = z_n + Q_\rho(z, z)$ with $z = (z_1, 0, \ldots, 0, z_n)$. Since $\frac{\partial h}{\partial z_n}(0) \neq 0$, the set $U = \{z \in \mathbb{C}^n : h(z) = 0, z_2 = \cdots = z_{n-1} = 0\}$ is a complex submanifold of dimension one in the neighborhood of 0. Let $B(0, \varepsilon)$ be the ball with center 0 and radius ε. Since $\frac{\partial^2 \rho}{\partial z_1 \partial \overline{z}_1}(0) < 0$, we deduce that there exists $\varepsilon > 0$ such that $U \cap B(0, \varepsilon) \smallsetminus \{0\} \subset M$. In the holomorphic coordinates $w_1(z) = h(z)$, $w_2 = z_2, \ldots, w_n = z_n$, U is a complex plane through 0. Moreover, for $\eta > 0$ small enough $\{w_1 = \ldots = w_{n-1} = 0, |w_n| < \eta\} \cap B(0, \varepsilon) \smallsetminus \{0\} \subset M$. The lemma follows now easily by a continuity argument. $\square$

Theorem 3.4.5. *Let X be an Andreotti-pseudoconcave manifold. Then for any holomorphic line bundle L on X, we have $\dim H^0(X, L) < \infty$. Moreover, there exists $C > 0$ such that*

$$\dim H^0(X, L^p) \leqslant C p^{\varrho_p}, \quad \textit{for } p \geqslant 1, \qquad (3.4.3)$$

where ϱ_p is the maximal rank of the Kodaira map $\Phi_p : X \smallsetminus \mathrm{Bl}_{|H^0(X,L^p)|} \to \mathbb{P}(H^0(X, L^p)^*)$. The field of meromorphic functions $\mathscr{M}_X(X)$ is an algebraic field of transcendence degree $a(X) \leqslant \dim X$.

Proof. First we choose an open set $M \Subset X$ as in Definition 3.4.3. As in the proof of Lemma 2.2.6, the set of points where the Kodaira map Φ_p has rank less than ϱ_p

is a proper analytic set of X, so $\{x \in X, \mathrm{rk}_x\, \Phi_p = \varrho_p \text{ for any } p \geqslant 1\}$ is dense in X. By Lemma 3.4.4, there exist $x_1, \ldots, x_m \in M$ as in Lemma 2.2.1 such that Φ_p has rank ϱ_p at each x_j for any $p \geqslant 1$, $\overline{M} \subset \cup_{i=1}^m P(x_i, r_i e^{-1})$, L is trivial on coordinate polydiscs $P(x_i, 2r_i)$ and the Silov boundary $S(P(x_i, r_i)) \subset M$, for $i = 1, \ldots, m$.

Then the proof goes through as in the compact case by observing that there exists $q \in \{1, \ldots, m\}$ such that for some $y \in S(P(x_q, r_q))$ with $|s_q(y)| = \|s\|$. Thus $y \in M$ and we can find $j \neq q$ such that $y \in P(x_j, r_j e^{-1})$. Now the same argument as in the proof of Lemma 2.2.6 gives (3.4.3). The proof of the last part is completely analogous to that of Theorem 2.2.11. □

Remark 3.4.6. Under the hypothesis of Theorem 3.4.5 we observe that Siegel's lemma 2.2.6 holds for the adjoint bundles, that is there exists $C > 0$ such that for and $p \in \mathbb{N}^*$,

$$\dim H^0(X, L^p \otimes K_X) \leqslant C p^{\varrho_p}, \tag{3.4.4}$$

where ϱ_p is the maximal rank of the Kodaira map Φ_p associated to $H^0(X, L^p \otimes K_X)$.

Theorem 3.4.5 allows us to extend the notion of Moishezon manifold for the case of Andreotti-pseudoconcave manifolds. Thus, an Andreotti-pseudoconcave manifold is called *Moishezon* if $a(X) = \dim X$.

Theorem 3.4.7. *Let X be an Andreotti-pseudoconcave manifold and L be a holomorphic line bundle over X. If there exists $p \in \mathbb{N}^*$ such that $\mathrm{rk}\, \Phi_p = \dim X$, where Φ_p is the Kodaira map associated to $H^0(X, L^p)$ or $H^0(X, L^p \otimes K_X)$, then X is Moishezon.*

Proof. By using Theorem 3.4.5, the proof is the same as that of Theorem 2.2.15. □

Theorem 3.4.8 (Levi's removable singularity theorem). *Let X be a reduced and irreducible complex space with an analytic subset $A \subset X$ of codimension at least 2 at every point. Then every meromorphic function $f \in \mathcal{M}_X(X \smallsetminus A)$ has a unique extension $\tilde{f} \in \mathcal{M}_X(X)$.*

After this study of the function theory on pseudoconcave manifolds, we return to the holomorphic Morse inequalities. As in Section 1.7, we will use the notation $X(q), X(\leqslant q), X(\geqslant q)$; when we need to make precise the metric h^L as in Definition 2.3.17, (2.3.47), we denote them by $X(q, c_1(L, h^L))$, $X(\leqslant q, c_1(L, h^L))$, $X(\geqslant q, c_1(L, h^L))$, respectively.

Theorem 3.4.9. *Let X be a hyperconcave manifold of dimension n carrying a holomorphic line bundle (L, h^L) which is semi-positive outside a compact set. Then, for $p \to \infty$*

$$\dim H^{n,0}_{(2)}(X, L^p) \geqslant \frac{p^n}{n!} \int_{X(\leqslant 1)} \left(\frac{\sqrt{-1}}{2\pi} R^L \right)^n + o(p^n), \tag{3.4.5}$$

where the L^2 condition is with respect to h^L on L and any metric on X.

Proof. Let us consider a proper smooth function $\varphi : X \longrightarrow]-\infty, 0[$ which is strictly plurisubharmonic outside a compact set $K \Subset X$. The fact that φ goes to $-\infty$ to the ideal boundary of X allows us to construct a complete metric g^{TX} on X. Let χ be the smooth function on X defined by

$$\chi = -\log(-\varphi). \tag{3.4.6}$$

Note that

$$\partial\bar{\partial}\chi = \frac{\partial\bar{\partial}\varphi}{-\varphi} + \frac{\partial\varphi \wedge \bar{\partial}\varphi}{\varphi^2}, \quad \text{and} \quad \frac{\partial\varphi \wedge \bar{\partial}\varphi}{\varphi^2} = \partial\chi \wedge \bar{\partial}\chi. \tag{3.4.7}$$

As $\sqrt{-1}\frac{\partial\bar{\partial}\varphi}{-\varphi}$ is positive and $\sqrt{-1}\partial\chi \wedge \bar{\partial}\chi$ is semi-positive on $X \smallsetminus K$, thus $\sqrt{-1}\partial\bar{\partial}\chi$ is positive on $X \smallsetminus K$. We can now patch $\partial\bar{\partial}\chi$ and an arbitrary positive $(1,1)$-form on X by using a smooth partition of unity, to get a positive $(1,1)$-form Θ on X such that

$$\Theta = \sqrt{-1}\partial\bar{\partial}\chi = -\sqrt{-1}\partial\bar{\partial}\log(-\varphi), \quad \text{on } X \smallsetminus K_1 \text{ with } K \Subset K_1 \Subset X. \tag{3.4.8}$$

The metric associated to Θ is the desired g^{TX}. In fact, from (3.4.7) and (3.4.8), we get

$$|d\chi|_{g^{TX}} \leqslant 2 \quad \text{on } X \smallsetminus K_1. \tag{3.4.9}$$

Since $\chi : X \longrightarrow \mathbb{R}$ is proper, (3.4.9) ensures that g^{TX} is complete. Indeed, (3.4.9) entails that χ is Lipschitz with respect to the geodesic distance induced by g^{TX}, so any geodesic ball must be relatively compact.

Note that g^{TX} is obviously Kähler on $X \smallsetminus K_1$. Let us assume $\sqrt{-1}R^L \geqslant 0$ on $X \smallsetminus K_1$ (we stretch K_1 if necessary). We equip L with the metric $h_\varepsilon^L = h^L \exp(-\varepsilon\chi)$ and the curvature relative to the new metric satisfies $c_1(L, h_\varepsilon^L) \geqslant \varepsilon\Theta$ on $X \smallsetminus K_1$. (L, h_ε^L) therefore in the conditions of Theorem 3.3.5. Since $h_\varepsilon^L \geqslant Ch^L$ for some $C > 0$, there is an injective morphism

$$H^{n,0}_{(2)}(X, L^p, \Theta, h_\varepsilon^L) \longrightarrow H^{n,0}_{(2)}(X, L^p, \Theta, h^L). \tag{3.4.10}$$

By Theorem 3.3.5 for the space $H^{n,0}_{(2)}(X, L^p, \Theta, h_\varepsilon^L)$ and (3.4.10), for any $K_1 \Subset U \Subset X$, as $c_1(L, h_\varepsilon^L))$ is positive on $X \smallsetminus U$, we get

$$\liminf_{p\to\infty} p^{-n} \dim H^{n,0}_{(2)}(X, L^p, \Theta, h^L) \geqslant \frac{1}{n!} \int_{U(\leqslant 1, c_1(L, h_\varepsilon^L))} c_1(L, h_\varepsilon^L)^n. \tag{3.4.11}$$

We let now $\varepsilon \to 0$ in (3.4.11); since h_ε^L converges uniformly together with its derivatives to h^L on compact sets, we see that we can replace h_ε^L with h^L in the right-hand side of (3.4.11). Now, $X(\leqslant 1) = X(0) \cup X(1)$. By hypothesis $X(1) \subset K$ and on $X(0)$ the integrand is positive. Hence we can let U exhaust X to get (3.4.5). $\qquad\square$

Theorem 3.4.9 implies the first main result of this section. Let us remark that Definition 2.2.8 and Theorem 2.2.11 carry over compact irreducible spaces. Such a space is called *Moishezon* if its algebraic and complex dimensions are equal.

Theorem 3.4.10. *Let X be a compact irreducible complex space of dimension $n \geqslant 2$ with at most isolated singularities, and let (L, h^L) be a holomorphic Hermitian line bundle on X_{reg} which is semi-positive in a deleted neighborhood of the singular locus X_{sing} and satisfies Demailly's condition:*

$$\int_{X_{\text{reg}}(\leqslant 1)} c_1(L, h^L)^n > 0. \qquad (3.4.12)$$

Then X is Moishezon.

Proof. By Example 3.4.2 (ii), X_{reg} here is a hyperconcave manifold. We deduce from Theorem 3.4.9 and (3.4.12) that $\dim H^0(X_{\text{reg}}, L^p \otimes K_X) \geqslant Cp^n$ for some $C > 0$ and p sufficiently large. By Theorem 3.4.5 and Remark 3.4.6, we deduce that the rank of the Kodaira map of $H^0(X_{\text{reg}}, L^p \otimes K_X)$ is maximal. Theorem 3.4.7 entails that there exists $\dim X$ independent meromorphic functions on X_{reg}.

By Theorem 3.4.8, we conclude that these functions extend to $\dim X$ independent meromorphic functions on X. Thus X is Moishezon.   □

Let us define the "adjoint" volume of a line bundle L over a complex manifold M of dimension n by

$$\text{vol}^*(L) = \limsup_{p \to \infty} n! \, p^{-n} \dim H^0(M, L^p \otimes K_M). \qquad (3.4.13)$$

Corollary 3.4.11. *Let L be a holomorphic line bundle over X_{reg}, where X is a compact complex space with only isolated singularities, $\dim X = n$.*

(i) *If L is semi-positive outside a compact set, then*

$$0 \leqslant \int_{X_{\text{reg}}(0)} c_1(L, h^L)^n \leqslant \text{vol}^*(L) - \int_{X_{\text{reg}}(1)} c_1(L, h^L)^n < +\infty. \qquad (3.4.14)$$

(ii) *If $\psi : X_{\text{reg}} \longrightarrow \mathbb{R}$ is a smooth function which is plurisubharmonic outside a compact set, then*

$$\int_{X_{\text{reg}}(0)} (\sqrt{-1}\partial\overline{\partial}\psi)^n \leqslant -\int_{X_{\text{reg}}(1)} (\sqrt{-1}\partial\overline{\partial}\psi)^n < +\infty, \qquad (3.4.15)$$

where $X_{\text{reg}}(0)$ is the open set where ψ is strictly plurisubharmonic.

Proof. By Example 3.4.2 (ii) and Remark 3.4.6, $\text{vol}^*(L) < +\infty$. Under the condition (i), since $X_{\text{reg}}(1)$ is relatively compact by the hypothesis on the semi-positivity of L, we have $-\infty < \int_{X_{\text{reg}}(1)} c_1(L, h^L)^n \leqslant 0$. Now (3.4.14) is from (3.4.5).

To prove (ii), we apply (i) to the trivial bundle L endowed with the metric $\exp(-\psi)$ and we use the obvious fact that $\text{vol}^*(L) = 0$, if L is trivial.   □

We recall now the notion of Hermitian form (metric) on a singular space.

Definition 3.4.12. Let us consider a covering $\{U_\alpha\}$ of a complex space X and embeddings $\iota_\alpha : U_\alpha \hookrightarrow \mathbb{C}^{N_\alpha}$. A *Hermitian form* on X is a Hermitian form Θ (i.e., Θ is a positive $(1,1)$-form) on X_{reg} which on every open set U_α as above is the pullback of a Hermitian form on the ambient space $\mathbb{C}^{N_\alpha}$, $\Theta = \iota_\alpha^* \Theta_\alpha$. The Hermitian form is called *distinguished*, if in the neighborhood of the singular points, Θ_α is the Euclidean Kähler form.

A Hermitian form on a singular space is constructed as usual by a partition of unity argument. If the singularities are isolated, we can assume that the Hermitian form is distinguished.

Definition 3.4.13. Let L be a holomorphic line bundle on a complex space X. Assume that $L|_{U_\alpha}$ is the inverse image by ι_α of the trivial line bundle $\mathbb{C}_\alpha$ on $\mathbb{C}^{N_\alpha}$. Consider Hermitian metrics $h_\alpha = e^{-2\varphi_\alpha}$ on $\mathbb{C}_\alpha$ such that $\iota_\alpha^* h_\alpha = \iota_\beta^* h_\beta$ on $U_\alpha \cap U_\beta \cap X_{\text{reg}}$. The system $h^L = \{\iota_\alpha^* h_\alpha\}$ is called a *Hermitian metric* on L. It clearly induces a Hermitian metric on $L|_{X_{\text{reg}}}$. We shall allow our metrics to be singular at the singular points, that is, $\varphi_\alpha \in L^1_{\text{loc}}(\mathbb{C}^{N_\alpha})$ and φ_α is smooth outside $\iota_\alpha(X_{\text{sing}})$. The curvature current $\sqrt{-1}R^L$ is given in U_α by $\iota_\alpha^*(2\sqrt{-1}\partial\bar\partial\varphi_\alpha)$, which on X_{reg} agrees with the curvature of the induced metric.

The following Theorem 3.4.14 is the second main result of this section. Theorems 3.4.10 and 3.4.14 show that Demailly's criterion generalizes to singular spaces with at most isolated singularities under mild growth conditions of the curvature near the singular set.

Theorem 3.4.14. *Let X be a compact irreducible complex space of dimension $n \geqslant 2$ and with isolated singularities. Assume that a holomorphic line bundle L is defined over all X, the Hermitian metric h^L on L may be singular at X_{sing}, but the curvature current R^L is dominated by the Euclidean metric near X_{sing} (cf. (3.4.18)) and moreover condition (3.4.12) is fulfilled on X_{reg}. Then X is Moishezon.*

Proof. In order to perform analysis on X_{reg}, we introduce first a good exhaustion function and a complete metric. Let us consider a coherent ideal $\mathcal{I} \subset \mathcal{O}_X$ with $\text{supp}(\mathcal{O}_X/\mathcal{I}) = X_{\text{sing}}$ such that the blow-up $\pi : \widetilde{X} \to X$ of the ideal $\mathcal{I}$ is a resolution of singularities of X. Let E be the exceptional divisor and E_i its smooth irreducible components such that $E = \sum_i c_i E_i$, where $c_i \in \mathbb{N}^*$. Then $\sum_i E_i$ has normal crossings and $\pi^{-1}(\mathcal{I})$ equals the invertible sheaf $\mathcal{O}_{\widetilde{X}}(F)$, where $F = \mathcal{O}_{\widetilde{X}}(-E)$.

Lemma 3.4.15. *There exists a Hermitian metric h^F on F which has positive curvature in a neighborhood $\widetilde{U}$ of E.*

Proof. Let us consider a small neighborhood U_y of $y \in X_{\text{sing}}$ which does not contain any other singular point and an embedding $\iota : U_y \longrightarrow \mathbb{C}^N$ as an analytic subset of a ball $B_\varepsilon^N = \{z \in \mathbb{C}^N : \sum_j |z_j|^2 < \varepsilon^2\}$. By definition of the blow-up, there exist sections $s_0, \ldots, s_k \in H^0(\widetilde{U}, F)$ which define a map $\Phi : \widetilde{U} \longrightarrow \mathbb{C}\mathbb{P}^k$, $\Phi(x) = [s_0(x), \ldots, s_k(x)]$, such that $\pi \times \Phi : \widetilde{U} \longrightarrow B_\varepsilon^N \times \mathbb{C}\mathbb{P}^k$ is an embedding. Moreover, $F = \Phi^* \mathscr{O}_{\mathbb{C}\mathbb{P}^k}(1) = (\pi \times \Phi)^* \sigma^* \mathscr{O}_{\mathbb{C}\mathbb{P}^k}(1)$, with $\sigma : B_\varepsilon^N \times \mathbb{C}\mathbb{P}^k \longrightarrow \mathbb{C}\mathbb{P}^k$ is the projection. We can endow $\sigma^* \mathscr{O}_{\mathbb{C}\mathbb{P}^k}(1)$ with a metric with positive curvature on $B_\varepsilon^N \times \mathbb{C}\mathbb{P}^k$ and pulling back this metric by $\pi \times \Phi$, we obtain the desired h^F. $\square$

Let us consider a canonical section s of $F = \mathscr{O}_{\widetilde{X}}(-E)$, and denote by $|s|_{h^F}$ the pointwise norm of s. By the Poincaré–Lelong formula (2.3.6),

$$\frac{\sqrt{-1}}{2\pi} \partial\overline{\partial} \log |s|_{h^F}^2 = [\mathrm{Div}(s)] - c_1(F, h^F). \tag{3.4.16}$$

Hence the function $\log |s|_{h^F}^2$ is strictly plurisubharmonic on $\widetilde{U} \smallsetminus E$ and converges to $-\infty$ on E. By using a smooth function on $\widetilde{X}$ with compact support in $\widetilde{U}$ which equals one near E, we construct a smooth function χ on $\widetilde{X} \smallsetminus E \simeq X_{\mathrm{reg}}$ such that $\chi = -\log(-\log |s|_{h^F}^2)$ on $\widetilde{U} \smallsetminus E$.

Let Θ be a distinguished Hermitian form on X, in particular Θ is Kähler near X_{sing}. We consider then the positive $(1,1)$-form on X_{reg},

$$\omega_0 = A_1 \Theta + \frac{\sqrt{-1}}{2\pi} \partial\overline{\partial}\chi, \tag{3.4.17}$$

where $A_1 > 0$ is chosen sufficiently large (to ensure that ω_0 is positive away from the open set where $\partial\overline{\partial}\chi$ is positive definite, cf. (3.4.7)). The metric g_0^{TX} associated to ω_0 is complete by the same argument as in the proof of Theorem 3.4.9 (see (3.4.9)).

Lemma 3.4.16. *The metric ω_0 has finite volume.*

Proof. Let $\pi : \widetilde{X} \to X$ as above. Observe first that the integral of $(\sqrt{-1}\partial\overline{\partial}\chi)^n$ on $\widetilde{U}$ is finite by Corollary 3.4.11 (ii). Since $\sqrt{-1}\partial\overline{\partial}\chi$ is a strictly positive current and $\pi^*(\Theta)$ is smooth on $\widetilde{U}$, there exists $C > 0$ such that $\pi^*(\Theta) \leqslant \sqrt{-1} C \partial\overline{\partial}\chi$ on $\widetilde{U} \smallsetminus E$. Therefore, ω_0^n is dominated on $\widetilde{U} \smallsetminus E$ by a multiple of $(\sqrt{-1}\partial\overline{\partial}\chi)^n$. The integral of ω_0^n on X_{reg} must be finite. $\square$

We shall suppose in the sequel that the curvature current of the Hermitian metric h^L on L is dominated by the Euclidean Kähler form, i.e., for φ_α in Definition 3.4.13, there exists $C > 0$ with

$$-C\omega_E \leqslant \sqrt{-1}\partial\overline{\partial}\varphi_\alpha \leqslant C\omega_E, \quad \omega_E = \frac{\sqrt{-1}}{2}\textstyle\sum_j dz_j \wedge d\overline{z}_j. \tag{3.4.18}$$

We assume that U is a small enough neighborhood of the singular set so that on U, there are well-defined potentials ρ, ρ_1 for $2\pi\Theta$, $\sqrt{-1}R^L$ from ambient spaces, i.e.,

$$\Theta = \frac{\sqrt{-1}}{2\pi}\partial\overline{\partial}\rho, \quad \sqrt{-1}R^L = \sqrt{-1}\partial\overline{\partial}\rho_1. \tag{3.4.19}$$

By suitably cutting-off, we may define a function $\psi \in \mathscr{C}^{\infty}(X_{\text{reg}})$ such that near X_{sing},

$$\psi = \chi - \rho_1 + A_1\,\rho. \tag{3.4.20}$$

Remark that, since $\sqrt{-1}R^L$ is bounded above by a continuous $(1,1)$-form near X_{sing}, the potential $-\rho_1$ is bounded above near the singular set. This holds true for ρ too (it is smooth) so that ψ tends to $-\infty$ at the singular set X_{sing}. Let us consider a smooth function $\gamma : \mathbb{R} \longrightarrow \mathbb{R}$ such that

$$\gamma(u) = \begin{cases} 0 & \text{if } u \geq 0, \\ u & \text{if } u \leq -1, \end{cases}$$

and the functions $\gamma_\nu : \mathbb{R} \longrightarrow \mathbb{R}$ given by $\gamma_\nu(u) = \gamma(u + \nu)$ for $\nu \in \mathbb{N}$.

The curvature $R^{(L,h_\nu^L)}$ of the metric $h_\nu^L = h^L \exp\left(-\gamma_\nu(\psi)\right)$ is

$$R^{(L,h_\nu^L)} = R^{(L,h^L)} + \gamma_\nu'(\psi)\partial\overline{\partial}\psi + \gamma_\nu''(\psi)\partial\psi \wedge \overline{\partial}\psi. \tag{3.4.21}$$

Since ψ goes to $-\infty$ when we approach the singular set, we may choose $\nu_0 \in \mathbb{N}$ such that for $\nu \geq \nu_0$ we have $\{\psi \leq -\nu - 1\} \subset U$. On the set $\{\psi \leq -\nu - 1\}$, we have $\gamma_\nu(\psi) = \psi + \nu$, so that $\gamma_\nu'(\psi) = 1$ and $\gamma_\nu''(\psi) = 0$, by (3.4.17) and (3.4.19),

$$c_1(L, h_\nu^L) = \frac{\sqrt{-1}}{2\pi}R^{(L,h_\nu^L)} = c_1(L, h^L) + \frac{\sqrt{-1}}{2\pi}\partial\overline{\partial}\psi = \omega_0. \tag{3.4.22}$$

Set

$$U_\nu' = \{\psi \geq -\nu\}, \ U_\nu'' = \{-\nu - 2 \leq \psi \leq -\nu\}, \ U_\nu = U_\nu' \cup U_\nu''. \tag{3.4.23}$$

Then $U_\nu = \{\psi \geq -\nu - 2\}$ is a compact set. By Theorem 3.3.5 and the same argument in (3.4.11), as $c_1(L, h_\nu^L)$ is positive on $X_{\text{reg}} \setminus U_\nu$, we get for $p \to \infty$,

$$\dim H^{n,0}(X_{\text{reg}}, L^p, h^L) \geq \frac{p^n}{n!} \int_{U_\nu(\leq 1, c_1(L,h_\nu^L))} c_1(L, h_\nu^L)^n + o(p^n). \tag{3.4.24}$$

On U_ν', we have $\gamma_\nu(\psi) = 0$ and $c_1(L, h_\nu^L) = c_1(L, h^L)$. We infer that

$$\int_{U_\nu'(\leq 1, c_1(L,h_\nu^L))} c_1(L, h_\nu^L)^n = \int_{X_{\text{reg}}(\leq 1, c_1(L,h^L))} 1_{U_\nu'}\, c_1(L, h^L)^n$$

$$\longrightarrow \int_{X_{\text{reg}}(\leq 1, c_1(L,h^L))} c_1(L, h^L)^n \quad \text{as } \nu \to \infty. \tag{3.4.25}$$

Thus it suffices to show that

$$\int_{U_\nu''(\leq 1, c_1(L,h_\nu^L))} c_1(L, h_\nu^L)^n \longrightarrow 0 \quad \text{as } \nu \to \infty. \tag{3.4.26}$$

As $\gamma'_\nu, \gamma''_\nu$ are bounded, from (3.4.17), (3.4.19) and (3.4.21), we know that $c_1(L, h_\nu^L)$ is dominated by ω_0 on U''_ν. Since $\mathrm{vol}\,(U''_\nu)$, the volume of U''_ν with respect to ω_0, tends to 0 as $\nu \to \infty$, we get (3.4.26).

Hence from (3.4.12), (3.4.24), (3.4.25) and (3.4.26), there exists $C > 0$ such that for p large enough

$$\dim H^0(X_{\mathrm{reg}}, L^p \otimes K_X) \geqslant Cp^n, \qquad (3.4.27)$$

so that X_{reg} has n independent meromorphic functions which can be extended to X by Theorem 3.4.8. The proof of Theorem 3.4.14 is finished. $\qquad\square$

3.5   Morse inequalities for q-convex manifolds

In this section, we establish the holomorphic Morse inequalities for the q-convex and weakly 1-complete manifolds.

We start by defining the Kodaira Laplacian with $\overline{\partial}$-Neumann boundary conditions. Let M be a relatively compact smooth domain in a complex manifold X of dimension n and let (E, h^E) be a holomorphic Hermitian vector bundle on X.

Let $\overline{\partial}^E : \Omega^{0,\bullet}(X, E) \to \Omega^{0,\bullet+1}(X, E)$ be the Dolbeault operator; we denote by $\overline{\partial}^{E,*}$ its formal adjoint. Let $\overline{\partial}^E : \mathrm{Dom}(\overline{\partial}^E) \subset L^2_{0,\bullet}(M, E) \to L^2_{0,\bullet+1}(M, E)$ be its maximal extension on $L^2_{0,\bullet}(M, E)$ (cf. Lemma 3.1.1). Let $\overline{\partial}^{E,*}_H$ be the Hilbert space adjoint of $\overline{\partial}^E$ on M. We introduced the space $B^{0,q}(M, E)$ in (1.4.70). By Proposition 1.4.19, we have

$$\langle \overline{\partial}^E s_1, s_2 \rangle = \langle s_1, \overline{\partial}^{E,*} s_2 \rangle, \quad \text{for } s_1 \in \Omega^{0,q}(\overline{M}, E), s_2 \in B^{0,q+1}(M, E). \quad (3.5.1)$$

We consider the operator

$$\mathrm{Dom}(\square^E) := \left\{ s \in B^{0,q}(M, E) : \overline{\partial}^E s \in B^{0,q+1}(M, E) \right\},$$
$$\square^E s = \overline{\partial}^E \overline{\partial}^{E,*} s + \overline{\partial}^{E,*} \overline{\partial}^E s, \quad \text{for } s \in \mathrm{Dom}(\square^E), \qquad (3.5.2)$$

which by (3.5.1) is positive.

Let $e_{\mathfrak{n}}$ be the inward pointing unit normal at ∂M. We decompose $e_{\mathfrak{n}}$ as $e_{\mathfrak{n}} = e_{\mathfrak{n}}^{(1,0)} + e_{\mathfrak{n}}^{(0,1)} \in T^{(1,0)}X \oplus T^{(0,1)}X$. Then the boundary conditions of $\mathrm{Dom}(\square^E)$ in (3.5.2) are called $\overline{\partial}$-Neumann boundary conditions and are given by

$$\mathrm{Dom}(\square^E) = \left\{ s \in \Omega^{0,\bullet}(\overline{M}, E); \; i_{e_{\mathfrak{n}}^{(0,1)}} s = i_{e_{\mathfrak{n}}^{(0,1)}} \overline{\partial}^E s = 0 \quad \text{on } \partial M \right\}. \qquad (3.5.3)$$

An extension of the associated quadratic form Q is

$$\mathrm{Dom}(Q) := B^{0,q}(M, E), \; Q(s_1, s_2) := \langle \overline{\partial}^E s_1, \overline{\partial}^E s_2 \rangle + \langle \overline{\partial}^{E,*} s_1, \overline{\partial}^{E,*} s_2 \rangle. \qquad (3.5.4)$$

It is easy to see that Q is closable and its closure is the form $\overline{Q}$ given by (cf. (C.1.7))

$$\mathrm{Dom}(\overline{Q}) = \{s \in L^2_{0,\bullet}(M, E) : \text{there exists } \{s_k\} \subset B^{0,\bullet}(M, E), \text{with}$$

$$\lim_{k \to \infty} s_k = s \text{ and } \{\|\overline{\partial}^E s_k\|_{L^2}\}, \{\|\overline{\partial}^{E,*} s_k\|_{L^2}\} \text{ are Cauchy sequences}\}, \quad (3.5.5)$$

$$\overline{Q}(s, s) = \lim_{k \to \infty} \left(\|\overline{\partial}^E s_k\|^2_{L^2} + \|\overline{\partial}^{E,*} s_k\|^2_{L^2}\right), \quad \text{for } s \in \mathrm{Dom}(\overline{Q}).$$

By Proposition C.1.5, the self-adjoint operator associated to $\overline{Q}$ is the Friedrichs extension of $\square^E$ and is called the *Kodaira Laplacian with $\overline{\partial}$-Neumann boundary conditions*. We still denote this operator by $\square^E$. The $\overline{\partial}$-Neumann problem for a domain M consists in proving Theorem 3.1.8 for the Kodaira Laplacian with $\overline{\partial}$-Neumann boundary conditions and establishing regularity results up to the boundary ∂M for the Green operator.

We have an analogue of the Andreotti–Vesentini lemma 3.3.1.

Lemma 3.5.1. $\Omega^{0,\bullet}(\overline{M}, E)$ *is dense in* $\mathrm{Dom}(\overline{\partial}^E)$ *in the graph-norm of* $\overline{\partial}^E$, *and* $B^{0,q}(M, E)$ *is dense in* $\mathrm{Dom}(\overline{\partial}^{E,*}_H)$ *and in* $\mathrm{Dom}(\overline{\partial}^E) \cap \mathrm{Dom}(\overline{\partial}^{E,*}_H)$ *in the graph-norms of* $\overline{\partial}^{E,*}$ *and* $\overline{\partial}^E + \overline{\partial}^{E,*}$, *respectively.*

The proof is again based on the Friedrichs lemma 3.1.3, but a more delicate convolution process in the tangential direction to ∂M is required.

Proposition 3.5.2. *The Kodaira Laplacian with $\overline{\partial}$-Neumann conditions on M coincides with the Gaffney extension* (3.1.4) *of the Kodaira Laplacian.*

Proof. By Proposition C.1.4, Definition C.1.7 and Proposition 3.1.2, it suffices to show that the quadratic forms (3.5.5) and (3.1.5) are the same. But this results immediately from Lemma 3.5.1. $\square$

From now until Theorem 3.5.9, we suppose that X is a q-convex manifold of dimension n and let $\varrho : X \longrightarrow \mathbb{R}$ be an exhaustion function which is q-convex outside a compact set $K \subset X$ (cf. Definition B.3.2). Let (E, h^E) be a holomorphic Hermitian vector bundle on X.

Let us consider a smooth sub-level set $X_c = \{\varrho < c\}$ such that $K \subset X_c$. We fix $u < c < v$ such that $K \subset X_u$. We choose now a convenient metric on X.

Lemma 3.5.3. *For any $C_1 > 0$ there exists a metric g^{TX} (with Hermitian form Θ) on X such that for any $j \geqslant q$, and any Hermitian vector bundle (E, h^E) on X,*

$$\langle (\partial \overline{\partial} \varrho)(w_l, \overline{w}_k) \overline{w}^k \wedge i_{\overline{w}_l} s, s \rangle \geqslant C_1 |s|^2, \quad s \in \Omega^{0,j}_0(X_v \smallsetminus \overline{X}_u, E). \quad (3.5.6)$$

Proof. Fix $x \in X_v \smallsetminus \overline{X}_u$. Consider local coordinates $(U, z_1, \ldots, z_n)$ centered at x, such that $\partial \overline{\partial} \varrho$ is positive definite on the subbundle of $TX|_U$ generated by $\frac{\partial}{\partial z_q}, \ldots, \frac{\partial}{\partial z_n}$. Let $C > 0$ be given. For $\varepsilon > 0$, consider the metric $g^{TX} = \sum_{i=1}^n \varepsilon_i \, dz_i \otimes d\overline{z}_i$, where $\varepsilon_i = \varepsilon^{-1}$ for $i < q$ and $\varepsilon_i = \varepsilon$ for $i \geqslant q$. Let $d_1 \leqslant \cdots \leqslant d_n$

be the eigenvalues of $\partial\overline{\partial}\varrho$ with respect to g^{TX}. Then for $\varepsilon > 0$ sufficiently small, we have $d_1 \geqslant -C^{-1}$ and $d_q \geqslant C$. For any $s \in \Omega_0^{0,j}(U, E)$, we obtain as in (1.5.19),

$$\langle(\partial\overline{\partial}\varrho)(w_l, \overline{w}_k)\overline{w}^k \wedge i_{\overline{w}_l}s, s\rangle \geqslant (d_1 + \cdots + d_j)|s|^2$$
$$\geqslant ((j - q + 1)C - (q - 1)C^{-1})|s|^2 . \tag{3.5.7}$$

By choosing $C > 0$ large enough, we obtain (3.5.6) for $s \in \Omega_0^{0,j}(U, E)$ and $j \geqslant q$. Patching metrics as one constructed above with the help of a partition of unity, we conclude the proof of Lemma 3.5.3. $\qquad\square$

From now on, we use the metric g^{TX} on X defined in Lemma 3.5.3. For a convex increasing function $\chi \in \mathscr{C}^\infty(\mathbb{R})$, we set

$$h_\chi^E = h^E e^{-\chi(\varrho)}, \quad E_\chi = (E, h_\chi^E). \tag{3.5.8}$$

Lemma 3.5.4. *For any $C_2 > 0$, there exists a constant $C_3 > 0$ such that for $\chi'(\varrho) \geqslant C_3$ on $X_v \smallsetminus \overline{X}_u$ and for any $s \in B^{0,j}(X_c, E)$, with $\mathrm{supp}(s) \subset X_v \smallsetminus \overline{X}_u$, $j \geqslant q$, we have*

$$\|s\|_{L^2}^2 \leqslant C_2\big(\|\overline{\partial}^E s\|_{L^2}^2 + \|\overline{\partial}^{E*} s\|_{L^2}^2\big), \tag{3.5.9}$$

where the L^2 norm is taken with respect to g^{TX} in Lemma 3.5.3 and h_χ^E on X_c.

Proof. To simplify the notation, set $M := X_c$. From (3.5.8),

$$R^{E_\chi} = R^E + \chi'(\varrho)\partial\overline{\partial}\varrho + \chi''(\varrho)\partial\varrho \wedge \overline{\partial}\varrho. \tag{3.5.10}$$

$\sqrt{-1}\chi''(\varrho)\partial\varrho \wedge \overline{\partial}\varrho$ is positive semi-definite. By Lemma 3.5.3 and (3.5.10), there exists $C_4, C_5 > 0$ such that for $s \in \Omega_0^{0,j}(X_v \smallsetminus \overline{X}_u, E)$, $j \geqslant q$,

$$\left\langle R^{E_\chi \otimes K_X^*}(w_l, \overline{w}_k)\overline{w}^k \wedge i_{\overline{w}_l}s, s\right\rangle \geqslant (\chi'(\varrho)C_4 - C_5)|s|^2 . \tag{3.5.11}$$

We denote simply by $|d\varrho| := |d\varrho|_{g^{T*}X}$, $\varrho_1 = (\varrho - c)/|d\varrho|$ near ∂M. Then the boundary term in (1.4.84) is $\mathscr{L}_{\varrho_1}(s, s)$. Note that

$$\partial\overline{\partial}\varrho_1 = \frac{1}{|d\varrho|}\partial\overline{\partial}\varrho + \partial\frac{1}{|d\varrho|} \wedge \overline{\partial}\varrho + \partial\varrho \wedge \overline{\partial}\frac{1}{|d\varrho|} + (\varrho - c)\partial\overline{\partial}\frac{1}{|d\varrho|}. \tag{3.5.12}$$

Thus for $s \in B^{0,j}(M, E)$, $j \geqslant q$, as $\varrho - c = i_{e_n^{(0,1)}}s = 0$ on ∂M, by (3.5.6) and (3.5.12), we get

$$\mathscr{L}_{\varrho_1}(s, s) = \frac{1}{|d\varrho|}\langle(\partial\overline{\partial}\varrho)(w_l, \overline{w}_k)\overline{w}^k \wedge i_{\overline{w}_l}s, s\rangle \geqslant \frac{C_1}{|d\varrho|}|s|^2 , \quad \text{on } \partial M. \tag{3.5.13}$$

From (3.5.13), we deduce that

$$\int_{\partial M} \mathscr{L}_{\varrho_1}(s, s)\, dv_{\partial M} \geqslant 0, \quad \text{for } s \in B^{0,j}(M, E), \ j \geqslant q. \tag{3.5.14}$$

By the Bochner–Kodaira–Nakano formula with boundary term (1.4.84) and (3.5.11), (3.5.14) and the Hermitian torsion $\mathcal{T}$ is bounded, we deduce for any $j \geqslant q$, $s \in B^{0,j}(M, E)$ with $\mathrm{supp}(s) \subset X_v \setminus \overline{X}_u$,

$$\frac{3}{2}\left(\|\overline{\partial}^E s\|^2_{L^2} + \|\overline{\partial}^{E*} s\|^2_{L^2}\right) \geqslant \int_{X_c \setminus \overline{X}_u} \left(\chi'(\varrho)C_4 - 2C_5\right)|s|^2 \, dv_X \,, \tag{3.5.15}$$

and Lemma 3.5.4 follows. $\qquad\square$

Theorem 3.5.5. *Let X be a q-convex manifold and let $c \in \mathbb{R}$ such that $K \subset X_c$. Then for any $j \geqslant q$, $\overline{\partial}^E : \mathrm{Dom}(\overline{\partial}^E) \cap L^2_{0,j}(X_c, E) \to L^2_{0,j+1}(X_c, E)$ has closed range and $\dim H^{0,j}_{(2)}(X_c, E) < \infty$.*

Proof. The L^2 condition in the definition of $H^{0,j}_{(2)}(X_c, E)$ is understood with respect to an arbitrary metric on X, since all metrics are equivalent on $\overline{X}_c$.

By applying (3.5.9) to ζs where $s \in B^{0,j}(X_c, E)$, $j \geqslant q$ and ζ is a cut-off function with $\zeta - 1$ near ∂X_c and $\mathrm{supp}(\zeta) \subset X_v \setminus X_u$, we deduce that the fundamental estimate (3.1.27) holds for any $s \in B^{0,j}(X_c, E)$. Since $B^{0,j}(X_c, E)$ is dense in $\mathrm{Dom}(\overline{\partial}^E) \cap \mathrm{Dom}(\overline{\partial}^{E*})$, we see that (3.1.27) is satisfied. Theorem 3.5.5 follows therefore from Theorem 3.1.8. $\qquad\square$

Using weight functions $\eta(\rho)$, with a rapidly growing convex function $\eta :]-\infty, v[\to \mathbb{R}$, in order to temper the growth of forms at the boundary ∂X_v, we can prove:

Theorem 3.5.6 (Hörmander). *Let X be a q-convex manifold, and let $c \in \mathbb{R}$ such that $K \subset X_c$. For $j \geqslant q$, and $c < v$, the restriction morphism*

$$H^{0,j}(X_v, E) \longrightarrow H^{0,j}_{(2)}(X_c, E) \tag{3.5.16}$$

is an isomorphism, where $H^{0,j}(X_v, E)$ is the Dolbeault cohomology.

For any holomorphic vector bundle E on X, the sheaf cohomology of $\mathscr{O}_X(E)$ over an open set $U \subset X$ is denoted by $H^j(U, E)$. By the Dolbeault isomorphism (Theorem B.4.4), we know that $H^j(X_v, E) \simeq H^{0,j}(X_v, E)$. Using similar methods as in Theorem 3.5.6, we get:

Theorem 3.5.7 (Andreotti–Grauert). *Let X be a q-convex manifold, and let $c \in \mathbb{R}$ such that $K \subset X_c$. The following assertions hold true for $j \geqslant q$:*

(i) *(approximation theorem) The restriction morphism $H^j(X, E) \longrightarrow H^j(X_c, E)$ is an isomorphism and consequently*

(ii) *(finiteness theorem) $\dim H^j(X, E) < +\infty$.*

We are now ready to prove the holomorphic Morse inequalities.

Theorem 3.5.8. *Let X be a q-convex manifold, and let (L, h^L), (E, h^E) be holomorphic Hermitian vector bundles with $\mathrm{rk}(L) = 1$.*

For any smooth sub-level set $X_c \supset K$, let $h^L_\chi = h^L e^{-\chi(\varrho)}$ be the metric on L defined in (3.5.8) with $\chi'(\varrho) \geq C_0 > 0$ on $X_v \setminus X_u$. Then as $p \to \infty$, for any $r \geq q$,

$$\sum_{j=r}^{n} (-1)^{j-r} \dim H^j(X, L^p \otimes E)$$

$$\leq \mathrm{rk}(E) \frac{p^n}{n!} \int_{X_c(\geq r, c_1(L, h^L_\chi))} (-1)^r c_1(L, h^L_\chi)^n + o(p^n). \quad (3.5.17)$$

If $c_1(L, h^L) = \frac{\sqrt{-1}}{2\pi} R^L$ is semi-positive outside a compact set, we can replace the right-hand side integral by $\int_{X(\geq r, c_1(L, h^L))} c_1(L, h^L)^n$.

Proof. By Theorems 3.5.6 and 3.5.7, for any $v > c$, $j \geq q$,

$$H^{0,j}_{(2)}(X_c, L^p \otimes E) \simeq H^j(X_v, L^p \otimes E) \simeq H^j(X, L^p \otimes E). \quad (3.5.18)$$

An easy modification of the proof of Lemma 3.5.4 shows that there exists $C_2 > 0$ such that for any $j \geq q$, $s \in B^{0,j}(X_c, L^p \otimes E)$ with $\mathrm{supp}(s) \subset X_v \setminus \overline{X}_u$,

$$\|s\|^2_{L^2} \leq \frac{C_2}{p} \left(\|\overline{\partial}^E_p s\|^2_{L^2} + \|\overline{\partial}^{E,*}_p s\|^2_{L^2} \right), \quad (3.5.19)$$

where the L^2 condition is taken with respect to the metrics g^{TX} on X_c and h^L_χ, h^E. We see thus that (3.2.2) is satisfied for any $j \geq q$, so by Remark 3.2.15 and (3.5.18), for any $r \geq q$, (3.5.17) follows.

In order to justify the last assertion of Theorem 3.5.8, let us choose $u < c$ such that L is semi-positive outside X_u and $\partial\overline{\partial}\varrho$ has $n - q + 1$ positive eigenvalues.

Let us choose χ in (3.5.8) such that $\chi = 0$ on $]-\infty, u_1[$ where $u < u_1 < c$. By (3.5.10), R^{L_χ} (i.e., $\dot{R}^{L_\chi}$) has then $n - q + 1$ positive eigenvalues in $X_c \setminus \overline{X}_u$, so that $X_c(j, c_1(L, h^L_\chi)) \subset X_{u_1}$ for $j \geq q$. But from (3.5.8), $h^L_\chi = h^L$ on X_{u_1}, thus $c_1(L, h^L_\chi) = c_1(L, h^L)$ on X_{u_1} and

$$X_c(j, c_1(L, h^L_\chi)) = X_c(j, c_1(L, h^L)) \quad \text{for } j \geq q. \quad (3.5.20)$$

From (3.5.17) and (3.5.20), we get the last part of Theorem 3.5.8. $\square$

We note also a related vanishing theorem.

Theorem 3.5.9. *Let X be a q-convex manifold. and let (L, h^L), (E, h^E) be holomorphic Hermitian vector bundles on X with $\mathrm{rk}(L) = 1$.*

(i) *If (L, h^L) is positive in a neighborhood of the exceptional set $K \subset X$ (cf. Definition B.3.2), then there exists $p_0 \in \mathbb{N}$ such that*

$$H^j(X, L^p \otimes E) = 0, \quad \text{for } j \geq q, \ p \geq p_0. \quad (3.5.21)$$

(ii) *If X is q-complete, then*

$$H^j(X, E) = 0, \quad \text{for } j \geqslant q. \tag{3.5.22}$$

Proof. (i). We can assume that (L, h^L) is positive on $\overline{X}_c$. Then, for $j \geqslant q$, and p sufficiently large, the estimate (3.5.19) holds for all $s \in B^{0,j}(X_c, L^p \otimes E)$, hence $H^{0,j}_{(2)}(X_c, L^p \otimes E) = 0$. From Theorem 3.5.7, we get (3.5.21).

(ii). If $K = \emptyset$, we can take $X_u = \emptyset$ in the proof of Theorem 3.5.5, so that (3.5.9) holds for any $s \in B^{0,j}(X_c, E)$, $j \geqslant q$. By Theorem 3.5.7, (3.5.22) holds. $\square$

In the same vein one can study the growth of the cohomology groups of pseudoconvex domains and weakly 1-complete manifolds.

Theorem 3.5.10. *Let $M \Subset X$ be a smooth pseudoconvex domain in a complex manifold X and let (L, h^L), (E, h^E) be holomorphic Hermitian vector bundles over X. Assume that $\mathrm{rk}(L) = 1$ and L is positive in a neighborhood of ∂M. Then*

$$\dim H^{0,j}_{(2)}(M, L^p \otimes E) < +\infty, \quad \text{for } p \gg 1, \ j \geqslant 1, \tag{3.5.23}$$

For $r \geqslant 1$, as $p \to \infty$,

$$\sum_{j=r}^{n} (-1)^{j-r} \dim H^{0,j}_{(2)}(M, L^p \otimes E) \leqslant \mathrm{rk}(E) \frac{p^n}{n!} \int_{M(\geqslant r)} (-1)^r c_1(L, h^L)^n + o(p^n). \tag{3.5.24}$$

Moreover

$$\dim H^0(M, L^p \otimes E) \geqslant \mathrm{rk}(E) \frac{p^n}{n!} \int_{M(\leqslant 1)} c_1(L, h^L)^n + o(p^n). \tag{3.5.25}$$

In particular, if L is positive over $\overline{M}$,

$$\dim H^0(M, L^p) \geqslant p^n \int_M c_1(L, h^L)^n / n! + o(p^n). \tag{3.5.26}$$

Proof. Since M is pseudoconvex, (3.5.14) holds for any $j \geqslant 1$, so by taking an arbitrary Hermitian metric g^{TX} on X, we deduce as in Lemma 3.5.4 that (3.5.19) holds for $j \geqslant 1$. We conclude by Remark 3.2.15 and Theorem 3.2.16. $\square$

We have the following variant of Theorems 3.5.6 and 3.5.7 for weakly 1-complete manifolds (cf. Definition B.3.14):

Theorem 3.5.11. *Let X be weakly 1-complete, L, E be holomorphic vector bundles over X, with $\mathrm{rk}(L) = 1$, and L be positive outside a compact set $K \subset X_c$. Then for $j \geqslant 1$ and $p \gg 1$,*

(i) (Takegoshi) *The canonical morphism*

$$\mathscr{H}^{0,j}(X_c, L^p \otimes E) \longrightarrow H^{0,j}(X_c, L^p \otimes E), \quad s \to [s], \tag{3.5.27}$$

(ii) (Ohsawa) *The restriction morphisms*

$$H^j(X, L^p \otimes E) \longrightarrow H^j(X_c, L^p \otimes E) \tag{3.5.28}$$

are isomorphisms.

Since $X_c = \{\varrho < c\}$ are smooth pseudoconvex domains for a regular value c of ϱ with $\varrho : X \to \mathbb{R}$ in Definition B.3.14, Theorems 3.5.10 and 3.5.11 immediately imply:

Theorem 3.5.12. *Let X be a weakly 1-complete manifold and let (L, h^L), (E, h^E) be holomorphic Hermitian vector bundles over X. Assume that $\mathrm{rk}(L) = 1$ and (L, h^L) is positive outside a compact set $K \subset X$. Then for $r \geqslant 1$, as $p \to \infty$,*

$$\sum_{j=r}^n (-1)^{j-r} \dim H^j(X, L^p \otimes E) \leqslant \mathrm{rk}(E) \frac{p^n}{n!} \int_{X(\geqslant r)} (-1)^r c_1(L, h^L)^n + o(p^n). \tag{3.5.29}$$

$$\dim H^0(X, L^p \otimes E) \geqslant \mathrm{rk}(E) \frac{p^n}{n!} \int_{X(\leqslant 1)} \left(\tfrac{\sqrt{-1}}{2\pi} R^L \right)^n + o(p^n). \tag{3.5.30}$$

Corollary 3.5.13. *Let X be a weakly 1-complete manifold and let (L, h^L) be a positive holomorphic Hermitian line bundle on X, then*

$$\lim_{p \to \infty} p^{-n} \dim H^0(X, L^p \otimes E) = +\infty. \tag{3.5.31}$$

This follows immediately from (3.5.10) and (3.5.25) by taking the liberty of modifying h^L to $h^L e^{-\chi(\varrho)}$, where $\chi : \mathbb{R} \to \mathbb{R}$ is smooth, rapidly increasing and convex.

This result together with the effective base point freeness methods, introduced in algebraic geometry by Angehrn–Siu, produce an answer to the conjecture of Nakano and Ohsawa about the embeddability of weakly 1-complete manifolds.

Theorem 3.5.14 (Takayama). *Let X be an n-dimensional weakly 1-complete manifold with a positive line bundle L. Then X is embeddable into $\mathbb{C}\mathbb{P}^{2n+1}$ by a linear subsystem of $|H^0(X, (K_X \otimes L^p)^{\otimes(n+2)})|$ for $p > n(n+1)/2$.*

Let us finally note the Nakano vanishing theorem.

Theorem 3.5.15 (Nakano). *Let X be a weakly 1-complete manifold, $\dim X = n$, and let (L, h^L), (E, h^E) be holomorphic Hermitian vector bundles over X. Assume that $\mathrm{rk}(L) = 1$ and (L, h^L) is positive on X. Then for $j \geqslant 1$ there exists $p_0 \in \mathbb{N}$ such that $H^j(X, L^p \otimes E) = 0$, for any $p \geqslant p_0$.*

Proof. Indeed, since L is positive on X, we can obtain the fundamental estimate (3.5.19) for any $s \in B^{0,j}(X_c, L^p \otimes E)$ and $p \gg 1$, so by (3.1.29) $\mathscr{H}^{0,j}(X_c, L^p \otimes E) = 0$. Taking into account Theorem 3.5.11, we obtain the conclusion. $\square$

3.6 Morse inequalities for coverings

This section is organized as follows. In Section 3.6.1, we present some generalities about von Neumann dimension on covering manifolds. In Section 3.6.2, we establish the holomorphic Morse inequalities for covering manifolds by adapting the argument in Section 1.7.

3.6.1 Covering manifolds, von Neumann dimension

Let $(\widetilde{X}, g^{T\widetilde{X}})$ be a Riemannian manifold on which a discrete group Γ acts freely (i.e., $g \cdot \widetilde{x} = \widetilde{x}$ for some $\widetilde{x} \in \widetilde{X}$ implies that g is the unit element of Γ) and properly discontinuously (i.e., the map $\Gamma \times \widetilde{X} \to \widetilde{X}, (g, \widetilde{x}) \mapsto g \cdot \widetilde{x}$ is proper, where Γ is endowed with the discrete topology) such that $g^{T\widetilde{X}}$ is Γ-invariant.

Let $X = \widetilde{X}/\Gamma$ be the quotient and $\pi_\Gamma : \widetilde{X} \longrightarrow X$ be the canonical projection. Then $\widetilde{X}$ is a Galois covering of X of Galois group Γ. We assume in what follows that X is *compact*. Moreover, we suppose that $\widetilde{X}$ is *paracompact* so that Γ will be countable. Since $g^{T\widetilde{X}}$ is Γ-invariant, there exists a Riemannian metric g^{TX} on X such that $\pi_\Gamma^* g^{TX} = g^{T\widetilde{X}}$. We denote by $dv_{\widetilde{X}}$ the Riemannian volume form of $g^{T\widetilde{X}}$. We call an open set $U \subset \widetilde{X}$ a fundamental domain of the action of Γ on $\widetilde{X}$ if the following conditions are satisfied:

(a) $\widetilde{X} = \cup_{\gamma \in \Gamma} \gamma(\overline{U})$,

(b) $\gamma_1(U) \cap \gamma_2(U) = \emptyset$ for $\gamma_1, \gamma_2 \in \Gamma$, $\gamma_1 \neq \gamma_2$ and

(c) $\overline{U} \smallsetminus U$ has zero measure.

We construct a fundamental domain in the following way. Let $\{U_k\}$ be a finite cover of X with open balls having the property that for each k, there exists an open set $\widetilde{U}_k \subset \widetilde{X}$ such that $\pi_\Gamma : \widetilde{U}_k \longrightarrow U_k$ is biholomorphic with inverse $\phi_k : U_k \longrightarrow \widetilde{U}_k$. Define $W_k = U_k \smallsetminus (\cup_{j<k} \overline{U}_j \cap U_k)$. Then $U := \cup_k \phi_k(W_k)$ is a fundamental domain.

Let $(\widetilde{E}, h^{\widetilde{E}})$ be a Hermitian vector bundle over $\widetilde{X}$ such that the action of Γ lifts to $\widetilde{E}$. Then there exists a Hermitian vector bundle (E, h^E) on X such that $(\widetilde{E}, h^{\widetilde{E}}) = (\pi_\Gamma^* E, \pi_\Gamma^* h^E)$. On the sections with compact support $\Omega_0(\widetilde{X}, \widetilde{E})$, we introduce the scalar product $\langle s_1, s_2 \rangle = \int_{\widetilde{X}} \langle s_1, s_2 \rangle_{h^{\widetilde{E}}} dv_{\widetilde{X}}$ and let $L^2(\widetilde{X}, \widetilde{E})$ be the corresponding L^2 space. It is easy to see that

$$L^2(\widetilde{X}, \widetilde{E}) \simeq L^2\Gamma \otimes L^2(U, \widetilde{E}) \simeq L^2\Gamma \otimes L^2(X, E), \qquad (3.6.1)$$

where U is a fundamental domain for the action of Γ. A basis for $L^2\Gamma$ is formed by the functions

$$\delta_\gamma(\gamma') = \begin{cases} 1 & \text{if} \quad \gamma = \gamma', \\ 0 & \text{if} \quad \gamma \neq \gamma'. \end{cases} \qquad (3.6.2)$$

Then for $f \in L^2(\widetilde{X}, \widetilde{E})$, the above identification is given by

$$f \simeq (f|_{\gamma U})_\gamma \simeq \sum_\gamma \delta_\gamma \otimes \gamma^{-1} \cdot (f|_{\gamma U}), \tag{3.6.3}$$

which means that $L^2(U, \widetilde{E})$ is identified with those sections of $\widetilde{E}$ which vanish outside U, and for any γ we identify $\mathbb{C}\delta_\gamma \otimes L^2(U, \widetilde{E})$ with those sections which vanish outside γU.

There is a unitary action of Γ by left translations on $L^2\Gamma$ by $t_\gamma \delta_\eta = \delta_{\gamma \eta}$. It is easy to check that actually $T_\gamma = t_\gamma \otimes \mathrm{Id}$ and that $\{T_\gamma : \gamma \in \Gamma\}$ defines a unitary action of Γ on $L^2(\widetilde{X}, \widetilde{E})$.

We give now the definition of the Γ-dimension of a Γ-module of $L^2(\widetilde{X}, \widetilde{E})$.

We consider the algebra $\mathscr{A}_\Gamma \subset \mathscr{L}(L^2\Gamma)$ of all operators that commute with all left translations and denote the unit element of Γ by e. We introduce a trace on $\mathscr{A}_\Gamma$ by

$$\mathrm{Tr}_\Gamma[A] = \langle A\delta_e, \delta_e \rangle, \quad A \in \mathscr{A}_\Gamma. \tag{3.6.4}$$

A left Γ-invariant subspace $V \subset L^2\Gamma$ is called a Γ-module. If V is a Γ-module, the orthogonal projection P_V on V belongs to $\mathscr{A}_\Gamma$. Imitating the usual definition of dimension as the trace of the projector of the space in question, we put

$$\dim_\Gamma V = \mathrm{Tr}_\Gamma[P_V] = \langle P_V \delta_e, \delta_e \rangle. \tag{3.6.5}$$

There is a useful formula for $\dim_\Gamma V$ in terms of an orthonormal basis $\{f_i\}$ of V. We complete this basis to an orthonormal basis $\{f_i, g_j\}$ of $L^2\Gamma$ and expanding $\delta_e = \sum \langle \delta_e, f_i \rangle f_i + \sum \langle \delta_e, g_j \rangle g_j = \sum \bar{f}_i(e) f_i + \sum \langle \delta_e, g_j \rangle g_j$, we get

$$\dim_\Gamma V = \langle P_V \delta_e, \delta_e \rangle = \left\langle \sum_i \bar{f}_i(e) f_i, \delta_e \right\rangle = \sum_i |f_i(e)|^2. \tag{3.6.6}$$

As before, a Γ-invariant space $V \subset L^2(\widetilde{X}, \widetilde{E})$ is called a Γ-module. We denote by $\mathscr{A}_\Gamma \subset \mathscr{L}(L^2(\widetilde{X}, \widetilde{E}))$ the algebra of all operators which commute with the action of Γ. Then to any operator $A \in \mathscr{L}(L^2(\widetilde{X}, \widetilde{E}))$, we can associate operators $a_{\gamma \eta} \in \mathscr{L}(L^2(U, \widetilde{E}))$ such that $a_{\gamma \eta}(f)$ is the projection of $A(\delta_\gamma \otimes f)$ on $\mathbb{C}\delta_\eta \otimes L^2(U, \widetilde{E})$. If moreover $A \in \mathscr{A}_\Gamma$, then the matrix $(a_{\gamma \eta})$ satisfies the relation $a_{\gamma \eta} = a_{e, \gamma^{-1}\eta}$ and we can define

$$\mathrm{Tr}_\Gamma[A] = \mathrm{Tr}[a_{ee}], \tag{3.6.7}$$

if the right-hand side is well defined. When $B \in \mathscr{L}(L^2(U, \widetilde{E}))$ is trace class (cf. Def. D.1.4), then

$$\mathrm{Tr}[B] := \sum_j \langle B h_j, h_j \rangle, \tag{3.6.8}$$

where $\{h_j\}$ is an orthonormal basis of $L^2(U, \widetilde{E})$. If B is a positive operator, certainly $\mathrm{Tr}[B]$ in (3.6.8) is also well defined.

If $A \in \mathscr{A}_\Gamma$ is positive, a_{ee} is positive too, so

$$\mathrm{Tr}_\Gamma[A] := \mathrm{Tr}[a_{ee}] \geqslant 0. \qquad (3.6.9)$$

If $V \subset L^2(\widetilde{X}, \widetilde{E})$ is a Γ-module then the projection $P_V \in \mathscr{A}_\Gamma$.

Definition 3.6.1. The *von Neumann dimension* or Γ-*dimension* of a Γ-module V is defined by

$$\dim_\Gamma V = \mathrm{Tr}_\Gamma[P_V]. \qquad (3.6.10)$$

Lemma 3.6.2. *Let $V \subset L^2(\widetilde{X}, \widetilde{E})$ be a Γ-module. Then*

$$\dim_\Gamma V = \sum_i \int_U |f_i(\widetilde{x})|^2 dv_{\widetilde{X}}(\widetilde{x}), \qquad (3.6.11)$$

where $\{f_i\}$ is an orthonormal basis in V. If $P_V(\widetilde{x}, \widetilde{y})$ is the Schwartz kernel of the projector P_V on V,

$$\dim_\Gamma V - \int_U \mathrm{Tr}[P_V(x, \ddot{x})] dv_{\widetilde{X}}(\widetilde{x}). \qquad (3.6.12)$$

Proof. We have

$$P_V(\widetilde{x}, \widetilde{y}) = \sum_i f_i(\widetilde{x}) \otimes f_i(\widetilde{y})^* \in \widetilde{E}_{\widetilde{x}} \otimes \widetilde{E}_{\widetilde{y}}^*,$$

and therefore $\mathrm{Tr}[P_V(\ddot{x}, \widetilde{x})] = \sum_i |f_i(\widetilde{x})|^2$. Formula (3.6.11) is proved as (3.6.6), cf. also Lemma D.1.5. $\qquad \square$

3.6.2 Holomorphic Morse inequalities

Now, we suppose that $(\widetilde{X}, \widetilde{J})$ is a paracompact complex manifold and the discrete group Γ acts holomorphically, freely and properly on $\widetilde{X}$ having a compact quotient $X = \widetilde{X}/\Gamma$. We assume also that the metric $g^{T\widetilde{X}}$ is Γ-equivariant and $\widetilde{J}$-compatible.

Let $(\widetilde{E}, h^{\widetilde{E}})$ be a Γ-equivariant holomorphic Hermitian vector bundle on $\widetilde{X}$. Then the Kodaira Laplacian $\square^{\widetilde{E}}$ is Γ-invariant, i.e., for any $\gamma \in \Gamma$, $T_\gamma \square^{\widetilde{E}} = \square^{\widetilde{E}} T_\gamma$ on $\Omega_0^{0,\bullet}(\widetilde{X}, \widetilde{E})$. It is easy to see that $g^{T\widetilde{X}}$ is complete. Thus $\square^{\widetilde{E}}$ is essentially self-adjoint (cf. Corollary 3.3.4), and we denote still by $\square^{\widetilde{E}}$ its self-adjoint extension, which turns out to commute to the action of Γ (Problem 3.8). This means that for any $\gamma \in \Gamma$,

$$T_\gamma(\mathrm{Dom}(\square^{\widetilde{E}})) \subset \mathrm{Dom}(\square^{\widetilde{E}}) \quad \text{and} \quad T_\gamma \square^{\widetilde{E}} = \square^{\widetilde{E}} T_\gamma \text{ on } \mathrm{Dom}(\square^{\widetilde{E}}). \qquad (3.6.13)$$

Consider the spectral resolution $E_\lambda^j(\square^{\widetilde{E}})$ of $\square^{\widetilde{E}}$ acting on $L_{0,j}^2(\widetilde{X}, \widetilde{E})$ (see Appendix C.2).

Lemma 3.6.3. *For any* $j = 0, 1, \ldots, n$ *and* $\lambda \in \mathbb{R}$, *we have* $E_\lambda^j(\square^{\widetilde{E}}) \in \mathscr{A}_\Gamma$, *its Schwartz kernel is smooth and*

$$\dim_\Gamma E_\lambda^j(\square^{\widetilde{E}}) < +\infty. \tag{3.6.14}$$

Proof. Since $\square^{\widetilde{E}}$ commutes to the action of Γ, we see with the help of the spectral theorem C.2.1 that $E_\lambda^j(\square^{\widetilde{E}}) \in \mathscr{A}_\Gamma$ (Problem 3.8). Using again the spectral theorem, we obtain for any $m \in \mathbb{N}$,

$$\operatorname{Im}(E_\lambda(\square^{\widetilde{E}})) \subset \operatorname{Dom}((\square^{\widetilde{E}})^m). \tag{3.6.15}$$

Set $\widetilde{\mathbf{E}} := \Lambda(T^{*(0,1)}\widetilde{X}) \otimes \widetilde{E}$. Note that $\operatorname{Dom}((\square^{\widetilde{E}})^m) \subset \boldsymbol{H}^{2m}(\widetilde{X}, \widetilde{\mathbf{E}}, \operatorname{loc})$, by the regularity theorem A.3.4. Then due to Sobolev's embedding theorem A.3.1,

$$\operatorname{Im}(E_\lambda(\square^{\widetilde{E}})) \subset \cap_{m \in \mathbb{N}} \boldsymbol{H}^{2m}(\widetilde{X}, \widetilde{\mathbf{E}}, \operatorname{loc}) \subset \mathscr{C}^\infty(\widetilde{X}, \widetilde{\mathbf{E}}). \tag{3.6.16}$$

Therefore $E_\lambda(\square^{\widetilde{E}}) : L^2(\widetilde{X}, \widetilde{\mathbf{E}}) \longrightarrow \mathscr{C}^\infty(\widetilde{X}, \widetilde{\mathbf{E}})$ is linearly continuous. By duality, $E_\lambda(\square^{\widetilde{E}}) : \Omega_0^{\prime n, n}(\widetilde{X}, \widetilde{\mathbf{E}}^*) \to L^2(\widetilde{X}, \widetilde{\mathbf{E}})$, where $\Omega_0^{\prime n, n}(\widetilde{X}, \widetilde{\mathbf{E}}^*)$ is the dual of $\mathscr{C}^\infty(\widetilde{X}, \widetilde{\mathbf{E}})$, which is the space of currents with compact support (this space can be identified with the space of (n, n)-forms with values in $\widetilde{\mathbf{E}}^*$ and distribution coefficients). Composing the two maps, we obtain $E_\lambda(\square^{\widetilde{E}}) : \Omega_0^{\prime n, n}(\widetilde{X}, \widetilde{\mathbf{E}}^*) \to L^2(\widetilde{X}, \widetilde{\mathbf{E}}) \to \mathscr{C}^\infty(\widetilde{X}, \widetilde{\mathbf{E}})$, hence $E_\lambda(\square^{\widetilde{E}})$ is a smoothing operator. The Schwartz kernel theorem B.2.7 implies that the kernel $E_\lambda(\square^{\widetilde{E}})(\widetilde{x}, \widetilde{y})$ of $E_\lambda(\square^{\widetilde{E}})$ with respect to $dv_{\widetilde{X}}$ is smooth. By (3.6.12),

$$\dim_\Gamma E_\lambda(\square^{\widetilde{E}}) = \int_U \operatorname{Tr}[E_\lambda(\square^{\widetilde{E}})(\widetilde{x}, \widetilde{x})] \, dv_{\widetilde{X}}(\widetilde{x}) < +\infty.$$

The proof of Lemma 3.6.3 is completed. □

The quantity $\dim_\Gamma E_\lambda^j(\square^{\widetilde{E}})$ is called the *von Neumann spectrum distribution function.*

Let $(\widetilde{L}, h^{\widetilde{L}})$ be a Γ-equivariant holomorphic Hermitian line bundle on $\widetilde{X}$. We use in the sequel the same notation as in Section 1.7 for the corresponding objects on X. Following Section 1.6.1, we introduce the Γ-invariant operator $\widetilde{D}_p$ on $\Omega^{0,\bullet}(\widetilde{X}, \widetilde{L}^p \otimes \widetilde{E})$ and the Γ-invariant Kodaira Laplacian $\widetilde{\square}_p$. Let $e^{-u\widetilde{D}_p^2}(\widetilde{x}, \widetilde{x}')$ $(\widetilde{x}, \widetilde{x}' \in \widetilde{X})$ be the smooth kernel of the operator $e^{-u\widetilde{D}_p^2}$ with respect to $dv_{\widetilde{X}}(\widetilde{x}')$.

Let $\overline{H}_{(2)}^{0,j}(\widetilde{X}, \widetilde{L}^p \otimes \widetilde{E})$ be the reduced L^2-Dolbeault cohomology group. By the weak Hodge decomposition (3.1.21), we have:

$$\operatorname{Ker}(\widetilde{D}_p^2) \cap L_{0,j}^2(\widetilde{X}, \widetilde{L}^p \otimes \widetilde{E}) \simeq \overline{H}_{(2)}^{0,j}(\widetilde{X}, \widetilde{L}^p \otimes \widetilde{E}). \tag{3.6.17}$$

The next result shows that the local behavior of the heat kernel on $\widetilde{X}$ is the same as on X.

Theorem 3.6.4. *For any $m \in \mathbb{N}$, $u_0 > 0, \varepsilon > 0$, there exists $C > 0$ such that for any $\widetilde{x} \in \widetilde{X}$, $p \in \mathbb{N}^*$, $u > u_0$,*

$$\left| e^{-\frac{u}{2p}\widetilde{D}_p^2}(\widetilde{x}, \widetilde{x}) - e^{-\frac{u}{2p}D_p^2}(\pi(\widetilde{x}), \pi(\widetilde{x})) \right|_{\mathscr{C}^m} \leqslant C p^{3m+8n+8} \exp\left(-\frac{\varepsilon^2 p}{16u}\right). \quad (3.6.18)$$

Proof. By the same proof as that of Proposition 1.6.4, we have the estimate (1.6.15) for $\mathbf{G}_{\frac{u}{p}}(\sqrt{u/p}\widetilde{D}_p)(\widetilde{x}, \widetilde{x}')$ with $\widetilde{x}, \widetilde{x}' \in \widetilde{X}$, here we use the covering of $\widetilde{X}$ induced by U_{x_i} (cf. also the proof of Theorem 6.1.4).

Now by using the finite propagation speed, Theorem D.2.1, $\mathbf{F}_{\frac{u}{p}}\left(\sqrt{\frac{u}{p}}\widetilde{D}_p\right)(\widetilde{x}, \cdot)$ is the pull-back of $\mathbf{F}_{\frac{u}{p}}\left(\sqrt{\frac{u}{p}}D_p\right)(\pi(\widetilde{x}), \cdot)$. Thus from (1.6.14), we get (3.6.18). $\qquad\square$

From (3.6.18) and Theorem 1.6.1, we deduce:

Corollary 3.6.5. *For each $u > 0$ fixed, for any $k \in \mathbb{N}$, under $\mathscr{C}^k$-norm on $\mathscr{C}^\infty(\widetilde{X}, \mathrm{End}(\Lambda(T^{*(0,1)}\widetilde{X}) \otimes \widetilde{E}))$, as $p \to \infty$, uniformly on $\widetilde{x} \in \widetilde{X}$, we have*

$$e^{-\frac{u}{p}\widetilde{D}_p^2}(\widetilde{x}, \widetilde{x}) = (2\pi)^{-n} \frac{\det(\dot{R}^L) \exp(2u\omega_d)}{\det(1 - \exp(-2u\dot{R}^L))}(\pi(\widetilde{x})) \otimes \mathrm{Id}_E \, p^n + o(p^n). \quad (3.6.19)$$

We denote by $\mathrm{Tr}_{\Gamma,q}$ the Γ-trace of operators acting on $L^2_{0,q}(\widetilde{X}, \widetilde{L}^p \otimes \widetilde{E})$; then we have the following analogue of Lemma 1.7.2.

Lemma 3.6.6. *For any $u > 0$, $p \in \mathbb{N}^*$, $0 \leqslant q \leqslant n$, we have*

$$\sum_{j=0}^{q}(-1)^{q-j} \dim_\Gamma \overline{H}^{0,j}_{(2)}(\widetilde{X}, \widetilde{L}^p \otimes \widetilde{E}) \leqslant \sum_{j=0}^{q}(-1)^{q-j} \mathrm{Tr}_{\Gamma,j}\left[\exp(-\frac{u}{p}\widetilde{D}_p^2)\right], \quad (3.6.20)$$

with equality for $q = n$.

Proof. Let $E^{j,p}_\lambda$ be the spectral resolution of $\widetilde{\square}_p$ acting on $L^2_{0,j}(\widetilde{X}, \widetilde{L}^p \otimes \widetilde{E})$. For $\lambda_2 > \lambda_1 \geqslant 0$, we consider the projectors $E^{j,p}(]\lambda_1, \lambda_2]) = E^{j,p}_{\lambda_2} - E^{j,p}_{\lambda_1}$. Then, by the Hodge decomposition (3.1.21), $\sum_{j=0}^{q}(-1)^{q-j}E^{j,p}(]\lambda_1, \lambda_2])$ is the projection on the range of $\overline{\partial}^{\widetilde{L}^p \otimes \widetilde{E}} E^{q,p}(]\lambda_1, \lambda_2])$ and thus a positive operator. Hence the Γ-invariant measure $\sum_{j=0}^{q}(-1)^{q-j}dE^{j,p}_\lambda$ is positive on $\{\lambda > 0\}$. It follows that

$$R := \int_{\lambda > 0} e^{-\frac{u}{p}\lambda} \sum_{j=0}^{q}(-1)^{q-j}dE^{j,p}_\lambda \geqslant 0, \quad \text{and } R \in \mathscr{A}_\Gamma. \quad (3.6.21)$$

On the other hand,

$$\mathrm{Tr}_{\Gamma,j}\left[\exp(-\frac{u}{2p}\widetilde{D}_p^2)\right] = \dim_\Gamma \overline{H}^{0,j}_{(2)}(\widetilde{X}, \widetilde{L}^p \otimes \widetilde{E}) + \mathrm{Tr}_\Gamma \int_{\lambda > 0} e^{-\frac{u}{p}\lambda}dE^{j,p}_\lambda. \quad (3.6.22)$$

From (3.6.9), (3.6.21) and (3.6.22), we obtain (3.6.20). $\qquad\square$

By (3.6.7) and (3.6.8), as in (1.7.4),

$$\mathrm{Tr}_{\Gamma,q}\left[\exp(-\frac{u}{p}\widetilde{D}_p^2)\right] = \int_U \mathrm{Tr}_q\left[e^{-\frac{u}{p}\widetilde{D}_p^2}(\widetilde{x},\widetilde{x})\right] dv_{\widetilde{X}}(\widetilde{x}). \qquad (3.6.23)$$

Using (3.6.19) and (3.6.20) as in the proof of Theorem 1.7.1, we get:

Theorem 3.6.7. *As $p \to \infty$, the following strong holomorphic Morse inequalities hold for every $q = 0, 1, \ldots, n$:*

$$\sum_{j=0}^{q}(-1)^{q-j} \dim_\Gamma \overline{H}_{(2)}^{0,j}(\widetilde{X}, \widetilde{L}^p \otimes \widetilde{E}) \leqslant \mathrm{rk}(\widetilde{E})\frac{p^n}{n!}\int_{X(\leqslant q)}(-1)^q c_1(L, h^L)^n + o(p^n), \qquad (3.6.24)$$

with equality for $q = n$.

Corollary 3.6.8. *Suppose that $C = \frac{1}{n!}\int_{X(\leqslant 1)} c_1(L, h^L)^n > 0$. Then as $p \to \infty$,*

$$\dim_\Gamma H_{(2)}^0(\widetilde{X}, \widetilde{L}^p \otimes \widetilde{E}) \geqslant C\,\mathrm{rk}(\widetilde{E})p^n + o(p^n). \qquad (3.6.25)$$

Problems

Problem 3.1. Prove Lemma 3.1.1 and show that if R' is any closed extension of R, then $R' \subset R_{\max}$. Especially, $R_{\max} = (R_H^*)_H^*$.

Problem 3.2. Prove that the quadratic form (3.1.5) is closed: If $\{s_k\}$ is a Cauchy sequence in the norm $\|\cdot\|_Q$ (cf. (C.1.5)), there exist elements s, v, w in L^2 such that $s_k \to s$, $\overline{\partial}^E s_k \to v$, $\overline{\partial}^{E,*} s_k \to w$. Show that $\overline{\partial}^E s_k \to \overline{\partial}^E s$ and $v = \overline{\partial}^E s$ in distribution sense, hence $\overline{\partial}^E s \in L^2$. Since $s_k \in \mathrm{Dom}(\overline{\partial}_H^{E,*})$, we have $(\overline{\partial}^E \widetilde{s}, s_k) = (\widetilde{s}, \overline{\partial}^{E,*} s_k)$ for any $\widetilde{s} \in \mathrm{Dom}(\overline{\partial}^E)$. By passing to the limit, show that $(\overline{\partial}^E \widetilde{s}, s) = (\widetilde{s}, w)$, thus $s \in \mathrm{Dom}(\overline{\partial}_H^{E,*})$ and $\overline{\partial}_H^{E,*} s = w$.

Problem 3.3. Let X be a complex Hermitian manifold and (E, h^E) be a holomorphic Hermitian vector bundle on X. Using Friedrichs lemma 3.1.3 show that for any $s \in L_{0,\bullet}^2(X, E)$ with compact support, s is contained in the domain of the quadratic form associated to the Gaffney extension (3.1.4) if and only if s is contained in the domain of the quadratic form associated to the Friedrichs extension (C.1.7), and if this is the case, the two quadratic forms coincide on (s, s).

Problem 3.4. Let A be a positive self-adjoint operator on a Hilbert space and (E_λ) be its spectral resolution. Let $\lambda \geqslant 0$ and $u \in \mathrm{Dom}(A)$, set $u = u_1 + u_2$ where $u_1 = E_\lambda u$, $u_2 = u - u_1$.

(a) Show that

$$Q_A(u, u) = Q_A(u_1, u_1) + Q_A(u_2, u_2),$$
$$Q_A(u_1, u_1) \leqslant \lambda\|u_1\|^2,$$
$$Q_A(u_2, u_2) \geqslant \lambda\|u_2\|^2.$$

(b) Assume that $Q_A(u, u) \leqslant \lambda\|u\|^2$ and $u_1 = 0$. Show that $u = 0$.

Problem 3.5. Prove that for any $f \in \mathscr{C}^{\infty}(\mathbb{R}_+)$, there exists $g \in \mathscr{C}^{\infty}(\mathbb{R})$ such that $f = g$ on $\mathbb{R}_+$. (Hint: The $\mathscr{C}^k$ version of the statement is (A.2.1). A good reference for the $\mathscr{C}^{\infty}$ version is [213].)

Problem 3.6. Let $A \leqslant 0$ and $B_1, \ldots, B_l > 0$ and $l, n \in \mathbb{N}$, $l \leqslant n$. Consider the function $G : \{u \in \mathbb{C} : \mathrm{Re}(u) > 0\} \to \mathbb{C}$,

$$G(u) = \frac{e^{uA}}{u^{n-l} \prod_{j=1}^{l}(1 - e^{-uB_j})}.$$

With the notation from (3.2.36), show that its inverse Laplace transform $L^{-1}(G)$ is

$$L^{-1}(G)(\lambda) = \begin{cases} \sum_{\alpha \in \mathbb{N}^n} \delta\left(\lambda + A - \sum_{j=1}^{n} \alpha_j B_j\right), & \text{if } l = n, \\ \frac{1}{(n-l-1)!} \sum_{\alpha \in \mathbb{N}^l} \left(t^{n-l-1} H_+(t)\right)\big|_{t=\lambda+A-\sum_{j=1}^{l} \alpha_j B_j}, & \text{if } l < n. \end{cases}$$

Here $\delta(t)$ represents the Dirac measure at $\{0\}$. Thus

$$\int_0^{\lambda} L^{-1}(G)(\lambda_0) \, d\lambda_0 = \frac{1}{(n-l)!} \sum_{\alpha \in \mathbb{N}^l} \left(t^{n-l} H_+(t)\right)\big|_{t=\lambda+A-\sum_{j=1}^{l} \alpha_j B_j}.$$

Problem 3.7. Show that the Γ-dimension (3.6.5) has the following properties:

1. $0 \leqslant \dim_{\Gamma} V \leqslant 1$.
2. $\dim_{\Gamma} V = 0$ if and only if $V = 0$.
3. $V \subset V'$ implies $\dim_{\Gamma} V \leqslant \dim_{\Gamma} V'$ with equality if and only if $V = V'$.

Problem 3.8. Let $(\widetilde{X}, g^{T\widetilde{X}}, \Gamma, X)$ be as in Section 3.6.2. Let $(\widetilde{E}, h^{\widetilde{E}})$ be a Γ-equivariant holomorphic Hermitian vector bundle on $\widetilde{X}$.

(a) Let A be one of the following operators: the closure of $\overline{\partial}^{\widetilde{E}}$ on $L^2_{0,\bullet}(\widetilde{X}, \widetilde{E})$, its Hilbert-space adjoint, or the unique self-adjoint extension Kodaira Laplacian $\square^{\widetilde{E}}$. Show that A commutes with the action of Γ in the following sense: for any $\gamma \in \Gamma$, $T_\gamma(\mathrm{Dom}(A)) \subset \mathrm{Dom}(A)$ and $T_\gamma A = A T_\gamma$ on $\mathrm{Dom}(A)$.

(b) Show that the spectral projections $E_\lambda(\square^{\widetilde{E}})$ commute with any T_γ, i.e., $E_\lambda(\square^{\widetilde{E}}) \in \mathscr{A}_\Gamma$.

3.7 Bibliographic notes

Section 3.1. The approach to L^2 Hodge theory is inspired by Hörmander [131], Kohn [143], Folland–Kohn [108] and Ohsawa [187]. Friedrichs lemma 3.1.3 can be found in [131, Prop. 1.2.4].

Section 3.2. Theorems 3.2.13 and 3.2.16 appeared in [169, 170] and [173, 244]. Holomorphic Morse inequalities on non-compact manifolds were proved in various contexts by Nadel–Tsuji [181], Bouche [47], Takayama [235] and recently by Berman [17].

The proof of Proposition 3.2.7 is inspired by [55, §3] where the contribution from the boundary in the local index theorem is also determined. Sources for the Laplace transform are [212] and [106, XIII]. Lemma 3.2.10 is [106, XIII.1, Th. 2a].

Section 3.3. The Andreotti–Vesentini lemma 3.3.1 appeared in [7, Lemma 4, p. 92–93]. Corollary 3.3.7 is [181, Th. 1.1]. For another approach see [83].

Section 3.4. The Siu–Demailly criterion for isolated singularities appeared in [171]. The metric constructed in (3.4.17) is called a metric of Saper-type, cf. [209, 113, 114]. For the notion of Andreotti-pseudoconcavity see [2, 6]. Levi's removable singularity theorem can be found at [107, p. 180].

Section 3.5. For the proof of Lemma 3.5.1, we refer to [131, Prop. 1.2.4]. The treatment of Andreotti–Grauert cohomology theory on q-convex manifolds is inspired by the paper of Hörmander [131] (where only the case of a trivial bundle E is treated; passing to the general case poses no difficulties). One can as well work with complete metrics; this is done in [7, 187]. Theorem 3.5.5 is essentially [131, Th. 3.4.1] and Theorem 3.5.6 is [131, Th. 3.4.9]. Lemma 3.5.3 appears in [7, Lemma 18]. The fundamental theorem 3.5.7 can be found in [4].

Theorem 3.5.8 was obtained by Th. Bouche [47]. His proof is based on the same principle of showing the fundamental estimate outside a compact set but he works with complete metrics.

Theorem 3.5.11 was proved by Takegoshi [236]. Theorem 3.5.12 is [47, Theorem 0.2], where actually a q-positive line bundle L is considered over a Kähler weakly 1-complete manifold X. The Kähler assumption was removed in [169] answering positively a question of Ohsawa [187, p. 218] about the polynomial growth of degree n with respect to p of $\dim H^j(X, L^p \otimes E)$, $j \geqslant 1$. Theorem 3.5.14 was proved by Takayama [235, Theorem 1.2].

Section 3.6. For facts about analysis on covering manifolds, the reader is referred to [9, §4], [160, 244].

The holomorphic Morse inequalities for coverings appeared in [244, 173]. The approach there was to determine the asymptotics of the von Neumann spectrum counting function $N_\Gamma^{j,p}(\lambda) = \dim_\Gamma E_\lambda^{j,p}$, and was inspired by the analytic proof of Shubin [222] of the Novikov–Shubin (classical) Morse inequalities on coverings.

There is a large literature about existence of holomorphic functions or sections on coverings of projective manifolds (cf. Corollary 3.6.8), motivated by the Shafarevich conjecture, see, e.g., [147] and the references therein.

Chapter 4

Asymptotic Expansion of the Bergman Kernel

In this chapter, we establish the asymptotic expansion of the Bergman kernel associated to high tensor powers of a positive line bundle on a compact complex manifold. Thanks to the spectral gap property of the Kodaira Laplacian, Theorem 1.5.5, we can use the finite propagation speed of solutions of hyperbolic equations, (Theorem D.2.1), to localize our problem to a problem on $\mathbb{R}^{2n}$. Comparing with Section 1.6, the key point here is that we need to extend the connection of the line bundle L such that its curvature becomes uniformly positive on $\mathbb{R}^{2n}$. Then we still have the spectral gap property on $\mathbb{R}^{2n}$. Thus we can instead study the Bergman kernel on $\mathbb{R}^{2n}$ (cf. (4.1.27)), and use various resolvent representations (4.1.59), (4.2.22) of the Bergman kernel on $\mathbb{R}^{2n}$. We conclude our results by employing functional analysis resolvent techniques.

This chapter is organized as follows: In Section 4.1, we obtain the near diagonal asymptotic expansion of the Bergman kernel, and give a general way to compute the coefficients. In Section 4.2, we establish the full off-diagonal asymptotic expansion of the Bergman kernel and give a relation between the coefficients of the asymptotic expansion of the Bergman kernel and the heat kernel.

We will use the notation from Section 1.6.1.

4.1 Near diagonal expansion of the Bergman kernel

In this section, we establish the near diagonal asymptotic expansion of the Bergman kernel, Theorem 4.1.24. As pointed out at the beginning of this chapter, we will localize our problem to a problem on $\mathbb{R}^{2n}$ by combining the spectral gap property and the finite propagation speed of solutions of hyperbolic equations. Then we will use functional analysis methods to study the resolvent in order to obtain our result. Moreover, in Sections 4.1.6, 4.1.7, we develop a formal power series technique to compute the coefficients in general.

This section is organized as follows. In Section 4.1.1, we state the diagonal asymptotic expansion of the Bergman kernel. In Section 4.1.2, we show that the problem is local. In Section 4.1.3, we construct in detail the extension of the Dirac operator from a small neighborhood of 0 to the whole $\mathbb{R}^{2n}$, such that the spectral gap property still holds, and we study the Taylor expansion of the operator. In Section 4.1.4, we study the Sobolev estimates of the resolvent $(\lambda - \mathscr{L}_2^t)^{-1}$. In Section 4.1.5, we establish the existence of the near diagonal expansion, (Theorem 4.1.24), without knowing information about the coefficients $\mathscr{F}_r$. In Section 4.1.6, we study the Bergman kernel of the limit operator. In Section 4.1.7, we develop a formal power series technique to compute the coefficients and we get the description of the coefficients $\boldsymbol{b}_r$ from Theorem 4.1.1. In Sections 4.1.8 and 4.1.9, we prove Theorems 4.1.2 and 4.1.3 by applying the results from Sections 4.1.6 and 4.1.7.

In this section, we use the notation from Section 1.6.1. Especially, (X, J) is a compact complex manifold with complex structure J and $\dim_{\mathbb{C}} X = n$. g^{TX} is a Riemannian metric on TX compatible with J. $(L, h^L), (E, h^E)$ are holomorphic Hermitian vector bundles on X and $\mathrm{rk}(L) = 1$.

We suppose that the positivity condition (1.5.21) holds for R^L.

4.1.1 Diagonal asymptotic expansion of the Bergman kernel

Let P_p be the orthogonal projection from $\Omega^{0,\bullet}(X, L^p \otimes E)$ onto $\mathrm{Ker}(D_p)$. Then for each $p \in \mathbb{N}$, the Bergman kernel $P_p(x, x')$ $(x, x' \in X)$, is the smooth kernel of P_p with respect to the Riemannian volume form $dv_X(x')$.

By Theorems 1.4.1 and 1.5.5 we have for p large enough,

$$\mathrm{Ker}(D_p) = \mathrm{Ker}(D_p^2|_{\Omega^{0,0}}) = H^0(X, L^p \otimes E). \tag{4.1.1}$$

As D_p^2 preserves the $\mathbb{Z}$-grading on $\Omega^{0,\bullet}(X, L^p \otimes E)$, we know that for p large enough,

$$P_p(x, x') \in (L^p \otimes E)_x \otimes (L^p \otimes E)_{x'}^*, \tag{4.1.2}$$

especially,

$$P_p(x, x) \in \mathrm{End}(L^p \otimes E)_x = \mathrm{End}(E)_x, \tag{4.1.3}$$

where we use the canonical identification $\mathrm{End}(L^p) = \mathbb{C}$ for any line bundle L on X.

Let $\{S_i^p\}_{i=1}^{d_p}$ $(d_p := \dim H^0(X, L^p \otimes E))$ be any orthonormal basis of $H^0(X, L^p \otimes E)$ with respect to the inner product (1.3.14) induced by g^{TX}, h^L and h^E. Then for p large enough,

$$P_p(x, x') = \sum_{i=1}^{d_p} S_i^p(x) \otimes (S_i^p(x'))^* \in (L^p \otimes E)_x \otimes (L^p \otimes E)_{x'}^*. \tag{4.1.4}$$

Recall that R^{TX}, $R^{T^{(1,0)}X}$ are the curvatures of the Levi–Civita connection ∇^{TX} on TX and of the holomorphic Hermitian connection $\nabla^{T^{(1,0)}X}$ on $T^{(1,0)}X$.

Let $\Lambda_\omega(R^E)$ be the contraction of R^E with respect to ω and $g_\omega^{TX} := \omega(\cdot, J\cdot)$. If $\{w_{\omega,j}\}$ is an orthonormal basis of $(T^{(1,0)}X, g_\omega^{TX})$, then

$$\sqrt{-1}\Lambda_\omega(R^E) = \sqrt{-1}\left\langle R^E, \omega \right\rangle_{g_\omega^{TX}} = R^E(w_{\omega,j}, \overline{w}_{\omega,j}),$$
$$nR^E \wedge \omega^{n-1} = \Lambda_\omega(R^E) \cdot \omega^n. \tag{4.1.5}$$

Theorem 4.1.1. *There exist smooth coefficients $\boldsymbol{b}_r(x) \in \mathrm{End}(E)_x$ which are polynomials in R^{TX}, R^E (and $d\Theta$, R^L) and their derivatives of order $\leqslant 2r-2$ (resp. $2r-1$, $2r$) and reciprocals of linear combinations of eigenvalues of $\dot{R}^L$ at x (in the sense of Lemmas 1.2.3 and 1.2.4), such that*

$$\boldsymbol{b}_0 = \det(\dot{R}^L/(2\pi))\,\mathrm{Id}_E, \tag{4.1.6}$$

and for any $k, l \in \mathbb{N}$, there exists $C_{k,l} > 0$ such that for any $p \in \mathbb{N}$,

$$\left| P_p(x,x) - \sum_{r=0}^{k} \boldsymbol{b}_r(x) p^{n-r} \right|_{\mathscr{C}^l(X)} \leqslant C_{k,l}\, p^{n-k-1}. \tag{4.1.7}$$

Moreover, the expansion is uniform in the following sense. For any fixed $k, l \in \mathbb{N}$, assume that the derivatives of g^{TX}, h^L, h^E of order $\leqslant 2n+2k+l+6$ run over a set bounded in the $\mathscr{C}^l$-norm taken with respect to the parameter $x \in X$ and, moreover, g^{TX} runs over a set bounded below. Then the constant $C_{k,l}$ is independent of g^{TX} and the $\mathscr{C}^l$-norm in (4.1.7) includes also the derivatives with respect to the parameters.

Theorem 4.1.2. *If $\omega = \Theta$ in (1.5.14), i.e., $\omega = \frac{\sqrt{-1}}{2\pi}R^L$ is the Kähler form of (TX, g^{TX}), then there exist smooth functions $\boldsymbol{b}_r(x) \in \mathrm{End}(E)_x$ such that (4.1.7) holds, and $\boldsymbol{b}_r$ are polynomials in R^{TX}, R^E and their derivatives of order $\leqslant 2r-2$ at x. Moreover,*

$$\boldsymbol{b}_0 = \mathrm{Id}_E, \quad \boldsymbol{b}_1 = \frac{1}{4\pi}\left[2\sqrt{-1}\Lambda_\omega(R^E) + \frac{1}{2}r^X\,\mathrm{Id}_E\right]. \tag{4.1.8}$$

Theorem 4.1.3. *The term $\boldsymbol{b}_1$ in the expansion (4.1.7) is given by*

$$\boldsymbol{b}_1 = \frac{1}{8\pi}\det\left(\frac{\dot{R}^L}{2\pi}\right)\left[r_\omega^X - 2\Delta_\omega\left(\log(\det(\dot{R}^L))\right) + 4\sqrt{-1}\Lambda_\omega(R^E)\right], \tag{4.1.9}$$

here r_ω^X, Δ_ω are the scalar curvature and the Bochner Laplacian as in (1.3.19) associated to g_ω^{TX}.

Remark 4.1.4. By the Riemann–Roch–Hirzebruch theorem 1.4.6, and relations (4.1.1) and (4.1.4), we have for p large enough,

$$\int_X \mathrm{Tr}_E[P_p(x,x)]dv_X(x) = \dim H^0(X, L^p \otimes E)$$

$$= \int_X \mathrm{Td}(T_hX)\,\mathrm{ch}(L^p \otimes E) \tag{4.1.10}$$

$$= \mathrm{rk}(E)\int_X \frac{c_1(L)^n}{n!}p^n + \int_X \left(c_1(E) + \frac{\mathrm{rk}(E)}{2}c_1(T_hX)\right)\frac{c_1(L)^{n-1}}{(n-1)!}p^{n-1} + \mathscr{O}(p^{n-2}).$$

By integrating the trace of (4.1.7), and using (4.1.8) and (4.1.9), we recover the last equation of (4.1.10). Thus we can consider (4.1.8) as a local version of (4.1.10) in the spirit of local index theory.

4.1.2 Localization of the problem

Recall that inj^X is the injectivity radius of (X, g^{TX}) and the function $f : \mathbb{R} \to [0,1]$ was defined in (1.6.12) for $0 < \varepsilon < \mathrm{inj}^X/4$. Set

$$F(a) = \left(\int_{-\infty}^{+\infty} f(v)dv\right)^{-1}\int_{-\infty}^{+\infty} e^{iva}f(v)dv. \tag{4.1.11}$$

Then $F(a)$ lies in Schwartz space $\mathcal{S}(\mathbb{R})$ and $F(0) = 1$.

Proposition 4.1.5. *For any $l, m \in \mathbb{N}$, $\varepsilon > 0$, there exists $C_{l,m,\varepsilon} > 0$ such that for $p \geqslant 1$, $x, x' \in X$,*

$$|F(D_p)(x,x') - P_p(x,x')|_{\mathscr{C}^m} \leqslant C_{l,m,\varepsilon}\, p^{-l},$$
$$|P_p(x,x')|_{\mathscr{C}^m} \leqslant C_{l,m,\varepsilon}\, p^{-l} \quad \text{if } d(x,x') \geqslant \varepsilon. \tag{4.1.12}$$

Here the $\mathscr{C}^m$ norm is induced by ∇^L, ∇^E and h^L, h^E, g^{TX}.

Proof. Recall that the constant μ_0 is defined in (1.5.26). For $a \in \mathbb{R}$, set

$$\phi_p(a) = 1_{[\sqrt{p\mu_0},+\infty[}(|a|)F(a). \tag{4.1.13}$$

Then by Theorem 1.5.8 (cf. also Theorem 1.5.5), for $p > C_L/\mu_0$,

$$F(D_p) - P_p = \phi_p(D_p). \tag{4.1.14}$$

By (4.1.11), for any $m \in \mathbb{N}$, there exists $C_m > 0$ such that

$$\sup_{a\in\mathbb{R}} |a|^m|F(a)| \leqslant C_m. \tag{4.1.15}$$

Let Q be a differential operator of order $m \in \mathbb{N}$ with scalar principal symbol and with compact support in U_{x_i} as in (1.6.17). From $\langle D_p^{m'}\phi_p(D_p)Qs, s'\rangle =$

$\langle s, Q^* \phi_p(D_p) D_p^{m'} s' \rangle$, (1.6.6), (4.1.13) and (4.1.15), we know that for $l, m' \in \mathbb{N}$, there exists $C_{l,m'} > 0$ such that for $p \geqslant 1$,

$$\|D_p^{m'} \phi_p(D_p) Q s\|_{L^2} \leqslant C_{l,m'} p^{-l+2m} \|s\|_{L^2}. \tag{4.1.16}$$

We deduce from (1.6.6) and (4.1.16) that if P, Q are differential operators of order m, m' with compact support in U_{x_i}, U_{x_j} respectively, then for any $l > 0$, there exists $C_l > 0$ such that for $p \geqslant 1$,

$$\|P \phi_p(D_p) Q s\|_{L^2} \leqslant C_l p^{-l} \|s\|_{L^2}. \tag{4.1.17}$$

On $U_{x_i} \times U_{x_j}$, by using Sobolev inequality and (4.1.14), we get the first inequality of (4.1.12).

By the finite propagation speed of solutions of hyperbolic equations, Theorem D.2.1, $F(D_p)(x, x')$ only depends on the restriction of D_p to $B^X(x, \varepsilon)$, and is zero if $d(x, x') \geqslant \varepsilon$. Thus we get the second inequality of (4.1.12). The proof of Proposition 4.1.5 is complete. $\qquad\square$

4.1.3   Rescaling and Taylor expansion of the operator D_p^2

Recall that $\nabla^{B, \Lambda^{0,\bullet}}$ is the connection on $\Lambda(T^{*(0,1)}X)$ defined by (1.4.27).

Trivialization 4.1. We fix $x_0 \in X$. From now on, we identify $B^{T_{x_0}X}(0, 4\varepsilon)$ with $B^X(x_0, 4\varepsilon)$ as in Section 1.2.1. For $Z \in B^{T_{x_0}X}(0, 4\varepsilon)$, we identify E_Z, L_Z, $\Lambda(T_Z^{*(0,1)}X)$ to E_{x_0}, L_{x_0}, $\Lambda(T_{x_0}^{*(0,1)}X)$ by parallel transport with respect to the connections ∇^E, ∇^L, $\nabla^{B, \Lambda^{0,\bullet}}$ along the curve $[0,1] \ni u \to uZ$, cf. Section 1.6.3. Thus on $B^X(x_0, 4\varepsilon)$, (E, h^E), (L, h^L), $(\Lambda(T^{*(0,1)}X), h^{\Lambda^{0,\bullet}})$ are identified to the trivial Hermitian bundles $(E_{x_0}, h^{E_{x_0}})$, $(L_{x_0}, h^{L_{x_0}})$, $(\Lambda(T_{x_0}^{*(0,1)}X), h^{\Lambda_{x_0}^{0,\bullet}})$. Let Γ^E, Γ^L, $\Gamma^{B, \Lambda^{0,\bullet}}$ be the corresponding connection forms of ∇^E, ∇^L and $\nabla^{B, \Lambda^{0,\bullet}}$ on $B^X(x_0, 4\varepsilon)$. Then Γ^E, Γ^L, $\Gamma^{B, \Lambda^{0,\bullet}}$ are skew-adjoint with respect to $h^{E_{x_0}}$, $h^{L_{x_0}}$, $h^{\Lambda_{x_0}^{0,\bullet}}$.

Denote by ∇_U the ordinary differentiation operator on $T_{x_0}X$ in the direction U. We will identify $\mathbb{R}^{2n}$ to $T_{x_0}X$ as in (1.6.22) by choosing an orthonormal basis $\{e_i\}$ of $T_{x_0}X$ with dual basis $\{e^i\}$. Let $\varphi_\varepsilon : \mathbb{R}^{2n} \to \mathbb{R}^{2n}$ be the map defined by $\varphi_\varepsilon(Z) = \rho(|Z|/\varepsilon)Z$ with ρ defined in (1.6.19). As in (1.2.12), let $\mathcal{R}$ be the radial vector field defined by

$$\mathcal{R} = \sum_{i=1}^{2n} Z_i e_i = Z. \tag{4.1.18}$$

For the trivial vector bundle $E_0 := (E_{x_0}, h^{E_{x_0}})$, we defined a Hermitian connection on $X_0 := T_{x_0}X$ by

$$\nabla^{E_0} = \nabla + \rho(|Z|/\varepsilon)\Gamma^E. \tag{4.1.19}$$

For the trivial line bundle $L_0 := (L_{x_0}, h^{L_{x_0}})$, we define the Hermitian connection on X_0 by

$$\nabla^{L_0}|_Z = \varphi_\varepsilon^* \nabla^L + \frac{1}{2}(1 - \rho^2(|Z|/\varepsilon)) R_{x_0}^L(\mathcal{R}, \cdot). \tag{4.1.20}$$

Then we calculate easily that its curvature $R^{L_0} := (\nabla^{L_0})^2$ is

$$R^{L_0}(Z) = \varphi_\varepsilon^* R^L + \frac{1}{2} d\Big((1 - \rho^2(|Z|/\varepsilon)) R_{x_0}^L(\mathcal{R}, \cdot) \Big)$$

$$= \Big(1 - \rho^2(|Z|/\varepsilon)\Big) R_{x_0}^L + \rho^2(|Z|/\varepsilon) R_{\varphi_\varepsilon(Z)}^L$$

$$- (\rho\rho')(|Z|/\varepsilon) \sum_i \frac{Z_i e^i}{\varepsilon|Z|} \wedge \Big[R_{x_0}^L(\mathcal{R}, \cdot) - R_{\varphi_\varepsilon(Z)}^L(\mathcal{R}, \cdot) \Big]. \tag{4.1.21}$$

Thus R^{L_0} is positive in the sense of (1.5.21) for ε small enough, and the corresponding constant μ_0 for R^{L_0} is bigger than $\frac{4}{5}\mu_0$. From now on, we fix ε as above.

Let $g^{TX_0}(Z) := g^{TX}(\varphi_\varepsilon(Z))$, $J_0(Z) := J(\varphi_\varepsilon(Z))$ be the metric and almost complex structure on X_0. Let $T^{*(0,1)}X_0$ be the anti-holomorphic cotangent bundle of (X_0, J_0). Then $T_{Z,J_0}^{*(0,1)}X_0$ is naturally identified with $T_{\varphi_\varepsilon(Z),J}^{*(0,1)}X_0$. We identify $\Lambda(T_Z^{*(0,1)}X_0)$ with $\Lambda(T_{x_0}^{*(0,1)}X)$ by identifying first $\Lambda(T_Z^{*(0,1)}X_0)$ with $\Lambda(T_{\varphi_\varepsilon(Z),J}^{*(0,1)}X_0)$, which in turn is identified with $\Lambda(T_{x_0}^{*(0,1)}X)$ by using parallel transport with respect to $\nabla^{B,\Lambda^{0,\bullet}}$ along $\gamma_u : [0,1] \ni u \to u\varphi_\varepsilon(Z)$. We trivialize $\Lambda(T^{*(0,1)}X_0)$ in this way.

We trivialize the Hermitian line bundle $\det(T^{(1,0)}X_0)$ by identifying at first $\det(T^{(1,0)}X_0)_Z$ to $\det(T^{(1,0)}X)_{\varphi_\varepsilon(Z)}$, and then to $\det(T^{(1,0)}X)_{x_0}$ by using parallel transport along γ_u with respect to the holomorphic Hermitian connection $\nabla^{\det}$ on $\det(T^{(1,0)}X)$. Let $\Gamma^{\det}$ be the corresponding connection form. Let $\nabla^{\det_0}$ be the Hermitian connection on $\det(T^{(1,0)}X)_{x_0}$ defined by

$$\nabla^{\det_0} = \nabla + \rho(|Z|/\varepsilon)\Gamma^{\det}. \tag{4.1.22}$$

Then g^{TX_0} and $\nabla^{\det_0}$ define a Clifford connection ∇^{Cl_0} on $\Lambda(T^{*(0,1)}X_0)$ as in (1.3.5). The connection form Γ^{Cl_0} associated to the above trivialization of $\Lambda(T^{*(0,1)}X_0)$ on $\mathbb{R}^{2n}$, satisfies

$$\Gamma^{\mathrm{Cl}_0} = 0 \qquad \text{for } |Z| > 4\varepsilon. \tag{4.1.23}$$

Let $D_{0,p}^c$ be the spin^c Dirac operator on $\mathbb{R}^{2n}$ acting on $E_{0,p} := \Lambda(T^{*(0,1)}X_0) \otimes L_0^p \otimes E_0$, and associated to $\nabla^{L_0}, \nabla^{E_0}, \nabla^{\mathrm{Cl}_0}$, as constructed in (1.3.15).

Recall that the 3-form T_{as} was defined in (1.2.48). Set

$$A_0 = -\frac{1}{4}\rho(|Z|/\varepsilon)T_{as,Z}. \tag{4.1.24}$$

Then the modified Dirac operator

$$D_p^{c,A_0} = D_{0,p}^c + {}^c(A_0) \tag{4.1.25}$$

coincides with D_p on $B^{T_{x_0}X}(0, 2\varepsilon)$. Observe that all tensors (except R^{L_0}) in (1.3.35) are 0 outside $B^{T_{x_0}X}(0, 4\varepsilon)$, thus from (4.1.21) and the proof of Theorems 1.5.7 and 1.5.8, we see that $(D_p^{c,A_0})^2$ has the following spectral gap:

$$\operatorname{Spec}((D_p^{c,A_0})^2) \subset \{0\} \cup \left] \frac{8}{5}\mu_0 p - C, +\infty \right[. \tag{4.1.26}$$

Let $P_{0,p}$ be the orthogonal projection from $L^2(X_0, E_{0,p})$ onto $\operatorname{Ker}(D_p^{c,A_0})^2$, and let $P_{0,p}(x, x')$ be the smooth kernel of $P_{0,p}$ with respect to the Riemannian volume form $dv_{X_0}(x')$.

Proposition 4.1.6. *For any $l, m \in \mathbb{N}$, there exists $C_{l,m} > 0$ such that for $x, x' \in B^{T_{x_0}X}(0, \varepsilon)$,*

$$\left|(P_{0,p} - P_p)(x, x')\right|_{\mathscr{C}^m} \leqslant C_{l,m} p^{-l}. \tag{4.1.27}$$

Proof. Using (4.1.14) and (4.1.26), we see that $P_{0,p} - F(D_p)$ verifies also (4.1.12) for $x, x' \in B^{T_{x_0}X}(0, \varepsilon)$, thus we get (4.1.27). $\qquad\square$

As in Section 1.6.3, by means of a unit vector S_L of I_{x_0}, we construct an isometry $E_{p,x_0} \simeq (\Lambda(T^{*(0,1)}X) \otimes E)_{x_0} =: \mathbf{E}_{x_0}$. Since the operator $(D_p^{c,A_0})^2$ takes values in $\operatorname{End}(E_{p,x_0}) = \operatorname{End}(\mathbf{E})_{x_0}$ under the natural identification $\operatorname{End}(L^p) \simeq \mathbb{C}$ (which does not depend on S_L), our formulas do not depend on the choice of S_L. Now, under this identification, we will consider $(D_p^{c,A_0})^2$ acting on $\mathscr{C}^\infty(X_0, \mathbf{E}_{x_0})$.

Let dv_{TX} be the Riemannian volume form of $(T_{x_0}X, g^{T_{x_0}X})$. Let $\kappa(Z)$ be the smooth positive function defined by the equation

$$dv_{X_0}(Z) = \kappa(Z)dv_{TX}(Z), \tag{4.1.28}$$

with $\kappa(0) = 1$.

Let ∇^{A_0} be the connection induced by ∇^{Clo} and A_0 on $\Lambda(T^{*(0,1)}X_0)$ as in (1.3.33). Let Γ^{A_0} be the connection form of ∇^{A_0}. Recall that ∇_0, L_2^0 were defined in (1.6.28).

For $s \in \mathscr{C}^\infty(\mathbb{R}^{2n}, \mathbf{E}_{x_0})$, $Z \in \mathbb{R}^{2n}$, and for $t = \frac{1}{\sqrt{p}}$, set

$$
\begin{aligned}
(S_t s)(Z) &:= s(Z/t), \\
\nabla_t &:= S_t^{-1} t\, \kappa^{1/2} \nabla^{A_0} \kappa^{-1/2} S_t, \\
\mathscr{L}_2^t &:= S_t^{-1} \kappa^{1/2}\, t^2 (D_p^{c,A_0})^2 \kappa^{-1/2} S_t, \\
\mathscr{L}_2^0 &:= L_2^0.
\end{aligned}
\tag{4.1.29}
$$

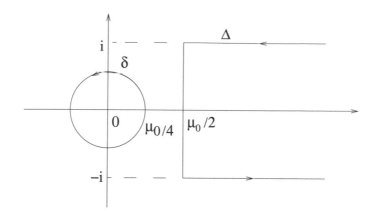

Figure 4.1.

By using Theorem 4.1.9, exactly in the same way as in proofs of Theorems 1.6.8–1.6.10, we obtain the following analogues which are uniform in $x_0 \in X$.

Theorem 4.1.10. *There exists $C > 0$ such that for $t \in]0, t_0]$, $\lambda \in \delta \cup \Delta$,*

$$\|(\lambda - \mathscr{L}_2^t)^{-1}\|_t^{0,0} \leqslant C,$$
$$\|(\lambda - \mathscr{L}_2^t)^{-1}\|_t^{-1,1} \leqslant C(1 + |\lambda|^2). \tag{4.1.41}$$

Proposition 4.1.11. *Take $m \in \mathbb{N}^*$. There exists $C_m > 0$ such that for $t \in]0, 1]$, $Q_1, \ldots, Q_m \in \{\nabla_{t,e_i}, Z_i\}_{i=1}^{2n}$ and $s, s' \in \mathscr{C}_0^\infty(\mathbb{R}^{2n}, \mathbf{E}_{x_0})$,*

$$\left| \langle [Q_1, [Q_2, \ldots, [Q_m, \mathscr{L}_2^t] \ldots]] s, s' \rangle_{t,0} \right| \leqslant C_m \|s\|_{t,1} \|s'\|_{t,1}. \tag{4.1.42}$$

Theorem 4.1.12. *For any $t \in]0, t_0]$, $\lambda \in \delta \cup \Delta$, $m \in \mathbb{N}$, the resolvent $(\lambda - \mathscr{L}_2^t)^{-1}$ maps $\mathbf{H}_t^m$ into $\mathbf{H}_t^{m+1}$. Moreover for any $\alpha \in \mathbb{N}^{2n}$, there exist $N \in \mathbb{N}$, $C_{\alpha,m} > 0$ such that for $t \in]0, t_0]$, $\lambda \in \delta \cup \Delta$, $s \in \mathscr{C}_0^\infty(X_0, \mathbf{E}_{x_0})$,*

$$\|Z^\alpha (\lambda - \mathscr{L}_2^t)^{-1} s\|_{t,m+1} \leqslant C_{\alpha,m} (1 + |\lambda|^2)^N \sum_{\alpha' \leqslant \alpha} \|Z^{\alpha'} s\|_{t,m}. \tag{4.1.43}$$

For $m \in \mathbb{N}$, let $\mathcal{Q}^m$ be the set of operators $\{\nabla_{t,e_{i_1}} \cdots \nabla_{t,e_{i_j}}\}_{j \leqslant m}$. For $k, r \in \mathbb{N}^*$, let

$$I_{k,r} = \left\{ (\mathbf{k}, \mathbf{r}) = (k_i, r_i) \middle| \sum_{i=0}^j k_i = k + j, \sum_{i=1}^j r_i = r, \ k_i, r_i \in \mathbb{N}^* \right\}. \tag{4.1.44}$$

For $(\mathbf{k}, \mathbf{r}) \in I_{k,r}$, $\lambda \in \delta \cup \Delta$, $t \in]0, t_0]$, set

$$A_\mathbf{r}^\mathbf{k}(\lambda, t) = (\lambda - \mathscr{L}_2^t)^{-k_0} \frac{\partial^{r_1} \mathscr{L}_2^t}{\partial t^{r_1}} (\lambda - \mathscr{L}_2^t)^{-k_1} \cdots \frac{\partial^{r_j} \mathscr{L}_2^t}{\partial t^{r_j}} (\lambda - \mathscr{L}_2^t)^{-k_j}. \tag{4.1.45}$$

Then there exist $a_{\mathbf{r}}^{\mathbf{k}} \in \mathbb{R}$ such that

$$\frac{\partial^r}{\partial t^r}(\lambda - \mathscr{L}_2^t)^{-k} = \sum_{(\mathbf{k},\mathbf{r}) \in I_{k,r}} a_{\mathbf{r}}^{\mathbf{k}} A_{\mathbf{r}}^{\mathbf{k}}(\lambda, t). \tag{4.1.46}$$

Theorem 4.1.13. *For any* $m \in \mathbb{N}$, $k > 2(m+r+1)$, $(\mathbf{k}, \mathbf{r}) \in I_{k,r}$, *there exist* $C > 0$, $N \in \mathbb{N}$ *such that for* $\lambda \in \delta \cup \Delta$, $t \in]0, t_0]$, $Q, Q' \in \mathcal{Q}^m$,

$$\|Q A_{\mathbf{r}}^{\mathbf{k}}(\lambda, t) Q' s\|_{t,0} \leqslant C(1 + |\lambda|)^N \sum_{|\beta| \leqslant 2r} \|Z^\beta s\|_{t,0}. \tag{4.1.47}$$

Proof. From Theorem 4.1.12, we deduce that if $Q \in \mathcal{Q}^m$, there are $C_m > 0$ and $N \in \mathbb{N}$ such that for any $\lambda \in \delta \cup \Delta$,

$$\|Q(\lambda - \mathscr{L}_2^t)^{-m}\|_t^{0,0} \leqslant C_m(1 + |\lambda|)^N. \tag{4.1.48}$$

Observe that $\mathscr{L}_2^t$ is self-adjoint with respect to $\|\cdot\|_{t,0}$, so after taking the adjoint of (4.1.48) we have for any $\lambda \in \delta \cup \Delta$,

$$\|(\lambda - \mathscr{L}_2^t)^{-m} Q\|_t^{0,0} \leqslant C_m(1 + |\lambda|)^N. \tag{4.1.49}$$

Thus from (4.1.48) and (4.1.49), we obtain (4.1.47) for $r = 0$.

Consider now $r > 0$. By (4.1.34) (cf. (1.6.42)), $\frac{\partial^r}{\partial t^r}\mathscr{L}_2^t$ is a combination of

$$\frac{\partial^{r_1}}{\partial t^{r_1}}(g^{ij}(tZ))\left(\frac{\partial^{r_2}}{\partial t^{r_2}}\nabla_{t,e_i}\right)\left(\frac{\partial^{r_3}}{\partial t^{r_3}}\nabla_{t,e_j}\right), \quad \frac{\partial^{r_1}}{\partial t^{r_1}}(d(tZ)), \quad \frac{\partial^{r_1}}{\partial t^{r_1}}(d_i(tZ))\left(\frac{\partial^{r_2}}{\partial t^{r_2}}\nabla_{t,e_i}\right)$$

where $d(Z), d_i(Z)$ and their derivatives in Z are uniformly bounded for $Z \in \mathbb{R}^{2n}$. Now $\frac{\partial^{r_1}}{\partial t^{r_1}}(d(tZ))$ (resp. $\frac{\partial^{r_1}}{\partial t^{r_1}}\nabla_{t,e_i}$) $(r_1 \geqslant 1)$, are functions of the type $d'(tZ)Z^\beta$, $|\beta| \leqslant r_1$ (resp. $r_1 + 1$) and $d'(Z)$ and its derivatives in Z are bounded smooth functions of Z.

Let $\mathscr{R}_t'$ be the family of operators of the type

$$\mathscr{R}_t' = \{[f_{j_1}Q_{j_1}, [f_{j_2}Q_{j_2}, \ldots [f_{j_l}Q_{j_l}, \mathscr{L}_2^t] \ldots]]\}$$

with f_{j_i} smooth bounded (with its derivatives) functions and $Q_{j_i} \in \{\nabla_{t,e_l}, Z_l\}_{l=1}^{2n}$.

We handle now the operator $A_{\mathbf{r}}^{\mathbf{k}}(\lambda, t)Q'$. We will move first all the terms Z^β in $d'(tZ)Z^\beta$ (defined above) to the right-hand side of this operator. To do so, we always use the commutator trick as in the proof of Theorem 1.6.10, i.e., each time, we perform only the commutation with Z_i (not directly with Z^β with $|\beta| > 1$). Then $A_{\mathbf{r}}^{\mathbf{k}}(\lambda, t)Q'$ is as the form $\sum_{|\beta| \leqslant 2r} L_\beta^t Q_\beta'' Z^\beta$, and Q_β'' is obtained from Q' and its commutation with Z^β. Now we move all the terms ∇_{t,e_i} in $\frac{\partial^{r_j}\mathscr{L}_2^t}{\partial t^{r_j}}$ to the right-hand side of the operator L_β^t. Then as in the proof of Theorem 4.1.12 (cf. Theorem 1.6.10), we get finally that $QA_{\mathbf{r}}^{\mathbf{k}}(\lambda, t)Q'$ is of the form $\sum_{|\beta| \leqslant 2r} \mathscr{L}_\beta^t Z^\beta$ where $\mathscr{L}_\beta^t$ is a linear combination of operators of the form

$$Q(\lambda - \mathscr{L}_2^t)^{-k_0'} R_1(\lambda - \mathscr{L}_2^t)^{-k_1'} R_2 \cdots R_{l'}(\lambda - \mathscr{L}_2^t)^{-k_{l'}'} Q''' Q'', \tag{4.1.50}$$

with $R_1, \ldots, R_{l'} \in \mathscr{R}'_t$, $Q''' \in \mathcal{Q}^{2r}$, $Q'' \in \mathcal{Q}^m$, $|\beta| \leqslant 2r$, and Q'' is obtained from Q' and its commutation with Z^β. Since $k > 2(m + r + 1)$, we can use the argument leading to (4.1.48) and (4.1.49), to split the above operator into two parts,

$$Q(\lambda - \mathscr{L}_2^t)^{-k'_0} R_1 (\lambda - \mathscr{L}_2^t)^{-k'_1} R_2 \cdots R_i (\lambda - \mathscr{L}_2^t)^{-k''_i},$$
$$(\lambda - \mathscr{L}_2^t)^{-(k'_i - k''_i)} \cdots R_{l'} (\lambda - \mathscr{L}_2^t)^{-k'_{l'}} Q''' Q'', \tag{4.1.51}$$

such that the $\|\cdot\|_t^{0,0}$-norm of each part is bounded by $C(1 + |\lambda|^2)^N$. Thus the proof of (4.1.47) is complete. □

Certainly, as $t \to 0$, the limit of $\|\cdot\|_{t,m}$ exists, and we denote it by $\|\cdot\|_{0,m}$.

Theorem 4.1.14. *For any $r \geqslant 0, k > 0$, there exist $C > 0$, $N \in \mathbb{N}$ such that for $t \in [0, t_0], \lambda \in \delta \cup \Delta$,*

$$\left\| \left(\frac{\partial^r \mathscr{L}_2^t}{\partial t^r} - \frac{\partial^r \mathscr{L}_2^t}{\partial t^r} \bigg|_{t=0} \right) s \right\|_{t,-1} \leqslant Ct \sum_{|\alpha| \leqslant r+3} \| Z^\alpha s \|_{0,1}, \tag{4.1.52}$$

$$\left\| \left(\frac{\partial^r}{\partial t^r} (\lambda - \mathscr{L}_2^t)^{-k} - \sum_{(\mathbf{k},r) \in I_{k,r}} a_{\mathbf{r}}^{\mathbf{k}} A_{\mathbf{r}}^{\mathbf{k}}(\lambda, 0) \right) s \right\|_{0,0} \leqslant Ct(1 + |\lambda|^2)^N \sum_{|\alpha| \leqslant 4r+3} \| Z^\alpha s \|_{0,0}.$$

Proof. Note that by (4.1.34) and (4.1.36), as in (1.6.57), for $t \in [0, 1]$, $k \geqslant 1$,

$$\| s \|_{t,k} \leqslant C \sum_{|\alpha| \leqslant k} \| Z^\alpha s \|_{0,k}. \tag{4.1.53}$$

An application of Taylor expansion for (4.1.34) leads to the following equation, if s, s' have compact support:

$$\left| \left\langle \left(\frac{\partial^r \mathscr{L}_2^t}{\partial t^r} - \frac{\partial^r \mathscr{L}_2^t}{\partial t^r} \bigg|_{t=0} \right) s, s' \right\rangle_{0,0} \right| \leqslant Ct \| s' \|_{t,1} \sum_{|\alpha| \leqslant r+3} \| Z^\alpha s \|_{0,1}. \tag{4.1.54}$$

Thus we get the first inequality of (4.1.52). Note that

$$(\lambda - \mathscr{L}_2^t)^{-1} - (\lambda - \mathscr{L}_2^0)^{-1} = (\lambda - \mathscr{L}_2^t)^{-1} (\mathscr{L}_2^t - \mathscr{L}_2^0)(\lambda - \mathscr{L}_2^0)^{-1}. \tag{4.1.55}$$

By passing to the limit we obtain that Theorems 4.1.10–4.1.12 still hold for $t = 0$. From Theorems 4.1.10, 4.1.12, (4.1.36), the first equation of (4.1.52) and (4.1.55), we get

$$\left\| \left((\lambda - \mathscr{L}_2^t)^{-1} - (\lambda - \mathscr{L}_2^0)^{-1} \right) s \right\|_{0,0} \leqslant Ct (1 + |\lambda|)^N \sum_{|\alpha| \leqslant 3} \| Z^\alpha s \|_{0,0}. \tag{4.1.56}$$

If we denote by $\mathscr{L}_{\lambda,t} = \lambda - \mathscr{L}_2^t$, then

$$A_{\mathbf{r}}^{\mathbf{k}}(\lambda, t) - A_{\mathbf{r}}^{\mathbf{k}}(\lambda, 0) = \sum_{i=1}^{j} \mathscr{L}_{\lambda,t}^{-k_0} \cdots \left(\frac{\partial^{r_i} \mathscr{L}_2^t}{\partial t^{r_i}} - \frac{\partial^{r_i} \mathscr{L}_2^t}{\partial t^{r_i}} \bigg|_{t=0} \right) \mathscr{L}_{\lambda,0}^{-k_i} \cdots \mathscr{L}_{\lambda,0}^{-k_j}$$

$$+ \sum_{i=0}^{j} \mathscr{L}_{\lambda,t}^{-k_0} \cdots \left(\mathscr{L}_{\lambda,t}^{-k_i} - \mathscr{L}_{\lambda,0}^{-k_i} \right) \left(\frac{\partial^{r_{i+1}} \mathscr{L}_2^t}{\partial t^{r_{i+1}}} \bigg|_{t=0} \right) \cdots \mathscr{L}_{\lambda,0}^{-k_j}. \quad (4.1.57)$$

Now from the first inequality of (4.1.52), (4.1.46), (4.1.56) and (4.1.57), we get the second equation of (4.1.52). $\qquad \square$

Remark 4.1.15. To get the near diagonal expansion, Theorem 4.1.24, we only need Theorems 4.1.10–4.1.14 for $\lambda \in \delta$. We need the part $\lambda \in \Delta$ in Section 4.2.

4.1.5 Uniform estimate on the Bergman kernel

The next step is to convert the estimates for the resolvent into estimates for the spectral projection $\mathcal{P}_{0,t} : (L^2(X_0, \mathbf{E}_{x_0}), \|\cdot\|_0) \to \mathrm{Ker}(\mathscr{L}_2^t)$. Let $\mathcal{P}_{0,t}(Z, Z') = \mathcal{P}_{0,t,x_0}(Z, Z')$, (with $Z, Z' \in X_0$) be the smooth kernel of $\mathcal{P}_{0,t}$ with respect to $dv_{TX}(Z')$.

Let $\pi : TX \times_X TX \to X$ be the natural projection from the fiberwise product of TX on X. Note that $\mathscr{L}_2^t$ is a family of differential operators on $T_{x_0}X$ with coefficients in $\mathrm{End}(\mathbf{E})_{x_0}$. Thus we can view $\mathcal{P}_{0,t}(Z, Z')$ as a smooth section of $\pi^*(\mathrm{End}(\mathbf{E}))$ over $TX \times_X TX$ by identifying a section $s \in \mathscr{C}^\infty(TX \times_X TX, \pi^* \mathrm{End}(\mathbf{E}))$ with the family $(s_x)_{x \in X}$, where $s_x = s|_{\pi^{-1}(x)}$. Let $\nabla^{\mathrm{End}(\mathbf{E})}$ be the connection on $\mathrm{End}(\mathbf{E})$ induced by $\nabla^{\mathbf{E}}$ (which is in turn induced by $\nabla^{B,\Lambda^{0,\bullet}}$ and ∇^E). Then $\nabla^{\pi^* \mathrm{End}(\mathbf{E})}$ induces naturally a $\mathscr{C}^m$-norm of s for the parameter $x_0 \in X$. In the rest of this chapter, we will denote by $\mathscr{C}^m(X)$ the $\mathscr{C}^m$-norm for the parameter $x_0 \in X$.

Theorem 4.1.16. *For any $m, m', r \in \mathbb{N}$, $q > 0$, there exists $C > 0$, such that for $t \in]0, t_0]$, $Z, Z' \in T_{x_0}X$, $|Z|, |Z'| \leq q$,*

$$\sup_{|\alpha| + |\alpha'| \leq m} \left| \frac{\partial^{|\alpha| + |\alpha'|}}{\partial Z^\alpha \partial Z'^{\alpha'}} \frac{\partial^r}{\partial t^r} \mathcal{P}_{0,t}(Z, Z') \right|_{\mathscr{C}^{m'}(X)} \leq C. \quad (4.1.58)$$

Proof. By (4.1.40), for any $k \in \mathbb{N}^*$,

$$\mathcal{P}_{0,t} = \frac{1}{2\pi i} \int_\delta \lambda^{k-1} (\lambda - \mathscr{L}_2^t)^{-k} d\lambda. \quad (4.1.59)$$

From (4.1.48), (4.1.49) and (4.1.59), we obtain

$$\|Q \mathcal{P}_{0,t} Q'\|_t^{0,0} \leq C_m, \quad \text{for } Q, Q' \in \mathcal{Q}^m. \quad (4.1.60)$$

Let $|\cdot|_{(q),m}$ be the usual Sobolev norm on $\mathscr{C}^\infty(B^{T_{x_0}X}(0, q+1), \mathbf{E}_{x_0})$ induced by $h^{\mathbf{E}_{x_0}}$ and the volume form $dv_{TX}(Z)$ as in (4.1.36). Observe that by (4.1.34) and (4.1.36), for $m > 0$, there exists $C_q > 0$ such that for $s \in \mathscr{C}^\infty(X_0, \mathbf{E}_{x_0})$, $\mathrm{supp}(s) \subset B^{T_{x_0}X}(0, q+1)$,

$$\frac{1}{C_q}\|s\|_{t,m} \leqslant |s|_{(q),m} \leqslant C_q\|s\|_{t,m}. \tag{4.1.61}$$

Now (4.1.60) and (4.1.61) together with Sobolev's inequalities imply

$$\sup_{|Z|,|Z'|\leqslant q} |Q_Z Q'_{Z'}\mathcal{P}_{0,t}(Z, Z')| \leqslant C, \quad \text{for } Q, Q' \in \mathcal{Q}^m. \tag{4.1.62}$$

Thanks to (4.1.34) and (4.1.62), estimate (4.1.58) holds for $r = m' = 0$. To obtain (4.1.58) for $r \geqslant 1$ and $m' = 0$, note that from (4.1.59),

$$\frac{\partial^r}{\partial t^r}\mathcal{P}_{0,t} = \frac{1}{2\pi i}\int_\delta \lambda^{k-1}\frac{\partial^r}{\partial t^r}(\lambda - \mathscr{L}_2^t)^{-k}d\lambda, \quad \text{for } k \geqslant 1. \tag{4.1.63}$$

By (4.1.46), (4.1.47), (4.1.63) and the above argument, we get the estimate (4.1.58) with $m' = 0$.

Finally, for any vector U on X,

$$\nabla_U^{\pi^*\,\mathrm{End}(\mathbf{E})}\mathcal{P}_{0,t} = \frac{1}{2\pi i}\int_\delta \lambda^{k-1}\nabla_U^{\pi^*\,\mathrm{End}(\mathbf{E})}(\lambda - \mathscr{L}_2^t)^{-k}d\lambda. \tag{4.1.64}$$

Now we use a similar formula as (4.1.46) for $\nabla_U^{\pi^*\,\mathrm{End}(\mathbf{E})}(\lambda - \mathscr{L}_2^t)^{-k}$ by replacing $\frac{\partial^{r_1}\mathscr{L}_2^t}{\partial t^{r_1}}$ by $\nabla_U^{\pi^*\,\mathrm{End}(\mathbf{E})}\mathscr{L}_2^t$, and remark that $\nabla_U^{\pi^*\,\mathrm{End}(\mathbf{E})}\mathscr{L}_2^t$ is a differential operator on $T_{x_0}X$ with the same structure as $\mathscr{L}_2^t$, i.e., it has the same type as (1.6.42). Then using the above argument, we conclude that (4.1.58) holds for $m' \geqslant 1$. The proof of Theorem 4.1.16 is complete. $\qquad\square$

For k large enough, set

$$\mathscr{F}_r = \frac{1}{2\pi i\, r!}\int_\delta \lambda^{k-1}\sum_{(\mathbf{k},\mathbf{r})\in I_{k,r}} a_{\mathbf{r}}^{\mathbf{k}} A_{\mathbf{r}}^{\mathbf{k}}(\lambda, 0)d\lambda,$$

$$\mathscr{F}_{r,t} = \frac{1}{r!}\frac{\partial^r}{\partial t^r}\mathcal{P}_{0,t} - \mathscr{F}_r. \tag{4.1.65}$$

Let $\mathscr{F}_r(Z, Z')$ $(Z, Z' \in T_{x_0}X)$ be the smooth kernel of $\mathscr{F}_r$ with respect to $dv_{TX}(Z')$. Then $\mathscr{F}_r \in \mathscr{C}^\infty(TX \times_X TX, \pi^*\mathrm{End}(\mathbf{E}))$.

Theorem 4.1.17. *For $q > 0$, there exists $C > 0$ such that for $t \in\,]0, 1]$, $Z, Z' \in T_{x_0}X$, $|Z|, |Z'| \leqslant q$,*

$$\left|\mathscr{F}_{r,t}(Z, Z')\right| \leqslant Ct^{1/(2n+1)}. \tag{4.1.66}$$

Proof. By (4.1.52), (4.1.63) and (4.1.65), there exists $C > 0$ such that for $t \in]0,1]$,

$$\|\mathscr{F}_{r,t}\|_{(q),0} \leqslant Ct. \tag{4.1.67}$$

By Theorem 4.1.16 and (4.1.67), the same as in (1.6.63) and (1.6.64), we get (4.1.66). $\qquad\square$

Finally, we obtain the following near diagonal estimate for the kernel of $\mathcal{P}_{0,t}$.

Theorem 4.1.18. *For any* $k, m, m' \in \mathbb{N}$, $q > 0$, *there exists* $C > 0$ *such that if* $t \in]0, t_0]$, $Z, Z' \in T_{x_0} X$, $|Z|, |Z'| \leqslant q$,

$$\sup_{|\alpha| + |\alpha'| \leqslant m} \left| \frac{\partial^{|\alpha| + |\alpha'|}}{\partial Z^\alpha \partial Z'^{\alpha'}} \left(\mathcal{P}_{0,t} - \sum_{r=0}^{k} \mathscr{F}_r t^r \right) (Z, Z') \right|_{\mathscr{C}^{m'}(X)} \leqslant Ct^{k+1}. \tag{4.1.68}$$

Proof. By (4.1.65) and (4.1.66),

$$\frac{1}{r!} \frac{\partial^r}{\partial t^r} \mathcal{P}_{0,t}|_{t=0} = \mathscr{F}_r, \tag{4.1.69}$$

Now by Theorem 4.1.16 and (4.1.65), $\mathscr{F}_r$ has the same estimate as $\frac{1}{r!} \frac{\partial^r}{\partial t^r} \mathcal{P}_{0,t}$ in (4.1.58). Again from (4.1.58), (4.1.65), and the Taylor expansion

$$G(t) - \sum_{r=0}^{k} \frac{1}{r!} \frac{\partial^r G}{\partial t^r}(0) t^r = \frac{1}{k!} \int_0^t (t - t_0)^k \frac{\partial^{k+1} G}{\partial t^{k+1}}(t_0) dt_0, \tag{4.1.70}$$

we have (4.1.68). $\qquad\square$

4.1.6   Bergman kernel of $\mathscr{L}$

Now we discuss the eigenvalues and eigenfunctions of $\mathscr{L}_2^0$ in detail.

We choose $\{w_i\}_{i=1}^n$ an orthonormal basis of $T_{x_0}^{(1,0)} X$, such that (1.5.18) holds. Let $\{w^j\}_{j=1}^n$ be its dual basis. Then $e_{2j-1} = \frac{1}{\sqrt{2}}(w_j + \overline{w}_j)$ and $e_{2j} = \frac{\sqrt{-1}}{\sqrt{2}}(w_j - \overline{w}_j)$, $j = 1, \ldots, n$ forms an orthonormal basis of $T_{x_0} X$. We use the coordinates on $T_{x_0} X \simeq \mathbb{R}^{2n}$ induced by $\{e_i\}$ as in (1.6.22) and in what follows we also introduce the complex coordinates $z = (z_1, \ldots, z_n)$ on $\mathbb{C}^n \simeq \mathbb{R}^{2n}$. Thus $Z = z + \overline{z}$, and $w_i = \sqrt{2} \frac{\partial}{\partial z_i}$, $\overline{w}_i = \sqrt{2} \frac{\partial}{\partial \overline{z}_i}$. We will also identify z to $\sum_i z_i \frac{\partial}{\partial z_i}$ and $\overline{z}$ to $\sum_i \overline{z}_i \frac{\partial}{\partial \overline{z}_i}$ when we consider z and $\overline{z}$ as vector fields. Remark that

$$\left| \frac{\partial}{\partial z_i} \right|^2 = \left| \frac{\partial}{\partial \overline{z}_i} \right|^2 = \frac{1}{2}, \quad \text{so that } |z|^2 = |\overline{z}|^2 = \frac{1}{2} |Z|^2. \tag{4.1.71}$$

From (1.6.28), set

$$\nabla_{0,\cdot} = \nabla_\cdot + \frac{1}{2} R_{x_0}^L(\mathcal{R}, \cdot), \quad \mathscr{L} = -\sum_i (\nabla_{0,e_i})^2 - \tau_{x_0}. \tag{4.1.72}$$

It is very useful to rewrite $\mathscr{L}$ by using the creation and annihilation operators. Set

$$b_i = -2\nabla_{0,\frac{\partial}{\partial z_i}}, \quad b_i^+ = 2\nabla_{0,\frac{\partial}{\partial \overline{z}_i}}, \quad b = (b_1, \dots, b_n). \tag{4.1.73}$$

Then by (1.5.18) and (4.1.72), we have

$$b_i = -2\frac{\partial}{\partial z_i} + \frac{1}{2}a_i \overline{z}_i, \quad b_i^+ = 2\frac{\partial}{\partial \overline{z}_i} + \frac{1}{2}a_i z_i, \tag{4.1.74}$$

and for any polynomial $g(z, \overline{z})$ on z and $\overline{z}$,

$$
\begin{aligned}
[b_i, b_j^+] &= b_i b_j^+ - b_j^+ b_i = -2a_i \delta_{ij}, \\
[b_i, b_j] &= [b_i^+, b_j^+] = 0, \\
[g(z, \overline{z}), b_j] &= 2\frac{\partial}{\partial z_j} g(z, \overline{z}), \\
[g(z, \overline{z}), b_j^+] &= -2\frac{\partial}{\partial \overline{z}_j} g(z, \overline{z}).
\end{aligned}
\tag{4.1.75}
$$

By (1.5.22), $\dot{R}^L \in \mathrm{End}(T^{(1,0)}X)$ is positive, thus a_i in (4.1.74) are strictly positive. From (1.5.19), (4.1.29), (4.1.72)–(4.1.75),

$$\mathscr{L} = \sum_i b_i b_i^+, \quad \mathscr{L}_2^0 = \mathscr{L} - 2\omega_{d,x_0} = \mathscr{L} + 2a_j \overline{w}^j \wedge i_{\overline{w}_j}. \tag{4.1.76}$$

As b_j is the formal adjoint of b_j^+, from (4.1.76), we get

$$\mathrm{Ker}(\mathscr{L}) = \cap_{j=1}^n \mathrm{Ker}(b_j^+). \tag{4.1.77}$$

Remark 4.1.19. Let $L = \mathbb{C}$ be the trivial holomorphic line bundle on $\mathbb{C}^n$ with the canonical frame $\mathbf{1}$, define by $\mathbb{C}^n \to L$, $z \mapsto (z, 1)$. Let h^L be the metric on L defined by

$$|\mathbf{1}|_{h^L}(z) := e^{-\frac{1}{4}\sum_{j=1}^n a_j |z_j|^2} = h(Z)$$

for $z \in \mathbb{C}^n$. Let $g^{T\mathbb{C}^n}$ be the canonical metric on $\mathbb{C}^n$. Then $\mathscr{L}$ is twice the corresponding Kodaira Laplacian $\overline{\partial}^{L*}\overline{\partial}^L$ under the trivialization of L by using the unit section $e^{\frac{1}{4}\sum_{j=1}^n a_j |z_j|^2}\mathbf{1}$. Let $\overline{\partial}^*$ be the adjoint of the Dolbeault operator $\overline{\partial}$ associated to L with the trivial metric on $(\mathbb{C}^n, g^{T\mathbb{C}^n})$. In fact, under the canonical trivialization by $\mathbf{1}$,

$$\overline{\partial}^L = \overline{\partial}, \quad \overline{\partial}^{L*} = h^{-2}\overline{\partial}^* h^2. \tag{4.1.78}$$

Set

$$\overline{\partial}_h = h\overline{\partial}h^{-1}, \quad \overline{\partial}_h^* = h^{-1}\overline{\partial}^* h. \tag{4.1.79}$$

Then

$$
\begin{aligned}
b_j^+ &= 2[i_{\frac{\partial}{\partial \overline{z}_j}}, \overline{\partial}_h], \quad b_j = [\overline{\partial}_h^*, d\overline{z}_j \wedge], \\
\overline{\partial}_h &= \frac{1}{2}\sum_j d\overline{z}_j \wedge b_j^+, \quad \overline{\partial}_h^* = \sum_j i_{\frac{\partial}{\partial \overline{z}_j}} b_j.
\end{aligned}
\tag{4.1.80}
$$

Under the trivialization by $h^{-1} \cdot \mathbf{1}$, we know the Kodaira Laplacian $\overline{\partial}^{L*} \overline{\partial}^{L} + \overline{\partial}^{L} \overline{\partial}^{L*}$ is $\overline{\partial}_h^* \overline{\partial}_h + \overline{\partial}_h \overline{\partial}_h^*$, and its restriction on functions is $\frac{1}{2}\mathscr{L}$.

Theorem 4.1.20. *The spectrum of the restriction of $\mathscr{L}$ on $L^2(\mathbb{R}^{2n})$, the space of square integrable functions on $\mathbb{R}^{2n}$, is given by*

$$\mathrm{Spec}(\mathscr{L}|_{L^2(\mathbb{R}^{2n})}) = \Big\{ 2\sum_{i=1}^{n} \alpha_i a_i \ : \ \alpha = (\alpha_1, \dots, \alpha_n) \in \mathbb{N}^n \Big\} \tag{4.1.81}$$

and an orthogonal basis of the eigenspace of $2\sum_{i=1}^{n} \alpha_i a_i$ is given by

$$b^\alpha \Big(z^\beta \exp\big(-\frac{1}{4}\sum_i a_i |z_i|^2\big)\Big), \quad \text{with } \beta \in \mathbb{N}^n. \tag{4.1.82}$$

Proof. At first $z^\beta \exp\big(-\frac{1}{4}\sum_i a_i |z_i|^2\big)$, $\beta \in \mathbb{N}^n$ are annihilated by b_i^+ $(1 \leqslant i \leqslant n)$, thus they are in the kernel of $\mathscr{L}|_{L^2(\mathbb{R}^{2n})}$. Now, by (4.1.75), (4.1.82) are eigenfunctions of $\mathscr{L}|_{L^2(\mathbb{R}^{2n})}$ with eigenvalue $2\sum_{i=1}^{n} \alpha_i a_i$. But the span of functions (4.1.82) includes all the rescaled Hermite polynomials multiplied by $\exp\big(-\frac{1}{4}\sum_i a_i |z_i|^2\big)$, which is an orthogonal basis of $L^2(\mathbb{R}^{2n})$ by Lemma E.1.3. Thus the eigenfunctions in (4.1.82) are all the eigenfunctions of $\mathscr{L}|_{L^2(\mathbb{R}^{2n})}$. The proof of Theorem 4.1.20 is complete. $\qquad\square$

Especially an orthonormal basis of $\mathrm{Ker}(\mathscr{L}|_{L^2(\mathbb{R}^{2n})})$ is

$$\Big(\frac{a^\beta}{(2\pi)^n 2^{|\beta|} \beta!} \prod_{i=1}^{n} a_i \Big)^{1/2} z^\beta \exp\Big(-\frac{1}{4}\sum_{j=1}^{n} a_j |z_j|^2 \Big), \quad \beta \in \mathbb{N}^n. \tag{4.1.83}$$

Let $\mathscr{P}(Z, Z')$ be the smooth kernel of $\mathscr{P}$, the orthogonal projection from $(L^2(\mathbb{R}^{2n}),$ $\|\cdot\|_0)$ onto $\mathrm{Ker}(\mathscr{L})$, with respect to $dv_{TX}(Z')$. From (4.1.83), we get

$$\mathscr{P}(Z, Z') = \prod_{i=1}^{n} \frac{a_i}{2\pi} \exp\Big(-\frac{1}{4}\sum_i a_i \big(|z_i|^2 + |z_i'|^2 - 2z_i \overline{z}_i'\big)\Big). \tag{4.1.84}$$

We denote by $I_{\mathbb{C}\otimes E}$ the projection from $\Lambda(T^{*(0,1)}X) \otimes E$ onto $\mathbb{C} \otimes E$ under the decomposition $\Lambda(T^{*(0,1)}X) = \mathbb{C} \oplus \Lambda^{>0}(T^{*(0,1)}X)$. Let P^N be the orthogonal projection from $(L^2(\mathbb{R}^{2n}, \mathbf{E}_{x_0}), \|\cdot\|_0 = \|\cdot\|_{t,0})$ onto $N = \mathrm{Ker}(\mathscr{L}_2^0)$, and $P^N(Z, Z')$ its smooth kernel with respect to $dv_{TX}(Z')$. Then from (4.1.76), we get

$$P^N(Z, Z') = \mathscr{P}(Z, Z') I_{\mathbb{C}\otimes E}. \tag{4.1.85}$$

4.1.7 Proof of Theorem 4.1.1

Let $f(\lambda, t)$ be a formal power series on t with values in $\mathrm{End}(L^2(\mathbb{R}^{2n}, \mathbf{E}_{x_0}))$,

$$f(\lambda, t) = \sum_{r=0}^{\infty} t^r f_r(\lambda), \quad f_r(\lambda) \in \mathrm{End}(L^2(\mathbb{R}^{2n}, \mathbf{E}_{x_0})). \tag{4.1.86}$$

By (4.1.31), consider the equation of formal power series for $\lambda \in \delta$,

$$(-\mathscr{L}_2^0 + \lambda - \sum_{r=1}^{\infty} t^r \mathcal{O}_r) f(\lambda, t) = \mathrm{Id}_{L^2(\mathbb{R}^{2n}, \mathbf{E}_{x_0})}. \qquad (4.1.87)$$

Let $N^\perp$ be the orthogonal space of $N = \mathrm{Ker}(\mathscr{L}_2^0)$ in $L^2(\mathbb{R}^{2n}, \mathbf{E}_{x_0})$, and $P^{N^\perp}$ be the orthogonal projection from $L^2(\mathbb{R}^{2n}, \mathbf{E}_{x_0})$ onto $N^\perp$. We decompose $f(\lambda, t)$ according to the splitting $L^2(\mathbb{R}^{2n}, \mathbf{E}_{x_0}) = N \oplus N^\perp$,

$$g_r(\lambda) = P^N f_r(\lambda), \quad f_r^\perp(\lambda) = P^{N^\perp} f_r(\lambda). \qquad (4.1.88)$$

Using (4.1.88) and identifying the powers of t in (4.1.87), we find that

$$g_0(\lambda) = \frac{1}{\lambda} P^N, \quad f_0^\perp(\lambda) = (\lambda - \mathscr{L}_2^0)^{-1} P^{N^\perp},$$

$$f_r^\perp(\lambda) = (\lambda - \mathscr{L}_2^0)^{-1} \sum_{j=1}^{r} P^{N^\perp} \mathcal{O}_j f_{r-j}(\lambda), \qquad (4.1.89)$$

$$g_r(\lambda) = \frac{1}{\lambda} \sum_{j=1}^{r} P^N \mathcal{O}_j f_{r-j}(\lambda).$$

Theorem 4.1.21. *There exist $J_r(Z, Z') \in \mathrm{End}(\Lambda(T^{*(0,1)}X) \otimes E)_{x_0}$ polynomials in Z, Z' with the same parity as r and $\deg J_r(Z, Z') \leqslant 3r$, whose coefficients are polynomials in R^{TX}, $R^{B, \Lambda^{0, \bullet}}$, $R^{\det}$, R^E ($d\Theta$ and R^L) and their derivatives of order $\leqslant r-2$ (resp. $r-1$, r), and reciprocals of linear combinations of eigenvalues of $\dot{R}^L$ at x_0, such that*

$$\mathscr{F}_r(Z, Z') = J_r(Z, Z') \mathscr{P}(Z, Z'), \quad J_0(Z, Z') = \mathrm{Id}_{\mathbb{C} \otimes E}. \qquad (4.1.90)$$

Proof. By (4.1.59), $\mathcal{P}_{0,t} = \frac{1}{2\pi i} \int_\delta (\lambda - \mathscr{L}_2^t)^{-1} d\lambda$. Thus by (4.1.63), (4.1.65), (4.1.69) and (4.1.88), we know that the operator $\mathscr{F}_r$ does not depend on the choice $k \geqslant 1$ in (4.1.66), thus

$$\mathscr{F}_r = \frac{1}{2\pi i} \int_\delta g_r(\lambda) d\lambda + \frac{1}{2\pi i} \int_\delta f_r^\perp(\lambda) d\lambda. \qquad (4.1.91)$$

From Theorem 4.1.20, (1.5.22) and (4.1.76), the only eigenvalue of $\mathscr{L}_2^0$ inside δ is 0. From (4.1.85), (4.1.89) and (4.1.91), we get

$$\mathscr{F}_0 = P^N = \mathscr{P} \, \mathrm{Id}_{\mathbb{C} \otimes E}. \qquad (4.1.92)$$

Generally, from Theorems 4.1.7, 4.1.20, Remark 4.1.8, (4.1.85), (4.1.89), (4.1.91) and the residue formula, we conclude that $\mathscr{F}_r$ has the form (4.1.90). $\qquad \square$

From Theorem 4.1.20, (4.1.76), (4.1.89), (4.1.91) and the residue formula, we can get $\mathscr{F}_r$ by using the operators $(\mathscr{L}_2^0)^{-1}$, P^N, $P^{N^\perp}$, $\mathcal{O}_k(k \leqslant r)$. This gives us a general method to compute $\mathscr{F}_r$ in view of Theorem 4.1.20. Especially, from (4.1.89) and (4.1.91), we get the following formula for $\mathscr{F}_1$, $\mathscr{F}_2$.

Theorem 4.1.22. *The following identities hold,*

$$\mathscr{F}_1 = -P^{N^\perp}(\mathscr{L}_2^0)^{-1}\mathcal{O}_1 P^N - P^N\mathcal{O}_1(\mathscr{L}_2^0)^{-1}P^{N^\perp},$$

$$\begin{aligned}
\mathscr{F}_2 = &(\mathscr{L}_2^0)^{-1}P^{N^\perp}\mathcal{O}_1(\mathscr{L}_2^0)^{-1}P^{N^\perp}\mathcal{O}_1 P^N - (\mathscr{L}_2^0)^{-1}P^{N^\perp}\mathcal{O}_2 P^N \\
&+ P^N\mathcal{O}_1(\mathscr{L}_2^0)^{-1}P^{N^\perp}\mathcal{O}_1(\mathscr{L}_2^0)^{-1}P^{N^\perp} - P^N\mathcal{O}_2(\mathscr{L}_2^0)^{-1}P^{N^\perp} \\
&+ (\mathscr{L}_2^0)^{-1}P^{N^\perp}\mathcal{O}_1 P^N\mathcal{O}_1(\mathscr{L}_2^0)^{-1}P^{N^\perp} - P^N\mathcal{O}_1(\mathscr{L}_2^0)^{-2}P^{N^\perp}\mathcal{O}_1 P^N \\
&- P^N\mathcal{O}_1 P^N\mathcal{O}_1(\mathscr{L}_2^0)^{-2}P^{N^\perp} - P^{N^\perp}(\mathscr{L}_2^0)^{-2}\mathcal{O}_1 P^N\mathcal{O}_1 P^N.
\end{aligned} \tag{4.1.93}$$

Remark 4.1.23. In Theorem 8.3.9 and Problem 8.5, we will show that even in the symplectic case, we have

$$P^N\mathcal{O}_1 P^N = 0. \tag{4.1.94}$$

Thus the last two terms in (4.1.93) are zero.

The following near diagonal expansion of the Bergman kernels is the main result of this section.

Theorem 4.1.24. *For any $k, m, m' \in \mathbb{N}$, $q > 0$, there exists $C > 0$ such that if $p \geqslant 1$, $Z, Z' \in T_{x_0}X$, $|Z|, |Z'| \leqslant q/\sqrt{p}$,*

$$\sup_{|\alpha|+|\alpha'|\leqslant m}\left| \frac{\partial^{|\alpha|+|\alpha'|}}{\partial Z^\alpha \partial Z'^{\alpha'}}\left(\frac{1}{p^n}P_p(Z, Z')\right.\right.$$

$$\left.\left. -\sum_{r=0}^{k}\mathscr{F}_r(\sqrt{p}Z, \sqrt{p}Z')\kappa^{-\frac{1}{2}}(Z)\kappa^{-\frac{1}{2}}(Z')p^{-\frac{r}{2}}\right)\right|_{\mathscr{C}^{m'}(X)} \leqslant Cp^{-\frac{k-m+1}{2}}. \tag{4.1.95}$$

Proof. By (4.1.28) and (4.1.29) as in (1.6.66), we have for $Z, Z' \in \mathbb{R}^{2n}$,

$$P_{0,p}(Z, Z') = t^{-2n}\kappa^{-\frac{1}{2}}(Z)P_{0,t}(Z/t, Z'/t)\kappa^{-\frac{1}{2}}(Z'). \tag{4.1.96}$$

By Proposition 4.1.6, Theorem 4.1.18 and (4.1.96), we get (4.1.95). $\qquad\square$

Proof of Theorem 4.1.1. Set now $Z = Z' = 0$ in (4.1.95). Since Theorem 4.1.21 implies $\mathscr{F}_{2r+1}(0,0) = 0$, we obtain (4.1.7) when we restrict to degree 0 and

$$\boldsymbol{b}_r(x_0) = I_{\mathbb{C}\otimes E}\mathscr{F}_{2r}(0,0)I_{\mathbb{C}\otimes E}. \tag{4.1.97}$$

Hence we obtain $\boldsymbol{b}_0$ from (4.1.84), (4.1.90) and (4.1.97).

By (1.2.38), (1.2.42), (1.2.48), (1.2.51) and (1.4.27), the curvatures $R^{B,\Lambda^{0,\bullet}}$ and $R^{\det}$ are linear combinations of R^{TX} and functions on $d\Theta$ and its first derivatives.

Now the statement about the structure of $\boldsymbol{b}_r$ follows from Theorem 4.1.21. In fact, we claim that for any $r \in \mathbb{N}$, $\mathscr{F}_r$ preserves the $\mathbb{Z}$-grading of $\mathbf{E}_{x_0}$, and

$$\mathscr{F}_r(Z, Z')|_{\Lambda^{>0}(T^{*(0,1)}X)\otimes E} = 0. \tag{4.1.98}$$

Recall that in Section 4.1.3 we use the connection $\nabla^{B,\Lambda^{0,\bullet}}$ to trivialize the bundle $\Lambda(T^{*0,1}X)$ on $B^X(x_0, 4\varepsilon)$, thus the trivialization preserves the $\mathbb{Z}$-grading of $\Lambda(T^{*0,1}X)$. As D_p^2 preserves the $\mathbb{Z}$-grading of $\Lambda(T^{*0,1}X)$, we know that on $B^{T_{x_0}X}(0, 2\varepsilon/t)$, $\mathscr{L}_2^t$ (thus $\mathscr{O}_r$) preserves the $\mathbb{Z}$-grading of $\mathbf{E}_{x_0} = (\Lambda(T^{*0,1}X) \otimes E)_{x_0}$. As a consequence, we know that $\mathscr{F}_r(Z, Z')$ also preserves the $\mathbb{Z}$-grading of $\mathbf{E}_{x_0}$. But from (4.1.76), $(\lambda - \mathscr{L}_2^0)^{-1}|_{\Lambda^{>0}(T^{*(0,1)}X)\otimes E}$ is holomorphic for $|\lambda| \leqslant \mu_0$, thus (4.1.89) and (4.1.91) yield (4.1.98).

To prove the uniformity part of Theorem 4.1.1, we notice that in the proof of Theorem 4.1.16, we only use the derivatives of the coefficients of $\mathscr{L}_2^t$ with order $\leqslant 2n + m + m' + r + 2$. Thus by (4.1.70), the constants in Theorems 4.1.16, 4.1.17, (resp. Theorem 4.1.18) are uniformly bounded, if with respect to a fixed metric g_0^{TX}, the $\mathscr{C}^{2n+m+m'+r+4}$ (resp. $\mathscr{C}^{2n+m+m'+k+5}$)-norms on X of the data (g^{TX}, h^L, h^E) are bounded, and g^{TX} is bounded below (as the coefficients of $\mathscr{L}_2^t$ are functions of g^{TX}, h^L, h^E and their derivatives with order $\leqslant 2$). Moreover, taking derivatives with respect to the parameters we obtain a similar equation as (4.1.64), where $x_0 \in X$ plays now a role of a parameter. Thus the $\mathscr{C}^{m'}$-norm in (4.1.95) can also include the parameters if the $\mathscr{C}^{m'}$-norms (with respect to the parameter $x_0 \in X$) of the derivatives of the above data with order $\leqslant 2n + m + k + 5$ are bounded. Thus we can take $C_{k,l}$ in (4.1.7) independent of g^{TX} under our condition (we apply (4.1.95) with k replaced by $2k+1$ to get (4.1.7)). This achieves the proof of Theorem 4.1.1. $\qquad\square$

4.1.8   The coefficient b_1: a proof of Theorem 4.1.2

From the end of Section 4.1.7, to compute the coefficient $\mathbf{b}_r$ in (4.1.7), we only need to do our computation on $\mathscr{C}^\infty(X, L^p \otimes E)$. Thus in this subsection, our operators $\mathscr{L}_t, \nabla_{t,\cdot}, \mathscr{L}_0, \mathscr{O}_r$ are the restriction of the operators $\mathscr{L}_2^t, \nabla_{t,\cdot}, \mathscr{L}_2^0, \mathscr{O}_r$ in (4.1.31) on $\mathscr{C}^\infty(\mathbb{R}^{2n}, \mathbf{E}_{x_0})$.

Recall that $\{e_i\}$ is an orthonormal basis of $T_{x_0}X$ as in Section 4.1.6. We assume here $\omega = \Theta$. Thus in (1.5.19) and (4.1.74), for $1 \leqslant j \leqslant n$,

$$a_j = 2\pi. \tag{4.1.99}$$

Theorem 4.1.25. *The following identities hold:*

$$\mathscr{L}_0 = \sum_j b_j b_j^+ = \mathscr{L}, \qquad \mathscr{O}_1 = 0,$$

$$\mathscr{O}_2 = \frac{1}{3}\left\langle R_{x_0}^{TX}(\mathcal{R}, e_i)\mathcal{R}, e_j \right\rangle \nabla_{0,e_i}\nabla_{0,e_j} - R_{x_0}^E(w_j, \overline{w}_j) - \frac{r_{x_0}^X}{6} \tag{4.1.100}$$
$$+ \left(\left\langle \frac{1}{3}R_{x_0}^{TX}(\mathcal{R}, e_k)e_k + \frac{\pi}{3}R_{x_0}^{TX}(z, \overline{z})\mathcal{R}, e_j \right\rangle - R_{x_0}^E(\mathcal{R}, e_j)\right)\nabla_{0,e_j}.$$

Proof. By (1.2.19) and (4.1.28),

$$\kappa(Z) = |\det(g_{ij}(Z))|^{1/2} = 1 + \frac{1}{6}\left\langle R_{x_0}^{TX}(\mathcal{R}, e_j)\mathcal{R}, e_j \right\rangle_{x_0} + \mathscr{O}(|Z|^3). \tag{4.1.101}$$

If Γ_{ij}^l are the Christoffel symbols of ∇^{TX} with respect to the frame $\{e_i\}$, then $(\nabla_{e_i}^{TX} e_j)(Z) = \Gamma_{ij}^l(Z)e_l$. By (1.2.1) and (1.2.19), with $\partial_j := \frac{\partial}{\partial Z_j}$,

$$\Gamma_{ij}^l(Z) = \frac{1}{2}\sum_k g^{lk}(\partial_i g_{jk} + \partial_j g_{ik} - \partial_k g_{ij})(Z) \tag{4.1.102}$$

$$= \frac{1}{3}\Big[\big\langle R_{x_0}^{TX}(\mathcal{R}, e_j)e_i, e_l\big\rangle_{x_0} + \big\langle R_{x_0}^{TX}(\mathcal{R}, e_i)e_j, e_l\big\rangle_{x_0}\Big] + \mathscr{O}(|Z|^2).$$

Observe that J is parallel with respect to ∇^{TX}, thus $\langle J\tilde{e}_i, \tilde{e}_j\rangle_Z = \langle Je_i, e_j\rangle_{x_0}$. By (1.2.12), (1.2.21) and (1.2.27),

$$\frac{\sqrt{-1}}{2\pi} R_Z^L(\mathcal{R}, e_l) = \theta_k^i(Z)\theta_l^j(Z)\,\langle J\tilde{e}_i, \tilde{e}_j\rangle_Z\, Z_k \tag{4.1.103}$$

$$= \langle J\mathcal{R}, e_l\rangle_{x_0} + \frac{1}{6}\big\langle R_{x_0}^{TX}(\mathcal{R}, J\mathcal{R})\mathcal{R}, e_l\big\rangle_{x_0} + \mathscr{O}(|Z|^4).$$

By (1.2.30), (4.1.29) and (4.1.103), for $t = \frac{1}{\sqrt{p}}$, we get

$$\nabla_{t,e_i} = \nabla_{e_i} + \Big(\frac{1}{t}\Gamma^L(e_i) + t\Gamma^E(e_i) - \frac{t}{2\kappa}\nabla_{e_i}\kappa\Big)(tZ) \tag{4.1.104}$$

$$= \nabla_{0,e_i} - \frac{t^2}{6}\big\langle \pi R_{x_0}^{TX}(z, \bar{z})\mathcal{R} - R_{x_0}^{TX}(\mathcal{R}, e_k)e_k, e_i\big\rangle + \frac{t^2}{2}R_{x_0}^E(\mathcal{R}, e_i) + \mathscr{O}(t^3).$$

Let $\{\tilde{w}_i(Z)\}_i$ be the parallel transport of $\{w_i\}_i$ along the curve $[0,1] \ni u \to uZ$. By (1.4.31), as in (4.1.34), when we restrict on $\mathscr{C}^\infty(\mathbb{R}^{2n}, E_{x_0})$, we get,

$$\mathscr{L}_t = -g^{ij}(tZ)\Big[\nabla_{t,e_i}\nabla_{t,e_j} - t\Gamma_{ij}^l(t\cdot)\nabla_{t,e_l}\Big](Z) - t^2 R^E(\tilde{w}_i, \overline{\tilde{w}}_i)(tZ) - 2\pi n. \tag{4.1.105}$$

From the fact that R^{TX} is a $(1,1)$-form with values in $\mathrm{End}(TX)$, we get

$$\nabla_{e_j}\big\langle R_{x_0}^{TX}(z, \bar{z})\mathcal{R}, e_j\big\rangle$$

$$= 2\Big(\frac{\partial}{\partial\bar{z}_j}\big\langle R_{x_0}^{TX}(z, \bar{z})\bar{z}, \frac{\partial}{\partial z_j}\big\rangle + \frac{\partial}{\partial z_j}\big\langle R_{x_0}^{TX}(z, \bar{z})z, \frac{\partial}{\partial\bar{z}_j}\big\rangle\Big) = 0. \tag{4.1.106}$$

From (1.2.5), (4.1.102), (4.1.104)–(4.1.106), we derive (4.1.100). $\qquad\square$

Proof of Theorem 4.1.2. Recall that we do all our computations on $\mathscr{C}^\infty(\mathbb{R}^{2n}, E_{x_0})$, thus here we still use $\mathscr{P}$ to denote the orthogonal projection from $L^2(\mathbb{R}^{2n}, E_{x_0})$ onto $\mathrm{Ker}(\mathscr{L})$ and set $\mathscr{P}^\perp = 1 - \mathscr{P}$. By Theorem 4.1.22 and (4.1.100), we know that

$$\mathscr{F}_2 = -\mathscr{L}^{-1}\mathscr{P}^\perp O_2\mathscr{P} - \mathscr{P}O_2\mathscr{L}^{-1}\mathscr{P}^\perp. \tag{4.1.107}$$

The second term is the adjoint of the first term by Remark 4.1.8. Thus we only need to compute the first term of (4.1.107).

Note that by (4.1.74), (4.1.77), (4.1.84) and (4.1.99),

$$(b_i^+ \mathscr{P})(Z, Z') = 0, \quad (b_i \mathscr{P})(Z, Z') = 2\pi(\overline{z}_i - \overline{z}_i') \mathscr{P}(Z, Z'). \tag{4.1.108}$$

Thus from Theorem 4.1.20, (4.1.73), (4.1.75), (4.1.100) and (4.1.108), and R^{TX} is a $(1, 1)$-form, we get

$$
\begin{aligned}
(\mathscr{P}^\perp \mathcal{O}_2 \mathscr{P})(Z, 0) &= \Big\{ \mathscr{P}^\perp \Big[\frac{1}{3} \Big\langle R^{TX}(\mathcal{R}, \tfrac{\partial}{\partial \overline{z}_i})\mathcal{R}, \tfrac{\partial}{\partial \overline{z}_j} \Big\rangle b_i b_j \\
&\quad - \frac{4\pi}{3} \Big\langle R^{TX}(\mathcal{R}, \tfrac{\partial}{\partial z_k})\mathcal{R}, \tfrac{\partial}{\partial \overline{z}_k} \Big\rangle - \frac{2}{3} \Big\langle R^{TX}(\mathcal{R}, \tfrac{\partial}{\partial \overline{z}_k})\tfrac{\partial}{\partial z_k}, \tfrac{\partial}{\partial \overline{z}_j} \Big\rangle b_j \\
&\quad - \frac{\pi}{3} \Big\langle R^{TX}(z, \overline{z})z, \tfrac{\partial}{\partial \overline{z}_j} \Big\rangle b_j + R^E(\mathcal{R}, \tfrac{\partial}{\partial \overline{z}_j})b_j \Big] \mathscr{P} \Big\}(Z, 0) \\
&= \Big\{ \mathscr{P}^\perp \Big[\frac{1}{6} \Big\langle R^{TX}(z, \tfrac{\partial}{\partial \overline{z}_i})z, \tfrac{\partial}{\partial \overline{z}_j} \Big\rangle b_i b_j + R^E(z, \tfrac{\partial}{\partial \overline{z}_j})b_j \Big] \mathscr{P} \Big\}(Z, 0) \\
&= \Big\{ \Big[\frac{b_i b_j}{6} \Big\langle R^{TX}(z, \tfrac{\partial}{\partial \overline{z}_i})z, \tfrac{\partial}{\partial \overline{z}_j} \Big\rangle + \frac{4b_i}{3} \Big\langle R^{TX}(z, \tfrac{\partial}{\partial \overline{z}_i})\tfrac{\partial}{\partial z_j}, \tfrac{\partial}{\partial \overline{z}_j} \Big\rangle \\
&\quad + b_j R^E(z, \tfrac{\partial}{\partial \overline{z}_j}) \Big] \mathscr{P} \Big\}(Z, 0).
\end{aligned}
\tag{4.1.109}
$$

Thus by Theorem 4.1.20, (1.2.5), (4.1.74), (4.1.84), (4.1.99) and (4.1.109), we obtain

$$
\begin{aligned}
-(\mathscr{L}^{-1} \mathscr{P}^\perp \mathcal{O}_2 \mathscr{P})(0, 0) &= -\Big\{ \Big[\frac{b_i b_j}{48\pi} \Big\langle R^{TX}(z, \tfrac{\partial}{\partial \overline{z}_i})z, \tfrac{\partial}{\partial \overline{z}_j} \Big\rangle \\
&\quad + \frac{b_i}{3\pi} \Big\langle R^{TX}(z, \tfrac{\partial}{\partial \overline{z}_i})\tfrac{\partial}{\partial z_j}, \tfrac{\partial}{\partial \overline{z}_j} \Big\rangle + \frac{b_j}{4\pi} R^E(z, \tfrac{\partial}{\partial \overline{z}_j}) \Big] \mathscr{P} \Big\}(0, 0) \\
&= \frac{1}{2\pi} \Big\langle R^{TX}(\tfrac{\partial}{\partial z_i}, \tfrac{\partial}{\partial \overline{z}_i})\tfrac{\partial}{\partial z_j}, \tfrac{\partial}{\partial \overline{z}_j} \Big\rangle + \frac{1}{2\pi} R^E(\tfrac{\partial}{\partial z_j}, \tfrac{\partial}{\partial \overline{z}_j}) \\
&= \frac{1}{16\pi} r_{x_0}^X + \frac{1}{2\pi} R^E(\tfrac{\partial}{\partial z_j}, \tfrac{\partial}{\partial \overline{z}_j}).
\end{aligned}
\tag{4.1.110}
$$

From (4.1.97), (4.1.107) and (4.1.110), we get (4.1.8).

Since $\omega = \Theta$, we have $d\Theta = 0$ and $R^{TX} = R^{T^{(1,0)}X}$. Hence the $2r$-order derivatives of R^L are the $(2r - 2)$-order derivatives of R^{TX}. This follows by (4.1.103), as J is parallel with respect to ∇^{TX}. Applying Theorem 4.1.1 we conclude the proof of Theorem 4.1.2. □

Remark 4.1.26. From Theorem 4.1.22 and (4.1.100), we see that under the hypothesis of Theorem 4.1.2, we have

$$\mathscr{F}_1 = 0. \tag{4.1.111}$$

4.1.9   Proof of Theorem 4.1.3

We define a metric on E by $h_\omega^E := \det(\frac{\dot{R}^L}{2\pi})^{-1} h^E$. Let R_ω^E be the curvature associated to the holomorphic Hermitian connection of (E, h_ω^E); then

$$R_\omega^E = R^E - \overline{\partial}\partial \log \det(\dot{R}^L). \tag{4.1.112}$$

Let $\langle \cdot, \cdot \rangle_\omega$ be the Hermitian product on $\mathscr{C}^\infty(X, L^p \otimes E)$ induced by $g_\omega^{TX}, h^L, h_\omega^E$. Therefore,

$$\begin{aligned}
(\mathscr{C}^\infty(X, L^p \otimes E), \langle \cdot, \cdot \rangle_\omega) &= (\mathscr{C}^\infty(X, L^p \otimes E), \langle \cdot, \cdot \rangle), \\
dv_{X,\omega} &= (2\pi)^{-n} \det(\dot{R}^L) dv_X.
\end{aligned} \tag{4.1.113}$$

Observe that $H^0(X, L^p \otimes E)$ does not depend on g^{TX}, h^L or h^E. If $P_{\omega,p}(x, x')$, $(x, x' \in X)$ denotes the smooth kernel of the orthogonal projection from $(\mathscr{C}^\infty(X, L^p \otimes E), \langle \cdot, \cdot \rangle_\omega)$ onto $H^0(X, L^p \otimes E)$ with respect to $dv_{X,\omega}(x)$, we have

$$P_p(x, x') = (2\pi)^{-n} \det(\dot{R}^L)(x') P_{\omega,p}(x, x'). \tag{4.1.114}$$

Now we can apply Theorem 4.1.2 for the kernel $P_{\omega,p}(x, x')$, since $g_\omega^{TX}(\cdot, \cdot) = \omega(\cdot, J\cdot)$ is a Kähler metric on TX. Hence (4.1.9) follows from (4.1.8), (4.1.112) and (4.1.114).

Remark 4.1.27. Since X is compact, (4.1.114) allowed us to reduce the general situation considered in Theorem 4.1.1 to the case $\omega = \Theta$ and apply Theorem 4.1.2. However, if X is not compact, the trick of using (4.1.114) does not work anymore, because the operator associated to $g_\omega^{TX}, h^L, h_\omega^E$ might not have a spectral gap (cf. Section 6.1). More generally, the approach here still holds if the curvature R^L is just non-degenerate and our argument extends to the symplectic case without any modification, cf. Chapter 8.

4.2   Off-diagonal expansion of the Bergman kernel

In this section, we study in detail the relation on the asymptotic expansion of the Bergman kernel and of the heat kernel. Especially, we establish the off-diagonal asymptotic expansion of the Bergman kernel by using the heat kernel. Again the spectral gap property plays an essential role in our approach.

This section is organized as follows. In Section 4.2.1, we explain the relation between the asymptotic expansions of the Bergman kernel and of the heat kernel, and we state the off-diagonal asymptotic expansion of the Bergman kernel. In Section 4.2.2, we study the uniform estimate on the heat kernel and the Bergman kernel. In Section 4.2.3, we prove the results stated in Section 4.2.1.

We use the notation from Sections 1.6.1, 4.1, and we suppose that the positivity condition (1.5.21) holds for R^L.

4.2.1   From heat kernel to Bergman kernel

In Section 4.1.3, on $B^X(x_0, 4\varepsilon) \simeq B^{T_{x_0}X}(0, 4\varepsilon)$, we trivialize E_p (cf. (1.6.1)) by using the parallel transport with respect to the connection ∇^{B, E_p} defined by (1.4.27) along the curve $[0, 1] \ni u \to uZ$. Recall that $\mathbf{E} = \Lambda(T^{*(0,1)}X) \otimes E$. Recall that we denote by $\mathscr{C}^m(X)$ the $\mathscr{C}^m$-norm for the parameter $x_0 \in X$.

Under this trivialization, for $Z, Z' \in T_{x_0}X$,

$$P_p(Z, Z') \in \mathrm{End}(E_{p,x_0}) = \mathrm{End}(\mathbf{E}_{x_0}).$$

Thus as in Section 4.1.5, we view $P_{p,x_0}(Z, Z') := P_p(Z, Z')$, $(Z, Z' \in T_{x_0}X$, $|Z|, |Z'| \leqslant 2\varepsilon)$, as a smooth section of $\pi^*(\mathrm{End}(\mathbf{E}))$ over $TX \times_X TX$.

Recall that $\mu_0, \kappa, \mathscr{F}_r$ were defined in (1.5.26), (4.1.28), (4.1.65) and (4.1.91). The following off-diagonal expansion of the Bergman kernel which extends Theorem 4.1.24, is the main result of this section.

Theorem 4.2.1. *There exists $C'' > 0$ such that for any $k, m, m' \in \mathbb{N}$, there exists $C > 0$ such that for $p \geqslant 1$, $Z, Z' \in T_{x_0}X, |Z|, |Z'| \leqslant 2\varepsilon$, $\alpha, \alpha' \in \mathbb{N}^{2n}, |\alpha|+|\alpha'| \leqslant m$, we have*

$$\left| \frac{\partial^{|\alpha|+|\alpha'|}}{\partial Z^\alpha \partial Z'^{\alpha'}} \left(\frac{1}{p^n} P_p(Z, Z') - \sum_{r=0}^{k} \mathscr{F}_r(\sqrt{p}Z, \sqrt{p}Z')\kappa^{-\frac{1}{2}}(Z)\kappa^{-\frac{1}{2}}(Z')p^{-\frac{r}{2}} \right) \right|_{\mathscr{C}^{m'}(X)}$$

$$\leqslant C p^{-(k+1-m)/2}(1 + \sqrt{p}|Z| + \sqrt{p}|Z'|)^{M_{k+1,m,m'}} \exp(-\sqrt{C''\mu_0}\sqrt{p}|Z - Z'|)$$
$$+ \mathcal{O}(p^{-\infty}), \tag{4.2.1}$$

with

$$M_{k,m,m'} = 2(n + k + m' + 1) + m. \tag{4.2.2}$$

The term $\mathcal{O}(p^{-\infty})$ means that for any $l, l_1 \in \mathbb{N}$, there exists $C_{l,l_1} > 0$ such that its $\mathscr{C}^{l_1}$-norm is dominated by $C_{l,l_1}p^{-l}$.

Remark 4.2.2. In (4.2.2), we make precise the exponent $M_{k+1,m,m'}$ in (4.2.1). For the most applications, we only need to know the existence of the constant $M_{k+1,m,m'}$.

The following result (and more generally (4.2.35)) relates the coefficients of the expansion of the Bergman kernel and the heat kernel.

Theorem 4.2.3. *There exist smooth sections $b_{r,u}$ of $\mathrm{End}(\Lambda(T^{*(0,1)}X) \otimes E)$ on X which are polynomials in R^{TX}, R^E ($d\Theta$ and R^L) and their derivatives with order $\leqslant 2r - 2$ (resp. $2r - 1$, $2r$) and functions on the eigenvalues of $\dot{R}^L$ at x, and*

$$b_{0,u} = (2\pi)^{-n} \frac{\det(\dot{R}^L) \exp(2u\omega_d)}{\det(1 - \exp(-2u\dot{R}^L))}, \tag{4.2.3}$$

such that for each $u > 0$ fixed, we have the asymptotic expansion in the sense of (4.1.7) as $p \to \infty$,

$$\exp(-\frac{u}{p}D_p^2)(x, x) = \sum_{r=0}^{k} b_{r,u}(x)p^{n-r} + \mathcal{O}(p^{n-k-1}). \tag{4.2.4}$$

Moreover, as $u \to +\infty$, with $\mathbf{b}_r$ in (4.1.7), we have

$$b_{r,u}(x) = \mathbf{b}_r(x) I_{\mathbb{C}\otimes E} + \mathcal{O}(e^{-\frac{1}{8}\mu_0 u}). \tag{4.2.5}$$

Remark 4.2.4. The formula (4.2.4) holds without the assumption (1.5.21) and $b_{0,u}$ is computed in Theorem 1.6.1. In fact, we only need to replace our contour $\delta \cup \Delta$ by Γ in Section 1.6.4, and Theorems 4.1.10–4.1.14 hold for $\lambda \in \Gamma$ without the assumption (1.5.21). Thus we still get the second equation of (4.2.30).

4.2.2 Uniform estimate on the heat kernel and the Bergman kernel

Let $e^{-u\mathscr{L}_2^t}(Z, Z')$, $(\mathscr{L}_2^t e^{-u\mathscr{L}_2^t})(Z, Z')$ $(Z, Z' \in T_{x_0}X)$ be the smooth kernels of the operators $e^{-u\mathscr{L}_2^t}$, $\mathscr{L}_2^t e^{-u\mathscr{L}_2^t}$ with respect to $dv_{TX}(Z')$. We view $e^{-u\mathscr{L}_2^t}(Z, Z')$, $(\mathscr{L}_2^t e^{-u\mathscr{L}_2^t})(Z, Z')$ as smooth sections of $\pi^*(\mathrm{End}(\mathbf{E}))$ on $TX \times_X TX$.

Theorem 4.2.5. *There exists $C'' > 0$ such that for any $m, m', r \in \mathbb{N}$, $u_0 > 0$, there exists $C > 0$ such that for $t \in]0, t_0]$, $u \geqslant u_0$, $Z, Z' \in T_{x_0}X$,*

$$\sup_{|\alpha|+|\alpha'|\leqslant m} \left| \frac{\partial^{|\alpha|+|\alpha'|}}{\partial Z^\alpha \partial Z'^{\alpha'}} \frac{\partial^r}{\partial t^r} e^{-u\mathscr{L}_2^t}(Z, Z') \right|_{\mathscr{C}^{m'}(X)}$$

$$\leqslant C(1 + |Z| + |Z'|)^{M_{r,m,m'}} \exp\left(\frac{1}{2}\mu_0 u - \frac{2C''}{u}|Z - Z'|^2 \right),$$

$$\sup_{|\alpha|+|\alpha'|\leqslant m} \left| \frac{\partial^{|\alpha|+|\alpha'|}}{\partial Z^\alpha \partial Z'^{\alpha'}} \frac{\partial^r}{\partial t^r} (\mathscr{L}_2^t e^{-u\mathscr{L}_2^t})(Z, Z') \right|_{\mathscr{C}^{m'}(X)}$$ (4.2.6)

$$\leqslant C(1 + |Z| + |Z'|)^{M_{r,m,m'}} \exp\left(-\frac{1}{4}\mu_0 u - \frac{2C''}{u}|Z - Z'|^2 \right).$$

Proof. By (4.1.40), for any $k \in \mathbb{N}^*$, $t \in]0, t_0]$,

$$e^{-u\mathscr{L}_2^t} = \frac{(-1)^{k-1}(k-1)!}{2\pi i u^{k-1}} \int_{\delta \cup \Delta} e^{-u\lambda}(\lambda - \mathscr{L}_2^t)^{-k} d\lambda,$$ (4.2.7)

$$\mathscr{L}_2^t e^{-u\mathscr{L}_2^t} = \frac{(-1)^{k-1}(k-1)!}{2\pi i u^{k-1}} \int_\Delta e^{-u\lambda} \left[\lambda(\lambda - \mathscr{L}_2^t)^{-k} - (\lambda - \mathscr{L}_2^t)^{-k+1} \right] d\lambda.$$

From (4.1.48), (4.1.49) and (4.2.7), there exists $C_m > 0$ such that for $u \geqslant u_0$, $Q, Q' \in \mathcal{Q}^m$, we have

$$\|Q e^{-u\mathscr{L}_2^t} Q'\|_t^{0,0} \leqslant C_m e^{\frac{1}{4}\mu_0 u},$$

$$\|Q(\mathscr{L}_2^t e^{-u\mathscr{L}_2^t})Q'\|_t^{0,0} \leqslant C_m e^{-\frac{1}{2}\mu_0 u}.$$ (4.2.8)

For $m \geqslant 0$, let $\| \ \|_m$ be the usual Sobolev norm on $\mathscr{C}^\infty(\mathbb{R}^{2n}, \mathbf{E}_{x_0})$ induced by $h^{\mathbf{E}_{x_0}} = h^{\Lambda(T_{x_0}^{*(0,1)}X)\otimes E_{x_0}}$ and the volume form $dv_{TX}(Z)$ as in (A.1.10). Observe that by (4.1.34) and (4.1.36), there exists $C > 0$ such that for $s \in \mathscr{C}^\infty(X_0, \mathbf{E}_{x_0})$, $\mathrm{supp}\, s \subset B^{T_{x_0}X}(0, q)$, $m \geqslant 0$,

$$\frac{1}{C}(1 + q)^{-m}\|s\|_{t,m} \leqslant \|s\|_m \leqslant C(1 + q)^m\|s\|_{t,m}.$$ (4.2.9)

Now (4.2.8), (4.2.9) together with Sobolev's inequalities implies that if $Q, Q' \in \mathcal{Q}^m$,

$$\sup_{|Z|,|Z'| \leqslant q} |Q_Z Q'_{Z'} e^{-u\mathcal{L}_2^t}(Z, Z')| \leqslant C(1+q)^{2n+2} \, e^{\frac{1}{4}\mu_0 u},$$
$$\sup_{|Z|,|Z'| \leqslant q} |Q_Z Q'_{Z'} (\mathcal{L}_2^t e^{-u\mathcal{L}_2^t})(Z, Z')| \leqslant C(1+q)^{2n+2} \, e^{-\frac{1}{2}\mu_0 u}. \tag{4.2.10}$$

Thus by (4.1.34) and (4.2.10), we derive (4.2.6) with the exponentials $e^{\frac{1}{4}\mu_0 u}$, $e^{-\frac{1}{2}\mu_0 u}$ for the case when $r = m' = 0$ and $C'' = 0$.

To obtain (4.2.6) in general, we proceed as follows. Note that the function f is defined in (1.6.12). For $h > 1$, put

$$K_{u,h}(a) = \int_{-\infty}^{+\infty} \exp(iv\sqrt{2u}a) \exp\left(-\frac{v^2}{2}\right)\left(1 - f(\frac{1}{h}\sqrt{2u}v)\right) \frac{dv}{\sqrt{2\pi}}. \tag{4.2.11}$$

Then there exist $C', C_1 > 0$ such that for any $c > 0$, $m, m' \in \mathbb{N}$, there is $C > 0$ such that for $u \geqslant u_0$, $h > 1$, $a \in \mathbb{C}, |\mathrm{Im}(a)| \leqslant c$, we have

$$|a|^m |K_{u,h}^{(m')}(a)| \leqslant C \exp\left(C'c^2 u - \frac{C_1}{u}h^2\right). \tag{4.2.12}$$

For any $c > 0$, let V_c be the images of $\{\lambda \in \mathbb{C}, |\mathrm{Im}(\lambda)| \leqslant c\}$ by the map $\lambda \to \lambda^2$. Then

$$V_c = \{\lambda \in \mathbb{C}, \mathrm{Re}(\lambda) \geqslant \frac{1}{4c^2}\mathrm{Im}(\lambda)^2 - c^2\}, \tag{4.2.13}$$

and $\delta \cup \Delta \subset V_c$ for c big enough. Let $\widetilde{K}_{u,h}$ be the holomorphic function such that $\widetilde{K}_{u,h}(a^2) = K_{u,h}(a)$. Then by (4.2.12), for $\lambda \in V_c$,

$$|\lambda|^m |\widetilde{K}_{u,h}^{(m')}(\lambda)| \leqslant C \exp\left(C'c^2 u - \frac{C_1}{u}h^2\right). \tag{4.2.14}$$

Let $\widetilde{K}_{u,h}(\mathcal{L}_2^t)(Z, Z')$ be the kernel of $\widetilde{K}_{u,h}(\mathcal{L}_2^t)$ with respect to $dv_{TX}(Z')$. Using finite propagation speed of solutions of hyperbolic equations, Theorem D.2.1, (D.2.17) and (4.2.11), we find that there exists a fixed constant $c' > 0$ (which depends on ε), such that

$$\widetilde{K}_{u,h}(\mathcal{L}_2^t)(Z, Z') = e^{-u\mathcal{L}_2^t}(Z, Z') \quad \text{if } |Z - Z'| \geqslant c'h. \tag{4.2.15}$$

By (4.2.14), we see that given $k \in \mathbb{N}$, there is a unique holomorphic function $\widetilde{K}_{u,h,k}(\lambda)$ defined on a neighborhood of V_c which verifies the same estimates as $\widetilde{K}_{u,h}$ in (4.2.14) and $\widetilde{K}_{u,h,k}(\lambda) \to 0$ as $\lambda \to +\infty$; moreover

$$\widetilde{K}_{u,h,k}^{(k-1)}(\lambda)/(k-1)! = \widetilde{K}_{u,h}(\lambda). \tag{4.2.16}$$

As in (4.2.7),

$$
\begin{aligned}
\widetilde{K}_{u,h}(\mathscr{L}_2^t) &= \frac{1}{2\pi i}\int_{\delta\cup\Delta}\widetilde{K}_{u,h,k}(\lambda)(\lambda-\mathscr{L}_2^t)^{-k}d\lambda, \\
\mathscr{L}_2^t\widetilde{K}_{u,h}(\mathscr{L}_2^t) &= \frac{1}{2\pi i}\int_{\Delta}\widetilde{K}_{u,h,k}(\lambda)\Big[\lambda(\lambda-\mathscr{L}_2^t)^{-k}-(\lambda-\mathscr{L}_2^t)^{-k+1}\Big]d\lambda.
\end{aligned}
\tag{4.2.17}
$$

Using (4.1.48), (4.1.49) and proceeding as in (4.2.8)–(4.2.10), we find that for $\mathbf{K}(a)=\widetilde{K}_{u,h}(a)$ or $a\widetilde{K}_{u,h}(a)$, for α,α' with $|\alpha|+|\alpha'|\leqslant m$, and $|Z|,|Z'|\leqslant q$,

$$
\left|\frac{\partial^{|\alpha|+|\alpha'|}}{\partial Z^\alpha\partial Z'^{\alpha'}}\mathbf{K}(\mathscr{L}_2^t)(Z,Z')\right|\leqslant C(1+q)^{2n+m+2}\exp\left(C'c^2u-\frac{C_1}{u}h^2\right). \tag{4.2.18}
$$

Setting $h=\frac{1}{c'}|Z-Z'|$ in (4.2.18), we get for any α,α' verifying $|\alpha|+|\alpha'|\leqslant m$,

$$
\left|\frac{\partial^{|\alpha|+|\alpha'|}}{\partial Z^\alpha\partial Z'^{\alpha'}}\mathbf{K}(\mathscr{L}_2^t)(Z,Z')\right|
$$
$$
\leqslant C(1+|Z|+|Z'|)^{2n+m+2}\exp\left(C'c^2u-\frac{C_1}{2c'^2u}|Z-Z'|^2\right). \tag{4.2.19}
$$

By (4.2.6) with the exponentials $e^{\frac{1}{4}\mu_0 u}$, $e^{-\frac{1}{2}\mu_0 u}$ for $r=m'=C''=0$, (4.2.15) and (4.2.19), we infer (4.2.6) for $r=m'=0$. To prove (4.2.6) for $r\geqslant 1$, note that (4.2.7) imply that for $k\geqslant 1$,

$$
\frac{\partial^r}{\partial t^r}e^{-u\mathscr{L}_2^t}=\frac{(-1)^{k-1}(k-1)!}{2\pi i u^{k-1}}\int_{\delta\cup\Delta}e^{-u\lambda}\frac{\partial^r}{\partial t^r}(\lambda-\mathscr{L}_2^t)^{-k}d\lambda. \tag{4.2.20}
$$

We have a similar equation for $\frac{\partial^r}{\partial t^r}(\mathscr{L}_2^t e^{-u\mathscr{L}_2^t})$.

By (4.1.46), (4.1.47) and (4.2.20), we get the similar estimates (4.2.6) with $m'=C''=0$, (4.2.19) for $\frac{\partial^r}{\partial t^r}e^{-u\mathscr{L}_2^t}$, $\frac{\partial^r}{\partial t^r}(\mathscr{L}_2^t e^{-u\mathscr{L}_2^t})$ with the weight $2n+2r+2+m$ instead of $2n+m+2$. Thus we get (4.2.6) for $m'=0$.

Finally, by using the argument after (4.1.64) and the above argument, we obtain (4.2.6) for $m'\geqslant 1$. The proof of Theorem 4.2.5 is complete. $\qquad\square$

Recall that $\mathcal{P}_{0,t}$ is the orthogonal projection from $\mathscr{C}^\infty(X_0,\mathbf{E}_{x_0})$ to $\mathrm{Ker}(\mathscr{L}_2^t)$ with respect to $\langle\,,\,\rangle_{t,0}$. Set

$$
F_u(\mathscr{L}_2^t)=\frac{1}{2\pi i}\int_\Delta e^{-u\lambda}(\lambda-\mathscr{L}_2^t)^{-1}d\lambda. \tag{4.2.21}
$$

Let $F_u(\mathscr{L}_2^t)(Z,Z')$ be the smooth kernel of $F_u(\mathscr{L}_2^t)$ with respect to $dv_{TX}(Z')$. Then by (4.1.40),

$$
F_u(\mathscr{L}_2^t)=e^{-u\mathscr{L}_2^t}-\mathcal{P}_{0,t}=\int_u^{+\infty}\mathscr{L}_2^t e^{-u_1\mathscr{L}_2^t}\,du_1. \tag{4.2.22}
$$

Corollary 4.2.6. *With the notation in Theorem 4.2.5,*

$$\sup_{|\alpha|+|\alpha'|\leqslant m}\left|\frac{\partial^{|\alpha|+|\alpha'|}}{\partial Z^\alpha\partial Z'^{\alpha'}}\frac{\partial^r}{\partial t^r}F_u(\mathscr{L}_2^t)(Z,Z')\right|_{\mathscr{C}^{m'}(X)}$$
$$\leqslant C(1+|Z|+|Z'|)^{M_{r,m,m'}}\exp(-\tfrac{1}{8}\mu_0 u-\sqrt{C''\mu_0}|Z-Z'|).\quad (4.2.23)$$

Proof. Note that $\frac{1}{8}\mu_0 u+\frac{2C''}{u}|Z-Z'|^2\geqslant\sqrt{C''\mu_0}|Z-Z'|$, thus

$$\int_u^{+\infty}e^{-\frac{1}{4}\mu_0 u_1-\frac{2C''}{u_1}|Z-Z'|^2}\,du_1\leqslant e^{-\sqrt{C''\mu_0}|Z-Z'|}\int_u^{+\infty}e^{-\frac{1}{8}\mu_0 u_1}\,du_1$$
$$=\frac{8}{\mu_0}e^{-\frac{1}{8}\mu_0 u-\sqrt{C''\mu_0}|Z-Z'|}.\quad (4.2.24)$$

By (4.2.6), (4.2.22) and (4.2.24), we get (4.2.23). □

In view of (4.1.46),we set for k large enough

$$F_{r,u}=\frac{(-1)^{k-1}(k-1)!}{2\pi i\,r!\,u^{k-1}}\int_\Delta e^{-u\lambda}\sum_{(\mathbf{k},\mathbf{r})\in I_{k,r}}a_{\mathbf{r}}^{\mathbf{k}}A_{\mathbf{r}}^{\mathbf{k}}(\lambda,0)d\lambda,$$

$$J_{r,u}=\frac{(-1)^{k-1}(k-1)!}{2\pi i\,r!\,u^{k-1}}\int_{\delta\cup\Delta} e^{-u\lambda}\sum_{(\mathbf{k},\mathbf{r})\in I_{k,r}}a_{\mathbf{r}}^{\mathbf{k}}A_{\mathbf{r}}^{\mathbf{k}}(\lambda,0)d\lambda,$$

$$F_{r,u,t}=\frac{1}{r!}\frac{\partial^r}{\partial t^r}F_u(\mathscr{L}_2^t)-F_{r,u},$$

$$J_{r,u,t}=\frac{1}{r!}\frac{\partial^r}{\partial t^r}e^{-u\mathscr{L}_2^t}-J_{r,u}.$$
$$(4.2.25)$$

Theorem 4.2.7. *There exist $C>0$, $N\in\mathbb{N}$ such that for $t\in]0,t_0]$, $u\geqslant u_0$, $q\in\mathbb{N}^*$, $Z,Z'\in T_{x_0}X$, $|Z|,|Z'|\leqslant q$,*

$$\left|F_{r,u,t}(Z,Z')\right|\leqslant Ct^{1/(2n+1)}(1+q)^N e^{-\frac{1}{8}\mu_0 u},$$
$$\left|J_{r,u,t}(Z,Z')\right|\leqslant Ct^{1/(2n+1)}(1+q)^N e^{\frac{1}{2}\mu_0 u}.$$
$$(4.2.26)$$

Proof. Let $J^0_{x_0,q}$ be the vector space of square integrable sections of $\mathbf{E}_{x_0}$ over $\{Z\in T_{x_0}X,|Z|\leqslant q+1\}$. Let $\|A\|_{(q)}$ be the operator norm of $A\in\mathscr{L}(J^0_{x_0,q})$ with respect to $\|\ \|_{(q),0}$ in (4.1.61). By (4.1.52), (4.2.20) and (4.2.25), we get: There exist $C>0$, $N\in\mathbb{N}$ such that for $t\in]0,t_0]$, $u\geqslant u_0$, $q>1$,

$$\|F_{r,u,t}\|_{(q)}\leqslant Ct(1+q)^N e^{-\frac{1}{2}\mu_0 u},$$
$$\|J_{r,u,t}\|_{(q)}\leqslant Ct(1+q)^N\;e^{\frac{1}{4}\mu_0 u}.$$
$$(4.2.27)$$

Let $\phi : \mathbb{R}^{2n} \to [0, 1]$ be a smooth function with compact support, equal 1 near 0, such that $\int_{T_{x_0} X} \phi(Z) dv_{TX}(Z) = 1$. Take $\nu \in]0, 1]$. By the proof of Theorem 4.2.5, $F_{r,u}$ verifies a similar inequality as (4.2.23). Thus by (4.2.23), there exists $C > 0$ such that if $|Z|, |Z'| \leqslant q$, $U, U' \in \mathbf{E}_{x_0}$,

$$
\left| \langle F_{r,u,t}(Z, Z')U, U' \rangle - \int_{T_{x_0} X \times T_{x_0} X} \langle F_{r,u,t}(Z - W, Z' - W')U, U' \rangle \frac{1}{\nu^{4n}} \phi(W/\nu) \right.
$$
$$
\left. \times \phi(W'/\nu) dv_{TX}(W) dv_{TX}(W') \right| \leqslant C\nu(1 + q)^N e^{-\frac{1}{8}\mu_0 u} |U||U'|. \quad (4.2.28)
$$

On the other hand, by (4.2.27), we have for $|Z|, |Z'| \leqslant q$,

$$
\left| \int_{T_{x_0} X \times T_{x_0} X} \langle F_{r,u,t}(Z - W, Z' - W')U, U' \rangle \frac{1}{\nu^{4n}} \phi(W/\nu) \right.
$$
$$
\left. \times \phi(W'/\nu) dv_{TX}(W) dv_{TX}(W') \right| \leqslant Ct \frac{1}{\nu^{2n}}(1 + q)^N e^{-\frac{1}{2}\mu_0 u} |U||U'|. \quad (4.2.29)
$$

By taking $\nu = t^{1/(2n+1)}$, we deduce the first estimate (4.2.26). In the same way, we obtain (4.2.26) for $J_{r,u,t}$. □

Theorem 4.2.8. *There exists $C'' > 0$ such that for any $k, m, m' \in \mathbb{N}$, there exists $C > 0$ such that if $t \in]0, t_0], u \geqslant u_0, Z, Z' \in T_{x_0} X$,*

$$
\sup_{|\alpha|+|\alpha'|\leqslant m} \left| \frac{\partial^{|\alpha|+|\alpha'|}}{\partial Z^{\alpha} \partial Z'^{\alpha'}} \left(F_u(\mathscr{L}_2^t) - \sum_{r=0}^{k} F_{r,u} t^r \right)(Z, Z') \right|_{\mathscr{C}^{m'}(X)}
$$
$$
\leqslant Ct^{k+1}(1 + |Z| + |Z'|)^{M_{k+1,m,m'}} \exp(-\frac{1}{8}\mu_0 u - \sqrt{C''\mu_0}|Z - Z'|),
$$
$$
\sup_{|\alpha|+|\alpha'|\leqslant m} \left| \frac{\partial^{|\alpha|+|\alpha'|}}{\partial Z^{\alpha} \partial Z'^{\alpha'}} \left(e^{-u\mathscr{L}_2^t} - \sum_{r=0}^{k} J_{r,u} t^r \right)(Z, Z') \right|_{\mathscr{C}^{m'}(X)}
$$
$$
\leqslant Ct^{k+1}(1 + |Z| + |Z'|)^{M_{k+1,m,m'}} \exp(\frac{1}{2}\mu_0 u - \frac{2C''}{u}|Z - Z'|^2). \quad (4.2.30)
$$

Proof. By (4.2.25) and (4.2.26),

$$
\frac{1}{r!} \frac{\partial^r}{\partial t^r} F_u(\mathscr{L}_2^t)|_{t=0} = F_{r,u},
$$
$$
\frac{1}{r!} \frac{\partial^r}{\partial t^r} e^{-u\mathscr{L}_2^t}|_{t=0} = J_{r,u}. \quad (4.2.31)
$$

Now by Theorem 4.2.5 and (4.2.25), $J_{r,u}$, $F_{r,u}$ have the same estimates as $\frac{\partial^r}{\partial t^r} e^{-u\mathscr{L}_2^t}$, $\frac{\partial^r}{\partial t^r} F_u(\mathscr{L}_2^t)$ in (4.2.6), (4.2.23). Again from the Taylor expansion (4.1.70), (4.2.6), (4.2.23) and (4.2.25), we get (4.2.30). □

4.2.3   Proof of Theorem 4.2.1

By (4.2.22) and (4.2.30), for any $u > 0$ fixed, there exists $C_u > 0$ such that for $l = \frac{1}{\sqrt{p}}$, $Z, Z' \in T_{x_0}X$, $x_0 \in X$, we have

$$
\sup_{|\alpha|+|\alpha'|\leqslant m} \left| \frac{\partial^{|\alpha|+|\alpha'|}}{\partial Z^\alpha \partial Z'^{\alpha'}} \left(\mathcal{P}_{0,t} - \sum_{r=0}^{k} t^r (J_{r,u} - F_{r,u}) \right)(Z, Z') \right|_{\mathscr{C}^{m'}(X)}
$$
$$
\leqslant C_u t^{k+1}(1+|Z|+|Z'|)^{M_{k+1,m,m'}} \exp(-\sqrt{C''\mu_0}|Z-Z'|). \quad (4.2.32)
$$

Comparing with Theorem 4.1.18, we get

$$
\mathscr{F}_r = J_{r,u} - F_{r,u}. \quad (4.2.33)
$$

By taken the limit of (4.2.23) as $t \to 0$,

$$
\left| F_{r,u}(Z, Z') \right| \leqslant C(1+|Z|+|Z'|)^{2(n+r+1)} e^{-\frac{1}{8}\mu_0 u - \sqrt{C''\mu_0}|Z-Z'|}. \quad (4.2.34)
$$

Thus from (4.2.33) and (4.2.34), when $u \to \infty$,

$$
J_{r,u}(Z, Z') = \mathscr{F}_r(Z, Z') + \mathcal{O}(e^{-\frac{1}{8}\mu_0 u}), \quad (4.2.35)
$$

uniformly on any compact set of $T_{x_0}X \times T_{x_0}X$.

We now observe that, as a consequence of (4.1.96), (4.2.32) and (4.2.33), we obtain the following important estimate.

Theorem 4.2.9. *There exists $C'' > 0$ such that for any $k, m, m' \in \mathbb{N}$, there exists $C > 0$ such that for $\alpha, \alpha' \in \mathbb{N}^n$, $|\alpha| + |\alpha'| \leqslant m$, $Z, Z' \in T_{x_0}X$, $|Z|, |Z'| \leqslant \varepsilon$, $x_0 \in X$, $p \geqslant 1$,*

$$
\left| \frac{\partial^{|\alpha|+|\alpha'|}}{\partial Z^\alpha \partial Z'^{\alpha'}} \left(\frac{1}{p^n} P_{0,p}(Z, Z') - \sum_{r=0}^{k} \mathscr{F}_r(\sqrt{p}Z, \sqrt{p}Z') \kappa^{-\frac{1}{2}}(Z) \kappa^{-\frac{1}{2}}(Z') p^{-\frac{r}{2}} \right) \right|_{\mathscr{C}^{m'}(X)}
$$
$$
\leqslant C p^{-(k+1-m)/2}(1+|\sqrt{p}Z|+|\sqrt{p}Z'|)^{M_{k+1,m,m'}} \exp(-\sqrt{C''\mu_0}\sqrt{p}|Z-Z'|). \quad (4.2.36)
$$

By Theorem 4.2.9 and (4.1.27), we conclude Theorem 4.2.1.

4.2.4   Proof of Theorem 4.2.3

Using (4.1.29) and proceeding as in (4.1.96), we obtain for $Z, Z' \in T_{x_0}X$,

$$
\exp\left(-\frac{u}{p}(D_p^{c,A_0})^2 \right)(Z, Z') = p^n e^{-u\mathscr{L}_2^t}(Z/t, Z'/t) \kappa^{-1/2}(Z) \kappa^{-1/2}(Z'). \quad (4.2.37)
$$

From Proposition 1.6.4, we deduce as in Lemma 1.6.5, that for any $u > 0$ fixed and for any $l \in \mathbb{N}$, there exists $C > 0$ such that for $x_0 \in X$, $Z, Z' \in T_{x_0}X$, $|Z|, |Z'| \leqslant \varepsilon$,

$$\left|\left(\exp\left(-\frac{u}{p}D_p^2\right) - \exp\left(-\frac{u}{p}(D_p^{c,A_0})^2\right)\right)(Z, Z')\right|_{\mathscr{C}^{m'}(X)} \leqslant Cp^{-l}. \tag{4.2.38}$$

Thus from (4.1.28), (4.2.30), (4.2.37) and (4.2.38) we get

$$\left|\frac{1}{p^n}\exp\left(-\frac{u}{p}D_p^2\right)(x_0, x_0) - \sum_{r=0}^{k} J_{r,u}(0,0)p^{-r/2}\right|_{\mathscr{C}^{m'}(X)} \leqslant Cp^{-\frac{k+1}{2}}. \tag{4.2.39}$$

By Theorem 4.1.7, (4.1.46), (4.2.25) and the discussion after (4.1.97), we know that $J_{r,u}(Z, Z')$ are polynomials in R^{TX}, R^E (resp. $d\Theta, R^L$) and their derivatives with order $\leqslant r - 2$ (resp $r - 1, r$) and functions on the eigenvalues of $\dot{R}^L$ at x_0.

If we can prove that for any $r \in \mathbb{N}$,

$$J_{2r+1,u}(0,0) = 0, \tag{4.2.40}$$

then we have (4.2.4) and at x_0,

$$b_{r,u} = J_{2r,u}(0,0). \tag{4.2.41}$$

Relations (4.1.97), (4.1.98), (4.2.35) and (4.2.41) imply (4.2.5).

In the rest, we give a proof of (4.2.40). Let $\{\sigma_k\}$ be an orthonormal basis of $\mathbf{E}_{x_0}$. We denote by

$$\psi_{\alpha,\beta,k} = c_{\alpha\beta}b^\alpha\left(z^\beta e^{-\frac{1}{4}\sum_i a_i|z_i|^2}\right)\sigma_k = \varphi_{\alpha\beta}(Z)e^{-\frac{1}{4}\sum_i a_i|z_i|^2}\sigma_k, \tag{4.2.42}$$

such that $\|\psi_{\alpha,\beta,k}\|_0 = 1$. Then $\varphi_{\alpha\beta}(Z)$ is a polynomial in Z with the same parity as $|\alpha| + |\beta|$. Set

$$\begin{aligned} F^{\text{even}} &= \text{span}\{\psi_{\alpha,\beta,k}; |\alpha| + |\beta| \text{ is even}\}, \\ F^{\text{odd}} &= \text{span}\{\psi_{\alpha,\beta,k}; |\alpha| + |\beta| \text{ is odd}\}. \end{aligned} \tag{4.2.43}$$

Then from Theorem 4.1.20, $L^2(\mathbb{R}^{2n}, \mathbf{E}_{x_0}) = F^{\text{even}} \oplus F^{\text{odd}}$.

From Theorems 4.1.7, 4.1.20, (4.2.25), we know that for r odd, $J_{r,u}$ exchanges F^{even} and F^{odd}. Thus for any $\alpha, \beta \in \mathbb{N}^n$, we have

$$(J_{r,u}\psi_{\alpha,\beta,k})(0)^* \psi_{\alpha,\beta,k}(0) = 0. \tag{4.2.44}$$

We claim that

$$J_{r,u}(Z, Z') = \sum_{\alpha,\beta,k}(J_{r,u}\psi_{\alpha,\beta,k})(Z')^*\psi_{\alpha,\beta,k}(Z), \tag{4.2.45}$$

converges uniformly on any compact set of $\mathbb{R}^{2n} \times \mathbb{R}^{2n}$.

To prove the uniform convergence of (4.2.45), set

$$\mathcal{I}_l := \{(\alpha, \beta); \alpha, \beta \in \mathbb{N}^n, |\alpha| + |\beta| \leqslant l\} \quad \text{for} \quad l \in \mathbb{N}.$$

For $s \in L^2(\mathbb{R}^{2n}, \mathbf{E}_{x_0})$, we define

$$J_{\mathcal{I}_l} s = \sum_{(\alpha,\beta)\in\mathcal{I}_l,k} \langle s, J_{r,u}\psi_{\alpha,\beta,k} \rangle \, \psi_{\alpha,\beta,k}. \tag{4.2.46}$$

Then the smooth kernel of $J_{\mathcal{I}_l}$ is

$$J_{\mathcal{I}_l}(Z, Z') = \sum_{(\alpha,\beta)\in\mathcal{I}_l,k} (J_{r,u}\psi_{\alpha,\beta,k})(Z')^* \psi_{\alpha,\beta,k}(Z). \tag{4.2.47}$$

From Theorem 4.1.20, we know that for $k_1, k_2 \in \mathbb{N}$, $s \in L^2(\mathbb{R}^{2n}, \mathbf{E}_{x_0})$,

$$(\mathscr{L}_2^0)^{k_1} J_{\mathcal{I}_l} (\mathscr{L}_2^0)^{k_2} s = \sum_{(\alpha,\beta)\in\mathcal{I}_l,k} \langle s, (\mathscr{L}_2^0)^{k_2} J_{r,u} (\mathscr{L}_2^0)^{k_1} \psi_{\alpha,\beta,k} \rangle \, \psi_{\alpha,\beta,k}$$

$$= \sum_{(\alpha,\beta)\in\mathcal{I}_l,k} \langle (\mathscr{L}_2^0)^{k_1} J_{\mathcal{I}_l} (\mathscr{L}_2^0)^{k_2} s, \psi_{\alpha,\beta,k} \rangle \, \psi_{\alpha,\beta,k}. \tag{4.2.48}$$

Thus for Theorem 4.1.13, for any $q > 1$, $k_1, k_2 \in \mathbb{N}$, there exists $C_q > 0$ such that for $s \in L^2(\mathbb{R}^{2n}, \mathbf{E}_{x_0})$, $\mathrm{supp}(s) \subset B^{T_{x_0}X}(0, q+1)$, $l \in \mathbb{N}$,

$$\|(\mathscr{L}_2^0)^{k_1} J_{\mathcal{I}_l} (\mathscr{L}_2^0)^{k_2} s\|_0 \leqslant \|(\mathscr{L}_2^0)^{k_1} J_{r,u} (\mathscr{L}_2^0)^{k_2} s\|_0 \leqslant C_q \|s\|_0. \tag{4.2.49}$$

Thus by Theorems A.1.6 and A.1.7, $J_{\mathcal{I}_l}(Z, Z')$ and its derivatives are uniformly bounded for $|Z|, |Z'| \leqslant q$. When $l \to \infty$, $J_{\mathcal{I}_l} \to J_{r,u}$ in the $|\cdot|_{(q),0}$-norm from (4.1.61). Applying the argument from the proof of Theorem 4.2.7, we get (4.2.45).

Now, (4.2.44) and (4.2.45) entail (4.2.40). The proof of Theorem 4.2.3 is complete.

Problems

Problem 4.1. Let (X, ω) be a Kähler manifold and let g^{TX} be the Kähler metric associated to the Kähler form ω. Verify that

$$r^X = 4\pi \left\langle c_1(T^{(1,0)}X, \nabla^{T^{(1,0)}X}), \omega \right\rangle.$$

Problem 4.2. Verify Remark 4.1.19 and Theorem 4.1.22.

Problem 4.3. Verify (4.2.12) and (4.2.15).

Problem 4.4 (cohomology of $\mathscr{O}(k)$). We continue here Problem 1.8. Let $n \geqslant 1$. Using Čech cohomology show that:

$$H^q(\mathbb{CP}^n, \mathscr{O}(k)) = 0,$$

for (a) $q \neq n$, $k < 0$ and (b) $q \neq 0$, $k > -n - 1$. Moreover for any $k \in \mathbb{N}$,

$$H^0(\mathbb{CP}^n, \mathscr{O}(k)) = \mathbb{C}\Big\{s_\alpha, \ |\alpha| = k, \text{ and } \alpha \in \mathbb{N}^{n+1}\Big\}.$$

Here the holomorphic section s_α of $\mathscr{O}(|\alpha|)$ is defined by the map $\mathbb{C}^{n+1} \ni z \to \prod_{j=0}^n z_j^{\alpha_j}$. Calculate the dimension of $H^0(\mathbb{CP}^n, \mathscr{O}(k))$.

Describe $H^n(\mathbb{CP}^n, \mathscr{O}(k))$, $k \leqslant -n - 1$. (One may consult [124, III.5]; for the general cohomology groups $H^q(\mathbb{CP}^n, \mathscr{O}^r_{\mathbb{CP}^n}(\mathscr{O}(k)))$ see [79, VII.10.7], [217, (4.3)])

Problem 4.5. We continue Problems 1.8 and 4.4. Verify that the Riemannian volume form on $(\mathbb{CP}^n, \omega_{FS})$ is

$$\frac{1}{n!}\omega_{FS}(z)^n = \left(\frac{\sqrt{-1}}{2\pi}\right)^n \frac{dz_1 d\bar{z}_1 \cdots dz_n d\bar{z}_n}{(1 + \sum_{j=1}^n |z_j|^2)^{n+1}}.$$

Show that for $\alpha \in \mathbb{N}^{n+1}$,

$$\|s_\alpha\|_{L^2}^2 = \frac{\alpha!}{(n + |\alpha|)!}.$$

Prove that the Bergman kernel $P_p(z, w)$ associated to $\mathscr{O}(p)$ is

$$P_p(z, w) = \sum_{\alpha \in \mathbb{N}^{n+1}, |\alpha|=p} \frac{(n+p)!}{\alpha!} s_\alpha(z) \otimes s_\alpha(w)^*.$$

Especially for any $z \in \mathbb{CP}^n$, we have

$$P_p(z, z) = (n+p)!/p!.$$

Problem 4.6 (moment map). In this problem, we use the notation from Problems 1.8 and 4.4, and we use the metric $g^{T\mathbb{CP}^1}$ associated to the Kähler form $\omega := 2\omega_{FS}$ on $\mathbb{CP}^1$.

Let K be the canonical basis of Lie $S^1 = \mathbb{R}$, i.e., for $t \in \mathbb{R}, \exp(tK) = e^{2\pi\sqrt{-1}t} \in S^1$.

We define an S^1-action on $\mathbb{CP}^1$ by $g \cdot [z_0, z_1] = [z_0, gz_1]$ for $g \in S^1$.

On our local coordinate U_0, $g \cdot z = gz$, and the vector field $K^{\mathbb{CP}^1}$ on $\mathbb{CP}^1$ induced by K is

$$K^{\mathbb{CP}^1}(z) := \tfrac{\partial}{\partial t} \exp(-tK) \cdot z|_{t=0} = -2\pi\sqrt{-1}\left(z\tfrac{\partial}{\partial z} - \bar{z}\tfrac{\partial}{\partial \bar{z}}\right).$$

We define the S^1-action on $(L, h^L) := (\mathscr{O}(2), h^{\mathscr{O}(2)})$ by $\exp(tK) \cdot s_{(2-j,j)} = e^{2\pi\sqrt{-1}(1-j)t} s_{(2-j,j)}$. Verify that it defines an S^1-action on L which preserves the metric h^L. (Hint: $(g \cdot s)(x) = g \cdot (s(g^{-1} \cdot x))$ for $g \in S^1$.)

Let $L_K s = \frac{\partial}{\partial t} \exp(tK) \cdot s$ be the associated Lie derivative for $s \in \mathscr{C}^\infty(\mathbb{CP}^1, L)$. Let ∇^L be the holomorphic Hermitian connection of (L, h^L). On $\mathbb{C}$, set

$$\mu(K) = 2|z|^2(1 + |z|^2)^{-1} - 1.$$

Prove that

$$2\pi\sqrt{-1}\mu(K) = \nabla^L_{K_{\mathbb{CP}^1}} - L_K,$$

and

$$d\mu(K) = i_{K_{\mathbb{CP}^1}}\omega.$$

The map μ is called the *moment map* associated to the S^1-action.

Let $P_p^{S^1}$ be the smooth kernel of the orthogonal projection from $\mathscr{C}^\infty(\mathbb{CP}^1, L^p)$ onto $H^0(\mathbb{CP}^1, L^p)^{S^1}$, the S^1-invariant sub-space of $H^0(\mathbb{CP}^1, L^p)$. Verify that

$$P_p^{S^1}(z, z) = \frac{(2p+1)!}{2\,(p!)^2} \frac{|z|^{2p}}{(1 + |z|^2)^{2p}}.$$

Use Stirling's formula,

$$p! = (2\pi p)^{1/2} p^p\, e^{-p} \left(1 + \mathscr{O}\left(\frac{1}{p}\right)\right) \qquad \text{as } p \to +\infty,$$

to show that for $z \in \mathbb{C}$ fixed, when $p \to +\infty$,

$$P_p^{S^1}(z, z) = \begin{cases} \sqrt{p/\pi}(1 + \mathscr{O}(\frac{1}{p})) & \text{for } |z| = 1, \\ \mathscr{O}(p^{-\infty}) & \text{for } |z| \neq 1. \end{cases}$$

Problem 4.7. We use the notation and assumption from Theorem 4.1.2.

Let dv be any volume form on X. Let η be the positive function on X defined by $dv_X = \eta\, dv$. The L^2-scalar product $\langle \cdot, \cdot \rangle_\nu$ on $\mathscr{C}^\infty(X, L^p)$ is given by

$$\langle \sigma_1, \sigma_2 \rangle_\nu := \int_X \langle \sigma_1(x), \sigma_2(x) \rangle_{L^p}\, d\nu(x).$$

Let $P_{\nu,p}(x, x')$ $(x, x' \in X)$ be the smooth kernel of the orthogonal projection from $(\mathscr{C}^\infty(X, L^p), \langle \cdot, \cdot \rangle_\nu)$ onto $H^0(X, L^p)$ with respect to $d\nu(x')$. Note that $P_{\nu,p}(x, x') \in L^p_x \otimes L^{p*}_{x'}$. Set

$$K_p(x, x') := |P_{\nu,p}(x, x')|^2_{h^{L^p}_x \otimes h^{L^{p*}}_{x'}},$$

$$R_{\nu,p} := (\dim H^0(X, L^p))/\operatorname{vol}(X, \nu),$$

where $\operatorname{vol}(X, \nu) := \int_X d\nu$. Set $\operatorname{vol}(X) := \int_X dv_X$.

Let Q_{K_p} be the integral operator associated to K_p which is defined for $f \in \mathscr{C}^\infty(X)$ by

$$Q_{K_p}(f)(x) := \frac{1}{R_{\nu,p}} \int_X K_p(x,y) f(y) d\nu(y).$$

Recall that Δ is the (positive) Laplace operator on (X, g^{TX}) acting on the functions on X. We denote by $\| \cdot \|_{L^2}$ the L^2-norm on the function on X with respect to dv_X.

Prove that there exists a constant $C > 0$ such that for any $f \in \mathscr{C}^\infty(X)$, $p \in \mathbb{N}^*$,

$$\left\| \left(Q_{K_p} - \frac{\mathrm{vol}(X,\nu)}{\mathrm{vol}(X)} \eta \exp\left(-\frac{\Delta}{4\pi p} \right) \right) f \right\|_{L^2} \leqslant \frac{C}{p} \|f\|_{L^2},$$

$$\left\| \left(\frac{\Delta}{p} Q_{K_p} - \frac{\mathrm{vol}(X,\nu)}{\mathrm{vol}(X)} \frac{\Delta}{p} \eta \exp\left(-\frac{\Delta}{4\pi p} \right) \right) f \right\|_{L^2} \leqslant \frac{C}{p} \|f\|_{L^2}.$$

(Hint: Use the heat kernel expansion in [15, Theorems 2.23, 2.26], Theorem 1.2.1 and (4.1.114)).

Problem 4.8 (Open problem). Let (L, h^L) be a holomorphic Hermitian line bundle on a compact connected complex Hermitian manifold (X, Θ). Donnelly [96] showed through an example that if $\sqrt{-1}R^L$ is just semipositive, the spectral gap property expressed in Theorem 1.5.5 cannot hold. This begs the question whether the asymptotic expansion of the Bergman kernel from Theorems 4.1.1, 4.1.24 and 4.2.1 still hold for semipositive line bundles?

4.3 Bibliographic notes

Section 4.1. The existence of the asymptotic expansion (4.1.7), started in the paper of Tian [241] (cf. also Bouche [48], Ruan [208]) following a suggestion of Yau [258], [259], was first established by Catlin [62] and Zelditch [261]. They computed also b_0.

Assume now $\omega = \frac{\sqrt{-1}}{2\pi} R^L$ is the Kähler form of (X, g^{TX}). Lu [156] obtained Theorem 4.1.2 when $E = \mathbb{C}$, i.e., more information on the coefficients b_r via R^{TX}, and he computed b_1, b_2, b_3, cf. also [157]; Wang [250] got also b_1 in (4.1.8) for general E. When $E = \mathbb{C}$, the existence of the asymptotic expansion similar to (4.1.95) was also obtained in [219, Th. 1] (cf. also [38]). For other versions of the asymptotic expansion (cf. also [135], [63], [18]).

In [62], [261], [63], [135] and [219], the Boutet de Monvel–Sjöstrand parametrix for the Szegö kernel [53], [105] was applied, and in [241], [156], [250] the coefficients were computed by the peak section trick in complex geometry [132].

The coefficient b_1 (when $E = \mathbb{C}$) can be also obtained from Bismut–Vasserot's result on the asymptotic of the analytic torsion [35] (cf. Section 5.5) and Donaldson's moment map picture [89].

Sections 4.1 and 4.2. The techniques here are basically from [69] and [161] which are inspired by local index theory, especially the analytic localization techniques of Bismut and Lebeau [33, §11], [29, §11]. In [69], by using the heat kernel method in Section 4.2, Dai–Liu–Ma established the full off-diagonal expansion of the Bergman kernel, Theorem 4.2.1, and the asymptotic relation between the Bergman kernel and the heat kernel, Theorem 4.2.3, for the spin^c Dirac operator on compact symplectic manifolds. By using the asymptotic expansion of the heat kernel, in [69, §5], they also compute b_1 in (4.1.8). In [161], by using the resolvent method presented in Section 4.1, we establish the near-diagonal asymptotic expansion of the generalized Bergman kernel, Theorem 4.1.18, associated to the renormalized Bochner Laplacian. Moreover, in [161], we found the formal power series trick in Section 4.1.7 which gives us a general and algorithmic way to compute the coefficients in various cases. In [165], Ma–Zhang have established the corresponding family version in the spirit of Bismut's family local index theory. They obtain further in [164] the asymptotic expansion of the G-invariant Bergman kernel with a Hamiltonian action of a Lie group. In both papers the spectral gap plays again an essential role.

The approach here is slightly different to what was explained in [161, §3.4] (To extend the operator $\overline{\partial} + \overline{\partial}^*$ from a local coordinate to $\mathbb{C}^n$, we need to use the holomorphic coordinate therein.) Here we work in normal coordinates, and we give a unified approach for the spin^c Dirac operators and the Kodaira Laplacian by using modified Dirac operators, thus we also work on the full algebra of differential forms; but the final result for the Kodaira Laplacian is in degree zero.

Problem 4.6 is taken from [164, §3.3]. Problem 4.7 is from [155] and Q_{K_p} was defined in Donaldson's paper [95].

Chapter 5

Kodaira Map

In this chapter we present some applications of the asymptotic expansion of the Bergman kernel.

We start with an analytic proof of the Kodaira embedding theorem in Section 5.1. In Section 5.2, we explain very briefly Donaldson's approach to the existence of Kähler metrics of constant scalar curvature and it's relation to the Bergman kernel. In Section 5.3, we give an introduction to the distribution of zeros of random sections. In Section 5.4, by applying the results in Chapter 4, we establish the asymptotic expansion of the Bergman kernel in the orbifold case, and then we study the analytic aspect of Baily's orbifold embedding theorem.

Finally, in Section 5.5, we explain the asymptotics of the Ray-Singer analytic torsion by Bismut and Vasserot. This Section is quite independent from other topics touched in this book. It can be read after Chapter 1; we only use the spectral gap property (Theorem 1.5.5) and the technique from Section 1.6.

We will use the notation from Chapter 4.

5.1 The Kodaira embedding theorem

In this section, as an application of the asymptotic expansion of the Bergman kernel, we present an analytic proof of the Kodaira embedding theorem and study the convergence of the induced Fubini–Study metrics.

In Section 5.1.1, we recall some facts on the projective spaces and Grassmannian manifolds. In Section 5.1.2, we present an analytic proof of the Kodaira embedding theorem, cf. also Section 8.3.5. In Section 5.1.3, we explain the classical proof of the Kodaira embedding theorem. In Section 5.1.4, we study the Grassmannian embedding.

5.1.1   Universal bundles

Let V be an m-dimensional complex vector space and let V^* be its dual.

The projective space $\mathbb{P}(V)$ is the set of complex lines through the origin in V. A line $l \subset V$ is determined by the hyperplane $\{f(l) = 0; f \in V^*\} \subset V^*$. A line $l \subset V$ is also determined by any $0 \neq u \in l$, so we can write $\mathbb{P}(V) = V \setminus \{0\}/\sim$; here $u \sim v$ if and only if there is $\lambda \in \mathbb{C}^*$ such that $u = \lambda v$. For $v \in V \setminus \{0\}$, we denote by $[v]$ the complex line through v, and $[v] \in \mathbb{P}(V)$ the corresponding point. In the following, we will use these different points of view about $\mathbb{P}(V)$.

Let $\mathscr{O}(-1)$ be the universal (tautological) line bundle on $\mathbb{P}(V^*)$, then

$$\mathscr{O}(-1) = \{(h, f) \in \mathbb{P}(V^*) \times V^*, f \in h \subset V^*\}. \tag{5.1.1}$$

For $k \in \mathbb{Z}$, set $\mathscr{O}(k) := \mathscr{O}(-1)^{-k}$, especially, $\mathscr{O}(1)$ is the dual bundle of $\mathscr{O}(-1)$.

Let h^V be a Hermitian metric on V; it induces naturally a Hermitian metric h^{V^*} on V^*, thus it induces a Hermitian metric $h^{\mathscr{O}(-1)}$ on $\mathscr{O}(-1)$, as a sub-bundle of the trivial bundle V^* on $\mathbb{P}(V^*)$. Let $h^{\mathscr{O}(1)}$ be the Hermitian metric on $\mathscr{O}(1)$ induced by $h^{\mathscr{O}(-1)}$.

For any $v \in V$, the linear map $V^* \ni f \to (f, v) \in \mathbb{C}$ defines naturally a holomorphic section σ_v of $\mathscr{O}(1)$ on $\mathbb{P}(V^*)$. By the definition, for $f \in V^* \setminus \{0\}$, at $[f] \in \mathbb{P}(V^*)$, we have

$$|\sigma_v([f])|^2_{h^{\mathscr{O}(1)}} = |(f, v)|^2/|f|^2_{h^{V^*}}. \tag{5.1.2}$$

Let $R^{\mathscr{O}(1)}$ be the curvature of the holomorphic Hermitian (Chern) connection $\nabla^{\mathscr{O}(1)}$ on $\mathscr{O}(1)$ on $\mathbb{P}(V^*)$. The *Fubini–Study form* ω_{FS} is the Kähler form associated to the *Fubini–Study metric* on $\mathbb{P}(V^*)$, and is defined by : for any $0 \neq v \in V$,

$$\omega_{FS} = \frac{\sqrt{-1}}{2\pi} R^{\mathscr{O}(1)} = \frac{\sqrt{-1}}{2\pi} \overline{\partial}\partial \log|\sigma_v|^2_{h^{\mathscr{O}(1)}} \quad \text{on } \{x \in \mathbb{P}(V^*), \sigma_v(x) \neq 0\}. \tag{5.1.3}$$

The Grassmannian $\mathbb{G}(k, V)$ is defined to be the set of k-dimensional complex linear subspaces of V; we write $\mathbb{G}(k, m)$ for $\mathbb{G}(k, \mathbb{C}^m)$. $\mathbb{G}(k, V)$ is also identified to $\mathbb{G}(m - k, V^*)$. The universal bundle U on $\mathbb{G}(k, V^*)$ is a k-dimensional holomorphic vector bundle defined by

$$U = \{(h, f) \in \mathbb{G}(k, V^*) \times V^*, f \in h \subset V^*\}. \tag{5.1.4}$$

Let h^U be the Hermitian metric on U induced by h^V, by considering U as a sub-bundle of the trivial bundle V^* on $\mathbb{G}(k, V^*)$. Let h^{U^*}, $h^{\det U^*}$ be the Hermitian metrics on U^*, $\det U^*$ induced by h^U.

For any $v \in V$, the linear map $V^* \ni f \to (f, v) \in \mathbb{C}$ defines naturally a holomorphic section σ_v of U^* on $\mathbb{G}(k, V^*)$.

At $[V_1^*] \in \mathbb{G}(k, V^*)$, assume that $\{f^1, \dots, f^k\}$ is a basis of V_1^*, $\{e^j\}$ is an orthonormal basis of (V^*, h^{V^*}), and $\{e_j\}$ its dual basis. Then there exist $A_{ij} \in \mathbb{C}$ such that $f^i = A_{ij}e^j$, $v = \sum_j v_j e_j$. Set $A = (A_{ij})_{\substack{1 \leqslant i \leqslant k \\ 1 \leqslant j \leqslant m}}$. Then $\widetilde{f}^i = ((AA^*)^{-1/2}A)_{ij}e^j$ is an orthonormal basis of V_1^*. Thus $\sigma_v([V_1^*]) = \sum_{i=1}^k \widetilde{f}_i(\widetilde{f}^i, v)$, where $\{\widetilde{f}_i\}$ is the dual basis of $\{\widetilde{f}^i\}$. Thus

$$|\sigma_v([V_1^*])|^2_{h^{U^*}} = \sum_{i=1}^k \left| \sum_{j=1}^m ((AA^*)^{-1/2}A)_{ij}v_j \right|^2 = (A^*(AA^*)^{-1}A)_{ij}v_j\overline{v}_i. \quad (5.1.5)$$

Let $R^{\det U^*}$ be the curvature of the holomorphic Hermitian connection of $\det U^* \to \mathbb{G}(k, V^*)$. Set

$$\omega_{FS} = \frac{\sqrt{-1}}{2\pi}R^{\det U^*}. \quad (5.1.6)$$

We verify that ω_{FS} is actually a positive $(1,1)$-form. At $[V_1^*] \in \mathbb{G}(k, V^*)$, we choose $\{e^j\}_{j=1}^m$ an orthonormal basis of V^* such that $\{e^j\}_{j=1}^k$ is an orthonormal basis of V_1^*, then $B = (B_{jl})_{\substack{1 \leqslant l \leqslant m-k \\ 1 \leqslant j \leqslant k}} \to \mathrm{span}\{e^j + B_{jl}e^{k+l}\}_j$ is a local chart of $\mathbb{G}(k, V^*)$ near $[V_1^*]$, and $A = (I, D)$, where I is the $k \times k$ identity matrix. Especially

$$A^*(AA^*)^{-1}A = \begin{pmatrix} (I + BB^*)^{-1} & (I + BB^*)^{-1}B \\ B^*(I + BB^*)^{-1} & B^*(I + BB^*)^{-1}B \end{pmatrix}. \quad (5.1.7)$$

From (5.1.5) and (5.1.7), we have

$$|\sigma_{e_1} \wedge \cdots \wedge \sigma_{e_k}|^2_{h^{\det U^*}} = \det(\langle \sigma_{e_i}, \sigma_{e_j}\rangle_{h^{U^*}})_{1 \leqslant i, j \leqslant k}$$
$$= \det((A^*(AA^*)^{-1}A)_{ij})_{1 \leqslant i, j \leqslant k} = \det(I + BB^*)^{-1}. \quad (5.1.8)$$

As $\sigma_{e_1} \wedge \cdots \wedge \sigma_{e_k}$ is a holomorphic frame of $\det U^*$, we get from (1.5.8) and (5.1.8),

$$\omega_{FS}([V_1^*]) = \frac{\sqrt{-1}}{2\pi}dB_{jl} \wedge d\overline{B}_{jl}. \quad (5.1.9)$$

In analogy to the projective space we call ω_{FS} and its associated Hermitian metric the *Fubini–Study form* and *Fubini–Study metric* on $\mathbb{G}(k, V^*)$, respectively.

5.1.2 Convergence of the induced Fubini–Study metrics

Let (X, J) be a compact complex manifold with complex structure J and $\dim X = n$. We introduce a Riemannian metric g^{TX} on TX compatible with J and denote by dv_X the Riemannian volume form on (X, g^{TX}).

We consider a holomorphic positive line bundle L on X, endowed with a Hermitian metric h^L such that the curvature R^L associated to h^L verifies (1.5.21). The metric h^L on L induces metrics h^{L^p} on L^p.

We denote by ω and Θ the $(1,1)$-forms on X associated to R^L and g^{TX} as in (1.5.14). Then $dv_X = \Theta^n/n!$.

On $\mathscr{C}^\infty(X, L^p)$ there is an L^2-scalar product $\langle\cdot, \cdot\rangle$ induced by h^L and g^{TX} as in (1.3.14); for $s, s' \in \mathscr{C}^\infty(X, L^p)$ we set

$$\langle s, s'\rangle = \int_X \langle s(x), s'(x)\rangle_{L^p}\, dv_X(x). \qquad (5.1.10)$$

The scalar product $\langle\cdot, \cdot\rangle$ induces an L^2-metric $h^{H^0(X, L^p)}$ on the space $H^0(X, L^p)$ of holomorphic sections of L^p on X.

Let $P_p(x, x')$, $(x, x' \in X)$, be the smooth kernel of the orthogonal projection P_p from $(\mathscr{C}^\infty(X, L^p), \langle\cdot, \cdot\rangle)$ onto $H^0(X, L^p)$, with respect to $dv_X(x')$.

If $\{S_i^p\}_{i=1}^{d_p}$, $(d_p = \dim H^0(X, L^p))$, is an orthonormal basis of $(H^0(X, L^p),$ $h^{H^0(X, L^p)})$, then by (4.1.4),

$$P_p(x, x) = \sum_{i=1}^{d_p} |S_i^p(x)|_{h^{L^p}}^2. \qquad (5.1.11)$$

The Kodaira map for L^p was defined in (2.2.10) and (2.2.11):

$$\begin{aligned}
\Phi_p &: X \smallsetminus \mathrm{Bl}_p \longrightarrow \mathbb{P}(H^0(X, L^p)^*), \\
\Phi_p(x) &= \{s \in H^0(X, L^p) : s(x) = 0\},
\end{aligned} \qquad (5.1.12)$$

where $\mathrm{Bl}_p = \{x \in X : s(x) = 0 \text{ for all } s \in H^0(X, L^p)\}$ is the base locus of $H^0(X, L^p)$.

Definition 5.1.1. Let F be a holomorphic line bundle over X. F is called *semi-ample* if there exists p_0 such that $\mathrm{Bl}_p = \emptyset$ for all $p \geqslant p_0$ (so that Φ_p are holomorphic maps on X for $p \geqslant p_0$). F is called *ample* if F is semi-ample and Φ_p is an embedding for p large enough. F is called *very ample* if $\mathrm{Bl}_1 = \emptyset$ and Φ_1 is an embedding.

Due to the Cartan–Serre–Grothendieck lemma 5.1.11, we can require in the definition of ampleness that Φ_p be an embedding for just one p. Thus, F is ample if and only if F^p is very ample for p large enough.

Lemma 5.1.2. *The positive line bundle L is semi-ample on X.*

Proof. By Theorem 4.1.1, we know that uniformly for $x \in X$,

$$P_p(x, x) = \boldsymbol{b}_0(x)p^n + \mathscr{O}(p^{n-1}), \quad \text{with } \boldsymbol{b}_0(x) = \det\left(\frac{\dot{R}^L}{2\pi}\right)(x) > 0. \qquad (5.1.13)$$

Since X is compact, there exists p_0 such that $P_p(x, x) \neq 0$ for all $p \geqslant p_0$ and $x \in X$. By (5.1.11), $\mathrm{Bl}_p = \emptyset$ for all $p \geqslant p_0$. $\qquad \square$

Theorem 5.1.3. *For $p \geqslant p_0$, $\Phi_p : X \longrightarrow \mathbb{P}(H^0(X, L^p)^*)$ is holomorphic and the map*

$$\begin{aligned}
\Psi_p &: \Phi_p^* \mathscr{O}(1) \to L^p, \\
\Psi_p((\Phi_p^* \sigma_s)(x)) &= s(x), \quad \text{for any } s \in H^0(X, L^p)
\end{aligned} \qquad (5.1.14)$$

defines a canonical isomorphism from $\Phi_p^* \mathcal{O}(1)$ to L^p on X, and under this isomorphism, we have

$$h^{\Phi_p^* \mathcal{O}(1)}(x) = P_p(x,x)^{-1} h^{L^p}(x). \tag{5.1.15}$$

Here $h^{\Phi_p^* \mathcal{O}(1)}$ is the metric on $\Phi_p^* \mathcal{O}(1)$ induced by $h^{\mathcal{O}(1)}$ on $\mathcal{O}(1)$ on $\mathbb{P}(H^0(X, L^p)^*)$.

Proof. Let $\{s_j\}$ be any basis of $H^0(X, L^p)$, and $\{s^j\}$ be its dual basis. Let e_L be any local frame of L near $x_0 \in X$, and write $s_j = f_j e_L^{\otimes p}$. Then for any $s \in H^0(X, L^p)$,

$$\Big(\sum_{j=1}^{d_p} f_j(x) s^j, s \Big) e_L^{\otimes p}(x) = \sum_{j=1}^{d_p} s_j(x)(s^j, s) = s(x). \tag{5.1.16}$$

By (5.1.12) and (5.1.16), we have

$$\Phi_p(x) = \Big[\sum_{j=1}^{d_p} f_j(x) s^j \Big] \in \mathbb{P}(H^0(X, L^p)^*). \tag{5.1.17}$$

Taking a holomorphic local frame e_L near x_0, then $f_j(x)$ are holomorphic; from (5.1.17), we know Φ_p is holomorphic.

By definition of σ_s and (5.1.17), for $\sum_{j=1}^{d_p} f_j(x) s^j \in \mathcal{O}(-1)_{\Phi_p(x)}$, we have

$$\Big((\Phi_p^* \sigma_s)(x), \sum_{j=1}^{d_p} f_j(x) s^j \Big) e_L^{\otimes p}(x) = \sum_{j=1}^{d_p} f_j(x)(s, s^j) e_L^{\otimes p}(x) = s(x). \tag{5.1.18}$$

Thus $(\Phi_p^* \sigma_s)(x) = 0$ is equivalent to $s(x) = 0$.

For any $\zeta \in (\Phi_p^* \mathcal{O}(1))_x$, from Lemma 5.1.2, take $s \in H^0(X, L^p)$ such that $\zeta = (\Phi_p^* \sigma_s)(x)$. Thus the map Ψ_p in (5.1.14) is well defined from (5.1.18).

As $(\Phi_p^* \sigma_s)$, s are holomorphic sections of $\Phi_p^* \mathcal{O}(1)$, L^p, from (5.1.14), we know Ψ_p is holomorphic.

To get (5.1.15), we take $\{s_j\}$ an orthonormal basis of $(H^0(X, L^p), h^{H^0(X, L^p)})$. Then by (5.1.2), (5.1.11), (5.1.16) and (5.1.17), for $\zeta \in (\Phi_p^* \mathcal{O}(1))_x$, there is $s \in H^0(X, L^p)$ such that $\zeta = (\Phi_p^* \sigma_s)(x)$, then under the isomorphism Ψ_p,

$$|\zeta|^2_{h^{\Phi_p^* \mathcal{O}(1)}} = |\sigma_s(\Phi_p(x))|^2_{h^{\mathcal{O}(1)}}$$

$$= \frac{|(\sum_{j=1}^{d_p} f_j(x) s^j, s)|^2}{|\sum_{j=1}^{d_p} f_j(x) s^j|^2_{h^{H^0(X, L^p)^*}}} = \frac{|s(x)|^2_{h^{L^p}}}{\sum_{j=1}^{d_p} |s_j(x)|^2_{h^{L^p}}} = P_p(x,x)^{-1} |\zeta|^2_{h^{L^p}}. \tag{5.1.19}$$

The proof of Theorem 5.1.3 is completed. $\qquad\square$

The next result shows that the Kodaira map tends to be isometric.

$H^0(X, L^l)$ such that $\sigma_1(x) \neq 0$, $\sigma_1(y) = 0$. Thus $\sigma_1 \otimes s \in H^0(X, L^{l+p} \otimes E)$ such that $(\sigma_1 \otimes s)(x) \neq 0, (\sigma_1 \otimes s)(y) = 0$. Thus $(x, y) \notin A_{l+p}$. By Lemma 5.1.9, there exists p_1 such that $A_p = A_{p_1}$ for $p \geqslant p_1$.

Summarizing, for any $p_2 \geqslant p_1$, we have

$$\mathrm{Diag}(X \times X) = \cap_p A_p = A_{p_2}. \tag{5.1.52}$$

Therefore Φ_p is injective for $p \geqslant p_1$. $\square$

5.2 Stability and Bergman kernel

In this section, we give an introduction to Donaldson's approach to the existence of Kähler metrics of constant scalar curvature (briefly, *csc Kähler metrics*) and especially, the relation with the Bergman kernel, which originates from his attempt to understand a conjecture of Yau.

In Section 5.2.1, we explain very briefly the history of the csc Kähler metrics. Section 5.2.2 is an introduction to Donaldson's approach on the existence of csc Kähler metrics. In Section 5.2.3, we explain the vector bundle version.

We will use the notation from Section 5.1.

5.2.1 Extremal Kähler metrics

The existence of special metrics is one of the central problems in differential geometry. Usually, once we know the existence of special metrics, it will have many interesting applications. In Kähler geometry, many of these questions have been instigated by the seminal work of Calabi.

Let (X, J, ω) be a compact Kähler manifold and $\dim X = n$. Let g^{TX} be the Riemannian metric on TX associated to ω. We call the cohomology class associated to ω the Kähler class of ω.

Let $R^{K_X^*}$ be the curvature of the holomorphic Hermitian connection on K_X^*, the dual of the canonical bundle $K_X := \det(T^{*(1,0)}X)$ of X, with metric induced by ω. Then the Ricci form Ric_ω of (X, ω) is a real $(1, 1)$-form on X given by (cf. Problem 1.7),

$$\mathrm{Ric}_\omega = \sqrt{-1} R^{K_X^*}. \tag{5.2.1}$$

We denote by $[\mathrm{Ric}_\omega]$ the cohomology class of Ric_ω, then

$$[\mathrm{Ric}_\omega] = 2\pi c_1(K_X^*) =: 2\pi c_1(X) \in H^2(X, \mathbb{R}). \tag{5.2.2}$$

Definition 5.2.1. ω is a *Kähler–Einstein metric* on X if there exists $\lambda \in \mathbb{R}$ such that

$$\mathrm{Ric}_\omega = \lambda\omega. \tag{5.2.3}$$

By (5.2.2), if ω is a Kähler–Einstein metric, it verifies the topological condition

$$\lambda[\omega] = 2\pi c_1(X) \in H^2(X, \mathbb{R}). \tag{5.2.4}$$

In the 1950s, Calabi initiated the study of Kähler–Einstein metrics. In the 1970s, the existence of Kähler–Einstein metrics was proved by Aubin and Yau independently, if $c_1(X) < 0$, i.e., K_X^* is negative. If $c_1(X) = 0$, Yau established the existence of the Ricci flat metric in each Kähler class on X: the renowned Calabi–Yau metrics. If $c_1(X) > 0$, X need not have a Kähler–Einstein metric. If $\dim X = 2$ and $c_1(X) > 0$, Tian solved completely this problem in the 1990s: X admits Kähler–Einstein metrics if and only if the algebra of holomorphic vector fields is reductive.

In the late 1980s, Yau conjectured the relation between the notions of stability of manifolds and existence of special metrics such as Kähler–Einstein metrics and csc Kähler metrics. Tian has made enormous progress towards understanding precisely when X has a Kähler–Einstein metrics if $c_1(X) > 0$. Tian introduced K-stability (later modified by Donaldson) which is thought to be the right notion of stability equivalent to the existence of csc Kähler metrics.

Now, we relax the topological condition (5.2.4) on the Kähler form ω on X; then the natural metrics we are looking for are the csc Kähler metrics. Certainly, a Kähler–Einstein metric has constant scalar curvature (cf. Problem 1.7).

Proposition 5.2.2. *If there exists $\lambda \in \mathbb{R}$ such that (5.2.4) holds and ω is a csc Kähler metric on X, then ω is a Kähler–Einstein metric.*

Proof. By the $\partial\bar{\partial}$-Lemma 1.5.1, there exists a real function f on X such that

$$\mathrm{Ric}_\omega - \lambda\omega = \sqrt{-1}\,\partial\bar{\partial}f. \tag{5.2.5}$$

Taking the trace of these $(1,1)$-forms, we get from Problem 4.1 and (5.2.2) that

$$\frac{1}{2}r^X - \lambda n = -\frac{1}{2}\Delta f. \tag{5.2.6}$$

Since $\frac{1}{2}r^X - \lambda n$ is a constant function, (5.2.6) implies

$$\int_X |\Delta f|^2 dv_X = -(r_\omega^X - 2\lambda n)\int_X \Delta f\, dv_X = 0. \tag{5.2.7}$$

This means $\Delta f = 0$. Thus f is constant on X (cf. Problem 1.9). We conclude that ω is a Kähler–Einstein metric. $\qquad\square$

More generally, in the early 1980s, Calabi introduced the extremal Kähler metrics which are critical points of the functional

$$\mathrm{Ca}(\omega) = \int_X (r_\omega^X)^2 \omega^n, \tag{5.2.8}$$

in a given Kähler class, where r_ω^X is the scalar curvature of (X, J, ω). Especially, a csc Kähler metric is an extremal Kähler metric.

The uniqueness problem for extremal metrics is well understood now. The uniqueness of Kähler–Einstein metrics of non-positive scalar curvature was already known to Calabi in the 1950s. In 1986, Bando and Mabuchi proved the uniqueness of Kähler–Einstein metrics of positive scalar curvature. In 2000, following a suggestion of Donaldson, Chen proved the uniqueness of csc Kähler metrics in any Kähler class on a compact Kähler manifold with non-positive first Chern class. Donaldson proved the uniqueness of csc Kähler metrics with rational Kähler class on any projective manifolds without non-trivial holomorphic vector fields and Mabuchi was able to extend Donaldson's result to some cases that holomorphic vector fields are not zero. Finally, Chen and Tian established the uniqueness of extremal metrics in the most general case.

5.2.2   Scalar curvature and projective embeddings

Let X be a compact complex manifold. Let L be a positive line bundle on X.

Following Donaldson, we consider the existence and uniqueness problem of csc Kähler metrics on X in the Kähler class $c_1(L)$.

By Proposition 1.5.3, the Kähler forms on X in the Kähler class $c_1(L)$ are given by the Hermitian metrics on L (unique up to multiplication by positive constants) such that the associated curvature is $-2\pi\sqrt{-1}$ times the Kähler form.

By Theorem 5.1.12, there exists $p_0 \in \mathbb{N}$ such that the Kodaira map $\Phi_p : X \longrightarrow \mathbb{P}(H^0(X, L^p)^*)$ is an embedding for any $p \geqslant p_0$. By Theorem 5.1.3, $\Phi_p^* \mathscr{O}(1)$ is canonically isomorphic to L^p.

Thus the Kähler metrics on X in the Kähler class $c_1(L)$ is closed related to the Hermitian metrics on $H^0(X, L^p)$. We explore this point now.

Set

$$\mathrm{vol}(X) = \int_X c_1(L)^n/n!, \quad d_p = \dim H^0(X, L^p), \quad R_p = \frac{d_p}{\mathrm{vol}(X)},$$

$$\bar{r}^X = \frac{4\pi}{\mathrm{vol}(X)} \int_X c_1(X) \frac{c_1(L)^{n-1}}{(n-1)!}. \tag{5.2.9}$$

Then $\bar{r}^X$ is the average of the scalar curvature of any Kähler metric in the Kähler class $c_1(L)$ (cf. Problem 4.1).

Set

$$\mathrm{Met}(L^p) = \{h^{L^p}; h^{L^p} \text{ is a metric on } L^p \text{ with associated curvature}$$

$$- 2\pi\sqrt{-1}p\,\omega_{h^{L^p}} \text{ where } \omega_{h^{L^p}} \text{ is a positive } (1,1)\text{-form}\}. \tag{5.2.10}$$

In this section, given $h^{L^p} \in \mathrm{Met}(L^p)$, we will always use the Riemannian metric on TX associated to $\omega_{h^{L^p}}$.

We assume from now on that $p \geqslant p_0$.

Definition 5.2.3. For $h^{L^p} \in \mathrm{Met}(L^p)$, set $\mathrm{Hilb}(h^{L^p})$ as the Hermitian metric on $H^0(X, L^p)$ defined by

$$\|s\|^2_{\mathrm{Hilb}(h^{L^p})} = R_p \int_X |s(x)|^2_{h^{L^p}} \omega^n_{h^{L^p}}/n!. \tag{5.2.11}$$

Definition 5.2.4. For a Hermitian metric h^H on $H^0(X, L^p)$, the metric $\mathrm{FS}(h^H)$ on L^p is defined as follows: For $\{s_j\}_{j=1}^{d_p}$ an orthonormal basis of $(H^0(X, L^p), h^H)$, we have

$$\sum_j |s_j(x)|^2_{\mathrm{FS}(h^H)} = 1 \quad \text{for any } x \in X. \tag{5.2.12}$$

In other words, let e_L be a local frame of L near $x_0 \in X$, and write $s_j = f_j e_L^{\otimes p}$, then $\sum_j |f_j(x)|^2 > 0$; as the evaluation map $H^0(X, L^p) \to L^p_x$ is surjective for all $x \in X$ and $p \geqslant p_0$, we define

$$|e_L^{\otimes p}(x)|^2_{\mathrm{FS}(h^H)} = \frac{1}{\sum_j |f_j(x)|^2}. \tag{5.2.13}$$

Let $h_H^{\mathscr{O}(1)}$ be the metric on $\mathscr{O}(1) \to \mathbb{P}(H^0(X, L^p)^*)$ induced by h^H. Then, using the same argument as in (5.1.19), we see that $\mathrm{FS}(h^H)$ is the metric induced by $h_H^{\mathscr{O}(1)}$ under the isomorphism (5.1.14).

We denote by $\omega_{\mathrm{FS}(h^H)}$ the Kähler form which is the restriction on X of the standard Fubini–Study form on $\mathbb{P}(H^0(X, L^p)^*)$ defined by h^H as in (5.1.3).

Definition 5.2.5. A pair (G, h^{L^p}) is *balanced* if

$$G = \mathrm{Hilb}(h^{L^p}), \quad \text{and } h^{L^p} = \mathrm{FS}(G). \tag{5.2.14}$$

In this situation, we will say the metrics h^{L^p} on L^p and G on $H^0(X, L^p)$ are balanced, or they are balanced metrics.

By (5.1.2), (5.2.11) and (5.2.12), if h^{L^p} is a balanced metric on L^p, then $C h^{L^p}$ is also a balanced metric for any constant $C > 0$.

Proposition 5.2.6. *Let $h^{L^p} \in \mathrm{Met}(L^p)$ be a metric on L^p and as in (5.1.11), we denote by $P_p(x, x)$ the diagonal of the Bergman kernel on L^p associated to h^{L^p}; then h^{L^p} is balanced if and only if*

$$P_p(x, x) = constant \text{ on } X. \tag{5.2.15}$$

Proof. If (5.2.15) holds, then

$$P_p(x, x) = \frac{1}{\mathrm{vol}(X)} \int_X P_p(x, x) dv_X(x) = \frac{d_p}{\mathrm{vol}(X)} = R_p. \tag{5.2.16}$$

If $\{s_j\}$ is an orthonormal basis of $(H^0(X, L^p), h^{H^0(X,L^p)})$, then by (5.2.11), $\{R_p^{-1/2} s_j\}$ is an orthonormal basis of $(H^0(X, L^p), \mathrm{Hilb}(h^{L^p}))$.

Thus from (5.2.12), $\mathrm{FS}(\mathrm{Hilb}(h^{L^p})) = h^{L^p}$ if and only if

$$R_p^{-1} P_p(x, x) = R_p^{-1} \sum |s_j(x)|^2_{h^{L^p}} = 1. \qquad \square$$

Let $\{s_j\}$ be a basis of $H^0(X, L^p)$, and $\{s^j\}$ its dual basis. Then we can define $\psi : H^0(X, L^p)^* \simeq \mathbb{C}^{d_p}$ by $\psi(z_j s^j) = (z_1, \ldots, z_{d_p})$. This gives us an embedding $\Phi_{p,\{s_j\}} : X \to \mathbb{C}\mathbb{P}^{d_p-1}$. If we choose another basis $\{\tilde{s}_j\}$, then there exists $\sigma \in \mathrm{SL}(d_p, \mathbb{C})$, $c \in \mathbb{C}$ such that $\tilde{s}_i = c\,\sigma_{ij} s_j$, thus

$$\Phi_{p,\{\tilde{s}_j\}} = \sigma \circ \Phi_{p,\{s_j\}}. \tag{5.2.17}$$

This inspires us to introduce the following notation.

Definition 5.2.7. Let $X \subset \mathbb{C}\mathbb{P}^N$ be a smooth compact complex manifold. We define $M(X)$ to be the $(N+1) \times (N+1)$ matrix with entries

$$M(X)_{ij} = \int_X \frac{z_i \bar{z}_j}{\sum_l |z_l|^2} \frac{\omega_{\mathrm{FS}}^n}{n!}. \tag{5.2.18}$$

X is said to be *balanced* if there exists $\sigma \in \mathrm{SL}(N+1, \mathbb{C})$ such that $M(\sigma(X))$ is a multiple of the identity matrix.

For a pth embedding $X \hookrightarrow \mathbb{C}\mathbb{P}^{d_p-1}$, we have two classical notions of stability in geometric invariant theory (G.I.T.): Chow poly-stable, corresponding to the Chow point associated to (X, L^p) in the Chow group of $\mathbb{C}\mathbb{P}^{d_p-1}$ with degree $p^n \int_X c_1(L)^n$; and Hilbert poly-stable, corresponding to the Hilbert point associated to (X, L^p) in the Hilbert scheme of $\mathbb{C}\mathbb{P}^{d_p-1}$ with Hilbert polynomial $\chi(m) := d_{mp}$. It would be quite long to give an elementary introduction to stability, thus we will only give some references.

The following result explains the relation between the balance condition and the Chow poly-stable condition in algebraic geometry.

Theorem 5.2.8 (Zhang). *(X, L^p) is Chow poly-stable if and only if the projective embedding $X \to \mathbb{C}\mathbb{P}^{d_p-1}$ induced by L^p can be balanced.*

Definition 5.2.9. The automorphism group $\mathrm{Aut}(X, L)$ of the pair (X, L) is

$$\mathrm{Aut}(X, L) = \{\varphi : L \to L; \varphi \text{ is a biholomorphic, fiberwise linear}$$
$$\text{map, and it induces an automorphism on } X\}. \tag{5.2.19}$$

Theorem 5.2.10 (Donaldson). *Suppose that $\mathrm{Aut}(X, L)$ is discrete. If X has a csc Kähler form ω in the Kähler class $c_1(L)$, then for large enough p there is a unique balanced metric on L^p inducing a Kähler form $p\omega_p$ on X, and $\omega_p \to \omega$ as $p \to \infty$. Conversely, if there are balanced metrics on L^p for all large p and the sequence ω_p converges to ω, then ω has constant scalar curvature.*

Proof. The first part is the hard and central part. Donaldson needs to combine Theorem 4.1.2 and his moment map picture to complete the proof. We refer the readers to Donaldson's original paper for the details.

The second part is a consequence of Theorems 4.1.1 and 4.1.2. In fact, if there are balanced metrics on L^p for all large p and the sequence ω_p converges to ω, we denote by $P_{p,\omega_p}(x,x)$ the diagonal of the Bergman kernel on L^p associated to the balanced metric on L^p and ω_p. Then from Theorem 4.1.1, we know there exists $C > 0$ such that for $p \in \mathbb{N}$, $x \in X$,

$$\left| P_{p,\omega_p}(x,x) - p^n - \frac{1}{8\pi} r^X_{\omega_p}(x) p^{n-1} \right| \leqslant C p^{n-2}. \tag{5.2.20}$$

But by (4.1.10), (5.2.9) and (5.2.16), we get

$$P_{p,\omega_p}(x,x) = R_p = p^n + \frac{1}{8\pi} \bar{r}^X p^{n-1} + \mathcal{O}(p^{n-2}). \tag{5.2.21}$$

Thus

$$|r^X_{\omega_p}(x) - \bar{r}^X| \leqslant C/p. \tag{5.2.22}$$

But $r^X_{\omega_p}(x)$ converges to r^X_ω when $p \to \infty$, since $\omega_p \to \omega$. Thus r^X_ω is constant. $\square$

5.2.3 Gieseker stability and Grassmannian embeddings

We explain the vector bundle version of Section 5.2.2 which was established by Wang.

Let X be an n-dimensional compact complex manifold, and L a positive line bundle.

Suppose L is very ample now. We fix a metric h^L on L such that $\sqrt{-1}R^L$ is a positive $(1,1)$-form, where R^L is the curvature associated to h^L. We also fix the Kähler form $\omega = \frac{\sqrt{-1}}{2\pi} R^L$ on X.

Let E be a holomorphic vector bundle on X and $\mathrm{rk}(E) = k$. Set

$$\deg(E) = \int_X c_1(E) c_1(L)^{n-1}. \tag{5.2.23}$$

As L is very ample, then for any $\mathscr{O}_X$-coherent sheaf $\mathscr{F}$, we have a projective resolution of vector bundles on X,

$$0 \to \mathscr{O}_X(E^n) \to \cdots \to \mathscr{O}_X(E^0) \to \mathscr{O}_X(\mathscr{F}) \to 0, \tag{5.2.24}$$

here E^i are holomorphic vector bundles on X and (5.2.24) is an exact sequence of $\mathscr{O}_X$-sheaves. Especially $c_1(\mathscr{F})$ and $\deg(\mathscr{F})$ are well defined and

$$c_1(\mathscr{F}) = \sum_i (-1)^i c_1(E^i), \quad \mathrm{ch}(\mathscr{F}) = \sum_i (-1)^i \mathrm{ch}(E^i). \tag{5.2.25}$$

Definition 5.2.11. We call E *Mumford stable* (resp. *semi-stable*) if for any torsion free proper sub-sheaf $\mathcal{F} \subset \mathscr{O}_X(E)$, we have

$$\frac{\deg(E)}{\mathrm{rk}(E)} > (\text{resp. } \geqslant)\frac{\deg(\mathcal{F})}{\mathrm{rk}(\mathcal{F})}. \tag{5.2.26}$$

Recall that the Euler characteristic number $\chi(X, E)$ is defined in (1.4.25).

Definition 5.2.12. We call E *Gieseker stable* (resp. *semi-stable*) if for any torsion free proper sub-sheaf $\mathcal{F} \subset \mathscr{O}_X(E)$, there exists $p_0 \in \mathbb{N}$ such that for any $p \geqslant p_0$, we have

$$\frac{\chi(X, L^p \otimes E)}{\mathrm{rk}(E)} > (\text{resp. } \geqslant)\frac{\chi(X, L^p \otimes \mathcal{F})}{\mathrm{rk}(\mathcal{F})}. \tag{5.2.27}$$

By the Riemann–Roch–Hirzebruch theorem 1.4.6, we know that

$$\text{Mumford stable} \Longrightarrow \text{Gieseker stable} \Longrightarrow \text{Mumford semi-stable}. \tag{5.2.28}$$

We assume from now on that $p \geqslant p_0$ as in Section 5.1.4. Set

$$d_{E,p} = \dim H^0(X, L^p \otimes E) = \chi(X, L^p \otimes E), \quad R_{E,p} = \frac{d_{E,p}}{\mathrm{vol}(X)k}. \tag{5.2.29}$$

Definition 5.2.13. Let h^E be a Hermitian metric on E and define the Hermitian metric $\mathrm{Hilb}(h^E)$ on $H^0(X, L^p \otimes E)$ by

$$\|s\|^2_{\mathrm{Hilb}(h^E)} = R_{E,p} \int_X |s(x)|^2_{h^{L^p \otimes E}} \frac{\omega^n}{n!}. \tag{5.2.30}$$

Definition 5.2.14. For a Hermitian metric h^H on $H^0(X, L^p \otimes E)$, the metric $\mathrm{FS}(h^H)$ on E is defined as follows: Let h^{U^*} be the Hermitian metric on U^* (the dual of the universal bundle on $\mathbb{G}(k, H^0(X, L^p \otimes E)^*)$), induced by h^H. Then $\mathrm{FS}(h^H)$ is the metric satisfying,

$$h^{L^p} \otimes \mathrm{FS}(h^H) = \Phi_p^* h^{U^*}, \tag{5.2.31}$$

where $\Phi_p^* h^{U^*}$ is the pull-back of h^{U^*} by the canonical map (5.1.39).

Definition 5.2.15. A pair (G, h^E) is *balanced* for the Kodaira map Φ_p if

$$G = \mathrm{Hilb}(h^E), \quad \text{and } h^E = \mathrm{FS}(G). \tag{5.2.32}$$

In this situation, we will say the metrics h^E on E and G on $H^0(X, L^p \otimes E)$ are balanced.

By (5.1.5), (5.2.30) and (5.2.31), if h^E is a balanced metric on E, then Ch^E is also a balanced metric for any constant $C > 0$.

Theorem 5.2.16. *Let h^E be a metric on E. h^E is balanced for Φ_p if and only if the corresponding Bergman kernel $P_p(x, x')$ on $L^p \otimes E$ verifies*

$$P_p(x, x) = R_{E,p} \operatorname{Id}_E. \tag{5.2.33}$$

Proof. By using (5.1.40) as in the proof of Proposition 5.2.6, we get Theorem 5.2.16. □

Theorem 5.2.17 (Wang). *E is Gieseker stable if and only if there exists $p_0 \in \mathbb{N}$ such that for all $p \geq p_0$, there exists h_p^E on E such that h_p^E is balanced for Φ_p.*

Theorem 5.2.18 (Wang). *Suppose E is Gieseker stable. If for $p \to +\infty$, h_p^E in Theorem 5.2.17 converges to h_∞^E, then the metric h_∞^E solves the following weakly Hermitian–Einstein equation,*

$$\frac{\sqrt{-1}}{2\pi} \Lambda_\omega(R_\infty^E) + \frac{1}{8\pi} r^X \operatorname{Id}_E = \left(\frac{\deg(E)}{\operatorname{vol}(X) k\,(n-1)!} + \frac{1}{8\pi} \overline{r}^X \right) \operatorname{Id}_E, \tag{5.2.34}$$

where R_∞^E is the curvature associated to h_∞^E on E.

Conversely, suppose there is a Hermitian metric h_∞^E on E solving (5.2.34), then there exist balanced metrics h_p^E for any $p \geq p_0$ such that $h_p^E \to h_\infty^E$ in $\mathscr{C}^l$ for any $l \in \mathbb{N}$.

Proof of Theorem 5.2.18. From (4.1.10) and (5.2.29), we have for $p \geq p_0$:

$$R_{E,p} = p^n + \left(\frac{\deg(E)}{\operatorname{vol}(X) k\,(n-1)!} + \frac{1}{8\pi} r^X \right) p^{n-1} + \mathcal{O}(p^{n-2}). \tag{5.2.35}$$

Now from Theorems 4.1.1 and 5.2.17, we get (5.2.34).

The converse is the hard part of the proof. One needs to adapt Donaldson's argument for vector bundles. We refer to Wang's original paper for the details. □

5.3 Distribution of zeros of random sections

As an application of Theorem 5.1.4, we study the distribution of zeros of random sections.

We use the same notation and assumption in Section 5.1.2. Especially, (X, J) is a compact complex manifold with complex structure J and $\dim X = n$. g^{TX} is a Riemannian metric on TX compatible with J. (L, h^L) is a positive holomorphic Hermitian line bundle on X. $\omega = \frac{\sqrt{-1}}{2\pi} R^L$, $\Theta = g^{TX}(J\cdot, \cdot)$ are positive $(1, 1)$-forms on X.

Let dS be the usual volume form on the $2m - 1$-dimensional unit sphere $S^{2m-1} := \{\lambda \in \mathbb{C}^m; |\lambda| = 1\}$; then

$$\operatorname{vol}(S^{2m-1}) = \int_{S^{2m-1}} dS = \frac{2\pi^m}{(m-1)!}. \tag{5.3.1}$$

By the same proof as in Sections 1.5.1, 1.5.2, we get vanishing results and the spectral gap property.

Theorem 5.4.9.

(a) $H^{r,q}(X, L) = 0$ if $r + q > n$; especially,

(b) $H^q(X, L \otimes K_X) = 0$ if $q > 1$.

(c) There exists $C > 0$ such that for any $p \in \mathbb{N}$,

$$\text{Spec}(D_p^2) \subset \{0\} \cup \,]4\pi p - C, +\infty[, \tag{5.4.10}$$

and $D_p^2|_{\Omega^{0,>0}}$ is invertible for p large enough. Especially, for p large enough,

$$H^q(X, L^p \otimes E) = 0 \quad \text{for any } q > 0. \tag{5.4.11}$$

By Theorems 5.4.6 and 5.4.9, we can define the Bergman kernel $P_p(x, x')$ $(x, x' \in X)$, for $p > \frac{C}{2\pi}$, as the smooth kernel of the orthogonal projection from $\mathscr{C}^\infty(X, L^p \otimes E)$ onto $H^0(X, L^p \otimes E)$, with respect to the Riemannian volume form $dv_X(x')$.

From now on, we assume $p > \frac{C}{2\pi}$. Let $\{S_i^p\}_{i=1}^{d_p}$, $(d_p = \dim H^0(X, L^p \otimes E))$, be any orthonormal basis of $H^0(X, L^p \otimes E)$ with respect to the inner product (1.3.14). In fact, in the local coordinates above, $\widetilde{S}_i^p(\widetilde{z})$ are G_x-invariant on $\widetilde{U}_x$, and

$$P_p(z, z') = \sum_{i=1}^{d_p} \widetilde{S}_i^p(\widetilde{z}) \otimes (\widetilde{S}_i^p(\widetilde{z'}))^*, \tag{5.4.12}$$

where we use $\widetilde{z}$ to denote the point in $\widetilde{U}_x$ representing $z \in U_x$.

The following analogue of Theorem 4.1.1 is true.

Theorem 5.4.10. There exist smooth coefficients $\boldsymbol{b}_r(x) \in \text{End}(E)_x$ which are polynomials in R^{TX}, R^E and their derivatives with order $\leqslant 2r - 2$ at x and $C_0 > 0$ such that for any $k, l \in \mathbb{N}$, there exist $C_{k,l} > 0$, $N \in \mathbb{N}$ with

$$\left| \frac{1}{p^n} P_p(x, x) - \sum_{r=0}^{k} \boldsymbol{b}_r(x) p^{-r} \right|_{\mathscr{C}^l}$$

$$\leqslant C_{k,l} \left(p^{-k-1} + p^{l/2}(1 + \sqrt{p}d(x, X_{\text{sing}}))^N e^{-\sqrt{C_0 p}\, d(x, X_{\text{sing}})} \right), \tag{5.4.13}$$

for any $x \in X$, $p \in \mathbb{N}^*$. Moreover $\boldsymbol{b}_0, \boldsymbol{b}_1$ are given in (4.1.8).

Now, we explain in detail the asymptotic expansion near X_{sing}.

Let $\{x_i\}_{i=1}^m$ be points of X_{sing} such that the corresponding local charts $(G_{x_i}, \widetilde{U}_{x_i})$ with $\widetilde{U}_{x_i} \subset \mathbb{C}^n$ verify

$$B^{\widetilde{U}_{x_i}}(0, 2\varepsilon) \subset \widetilde{U}_{x_i}, \text{ and } X_{\text{sing}} \subset W := \cup_{i=1}^m B^{\widetilde{U}_{x_i}}\left(0, \frac{1}{4}\varepsilon\right)/G_{x_i}.$$

Let $\widetilde{U}_{\widetilde{x}_i}^g$ be the fixed point-set of $g \in G_{x_i}$ in $\widetilde{U}_{x_i}$, and let $\widetilde{N}_{x_i,g}$ be the normal bundle of $\widetilde{U}_{\widetilde{x}_i}^g$ in $\widetilde{U}_{x_i}$. For each $g \in G_{x_i}$, the exponential map $\widetilde{N}_{x_i,g,\widetilde{x}} \ni Y \rightarrow \exp_{\widetilde{x}}^{\widetilde{U}_{x_i}}(Y)$ identifies a neighborhood of $\widetilde{U}_{\widetilde{x}_i}^g$ to $\widetilde{W}_{x_i,g} = \{Y \in \widetilde{N}_{x_i,g}, |Y| \leqslant \varepsilon\}$. We identify $L|_{\widetilde{W}_{x_i,g}}, E|_{\widetilde{W}_{x_i,g}}$ to $L|_{\widetilde{U}_{\widetilde{x}_i}^g}, E|_{\widetilde{U}_{\widetilde{x}_i}^g}$ by using the parallel transport along the above exponential map. Then the g-action on $L|_{\widetilde{W}_{x_i,g}}$ is the multiplication by $e^{i\theta_g}$, and θ_g is locally constant on $\widetilde{U}_{\widetilde{x}_i}^g$. Likewise, the g-action on $E|_{\widetilde{W}_{x_i,g}}$ is $g^E \in \mathscr{C}^\infty(\widetilde{U}_{\widetilde{x}_i}^g, \operatorname{End}(E))$, which is parallel with respect to the connection ∇^E.

Note that there exists ε_0 such that for any $\widetilde{x} \in \widetilde{U}_{x_i} \smallsetminus \widetilde{W}_{x_i,g}$, we have $d(g^{-1}\widetilde{x}, \widetilde{x}) > \varepsilon_0$. If $\widetilde{Z} \in \widetilde{W}_{x_i,g}$, we write $\widetilde{Z} = (\widetilde{Z}_{1,g}, \widetilde{Z}_{2,g})$ with $\widetilde{Z}_{1,g} \in \widetilde{U}_{\widetilde{x}_i}^g$, $\widetilde{Z}_{2,g} \in \widetilde{N}_{x_i,g}$.

Theorem 5.4.11. *On $\widetilde{U}_{x_i}$ as above, there exist polynomials $\mathscr{K}_{r,\widetilde{Z}_{1,g}}(\widetilde{Z}_{2,g})$ in $\widetilde{Z}_{2,g}$ of degree $\leqslant 3r$, of the same parity as r, whose coefficients are polynomials in R^{TX}, R^E and their derivatives of order $\leqslant r-2$, and a constant $C_0 > 0$ such that for any $k, l \in \mathbb{N}$, there exists $C_{k,l} > 0$, $N \in \mathbb{N}$ such that*

$$
\left| \frac{1}{p^n} P_p(\widetilde{Z}, \widetilde{Z}) - \sum_{r=0}^{k} \boldsymbol{b}_r(\widetilde{Z}) p^{-r} \right. \tag{5.4.14}
$$

$$
\left. - \sum_{r=0}^{2k} p^{-\frac{r}{2}} \sum_{1 \neq y \in G_{x_0}} e^{-i\theta_g p} g^E(\widetilde{Z}_{1,g}) \mathscr{K}_{r,\widetilde{Z}_{1,g}}(\sqrt{p}\widetilde{Z}_{2,g}) e^{-2\pi p \langle (1-g^{-1})\widetilde{z}_{2,g}, \overline{\widetilde{z}_{2,g}} \rangle} \right|_{\mathscr{C}^l}
$$

$$
\leqslant C_{k,l} \left(p^{-k-1} + p^{-k+\frac{l-1}{2}} (1 + \sqrt{p} d(Z, X_{\mathrm{sing}}))^N \exp\left(-\sqrt{C_0 p}\, d(Z, X_{\mathrm{sing}})\right) \right),
$$

for any $|\widetilde{Z}| \leqslant \varepsilon/2$, $p \in \mathbb{N}^*$, with $\boldsymbol{b}_r(\widetilde{Z})$ as in Theorem 5.4.10 and

$$
\mathscr{K}_{0,\widetilde{Z}_{1,g}}(\widetilde{Z}_{2,g}) = \operatorname{Id}_E, \qquad \boldsymbol{b}_0 = \operatorname{Id}_E. \tag{5.4.15}
$$

Remark 5.4.12. a) The important feature in (5.4.14) is that even for the diagonal asymptotic expansion, we have the exponential decay near the singularity, which is very similar to the off-diagonal expansion. In fact, we use the off-diagonal expansion to get it.

b) By the argument in Section 4.1.9, we can reduce the situation for a general metric g^{TX} to the metric associated to ω, because X is compact. Certainly, our argument can be carried out to a certain complete orbifold version as done in Section 6.1 for a general metric on X.

Proof of Theorems 5.4.10 and 5.4.11. At first, from the spectral gap property (5.4.10), we have the analogue of Proposition 4.1.5,

$$
|P_p(x, x') - F(D_p)(x, x')|_{\mathscr{C}^m(X \times X)} \leqslant C_{l,m,\varepsilon} p^{-l}. \tag{5.4.16}
$$

To prove (5.4.16), we work on $\widetilde{U}_{x_i}$, and the Sobolev norm in (1.6.5) is summed over the $\widetilde{U}_{x_i}$.

Note that on an orbifold, the property of the finite propagation speed of solutions of hyperbolic equations still holds (see the proof in Appendix D.2). Thus
for $x, x' \in X$, $F(D_p)(x, x') = 0$, if $d(x, x') \geqslant \varepsilon$. Likewise, given $x \in X$, $F(D_p)(x, \cdot)$
only depends on the restriction of D_p to $B^X(x, \varepsilon)$. Thus the problem of the asymptotic expansion of $P_p(x, \cdot)$ is local.

For any compact set $K \subset X \setminus X_{\mathrm{sing}}$, we get the uniform estimate (5.4.13)
from Theorems 4.1.2, 4.1.18, Prop. 4.1.6, (4.1.96) and (5.4.16), as in Section 4.2,
since $G_x = \{1\}$ for $x \in K$.

Now, working near x_i, we replace X by $\mathbb{R}^{2n}/G_{x_i}$, by the above argument.
Let $\widetilde{L}, \widetilde{E}$ be the G_{x_i}-equivariant vector bundles on $\widetilde{U}_{x_i} \subset \mathbb{R}^{2n}$ corresponding to
L, E on $\widetilde{U}_{x_i}/G_{x_i}$. In particular, G_{x_i} acts linearly and effectively on $\mathbb{R}^{2n}$. We will
add a superscript $\sim$ to indicate the corresponding objects on $\mathbb{R}^{2n}$.

Now for $Z, Z' \in \mathbb{R}^{2n}/G_{x_i}$, $|\widetilde{Z}|, |\widetilde{Z}'| \leqslant \varepsilon/2$ and $\widetilde{Z}, \widetilde{Z}' \in \mathbb{R}^{2n}$ representing Z, Z',
we have

$$F(D_p)(Z, Z') = \sum_{g \in G_{x_i}} (g, 1) F(\widetilde{D}_p)(g^{-1}\widetilde{Z}, \widetilde{Z}'). \tag{5.4.17}$$

Here (g_1, g_2) acts on $E_{p,x} \times E_{p,x'}^*$ by $(g_1, g_2)(\xi_1, \xi_2) = (g_1\xi_1, g_2\xi_2)$. Indeed, if $s \in$
$\mathscr{C}^\infty(\mathbb{R}^{2n}/G_{x_i}, E_p)$ has support in $B^{\widetilde{U}_{x_i}}(0, \varepsilon) \subset \widetilde{U}_{x_i}$, then s is a G_x-invariant section
of $\widetilde{E}_p$ on $\mathbb{R}^{2n}$. By definition and the relation $g \cdot F(\widetilde{D}_p) = F(\widetilde{D}_p)g$, we have

$$\begin{aligned}
(F(D_p)s)(z) &= \int_{\mathbb{R}^{2n}} F(\widetilde{D}_p)(\widetilde{z}, \widetilde{z}')s(\widetilde{z}')dv_{\widetilde{U}_{x_i}}(\widetilde{z}') \\
&= \frac{1}{|G_{x_i}|} \sum_{g \in G_{x_i}} \int_{\mathbb{R}^{2n}} (g, 1) F(\widetilde{D}_p)(g^{-1}\widetilde{z}, \widetilde{z}')s(\widetilde{z}')dv_{\widetilde{U}_{x_i}}(\widetilde{z}') \\
&= \int_{\mathbb{R}^{2n}/G_{x_i}} \sum_{g \in G_{x_i}} (g, 1) F(\widetilde{D}_p)(g^{-1}\widetilde{z}, z')s(z')dv_{U_{x_i}}(z'). \tag{5.4.18}
\end{aligned}$$

By (4.1.12) and (4.1.27), for any $l, m \in \mathbb{N}$, there exists $C_{l,m,\varepsilon} > 0$ such that
for $p \geqslant 1$, $|\widetilde{Z}|, |\widetilde{Z}'| \leqslant \varepsilon/2$,

$$|F(\widetilde{D}_p)(\widetilde{Z}, \widetilde{Z}') - \widetilde{P}_{0,p}(\widetilde{Z}, \widetilde{Z}')|_{\mathscr{C}^m} \leqslant C_{l,m,\varepsilon} p^{-l}. \tag{5.4.19}$$

We use now the notation κ as in (4.1.28), $\mathscr{F}_r$ as in (4.1.95). By $\kappa_x, \mathscr{F}_{r,x}$ we indicate
the base point x. For $t = \frac{1}{\sqrt{p}}$, we have by (4.1.96),

$$\widetilde{P}_{0,p}((\widetilde{Z}, \widetilde{Z}') = \frac{1}{t^{2n}} \widetilde{P}_{0,t}(\widetilde{Z}/t, \widetilde{Z}'/t)\kappa^{-1/2}(Z)\kappa^{-1/2}(Z'). \tag{5.4.20}$$

By (4.2.30), (4.2.36) and (5.4.20), we have for $|\widetilde{Z}| \leqslant \varepsilon/2$, α, α' with $|\alpha'| \leqslant m$,

$$\left| \frac{\partial^{|\alpha|}}{\partial \widetilde{Z}_{1,g}^{\alpha}} \frac{\partial^{|\alpha'|}}{\partial \widetilde{Z}_{2,g}^{\alpha'}} \left(\frac{1}{p^n} \widetilde{P}_{0,p}(g^{-1}\widetilde{Z}, \widetilde{Z}) - \sum_{r=0}^{k} t^r \mathscr{F}_{r,\widetilde{Z}_{1,g}}(\sqrt{p}g^{-1}\widetilde{Z}_{2,g}, \sqrt{p}\widetilde{Z}_{2,g})\kappa_{\widetilde{Z}_{1,g}}^{-1}(\widetilde{Z}_{2,g}) \right) \right|$$

$$\leqslant Ct^{k+1-m'}(1 + \sqrt{p}|\widetilde{Z}_{2,g}|)^N \exp(-\sqrt{C_0 p}|\widetilde{Z}_{2,g}|). \quad (5.4.21)$$

Especially, from Theorem 4.1.21, (4.1.68) (or (5.4.21)), we obtain for $Z \in \mathbb{R}^{2n}/G_{x_i}$, $|\widetilde{Z}| \leqslant \varepsilon/2$,

$$\sup_{|\alpha| \leqslant m} \left| \frac{\partial^{|\alpha|}}{\partial \widetilde{Z}^{\alpha}} \left(\frac{1}{p^n} \widetilde{P}_{0,p}(\widetilde{Z}, \widetilde{Z}) - \sum_{r=0}^{k} p^{-r} \boldsymbol{b}_r(\widetilde{Z}) \right) \right| \leqslant Cp^{-k-1}. \quad (5.4.22)$$

Thus from (5.4.16)–(5.4.22), we get for $|\widetilde{Z}| \leqslant \varepsilon/2$,

$$\sup_{|\alpha| \leqslant m} \left| \frac{\partial^{|\alpha|}}{\partial \widetilde{Z}^{\alpha}} \left(\frac{1}{p^n} P_p(\widetilde{Z}, \widetilde{Z}) - \sum_{r=0}^{k} \boldsymbol{b}_r(\widetilde{Z})p^{-r} \right. \right.$$

$$\left. \left. - \sum_{r=0}^{2k} p^{-\frac{r}{2}} \sum_{1 \neq g \in G_{x_i}} (g,1)\mathscr{F}_{r,\widetilde{Z}_{1,g}}(\sqrt{p}g^{-1}\widetilde{Z}_{2,g}, \sqrt{p}\widetilde{Z}_{2,g})\kappa_{\widetilde{Z}_{1,g}}^{-1}(\widetilde{Z}_{2,g}) \right) \right|$$

$$\leqslant C \left(p^{-k-1} + p^{-k+\frac{m-1}{2}} (1 + \sqrt{p}d(Z, X_{\text{sing}}))^N \exp\left(-\sqrt{C_0 p}\, d(Z, X_{\text{sing}})\right) \right). \quad (5.4.23)$$

By our identification, from the notation in Theorem 4.1.21 and (4.1.71),

$$(g,1)\mathscr{F}_{r,\widetilde{Z}_{1,g}}(\sqrt{p}g^{-1}\widetilde{Z}_{2,g}, \sqrt{p}\widetilde{Z}_{2,g})$$

$$= e^{i\theta_g p} g^E(\widetilde{Z}_{1,g}) J_{r,\widetilde{Z}_{1,g}}(\sqrt{p}\widetilde{Z}_{2,g}) e^{-\pi p(|\widetilde{Z}_{2,g}|^2 - 2\langle g^{-1}\widetilde{z}_{2,g}, \overline{\widetilde{z}}_{2,g} \rangle)}$$

$$= e^{i\theta_g p} g^E(\widetilde{Z}_{1,g}) J_{r,\widetilde{Z}_{1,g}}(\sqrt{p}\widetilde{Z}_{2,g}) e^{-2\pi p\langle (1-g^{-1})\widetilde{z}_{2,g}, \overline{\widetilde{z}}_{2,g} \rangle}, \quad (5.4.24)$$

where $J_{0,\widetilde{Z}_{1,g}}(\widetilde{Z}_{2,g}) = \mathrm{Id}_E$.

Now, in (5.4.23), if we take also the Taylor expansion for $\kappa_{\widetilde{Z}_{1,g}}^{-1}(\widetilde{Z}_{2,g})$ in $\widetilde{Z}_{2,g}$, then we get (5.4.14) and the information on the polynomials $\mathscr{K}_{r,\widetilde{Z}_{1,g}}$ on $\widetilde{Z}_{2,g}$.

The proof of Theorems 5.4.10 and 5.4.11 is complete. $\qquad\square$

Remark 5.4.13. In the same way, by Theorem 4.2.1, (5.4.17)–(5.4.19), we get the full off-diagonal expansion of the Bergman kernel on the orbifolds as in Theorem 4.2.1.

Note that if $x_0 \in X_{\text{sing}}$, then $|G_{x_0}| > 1$. Now, if in addition, L and E are usual vector bundles, i.e., G_{x_0} acts on both L_{x_0} and E_{x_0} as identity, then by (5.4.23),

$$\left| \frac{1}{p^n} P_p(x_0, x_0) - |G_{x_0}| \boldsymbol{b}_0(x_0) \right| \leqslant Cp^{-1/2}. \quad (5.4.25)$$

Thus we can never have a uniform asymptotic expansion as (4.1.7) on X if X_{sing} is not empty.

In the spirit of (5.4.17), we study now the Bergman kernel on our model space $\mathbb{C}^n/G$ (compare Section 4.1.6).

Lemma 5.4.14. *Let $G \subset U(n)$ be a finite subgroup acting $\mathbb{C}$-linearly and isometric on $\mathbb{C}^n$, then there exists $C_{G,n} > 0$ such that for any $Z \in \mathbb{C}^n$,*

$$|G| \geqslant f_G(Z) := 1 + \sum_{1 \neq g \in G} e^{-2\pi \langle (1 - g^{-1})z, \overline{z} \rangle} \geqslant C_{G,n}. \qquad (5.4.26)$$

Proof. As $\operatorname{Re} \langle (1 - g^{-1})z, \overline{z} \rangle \geqslant 0$ for any $g \in G, z \in \mathbb{C}^n$, we know $|f_G(Z)| \leqslant |G|$.

We recall now some notation from Remark 4.1.19. Let (L, h^L) be the trivial holomorphic line bundle on $\mathbb{C}^n$ with the canonical section $\mathbf{1}$, and $|\mathbf{1}|_{h^L}(z) := e^{-\pi|z|^2}$ (Here under our notation (4.1.71), $|z|^2 = \frac{1}{2}|Z|^2$.) Then by (4.1.84), the Bergman kernel of (L, h^L) under the trivialization of L by using the unit section $e^{\pi|z|^2}\mathbf{1}$ is

$$\mathscr{P}(Z, Z') = \exp\left(-\pi(|z|^2 + |z'|^2 - 2\langle z, \overline{z}' \rangle)\right). \qquad (5.4.27)$$

We define the action of G on L by $g \cdot \mathbf{1} = \mathbf{1}$ for $g \in G$. Then the Bergman kernel $\mathscr{P}_{\mathbb{C}^n/G}$ of L on $\mathbb{C}^n/G$ is (cf. (5.4.17)),

$$\mathscr{P}_{\mathbb{C}^n/G}(Z, Z') = \sum_{g \in G} \mathscr{P}(g^{-1}Z, Z'). \qquad (5.4.28)$$

Now, the L^2-holomorphic sections of L on $\mathbb{C}^n/G$ are $H^0(\mathbb{C}^n, L)^G$, the G-invariant L^2-holomorphic sections of L on $\mathbb{C}^n$. Let $\{\varphi_j e^{-\pi|z|^2}\}_{j=1}^{\infty}$ be an orthonormal basis of $H^0(\mathbb{C}^n, L)^G$ with G-invariant homogeneous polynomials φ_j on z and $\varphi_1 = 1$. Thus from (5.4.27) and (5.4.28), we get for any $Z \in \mathbb{C}^n$,

$$f_G(Z) = \mathscr{P}_{\mathbb{C}^n/G}(Z, Z) = \left(1 + \sum_{j=2}^{\infty} |\varphi_j(z)|^2\right) e^{-2\pi|z|^2} > 0. \qquad (5.4.29)$$

To get the uniformly positivity of $f_G(Z)$, we will prove it by recurrence. Set

$$C_{G,n} = \inf_{Z \in \mathbb{C}^n} f_G(Z), \qquad (5.4.30)$$

and for $g \in G$, and $H \subset G$ a subgroup of G, set

$$V_g = \{z \in \mathbb{C}^n; gz = z\}, \quad V_H = \{z \in \mathbb{C}^n; hz = z \text{ for any } h \in H\}, \qquad (5.4.31)$$

and denote by $V_g^{\perp}, V_H^{\perp}$ their orthogonal complements in $\mathbb{C}^n$.

If $n = 1$, and if g does not act as identity on $\mathbb{C}$, then $e^{-2\pi \langle (1 - g^{-1})z, \overline{z} \rangle} \to 0$ as $|z| \to \infty$. Thus for any finite group G, we get $C_{G,1} > 0$.

Assume that for $n \leqslant k$, $C_{G,n} > 0$ for any finite group G. For $n = k+1$, let G be a finite group as above. If $V_g = \{0\}$ for any $g \in G$, then for each $g \in G$, there exists an orthonormal basis $\{w_j\}$ of $\mathbb{C}^n$ such that $g w_j = e^{i\theta_j} w_j$ and $0 < \theta_j < 2\pi$. Hence for $z = \sum_j v_j w_j$,

$$2\langle (1-g^{-1})z, \bar{z}\rangle = \sum_j (1 - e^{-i\theta_j}) |v_j|^2. \tag{5.4.32}$$

From (5.4.32), we know that for any $g \in G$, $e^{-2\pi\langle (1-g^{-1})z, \bar{z}\rangle} \to 0$ as $|z| \to \infty$. We get thus $C_{G,k+1} > 0$ from (5.4.26) and (5.4.29).

Assume now there exists $g \in G$ such that $V_g \neq \{0\}$. Let R be large enough such that for any $g \in G$, $Z \in V_g^{\perp}$, $|Z| > R$, we have

$$\left| e^{-2\pi\langle (1-g^{-1})z, \bar{z}\rangle} \right| < \frac{1}{4|G|} \inf_{H \subset G, V_H \neq \{0\}} \{ C_{H,k+1-\dim V_H}, 1\} =: \widetilde{C}_{G,k+1}. \tag{5.4.33}$$

Here by our recurrence hypothesis, $\widetilde{C}_{G,k+1} > 0$. Then from (5.4.26), we have

$$f_G(Z) > \frac{3}{4} \quad \text{for } Z \in \mathbb{C}^n \setminus \cup_g V_g \times B^{V_g^{\perp}}(0, R). \tag{5.4.34}$$

We will prove that $f_G(Z)$ is uniformly positive on $\cup_g V_g \times B^{V_g^{\perp}}(0, R)$ again by recurrence.

Note that $V_G = \{0\}$. If H_0 is a maximal subgroup of G such that $V_{H_0} \neq \{0\}$, then for any $g \notin H_0$, $V_g \cap V_{H_0} = \{0\}$. Thus for any $g \notin H_0$, $V_g \times B^{V_g^{\perp}}(0, R) \cap V_{H_0} \times B^{V_{H_0}^{\perp}}(0, R)$ is bounded. Thus there exists $R_1 > 0$ such that for $g \notin H_0$, $Z \in V_{H_0} \times B^{V_{H_0}^{\perp}}(0, R) \setminus B(0, R_1)$, (5.4.33) holds. But for $Z = (Z_0, Z_0^{\perp}) \in V_{H_0} \times V_{H_0}^{\perp}$, we have

$$f_{H_0}(Z) = 1 + \sum_{1 \neq g \in H_0} e^{-2\pi\langle (1-g^{-1})z, \bar{z}\rangle} = f_{H_0}(Z_0^{\perp}) \geqslant C_{H_0, k+1-\dim V_{H_0}}. \tag{5.4.35}$$

Thus $f_G(Z) > \frac{3}{4} C_{H_0, k+1-\dim V_{H_0}}$ on $V_{H_0} \times B^{V_{H_0}^{\perp}}(0, R) \setminus B(0, R_1)$. From (5.4.29), $f_G(Z) > 0$ on $B(0, R_1)$, thus $f_G(Z) > C > 0$ on $V_{H_0} \times B^{V_{H_0}^{\perp}}(0, R)$.

Now assume that $H_1 \subset G$ is a subgroup such that $V_{H_1} \neq \{0\}$ and for any subgroup $H \subset G$ such that $H_1 \subset H$ and $H_1 \neq H$, then $f_G(Z) > C > 0$ on $V_H \times B^{V_H^{\perp}}(0, R)$. Then we claim that there exists $C > 0$ such that $f_G(Z) > C$ on $V_{H_1} \times B^{V_{H_1}^{\perp}}(0, R)$. In fact, for $g \in G, g \notin H_1$, we denote by gH_1 the group generated by g and H_1. Then

$$V_{H_1} \times B^{V_{H_1}^{\perp}}(0, R) \cap V_g \times B^{V_g^{\perp}}(0, R) = V_{gH_1} \times B^{V_{gH_1}^{\perp}}(0, R). \tag{5.4.36}$$

For any $x \in X$, the Kodaira map Φ_p induces a G_x-equivariant map Ψ_p from $\widetilde{U}_x$ to $\mathbb{P}(H^0(X, L^p)^*)$. Thus, if $|G_x| > 1$, the differential of Ψ_p, $(d\Psi_p)_{\widetilde{x}}$ cannot be injective (otherwise, Ψ_p is an embedding near $\widetilde{x}$, which contradicts $|G_x| > 1$). Hence the induced Fubini–Study form $\Phi_p^* \omega_{FS}$ must be degenerate on the singular set X_{sing}. The following result tells us that we can approximate the original metric on L by using the metric induced by Φ_p, and we can still approximate the original Kähler form by using the induced Fubini–Study form away from the singular set.

Theorem 5.4.19. As $p \to \infty$, $\frac{1}{p} \log P_p(x, x) \to 0$ in $\mathscr{C}^{1,\alpha}$ for any $0 \leqslant \alpha < 1$. Especially, the metric $(h^{\Phi_p^* \mathscr{O}(1)})^{1/p}$ on L converges to h^L in $\mathscr{C}^0$-norm as $p \to \infty$. For any $l \in \mathbb{N}$, there exists $C_l > 0$ such that for $p \geqslant p_0$,

$$\left| \frac{1}{p} (\Phi_p^* \omega_{FS})(x) - \omega(x) \right|_{\mathscr{C}^l} \leqslant C_l \left(\frac{1}{p} + p^{\frac{l}{2}} e^{-c\sqrt{p} d(x, X_{\mathrm{sing}})} \right). \tag{5.4.46}$$

Proof. By (5.4.23), (5.4.41), (5.4.42) and (5.4.44), we get the first part of our theorem.

By (5.4.41), the analogue equation (5.1.21) still holds here. From (5.1.21), (5.4.14) and (5.4.44), we get (5.4.46). $\qquad\square$

The following result is the orbifold version of the Kodaira embedding theorem.

Theorem 5.4.20 (Baily). *There exists $p_0 \in \mathbb{N}$ such that for $p \geqslant p_0$, $\Phi_p(X)$ is an analytic subset of $\mathbb{P}(H^0(X, L^p)^*)$. The Kodaira map Φ_p is injective and for any $x \in X$, Φ_p induces an isomorphism from $\mathscr{O}_{\Phi_p(X), \Phi_p(x)}$ to $\mathscr{O}_{X,x}$. Thus $\Phi_p(X)$ is a normal algebraic variety in $\mathbb{P}(H^0(X, L^p)^*)$ for $p \geqslant p_0$.*

Before starting the proof, let us mention that we can use the asymptotic expansion of the Bergman kernel to prove Φ_p is injective as in Lemma 5.1.8, but it does not seem clear how to get the corresponding version of Lemma 5.1.6 by using the Bergman kernel. Thus we need to go back to the original proof of Baily which extends Kodaira's proof of Theorem 5.1.12 to the orbifold case. Namely, we use the vanishing Theorem 5.4.9 (b) and the blow-up technique to get the precise information about Φ_p near the singular set X_{sing} such that we can apply Lemma 5.4.17.

Proof. In what follows, for $x \in X$, $(\widetilde{U}_x, G_x)$ is a local coordinate of x as in Lemma 5.4.3 such that G_x acts $\mathbb{C}$-linearly on $\widetilde{U}_x \subset \mathbb{C}^n$ and $0 \in \widetilde{U}_x$ represents x.

Let $W_x \to \widetilde{U}_x$ be the blow-up of $\widetilde{U}_x$ with center 0; then G_x acts naturally on W_x, and under the notation in (2.1.6), for $(z, [t]) \in W_x$, $g \cdot (z, [t]) = (g \cdot z, [g \cdot t])$, and the exceptional divisor E defines a G_x-equivariant line bundle $\mathscr{O}_{W_x}(E)$ over W_x as explained after (2.1.8). Thus we define by $\widetilde{X} := \widetilde{X}_x = X \smallsetminus \{x\} \sqcup_\pi W_x/G_x$ the blow-up of x, and still by $\mathscr{O}_{\widetilde{X}}(E)$ the orbifold line bundle on $\widetilde{X}$ induced by $\mathscr{O}_{W_x}(E)$. (Locally, let f_α be the definition function of the divisor E on $\widetilde{U}_\alpha$ and S_E the corresponding canonical section of $\mathscr{O}_{W_x}(E)$. Then the orbifold bundle $\mathscr{O}_{\widetilde{X}}(E)$ is induced by the action $g \cdot (z, \xi) = (g \cdot z, \frac{f_\alpha(g \cdot z)}{f_\alpha(z)} \xi)$ on the trivialization of $\mathscr{O}_{W_x}(E)$ on $\widetilde{U}_\alpha$ defined by f_α, and $g \in G_{x_i}$ verifies $g(\widetilde{U}_\alpha) = \widetilde{U}_\alpha$. Moreover, S_E is g-invariant.) To obtain the result on $\widetilde{X}$, we always work on $\widetilde{U}_x$. Thus from the proof of Proposition 2.1.7, we get an isomorphism of holomorphic orbifold line bundles

$$K_{\widetilde{X}} = \pi^*(K_X) \otimes \mathscr{O}_{\widetilde{X}}((n-1)E). \qquad (5.4.47)$$

From the proof of Proposition 2.1.8, for any $x \in X$, there exists $p_0(x)$ such that $\pi^* L^p \otimes \mathscr{O}_{\widetilde{X}}(-E)$ is positive for $p \geqslant p_0(x)$. Moreover, there exists a neighborhood U_x of x such that $p_0(y) = p_0(x)$, for any $y \in U_x$.

For $(x, y) \in X \times X$, we define as in Section 5.1.3 the blow-up of $\pi : \widetilde{X} = \widetilde{X}_{x,y} \to X$ of X with center $\{x, y\}$. Then by Theorem 5.4.9 (b) and the above arguments, we know that Lemmas 5.1.13, 5.1.14 and (5.1.34) still hold in our case. Problem 5.11 shows the existence of $k_0 \in \mathbb{N}^*$ such that $|G_x| | k_0$ for all $x \in X$. There exists therefore a canonical section $S_{k_0 E_{x,y}}$ of $\mathscr{O}_{\widetilde{X}}(k_0 E_{x,y})$ which is a usual bundle on $\widetilde{X}$ (if S_{E_x} is the canonical section of $\mathscr{O}_{W_x}(E_x)$ on $\widetilde{U}_x$ as in Definition 2.1.3, then $\prod_{g \in G_x} g \cdot S_{E_x}$ is a G_x-invariant section of $\mathscr{O}_{\widetilde{X}}(|G_x| E_x)$). Let $\mathscr{I}_{E_{x,y}}$ be the ideal of holomorphic functions vanishing on $E_{x,y}$. Now, as in Proposition 2.1.4, the map

$$\mathscr{I}_{E_{x,y}}^{mk_0} \longrightarrow \mathscr{O}_{\widetilde{X}}(-mk_0 E_{x,y}), \quad f \longrightarrow f \cdot S_{k_0 E_{x,y}}^{-m},$$

is an isomorphism. Finally, from the argument after (5.1.35), we conclude that for $m \in \mathbb{N}$, there exists $p_0 \in \mathbb{N}$ such that for all $p \geqslant p_0$ and all $(x, y) \in X \times X$,

$$H^0(X, L^p) \xrightarrow{\ \beta\ } H^0(X, L^p \otimes \mathscr{O}_X/\mathscr{I}(x,y)^{mk_0}) \quad \text{is surjective .} \qquad (5.4.48)$$

For $x \neq y$, $m = 1$, as G_x acts trivially on L_x, we know

$$H^0(X, L^p \otimes \mathscr{O}_X/\mathscr{I}(x,y)^{k_0}) \supset L_x^p \oplus L_y^p. \qquad (5.4.49)$$

From (5.4.48) and (5.4.49), there exists $p_0 \in \mathbb{N}$ such that Φ_p is injective for $p \geqslant p_0$.

Now we take $x = y$, then

$$H^0(X, L^p \otimes \mathscr{O}_X/\mathscr{I}(x,y)^{mk_0}) \simeq \{Q \in \mathbb{C}[z_1, \ldots, z_n]^{G_x}; \ \deg Q \leqslant 2mk_0\}. \qquad (5.4.50)$$

Let $\{(\widetilde{U}_{x_i}, G_{x_i})\}_{i=1}^k$ be local coordinates of X such that $\widetilde{U}_{x_i}/G_{x_i}$ is a covering of X. Let $\{G_i^j\}_j$ be the subgroups of G_{x_i} and we denote by $d(G_i^j)$ the maximum degree of Q in Lemma 5.4.17 for G_i^j. Let $N = \sup_{i,j} d(G_i^j)$, and take $m > N/k_0$. By (5.4.48) and (5.4.50), for any $x \in X$, let $Q_{0,x} = 1, Q_{1,x}, \ldots, Q_{q,x}$ be G_x-invariant linear independent homogeneous polynomials which generate S_{G_x}; then there exist $s_x^0, \ldots, s_x^q$ linear independent sections of $H^0(X, L^p)$ such that $\beta(s_x^l) = Q_{l,x}$. Then we complete them as a basis of $H^0(X, L^p)$; from Lemma 5.4.17 and (5.1.17), we know that Φ_p induces an isomorphism from $\mathscr{O}_{\Phi_p(X), \Phi_p(x)}$ to $\mathscr{O}_{X,x}$. $\qquad \square$

5.5   The asymptotic of the analytic torsion

The holomorphic analytic torsion of Ray-Singer is obtained as the regularized determinant of the Kodaira Laplacian of holomorphic Hermitian vector bundles on a compact complex manifold. The Quillen metric is a natural metric on the determinant of the cohomology of a holomorphic Hermitian vector bundle, which one constructs by using the Ray-Singer analytic torsion.

One of its remarkable properties, established by Bismut–Gillet–Soulé, is that the Quillen metric on the determinant of the fiberwise cohomology of a holomorphic fibration is a smooth metric and the curvature of the corresponding holomorphic Hermitian connection is given by an explicit local formula, which is compatible with the Grothendieck–Riemann–Roch theorem at the level of differential forms.

The analytic torsion has also equivariant and family extensions. Especially, the analytic torsion forms of Bismut–Gillet–Soulé and Bismut–Köhler are differential forms on the base manifold for a holomorphic fibration. Its 0-degree component is the analytic torsion of Ray-Singer along the fiber. The analytic torsion and its extensions were studied extensively by Bismut and his coauthors in the last two decades.

The holomorphic analytic torsions have found a lot of applications, especially as the analytic counterpart of the direct image in Arakelov geometry.

In this section, we will study the asymptotic of the analytic torsion when the power of the line bundle tends to ∞. In the whole theory on the analytic torsion, to get explicit local terms as in Atiyah–Singer index theory (here we need to use the secondary classes of Bott-Chern type), in other words, to apply the local index type computations, we need to assume the base metrics are Kähler. However, here we get the local term without this assumption as the power tends to ∞.

This section is organized as follows. After briefly recalling the Mellin transformation in Section 5.5.1, we define the holomorphic analytic torsion in Section 5.5.2 by using the heat kernel. In Section 5.5.3, we study the dependence of the analytic torsion on a change of the metrics. In Section 5.5.4, we study the asymptotic of the analytic torsion when the power of the line bundle tends to ∞. In Section 5.5.5, we establish the corresponding version for L^2 metrics by combining Sections 5.5.3 and 5.5.4. In Section 5.5.6, we study certain technical results on heat kernels.

We use the notation in Sections 1.4.1, 1.6.1.

5.5.1   Mellin transformation

Let $\Gamma(z)$ be the Gamma function on $\mathbb{C}$. Then for $\mathrm{Re}(z) > 0$, we have

$$\Gamma(z) = \int_0^\infty e^{-u} u^{z-1} du. \tag{5.5.1}$$

We suppose that $f \in \mathscr{C}^\infty(\mathbb{R}_+^*)$ verifies the following two conditions:

(i) There exist $m \in \mathbb{N}$, $f_j \in \mathbb{C}$, $(j \geqslant -m)$ such that for any $k \in \mathbb{N}$, there exists $C_k > 0$ such that as $u \to 0$,

$$\left| f(u) - \sum_{j=-m}^{k} f_j u^j \right| \leqslant C_k u^{k+1}. \tag{5.5.2}$$

(ii) There exist $c, C > 0$ such that as $u \to +\infty$,

$$|f(u)| \leqslant Ce^{-cu}. \tag{5.5.3}$$

Definition 5.5.1. The *Mellin transformation* of f is the function defined for $\mathrm{Re}(z) > m$,

$$M[f](z) = \frac{1}{\Gamma(z)} \int_0^\infty f(u) u^{z-1} du. \tag{5.5.4}$$

Lemma 5.5.2. $M[f]$ *extends to a meromorphic function on* $\mathbb{C}$ *with poles contained in the set* $m - \mathbb{N}$, *and its possible poles are simple.* $M[f]$ *is holomorphic at 0 and*

$$M[f](0) = f_0. \tag{5.5.5}$$

Proof. By (5.5.3), the function $\int_1^\infty f(u) u^{z-1} du$ is an entire function on $z \in \mathbb{C}$. For any $k \in \mathbb{N}$, we have

$$\int_0^1 f(u) u^{z-1} du = \sum_{j=-m}^{k} f_j \int_0^1 u^{j+z-1} du + \int_0^1 \mathscr{O}(u^{k+z}) du$$

$$= \sum_{j=-m}^{k} \frac{f_j}{z+j} + R(z), \tag{5.5.6}$$

where $R(z)$ is a holomorphic function for $\mathrm{Re}(z) > -k - 1$.

Observe that the inverse of the Gamma function $\Gamma(z)^{-1}$ is entire on $\mathbb{C}$ and $\Gamma(z)^{-1} = z + \mathscr{O}(z^2)$ near $z = 0$. Hence we get Lemma 5.5.2. $\square$

5.5.2 Definition of the analytic torsion

For any $\mathbb{Z}_2$-graded finite-dimensional vector space $V = V^+ \oplus V^-$, let $\tau \in \mathrm{End}(V)$ be defined by $\tau = \mathrm{Id}$ on V^+, and $\tau = -\mathrm{Id}$ on V^-. For any $B \in \mathrm{End}(V)$, we define the supertrace Tr_s on $\mathrm{End}(V)$ by

$$\mathrm{Tr}_s[B] = \mathrm{Tr}[\tau B]. \tag{5.5.7}$$

Remember that $[B, C]$ denotes the supercommutator of B and C (cf. (1.3.30)). Then by a direct check, we get for any $B, C \in \mathrm{End}(V)$,

$$\mathrm{Tr}_s[[B, C]] = 0. \tag{5.5.8}$$

Let (X, J) be a compact complex manifold with complex structure J and $\dim X = n$. Let g^{TX} be any Riemannian metric on TX compatible with J. Let (E, h^E) be a holomorphic Hermitian vector bundle on X. Let ∇^E be the holomorphic Hermitian connection on (E, h^E) with curvature R^E.

Recall that the operators D, $\square^E$ were defined in (1.4.3). The operator P is the orthogonal projection from $(L^2(X, \Lambda(T^{*(0,1)}X) \otimes E), \langle \cdot, \cdot \rangle)$ onto $\text{Ker}(D)$. Set $P^{\perp} = \text{Id} - P$. Let N be the number operator on $\Lambda(T^{*(0,1)}X)$, i.e., N acts on $\Lambda^q(T^{*(0,1)}X)$ by multiplication by q.

The $\mathbb{Z}_2$-grading on $\Lambda(T^{*(0,1)}X) \otimes E$ and on $\Omega^{0,\bullet}(X, E)$ is defined by $\tau = (-1)^N$ (induced by the $\mathbb{Z}$-grading).

The following lemma is useful in computation of the supertrace.

Lemma 5.5.3. *For any operators K_1, K_2 on $\Omega^{0,\bullet}(X, E)$ with smooth kernels $K_1(x, y)$, $K_2(x, y)$ associated to $dv_X(y)$, we have*

$$\text{Tr}_s[K_1, K_2] = 0. \tag{5.5.9}$$

Proof. In fact, if K_1 or K_2 preserves the $\mathbb{Z}_2$-grading of $\Omega^{0,\bullet}(X, E)$, then by (5.5.8),

$$\text{Tr}_s[K_1 K_2] = \int_X \text{Tr}_s \left[\int_X K_1(x, y) K_2(y, x) dv_X(y) \right] dv_X(x)$$
$$= \int_X \text{Tr}_s \left[\int_X K_2(y, x) K_1(x, y) dv_X(x) \right] dv_X(y) = \text{Tr}_s[K_2 K_1].$$

In the same way, we can verify (5.5.9) if K_1 and K_2 exchange the $\mathbb{Z}_2$-grading of $\Omega^{0,\bullet}(X, E)$. $\qquad\square$

By (D.1.17) and (D.1.24), we have the following expansion in the sense of (5.5.2), for any $k \in \mathbb{N}$,

$$\text{Tr}_s \left[N e^{-\frac{u}{2} D^2} \right] = \sum_{j=-n}^{k} c_j u^j + \mathscr{O}(u^{k+1}), \tag{5.5.10}$$

when $u \to 0$. Now let $0 = \lambda_1 = \cdots = \lambda_m < \lambda_{m+1} \leqslant \cdots$ be the eigenvalues of D^2 with eigenfunctions φ_{λ_j}. Then by (D.1.17), for $u > 2$

$$\left| (e^{-\frac{u}{2} D^2} P^{\perp})(x, x) \right| \leqslant \sum_{k=m+1}^{\infty} e^{-\lambda_k u/2} |\varphi_{\lambda_k}(x) \otimes \varphi_{\lambda_k}(x)^*|$$

$$\leqslant C \sum_{k=m+1}^{\infty} e^{-\lambda_k u/2} |\varphi_{\lambda_k}(x)|^2 \leqslant C e^{-\lambda_{m+1}(u-2)/2} \sum_{k=m+1}^{\infty} e^{-\lambda_k} |\varphi_{\lambda_k}(x)|^2$$

$$= C e^{-\lambda_{m+1}(u-2)/2} \text{Tr}[(e^{-D^2} P^{\perp})(x, x)]. \tag{5.5.11}$$

By (D.1.17) and (5.5.11), the function $\text{Tr}_s \left[N e^{-\frac{u}{2} D^2} P^{\perp} \right]$ verifies also (5.5.3).

In view of the above discussion and Lemma 5.5.2, set

$$\theta(z) = -M\left[\text{Tr}_s[Ne^{-\frac{u}{2}D^2}P^{\perp}]\right](z) = -\text{Tr}_s[N(\square^E)^{-z}P^{\perp}]. \tag{5.5.12}$$

Then $\theta(z)$ is a meromorphic function on $\mathbb{C}$ with poles contained in the set $n - \mathbb{N}$, its possible poles are simple, and $\theta(z)$ is holomorphic at 0. Moreover,

$$\theta'(0) := \frac{\partial\theta}{\partial z}(0) = -\int_0^1 \left\{\text{Tr}_s\left[N\exp(-\frac{u}{2}D^2)\right] - \sum_{j=-n}^0 c_j u^j\right\}\frac{du}{u}$$

$$-\int_1^{\infty}\text{Tr}_s\left[N\exp(-\frac{u}{2}D^2)P^{\perp}\right]\frac{du}{u} - \sum_{j=-n}^{-1}\frac{c_j}{j} + \Gamma'(1)(c_0 - \text{Tr}_s[NP]). \tag{5.5.13}$$

Definition 5.5.4. The *analytic torsion of Ray–Singer* for the vector bundle E is $\exp(-\frac{1}{2}\theta'(0))$.

The determinant line of the cohomology $H^{\bullet}(X, E)$ is the complex line given by $\det H^{\bullet}(X, E) = \bigotimes_{q=0}^n (\det H^q(X, E))^{(-1)^q}$. We define

$$\lambda(E) = (\det H^{\bullet}(X, E))^{-1}. \tag{5.5.14}$$

Let $h^{H(X,E)}$ be the L^2-metric on $H^{\bullet}(X, E)$ induced by the canonical isomorphism (1.4.6) and the L^2-scalar product on $\Omega^{0,\bullet}(X, E)$. Let $|\cdot|_{\lambda(E)}$ be the L^2-metric on $\lambda(E)$ induced by $h^{H(X,E)}$.

Definition 5.5.5. The *Quillen metric* $\|\cdot\|_{\lambda(E)}$ on the complex line $\lambda(E)$ is defined by

$$\|\cdot\|_{\lambda(E)} = |\cdot|_{\lambda(E)}\exp(-\frac{1}{2}\theta'(0)). \tag{5.5.15}$$

The Quillen metric $\|\cdot\|_{\lambda(E)}$ depends on the choice of the metrics g^{TX} and h^E. We will study their dependence now.

5.5.3 Anomaly formula

We denote by $*^E : \Lambda(T^{*(0,1)}X) \otimes E \to \Lambda(T^*X \otimes_{\mathbb{R}} \mathbb{C}) \otimes E^*$ the Hodge $*$-operators defined by for $\sigma, \sigma' \in \Lambda(T^{*(0,1)}X) \otimes E$ (cf. (1.3.14)),

$$\langle\sigma, \sigma'\rangle_{\Lambda^{0,\bullet}\otimes E}dv_X = (\sigma \wedge *^E\sigma')_E. \tag{5.5.16}$$

We denote $*$ the corresponding Hodge operator for $E = \mathbb{C}$.

Let $\mathbb{R} \ni v \mapsto (g_v^{TX}, h_v^E)$ be a smooth family of metrics on TX and E and g_v^{TX} compatible with J. Let $*_v, *_v^E$ be the Hodge $*$-operators associated to the metrics g_v^{TX}, (g_v^{TX}, h_v^E). Let D_v be the operator D defined in (1.4.3) attached to the metrics (g_v^{TX}, h_v^E). Let $\|\cdot\|_{\lambda(E),v}$ be the corresponding Quillen metric on $\lambda(E)$. Set

$$Q_v = -(*_v^E)^{-1}\frac{\partial *_v^E}{\partial v} = -\left(*_v^{-1}\frac{\partial *_v}{\partial v} + (h_v^E)^{-1}\frac{\partial h_v^E}{\partial v}\right). \tag{5.5.17}$$

Theorem 5.5.6. *As $u \to 0$, for any $k \in \mathbb{N}$, there is an asymptotic expansion*

$$\mathrm{Tr}_s\left[Q_v \exp(-\frac{u}{2}D_v^2)\right] = \sum_{j=-n}^{k} M_{j,v}u^j + \mathcal{O}(u^{k+1}), \tag{5.5.18}$$

with

$$M_{0,v} = \frac{\partial}{\partial v}\log\left(\|\cdot\|_{\lambda(E),v}^2\right). \tag{5.5.19}$$

Proof. From (D.1.24), we get (5.5.18).

Moreover, we have

$$D_v = \sqrt{2}(\overline{\partial}^E + \overline{\partial}_v^{E,*}), \quad [D_v, N] = \sqrt{2}(-\overline{\partial}^E + \overline{\partial}_v^{E,*}). \tag{5.5.20}$$

For $s \in \Omega^{0,q}(X, E)$, we have

$$\overline{\partial}_v^{E,*} s = (-1)^q (*_v^E)^{-1}\overline{\partial}^{E^*} *_v^E s. \tag{5.5.21}$$

In fact, $\left\langle s_1, \overline{\partial}_v^{E,*} s\right\rangle_v = \left\langle \overline{\partial}^E s_1, s\right\rangle_v = \int_X (\overline{\partial}^E s_1 \wedge *_v^E s)_E = (-1)^q \int_X (s_1 \wedge \overline{\partial}^{E^*} *_v^E s)_E.$
Then by (5.5.17) and (5.5.21),

$$\frac{1}{\sqrt{2}}\frac{\partial}{\partial v}D_v = \frac{\partial}{\partial v}\overline{\partial}_v^{E,*} = -\left[\overline{\partial}_v^{E,*}, Q_v\right], \quad \frac{\partial}{\partial v}D_v^2 = \left[D_v, \frac{\partial}{\partial v}D_v\right]. \tag{5.5.22}$$

In what follows, we omit often the subscript v. By (D.1.19), we have

$$\frac{\partial}{\partial v}e^{-uD^2} = \int_0^u e^{-(u-u_1)D^2}\left(-\frac{\partial}{\partial v}D^2\right)e^{-u_1 D^2}\,du_1. \tag{5.5.23}$$

Observe that D^2 preserves the $\mathbb{Z}$-grading of $\Omega^{0,\bullet}(X, E)$, and thus commutes with N, and $\left(-\frac{\partial}{\partial v}D^2\right)e^{-uD^2}$ is an operator with smooth kernel. By (5.5.22) and (5.5.23), we have

$$\frac{\partial}{\partial v}\mathrm{Tr}_s[Ne^{-uD^2}] = \int_0^u \mathrm{Tr}_s\left[\left(-\frac{\partial}{\partial v}D^2\right)e^{-u_1 D^2}Ne^{-(u-u_1)D^2}\right]du_1$$

$$= -u\,\mathrm{Tr}_s\left[\left[D, \frac{\partial D}{\partial v}\right]Ne^{-uD^2}\right]. \tag{5.5.24}$$

By using (5.5.9), that D is an odd operator, (i.e., it changes the $\mathbb{Z}_2$-grading of $\Omega^{0,\bullet}(X, E)$), and that D commutes with e^{-uD^2}, we get for $0 < u_1 < u$,

$$\mathrm{Tr}_s\left[D\frac{\partial D}{\partial v}Ne^{-uD^2}\right] = \mathrm{Tr}_s\left[e^{-(u-u_1)D^2}D\frac{\partial D}{\partial v}Ne^{-u_1 D^2}\right]$$

$$= \mathrm{Tr}_s\left[De^{-(u-u_1)D^2}\frac{\partial D}{\partial v}Ne^{-u_1 D^2}\right]$$

$$= -\mathrm{Tr}_s\left[\frac{\partial D}{\partial v}Ne^{-u_1 D^2}De^{-(u-u_1)D^2}\right] = -\mathrm{Tr}_s\left[\frac{\partial D}{\partial v}NDe^{-uD^2}\right]. \tag{5.5.25}$$

(5.5.25) entails

$$\text{Tr}_s\left[\left[D, \frac{\partial D}{\partial v}\right]Ne^{-uD^2}\right] = \text{Tr}_s\left[\frac{\partial D}{\partial v}[D, N]e^{-uD^2}\right]. \tag{5.5.26}$$

Using that $\overline{\partial}^{E,*}$ commutes with $e^{-\frac{u}{2}D^2}$, and relations (5.5.20), (5.5.22), (5.5.24) and (5.5.26), we get by using the same trick as in (5.5.25):

$$\begin{aligned}
\frac{\partial}{\partial v}\text{Tr}_s\left[Ne^{-\frac{u}{2}D^2}\right] &= -u\,\text{Tr}_s\left[Q_v[\overline{\partial}^{E,*}, -\overline{\partial}^E + \overline{\partial}^{E,*}]e^{-\frac{u}{2}D^2}\right] \\
&= \frac{u}{2}\text{Tr}_s\left[Q_vD^2e^{-\frac{u}{2}D^2}\right] = -u\frac{\partial}{\partial u}\text{Tr}_s\left[Q_v\,e^{-\frac{u}{2}D^2}\right].
\end{aligned} \tag{5.5.27}$$

Let P_v be the orthogonal projection operator from $\Omega^{0,\bullet}(X, E)$ on $\text{Ker}(D_v)$ for the Hermitian product $\langle\ \rangle_v$ on $\Omega^{0,\bullet}(X, E)$ associated with g_v^{TX} and h_v^E defined in (1.3.14). From (5.5.11), $\text{Tr}_s[Q_ve^{-\frac{u}{2}D_v^2}P^\perp]$ decays exponentially when $u \to +\infty$. From (5.5.27), we obtain for $\text{Re}(z)$ large enough,

$$\begin{aligned}
\frac{\partial}{\partial v}\theta_v(z) &= \frac{1}{\Gamma(z)}\int_0^\infty u^z\frac{\partial}{\partial u}\text{Tr}_s\left[Q_ve^{-\frac{u}{2}D_v^2}\right]du \\
&= \frac{-z}{\Gamma(z)}\int_0^\infty u^{z-1}\text{Tr}_s\left[Q_ve^{-\frac{u}{2}D_v^2}P_v^\perp\right]du.
\end{aligned} \tag{5.5.28}$$

Using (5.5.5) and (5.5.28), we find

$$\frac{\partial}{\partial v}(\frac{\partial}{\partial z}\theta_v)(0) = -M_{0,v} + \text{Tr}_s[Q_vP_v]. \tag{5.5.29}$$

The line bundle $\lambda(E)$ is canonically identified to

$$\lambda_v = \otimes_{q=0}^n(\det\text{Ker}(D_v^2|_{\Omega^{0,q}}))^{(-1)^{q+1}}.$$

Under this identification, the canonical isomorphism of the line bundle $\phi_v : \lambda_0 \to \lambda_v$ is defined by $\phi_v(\sigma) = P_v(\sigma)$ for $\sigma \in \lambda_0$.

If σ and σ' are forms in the kernel of D_v^2, we have by definition

$$\langle P_v\sigma, P_v\sigma'\rangle_v = \int_X (P_v\sigma \wedge *_v^E P_v\sigma')_E. \tag{5.5.30}$$

From $P_v^2 = P_v$, we get $(\frac{\partial}{\partial v}P_v)P_v + P_v(\frac{\partial}{\partial v}P_v) = \frac{\partial}{\partial v}P_v$, thus the operator $\frac{\partial}{\partial v}P_v$ sends $\text{Ker}(D_v)$ to its orthogonal complement for $\langle\cdot,\cdot\rangle_v$. Therefore, from (5.5.30), we get

$$\frac{\partial}{\partial v}\langle P_v\sigma, P_v\sigma'\rangle_v = \int_X (P_v\sigma \wedge (\frac{\partial}{\partial v}*_v^E)P_v\sigma')_E = -\langle P_v\sigma, Q_vP_v\sigma'\rangle_v. \tag{5.5.31}$$

Thus from (5.5.14) and (5.5.31), we get

$$\frac{\partial}{\partial v}\log(|\phi_v(\sigma)|^2_{\lambda(E),v}) = \text{Tr}_s[Q_vP_v]. \tag{5.5.32}$$

From (5.5.15), (5.5.29) and (5.5.32), we get (5.5.19). $\qquad\square$

Remark 5.5.7. As the asymptotic of the heat kernel is local (as explained in (D.1.24)), $M_{0,v}$ is an integral of terms with local character. On the other hand, the Quillen metric $\|\cdot\|_{\lambda(E)}$ is defined by using the spectrum of the Kodaira Laplacian, so it is a global invariant of the manifold. Even if we do not compute $M_{0,v}$ here, it is remarkable that (5.5.7) shows that the variation of the Quillen metrics has local character. Actually, when g_v^{TX} are Kähler, Bismut–Gillet–Soulé gave a precise formula for $M_{0,v}$, known as the anomaly formula for Quillen metrics.

5.5.4   The asymptotics of the analytic torsion

We resume now the discussion from Section 5.5.2. Let L be a positive holomorphic line bundle on X. Let h^L be a Hermitian metric on L such that the curvature R^L associated to h^L verifies (1.5.21), i.e., $\sqrt{-1}R^L$ is a positive $(1,1)$-form.

We use the notation in Section 1.5.1 now, especially $D_p, \Box_p$ were defined in (1.5.20), and $D_p^2 = 2\Box_p$ preserves the $\mathbb{Z}$-grading of $\Omega^{0,\bullet}(X, L^p \otimes E)$, and $\omega = \frac{\sqrt{-1}}{2\pi}R^L$. We denote by $\mathrm{Tr}_q\left[\exp(-\frac{u}{p}D_p^2)\right]$ the trace of $\exp(-\frac{u}{p}D_p^2)$ acting on $\Omega^{0,q}(X, L^p \otimes E)$.

We denote by $\theta_p(z)$ the function associated to $L^p \otimes E$ as in (5.5.12).

The following result is the main result of this section.

Theorem 5.5.8. *As $p \to \infty$, we have*

$$\theta_p'(0) = \frac{1}{2}\,\mathrm{rk}(E)\int_X \log\left(\det\left(\frac{p\dot{R}^L}{2\pi}\right)\right)\,e^{p\,\omega} + o(p^n). \tag{5.5.33}$$

Now we state two intermediate results which will be used in the proof of Theorem 5.5.8. Note that Theorem 5.5.9 holds without the assumption that $\sqrt{-1}R^L$ is positive, but in Theorem 5.5.11 we have to assume the positivity of $\sqrt{-1}R^L$.

Let $A_j \in \mathscr{C}^\infty(X, \mathrm{End}(\Lambda(T^{*(0,1)}X)))$ such that as $u \to 0$,

$$(2\pi)^{-n}\frac{\det(\dot{R}^L)\exp(2u\omega_d)}{\det(1 - \exp(-2u\dot{R}^L))} = \sum_{j=-n}^{k} A_j(x)u^j + \mathscr{O}(u^{k+1}). \tag{5.5.34}$$

Theorem 5.5.9. *There exist $A_{p,j} \in \mathscr{C}^\infty(X, \mathrm{End}(\Lambda(T^{*(0,1)}X) \otimes E))$ such that for any $k, l \in \mathbb{N}$, there exists $C > 0$ such that for any $u \in]0, 1]$, $p \in \mathbb{N}^*$, we have the asymptotic expansion*

$$\left|p^{-n}\exp\left(-\frac{u}{p}D_p^2\right)(x, x) - \sum_{j=-n}^{k} A_{p,j}(x)u^j\right|_{\mathscr{C}^l(X)} \leq Cu^k. \tag{5.5.35}$$

For any $j \geq -n$, as $p \to \infty$, we have

$$A_{p,j}(x) = A_j(x) \otimes \mathrm{Id}_E + \mathscr{O}(p^{-1/2}). \tag{5.5.36}$$

We will prove Theorem 5.5.9 in Section 5.5.6, and we only use (5.5.35) with $\mathscr{C}^0$-norm (i.e., $l = 0$) in the proof of Theorem 5.5.8. Now we establish Theorem 5.5.10 as a consequence of Theorem 5.5.9.

For $x \in X$, $u > 0$, set

$$R_u(x) = \det\left(\frac{\dot{R}^L}{2\pi}\right) \operatorname{Tr}\left[(\operatorname{Id} - \exp(u\dot{R}^L))^{-1}\right]. \tag{5.5.37}$$

Observe that from (1.5.14),

$$dv_X = \Theta^n/n!, \quad \det\left(\frac{\dot{R}^L}{2\pi}\right) dv_X(x) = \omega^n/n!. \tag{5.5.38}$$

From (5.5.37) and (5.5.38), for any $k \in \mathbb{N}$, we have the following asymptotic expansion for $u \to 0$,

$$R_u(x) = \sum_{j=-1}^{k} \tilde{A}_j(x)u^j + \mathscr{O}(u^{k+1}), \tag{5.5.39}$$

with

$$\tilde{A}_j = 0 \quad \text{for } j \leqslant -2, \quad \tilde{A}_{-1} dv_X = -\frac{\Theta}{2\pi}\frac{\omega^{n-1}}{(n-1)!}, \quad \tilde{A}_0 dv_X = \frac{n}{2}\frac{\omega^n}{n!}. \tag{5.5.40}$$

For $j \geqslant -n$, set

$$B_{p,j} = 2^{-j}\int_X \operatorname{Tr}_s\left[NA_{p,j}(x)\right] dv_X(x),$$
$$B_j = \int_X \tilde{A}_j(x)dv_X(x). \tag{5.5.41}$$

Theorem 5.5.10. *For any $k, l \in \mathbb{N}$, there exists $C > 0$ such that for any $u \in\,]0, 1]$, $p \in \mathbb{N}^*$,*

$$\left|p^{-n}\operatorname{Tr}_s\left[N\exp(-\frac{u}{2p}D_p^2)\right] - \sum_{j=-n}^{k} B_{p,j}u^j\right| \leqslant Cu^k. \tag{5.5.42}$$

For any $j \geqslant -n$, as $p \to \infty$, we have

$$B_{p,j} = \begin{cases} \mathscr{O}(p^{-1/2}) & \text{for } j \leqslant -2, \\ \operatorname{rk}(E)B_j + \mathscr{O}(p^{-1/2}) & \text{for } j \geqslant -1. \end{cases} \tag{5.5.43}$$

Proof. By (1.5.19), we verify directly

$$\operatorname{Tr}_s\left[Ne^{u\omega_d}\right] = \frac{\partial}{\partial c}\left[\det(\operatorname{Id} - e^c\exp(-u\dot{R}^L))\right]\Big|_{c=0}. \tag{5.5.44}$$

From (5.5.37) and (5.5.44), we deduce that

$$(2\pi)^{-n} \frac{\det(\dot{R}^L) \operatorname{Tr}_s [Ne^{u\omega_d}]}{\det(1 - \exp(-u\dot{R}^L))} = R_u(x). \tag{5.5.45}$$

From (5.5.34), (5.5.39) and (5.5.45), we get

$$2^{-j} \operatorname{Tr}_s [NA_j(x)] = \tilde{A}_j(x). \tag{5.5.46}$$

By Theorem 5.5.9, for any $k, l \in \mathbb{N}$, there exists $C > 0$ such that for any $u \in]0, 1]$, $p \in \mathbb{N}^*$, we have

$$\left| p^{-n} \operatorname{Tr}_s \left[N \exp(-\frac{u}{2p} D_p^2)(x, x) \right] - \sum_{j=-n}^{k} \operatorname{Tr}_s [NA_{p,j}(x)] \frac{u^j}{2^j} \right|_{\mathscr{C}^l(X)} \leqslant Cu^k. \tag{5.5.47}$$

From (5.5.40), (5.5.41), (5.5.46) and (5.5.47), we get Theorem 5.5.10. $\square$

Theorem 5.5.11. *There exist* $C, c, c' > 0$ *such that for any* $q \geqslant 1$, $u \geqslant 1$, $p \in \mathbb{N}$, *we have*

$$p^{-n} \operatorname{Tr}_q \left[\exp(-\frac{u}{2p} D_p^2) \right] \leqslant C \exp \left(-(c - c'/p)u \right). \tag{5.5.48}$$

Proof. By Theorem 1.5.5, for $u \geqslant 1$, $q \geqslant 1$, we have

$$\operatorname{Tr}_q \left[\exp(-\frac{u}{2p} D_p^2) \right] \leqslant \exp \left(-\frac{(u-1)}{2p}(2C_0 p - C_L) \right) \operatorname{Tr}_q \left[\exp(-\frac{1}{p} D_p^2) \right]. \tag{5.5.49}$$

By (1.6.4), we know that $p^{-n} \operatorname{Tr}_q \left[\exp(-\frac{1}{p} D_p^2) \right]$ has a finite limit as $p \to \infty$.
From (5.5.49), we get (5.5.48). $\square$

Proof of Theorem 5.5.8. For $p \geqslant C_L/C_0$ in Theorem 1.5.5, set

$$\tilde{\theta}_p(z) = -\boldsymbol{M} \left[p^{-n} \operatorname{Tr}_s \left[N \exp(-\frac{u}{2p} D_p^2)(1 - P_p) \right] \right](z). \tag{5.5.50}$$

Clearly

$$p^{-n} \theta_p(z) = p^{-z} \tilde{\theta}_p(z). \tag{5.5.51}$$

Thus from (1.4.23), (5.5.5), (5.5.42) and (5.5.51), we get for $p \geq C_L/C_0$

$$\begin{aligned} p^{-n} \theta_p'(0) &= -\log(p) \tilde{\theta}_p(0) + \tilde{\theta}_p'(0), \\ \tilde{\theta}_p(0) &= -B_{p,0}, \quad \operatorname{Tr}_s[NP_p] = 0. \end{aligned} \tag{5.5.52}$$

By Lemma 5.5.2, (5.5.13) and (5.5.52), we get for $p \geq C_L/C_0$

$$\tilde{\theta}'_p(0) = -\int_0^1 \left\{ p^{-n} \operatorname{Tr}_s \left[N \exp(-\frac{u}{2p} D_p^2) \right] - \sum_{j=-n}^0 B_{p,j} u^j \right\} \frac{du}{u}$$

$$- \int_1^\infty p^{-n} \operatorname{Tr}_s \left[N \exp(-\frac{u}{2p} D_p^2) \right] \frac{du}{u} - \sum_{j=-n}^{-1} \frac{B_{p,j}}{j} + \Gamma'(1) B_{p,0}. \quad (5.5.53)$$

From Theorem 1.6.1, and (5.5.45), for any $u > 0$,

$$\lim_{p \to \infty} p^{-n} \operatorname{Tr}_s [N \exp(-\frac{u}{2p} D_p^2)(x,x)] = \operatorname{rk}(E) R_u(x), \quad (5.5.54)$$

and the convergence is uniform for $x \in X$ and for u varying in compact subsets of $\mathbb{R}_+^*$.

For $z \in \mathbb{C}$, $\operatorname{Re}(z) > 1$, set

$$\tilde{\zeta}(z) = -\boldsymbol{M}\left[\int_X R_u(x) dv_X(x) \right](z). \quad (5.5.55)$$

By Theorems 5.5.10, 5.5.11, (5.5.53), (5.5.54) and (5.5.55), as $p \to \infty$,

$$\tilde{\theta}'_p(0) \longrightarrow \eta = -\operatorname{rk}(E) \int_0^1 \left\{ \int_X R_u(x) dv_X(x) - \frac{B_{-1}}{u} - B_0 \right\} \frac{du}{u}$$

$$- \operatorname{rk}(E) \int_1^\infty \int_X R_u(x) dv_X(x) \frac{du}{u} + \operatorname{rk}(E)(B_{-1} + \Gamma'(1) B_0)$$

$$= \operatorname{rk}(E) \tilde{\zeta}'(0). \quad (5.5.56)$$

Since $\dot{R}^L \in \operatorname{End}(T^{(1,0)} X)$ has positive eigenvalues, we find that for $\operatorname{Re}(z) > 1$,

$$\tilde{\zeta}(z) = \left(\int_X \det \left(\frac{\dot{R}^L}{2\pi} \right) \operatorname{Tr}[(\dot{R}^L)^{-z}] dv_X(x) \right) \frac{1}{\Gamma(z)} \int_0^\infty u^{z-1} \frac{e^{-u}}{1 - e^{-u}} du. \quad (5.5.57)$$

Let $\zeta(z) = \sum_{k=1}^\infty \frac{1}{n^z}$ be the Riemann zeta function. Classically, for $\operatorname{Re}(z) > 1$,

$$\zeta(z) = \frac{1}{\Gamma(z)} \int_0^\infty u^{z-1} \frac{e^{-u}}{1 - e^{-u}} du. \quad (5.5.58)$$

Moreover,

$$\zeta(0) = -\frac{1}{2}, \quad \zeta'(0) = -\frac{1}{2} \log(2\pi). \quad (5.5.59)$$

From (5.5.57), (5.5.58) and (5.5.58),

$$\widetilde{\zeta}'(0) = -\int_X \det\left(\frac{\dot{R}^L}{2\pi}\right) \mathrm{Tr}\left[\log \dot{R}^L\right] dv_X(x)\,\zeta(0) + n\int_X \det\left(\frac{\dot{R}^L}{2\pi}\right) dv_X(x)\,\zeta'(0)$$

$$= \frac{1}{2}\int_X \det\left(\frac{\dot{R}^L}{2\pi}\right)\log\left(\det\left(\frac{\dot{R}^L}{2\pi}\right)\right) dv_X(x). \quad (5.5.60)$$

By (5.5.40), (5.5.43), (5.5.52), (5.5.56) and (5.5.60), we get (5.5.33). $\qquad\square$

5.5.5 Asymptotic anomaly formula for the L^2-metric

Let now g_0^{TX}, g_1^{TX} be two metrics on TX compatible with J. We keep the metrics on L, E fixed.

Let $|\cdot|_{K_X^*,0}, |\cdot|_{K_X^*,1}$ be the metrics on $K_X^* = \det(T^{(1,0)}X)$, the dual of the canonical line bundle K_X on X, induced by g_0^{TX}, g_1^{TX} respectively.

Let $\|\cdot\|_{p,0}, \|\cdot\|_{p,1}$ (resp. $|\cdot|_{p,0}, |\cdot|_{p,1}$) the Quillen (resp. L^2) metrics on $\lambda(L^p \otimes E)$ induced by g_0^{TX}, g_1^{TX} respectively, and with the given metrics on L, E.

Theorem 5.5.12. *As $p \to \infty$, we have*

$$\log\left(\frac{|\cdot|_{p,1}^2}{|\cdot|_{p,0}^2}\right) = -\mathrm{rk}(E)\int_X \log\left(\frac{|\cdot|_{K_X^*,1}^2}{|\cdot|_{K_X^*,0}^2}\right)\frac{\omega^n}{n!}\,p^n + o(p^n). \quad (5.5.61)$$

Proof. Let $\theta'_{p,0}(0), \theta'_{p,1}(0)$ be the real numbers in Theorem 5.5.8 associated to g_0^{TX}, g_1^{TX} respectively, and with the given metrics on L, E. By Theorem 5.5.8 and (1.5.15), as $|\sigma|_{K^*,1}^2 = |\Theta_1^n(\sigma,\overline{\sigma})|/n!$, we get

$$\theta'_{p,1}(0) - \theta'_{p,0}(0) = -\frac{1}{2}\,\mathrm{rk}(E)\int_X \log\left(\frac{|\cdot|_{K_X^*,1}^2}{|\cdot|_{K_X^*,0}^2}\right)\frac{\omega^n}{n!}\,p^n + o(p^n). \quad (5.5.62)$$

Now, we choose a path of metrics g_v^{TX} for $v \in [0,1]$ connected by g_0^{TX}, g_1^{TX}. We denote the objects associated to g_v^{TX} with a subscript v. Then by Theorems 5.5.6 and 5.5.9, we have

$$\frac{\partial}{\partial v}\log\left(\|\cdot\|_{\lambda(L^p\otimes E),v}^2\right) = p^n\int_X \mathrm{Tr}_s\left[Q_v A_{0,v}(x)\otimes \mathrm{Id}_E\right] dv_{X,v}(x) + \mathscr{O}(p^{n-1/2}).$$

$$\qquad (5.5.63)$$

Now, let $w_1, \ldots, w_n$ be an orthonormal basis of $T^{(1,0)}X$ for the metric g_v^{TX}, $\overline{w}_1, \ldots, \overline{w}_n$ the conjugate basis. Then we get

$$Q_v = -*_v^{-1}\frac{\partial *_v}{\partial v} = -\frac{\partial g_v^{TX}}{\partial v}(w_i, \overline{w}_j)i_{\overline{w}_i}\wedge \overline{w}^j \wedge . \quad (5.5.64)$$

In fact, let $h_v^{\Lambda^{0,\bullet}}$ be the metric on $\Lambda^\bullet(T^{*(0,1)}X)$ induced by g_v^{TX}. Then, (5.5.16) entails

$$*_v^{-1}\frac{\partial *_v}{\partial v} = (h_v^{\Lambda^{0,\bullet}})^{-1}\frac{\partial}{\partial v}h_v^{\Lambda^{0,\bullet}} + \frac{\partial(dv_{X,v})}{\partial v}/(dv_{X,v}). \qquad (5.5.65)$$

But

$$(h_v^{\Lambda^{0,\bullet}})^{-1}\frac{\partial}{\partial v}h_v^{\Lambda^{0,\bullet}} = -\left\langle(g_v^{TX})^{-1}\left(\frac{\partial}{\partial v}g_v^{TX}\right)w_i,\overline{w}_j\right\rangle\overline{w}^j\wedge i_{\overline{w}_i},$$

and

$$\frac{\partial(dv_{X,v})}{\partial v}/(dv_{X,v}) = \left\langle(g_v^{TX})^{-1}\left(\frac{\partial}{\partial v}g_v^{TX}\right)w_i,\overline{w}_i\right\rangle.$$

Let $\widetilde{Q}_v \in \mathrm{End}(T^{(1,0)}X)$ be defined by $\left\langle\widetilde{Q}_v w_i,\overline{w}_j\right\rangle_v = -\frac{\partial g_v^{TX}}{\partial v}(w_i,\overline{w}_j)$. From (5.5.64), we get

$$\mathrm{Tr}_s\left[Q_v e^{2u\omega_{d,v}}\right] = \frac{\partial}{\partial c}\left[\det(\mathrm{Id} - e^{c\widetilde{Q}_v}\exp(-2u\dot{R}_v^L))\right]\Big|_{c=0}. \qquad (5.5.66)$$

As in (5.5.45), we obtain

$$\frac{\mathrm{Tr}_s\left[Q_v e^{2u\omega_{d,v}}\right]}{\det(1-\exp(-2u\dot{R}_v^L))} = \mathrm{Tr}\left[\widetilde{Q}_v(\mathrm{Id}-\exp(2u\dot{R}_v^L))^{-1}\right]. \qquad (5.5.67)$$

Thus from (5.5.34), (5.5.38), we get

$$\int_X \mathrm{Tr}_s[Q_v A_{0,v}]dv_{X,v} = \frac{1}{2}\int_X\mathrm{Tr}[\widetilde{Q}_v]\frac{\omega^n}{n!} = -\frac{1}{2}\int_X\mathrm{Tr}\,|_{T^{(1,0)}X}\left[\frac{\partial g_v^{TX}}{\partial v}\right]\frac{\omega^n}{n!}. \qquad (5.5.68)$$

Now from (5.5.63) and (5.5.68), we get

$$\log\left(\frac{\|\cdot\|_{p,1}^2}{\|\cdot\|_{p,0}^2}\right) = -\frac{1}{2}\,\mathrm{rk}(E)\int_X\log\left(\frac{|\cdot|_{K_X^*,1}^2}{|\cdot|_{K_X^*,0}^2}\right)\frac{\omega^n}{n!}\,p^n + \mathcal{O}(p^{n-1/2}). \qquad (5.5.69)$$

Finally, by the definition of the Quillen metric, we get

$$\log\left(\frac{|\cdot|_{p,1}^2}{|\cdot|_{p,0}^2}\right) = \theta'_{p,1}(0) - \theta'_{p,0}(0) + \log\left(\frac{\|\cdot\|_{p,1}^2}{\|\cdot\|_{p,0}^2}\right). \qquad (5.5.70)$$

From (5.5.62), (5.5.69) and (5.5.70), we get (5.5.61). $\qquad\square$

5.5.6   Uniform asymptotic of the heat kernel

The results in this section work without the positivity assumption on R^L.

Proof of Theorem 5.5.9. By (1.6.13), for any $m \in \mathbb{N}$, there exists $C_m > 0$ (which depends on ε) such that for $0 < u \leqslant 1$,

$$\sup_{a \in \mathbb{R}} |a|^m |\mathbf{G}_u(\sqrt{u}a)| \leqslant C_m \exp(-\frac{\varepsilon^2}{16u}). \qquad (5.5.71)$$

Thus from (5.5.71), as in (1.6.17) and (1.6.18), we get that (1.6.17) still holds for $p \in \mathbb{N}^*, 0 < u \leqslant 1$, especially, for any $m \in \mathbb{N}$, $\varepsilon > 0$, there exists $C > 0$ such that for any $x, x' \in X$, $p \in \mathbb{N}^*$, $0 < u \leqslant 1$,

$$\left| \mathbf{G}_{\frac{u}{p}}(\sqrt{u/p}D_p)(x, x') \right|_{\mathscr{C}^m} \leqslant C \exp(-\frac{\varepsilon^2 p}{32u}). \qquad (5.5.72)$$

But as explained after (1.6.18), $\mathbf{F}_{\frac{u}{p}}(\sqrt{\frac{u}{p}}D_p)(x, x')$ only depends on the restriction of D_p to $B^X(x, \varepsilon)$, and is zero if $d(x, x') \geqslant \varepsilon$.

Now, by (5.5.72), we have the analogue of Lemma 1.6.5: For any $m \in \mathbb{N}$, there exists $C > 0$ such that for any $p \in \mathbb{N}^*$, $0 < u \leqslant 1$,

$$\left| \exp(-\frac{u}{2p}D_p^2)(x_0, x_0) - \exp(-\frac{u}{2p}L_{p,x_0})(0,0) \right|_{\mathscr{C}^m(X)} \leqslant C \exp(-\frac{\varepsilon^2 p}{32u}). \quad (5.5.73)$$

We denote L_2^t defined in (1.6.27) by L_{2,x_0}^t with $t = \frac{1}{\sqrt{p}}$. By (1.6.66), we have

$$\exp(-\frac{u}{p}L_{p,x_0})(0,0) = p^n e^{-uL_{2,x_0}^t}(0,0). \qquad (5.5.74)$$

As explained above in Theorem 1.6.11, L_{2,x_0}^t are families of differential operators with coefficients in $\text{End}(\mathbf{E}_{x_0}) = \text{End}(\Lambda(T^{*(0,1)}X) \otimes E)_{x_0}$, and we view $e^{-uL_{2,x_0}^t}(Z, Z')$ as a smooth section of $\pi^*(\text{End}(\Lambda(T^{*(0,1)}X) \otimes E))$ on $TX \times_X TX$.

Now we need to study the asymptotic of $e^{-uL_{2,x_0}^t}(0,0)$ as $u \to 0$, with parameters $x_0 \in X, t \in [0,1]$. Recall that the holomorphic functions $\widetilde{\mathbf{F}}_u, \widetilde{\mathbf{G}}_u$ were defined by (1.6.26). By (1.6.14), (5.5.71), we get again the analogue of (5.5.72):

$$\left| \widetilde{\mathbf{G}}_u(uL_{2,x_0}^t)(0,0) \right|_{\mathscr{C}^m(X \times [0,1])} \leqslant C_m \exp(-\frac{\varepsilon^2}{32u}). \qquad (5.5.75)$$

Here the $\mathscr{C}^m$-norm is for the parameters $x_0 \in X, t \in [0,1]$. In fact, for the $\mathscr{C}^0$-norm, it is from the argument in (1.6.17). Using (5.5.71) and proceeding as in (4.2.16), we show there exists a unique holomorphic function $\widetilde{H}_{u,1,k}$ defined on a neighborhood of V_c as in (4.2.13) which verifies the same estimates as $\widetilde{\mathbf{G}}_u(u\lambda)$. Moreover, $\widetilde{H}_{u,1,k}(\lambda) \to 0$ as $\lambda \to +\infty$ and

$$\widetilde{H}_{u,1,k}^{(k-1)}(\lambda)/(k-1)! = \widetilde{\mathbf{G}}_u(u\lambda). \qquad (5.5.76)$$

Then as in (1.6.55), we get

$$\nabla_U^{\pi^* \operatorname{End}(\mathbf{E})} \widetilde{\mathbf{G}}_u(uL_{2,x_0}^t) = \frac{1}{2\pi i} \int_\Gamma \widetilde{H}_{u,1,k}(\lambda) \nabla_U^{\pi^* \operatorname{End}(\mathbf{E})} (\lambda - L_{2,x_0}^t)^{-k} d\lambda. \quad (5.5.77)$$

As in Theorem 1.6.11, we obtain the estimate for the $\mathscr{C}^m$-norm in (5.5.75).

Now by the finite propagation speed, Theorem D.2.1 and (1.6.31), for t small, $\widetilde{\mathbf{F}}_u(uL_{2,x_0}^t)(0, \cdot)$ only depends on the restriction of L_{2,x_0}^t on $B^{T_{x_0}X}(0, 2\varepsilon)$ and

$$\operatorname{supp}\left(\widetilde{\mathbf{F}}_u(uL_{2,x_0}^t)(0, \cdot)\right) \subset B^{T_{x_0}X}(0, 2\varepsilon). \quad (5.5.78)$$

Consider a sphere bundle $V = \{(z, c) \in T_{x_0}X \times \mathbb{R}; |z|^2 + c^2 = 1\}$ on X with fiber V_{x_0} for $x_0 \in X$. We embed naturally $B^{T_{x_0}X}(0, 2\varepsilon)$ into V_{x_0} by sending z to (z, c) with $c > 0$, and extend the operator L_{2,x_0}^t to a generalized Laplacian $\widetilde{L}_{3,x_0}^t$ on V_{x_0} with values in $\pi^*(\operatorname{End}(\Lambda(T^{*(0,1)}X) \otimes E))$. By repeating the argument as in (5.5.75), we obtain

$$\left| \exp(-uL_{2,x_0}^t)(0, 0) - \exp(-u\widetilde{L}_{3,x_0}^t)(0, 0) \right|_{\mathscr{C}^m(X \times [0,1])} \leqslant C \exp(-\frac{\varepsilon^2}{32u}). \quad (5.5.79)$$

But now we can apply for $\exp(-u\widetilde{L}_{3,x_0}^t)(0, 0)$ the asymptotic expansion of the heat kernel stated in (D.1.24), as the total space is compact.

By (5.5.73), (5.5.74) and (5.5.79), we get (5.5.35) and $A_{p,j} = A_{\infty,j} + \mathscr{O}(p^{-1/2})$. Moreover, (1.6.68) entails $A_{\infty,j} = A_j \otimes \operatorname{Id}_E$. This completes the proof of Theorem 5.5.9. $\square$

We now explain how to use the argument in Section 4.1.4 to get a proof of (5.5.35) and (D.1.24).

Let $\Delta^{T_{x_0}X}$ be the usual Bochner Laplacian on $T_{x_0}X$. Set (with ρ in (1.6.19))

$$L_{3,x_0}^t = \rho(|Z|/\varepsilon) L_{2,x_0}^t + (1 - \rho(|Z|/\varepsilon)) \Delta^{T_{x_0}X}. \quad (5.5.80)$$

Then by (5.5.75), as in (5.5.79),

$$\left| \exp(-uL_{2,x_0}^t)(0, 0) - \exp(-uL_{3,x_0}^t)(0, 0) \right|_{\mathscr{C}^m(X \times [0,1])} \leqslant C \exp(-\frac{\varepsilon^2}{32u}). \quad (5.5.81)$$

Set (with S_v in (1.6.27))

$$L_{4,x_0}^{t,v} = S_v^{-1} u L_{3,x_0}^t S_v \quad \text{with } v = \sqrt{u}. \quad (5.5.82)$$

Then as in (1.6.66),

$$\exp(-uL_{3,x_0}^t)(0, 0) = u^{-n} \exp(-L_{4,x_0}^{t,v})(0, 0). \quad (5.5.83)$$

Now we use the usual Sobolev norm $\| \ \|_m$ on $\mathscr{C}^\infty(\mathbb{R}^{2n}, \mathbf{E}_{x_0})$ as in (1.6.53). In Section 4.1.4, we replace ∇_{t,e_i}, $\| \ \|_{t,m}$, $\mathscr{L}_2^t$ by ∇_{e_i}, $\| \ \|_m$, $L_{4,x_0}^{t,v}$, and the contour $\Delta \cup \delta$ by ∂V_c in (4.2.13), $\frac{\partial}{\partial t}$, t in (4.1.45) by $\frac{\partial}{\partial v}$, v. Then we get the analogue of Theorem 4.1.16,

$$\left| \exp(-L_{4,x_0}^{t,v})(0,0) \right|_{\mathscr{C}^m(X\times[0,1]\times[0,1])} \leqslant C, \tag{5.5.84}$$

by using the analogue of (4.1.59)

$$\exp(-L_{4,x_0}^{t,v}) = \frac{(-1)^{k-1}(k-1)!}{2\pi i} \int_{\partial V_c} e^{-\lambda}(\lambda - L_{4,x_0}^{t,v})^{-k} d\lambda. \tag{5.5.85}$$

For k large enough, set

$$\mathscr{B}_{r,x_0,t} = \frac{(-1)^{k-1}(k-1)!}{2\pi i\, r!} \int_{\partial V_c} e^{-\lambda} \sum_{(\mathbf{k},\mathbf{r})\in I_{k,r}} a_{\mathbf{r}}^{\mathbf{k}} A_{\mathbf{r}}^{\mathbf{k}}(\lambda,0) d\lambda, \tag{5.5.86}$$

with corresponding $A_{\mathbf{r}}^{\mathbf{k}}(\lambda,0)$ here. Then we get the analogue of (4.1.68): For any $m' \in \mathbb{N}$, there exists $C > 0$ such that if $v \in [0,1]$,

$$\left| \left(\exp(-L_{4,x_0}^{t,v}) - \sum_{r=0}^{k} \mathscr{B}_{r,x_0,t}v^r \right)(0,0) \right|_{\mathscr{C}^{m'}(X\times[0,1])} \leqslant Cv^{k+1}. \tag{5.5.87}$$

The analogue of Theorem 4.1.21 is

$$\mathscr{B}_{r,x_0,t}(Z,Z') = (-1)^r \mathscr{B}_{r,x_0,t}(-Z,-Z'). \tag{5.5.88}$$

In fact, let ψ be the operation defined by $(\psi s)(Z) = s(-Z)$; as $L_{4,x_0}^{t,0} = \Delta^{T_{x_0}X}$, we get

$$\left((\lambda - L_{4,x_0}^{t,0})^{-1}(\psi s) \right)(Z) = \left(\psi\left((\lambda - L_{4,x_0}^{t,0})^{-1}s \right) \right)(Z). \tag{5.5.89}$$

By using the expansion of $L_{4,x_0}^{t,v}$ on v as in Theorem 4.1.7 and (5.5.86), we get

$$\left(\mathscr{B}_{r,x_0,t}(\psi s) \right)(Z) = (-1)^r \left(\psi(\mathscr{B}_{r,x_0,t}s) \right)(Z). \tag{5.5.90}$$

From (5.5.90), we get (5.5.88).

From (5.5.81), (5.5.83), (5.5.87) and (5.5.88), we get

$$\left| u^n \exp(-uL_{2,x_0}^t)(0,0) - \sum_{r=0}^{k} \mathscr{B}_{2r}(0,0)u^r \right|_{\mathscr{C}^{m'}(X\times[0,1])} \leqslant Cu^{k+1}. \tag{5.5.91}$$

From (5.5.73), (5.5.74) and (5.5.91), we get also (5.5.35).

If we take $t = 1$ in (5.5.91), then we get also a proof of (D.1.24).

Problems

Problem 5.1 (Grauert's proof of the Kodaira embedding theorem [116]). Let X be a compact complex manifold and L be a Grauert positive line bundle on X (cf. Definition B.3.12). Let $Z(L^*) \subset L^*$ be the zero-section of L^*. We identify $Z(L^*)$ and X by means of the projection of L^*. Let $x_0 \in Z(L^*)$. Using this identification we denote by $\mathscr{O}_{L^*,x_0}$, $\mathscr{O}_{X,x_0}$ the rings of germs of holomorphic functions on L^* and X, respectively, at x_0. Let $e : U \longrightarrow L$, $e^* : U \longrightarrow L^*$ be dual holomorphic frames near x_0. Every germ $s \in \mathscr{O}_{X,x_0}(L^p)$ has a representative $f e^{\otimes p}$ with f holomorphic near x_0. We make the following identifications:

(i) $L|_U \simeq U \times \mathbb{C}$, $L^*|_U \simeq U \times \mathbb{C}$ via the frames e, e^*,

(ii) s with the polynomial $f(x) z^p$, with $x \in U$, $z \in \mathbb{C}$,

(iii) any germ $g \in \mathscr{O}_{L,x_0}$ with a Taylor series $\sum_{p=0}^{\infty} g_p(x) z^p$ converging in some neighborhood $V \times \{|z| < r\}$, $V \subset U$.

By using the identifications (ii) and (iii), define the maps

$$\Phi_k : \bigoplus_{p=0}^{k} \mathscr{O}_{X,x_0}(L^p) \to \mathscr{O}_{L^*,x_0}, \quad (s_0, \dots, s_k) \mapsto \sum_{p=0}^{k} f_p(x) z^p,$$

$$\Psi_k : \mathscr{O}_{L^*,x_0} \to \bigoplus_{p=0}^{k} \mathscr{O}_{X,x_0}(L^p), \quad \sum_{\mu=0}^{\infty} f_p(x) z^p \mapsto (s_0, \dots, s_k).$$

(a) Show that $\Psi_k \circ \Phi_k = \mathrm{Id}$, so Φ_k is injective.

(b) Let $\mathscr{F}$ be a coherent analytic sheaf on X and $\widetilde{\mathscr{F}}$ the corresponding analytic inverse image sheaf on L^* (cf. Appendix B). Show that Φ_k induces morphisms $\widetilde{\Phi}_k : \mathscr{F} \otimes_{\mathscr{O}_X} \oplus_{p=0}^{k} \mathscr{O}_X(L^p) \to \mathscr{F} \otimes_{\mathscr{O}_X} \mathscr{O}_{L^*} = \widetilde{\mathscr{F}}|_{Z(L^*)}$ and $\widetilde{\Phi}_k^* :$ $H^q(X, \oplus_{p=0}^{k} \mathscr{F} \otimes L^p) \to H^q(Z(L^*), \widetilde{\mathscr{F}})$.

Show further that Ψ_k induces a morphism $\widetilde{\Psi}_k : \widetilde{\mathscr{F}} \to \mathscr{F} \otimes_{\mathscr{O}_X} \oplus_{p=0}^{k} \mathscr{O}_X(L^p)$ with induced morphism $\widetilde{\Psi}_k^* : H^q(Z(L^*), \widetilde{\mathscr{F}}) \to H^q(X, \oplus_{p=0}^{k} \mathscr{F} \otimes L^p)$. Deduce that $\widetilde{\Psi}_k^* \circ \widetilde{\Phi}_k^* = \mathrm{Id}$, hence $\widetilde{\Phi}_k^*$ is injective.

(c) Consider a 1-convex neighborhood M of $Z(L^*)$. Let $j : Z(L^*) \to M$ be the inclusion and $\pi|_M : M \to Z(L^*)$ the restriction of the projection π. Show that these maps induce a commutative diagram

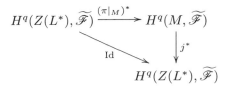

The space of holomorphic sections of $L^p \otimes E$ which are L^2 with respect to the norm given by (1.3.14) is denoted by $H^0_{(2)}(X, L^p \otimes E)$. Let $P_p(x, x')$, $(x, x' \in X)$ be the Schwartz kernel of the orthogonal projection P_p, from the space of L^2 sections of $L^p \otimes E$ onto $H^0_{(2)}(X, L^p \otimes E)$, with respect to the Riemannian volume form $dv_X(x')$ associated to (X, g^{TX}). Then by Remark 1.4.3, we know $P_p(x, x')$ is $\mathscr{C}^\infty$.

We recall at first some notation. $\Omega^{0,\bullet}_0(X, L^p \otimes E)$ is the space of smooth compactly supported $(0, \bullet)$- forms with values in $L^p \otimes E$, and by $L^2_{0,\bullet}(X, L^p \otimes E)$ the corresponding L^2-completion with Hermitian product $\langle \, \rangle$ defined in (1.3.14) induced by h^L, h^E and g^{TX}. As in (3.2.1), we denote the Dolbeault operator $\overline{\partial}^{L^p \otimes E}$ by $\overline{\partial}^E_p$ and by $\overline{\partial}^{E,*}_p$ its adjoint.

Theorem 6.1.1. *Suppose that there exist $\varepsilon > 0$, $C > 0$ such that:*

$$\sqrt{-1}R^L > \varepsilon\Theta, \quad \sqrt{-1}(R^{\det} + R^E) > -C\Theta\operatorname{Id}_E, \quad |\partial\Theta|_{g^{TX}} < C, \qquad (6.1.1)$$

then the kernel $P_p(x, x')$ has a full off-diagonal asymptotic expansion analogous to Proposition 4.1.5 and Theorem 4.2.1 with $\mathscr{F}_r(Z, Z') \in \operatorname{End}(E)_{x_0}$ given by (4.1.92) as $p \to \infty$, uniformly for any $x, x' \in K$, a compact set of X. Especially there exist coefficients $\boldsymbol{b}_r \in \mathscr{C}^\infty(X, \operatorname{End}(E))$, $r \in \mathbb{N}$, such that for any compact set $K \subset X$, any $k, l \in \mathbb{N}$, there exists $C_{k,l,K} > 0$ such that for $p \in \mathbb{N}^*$,

$$\left| \frac{1}{p^n} P_p(x, x) - \sum_{r=0}^k \boldsymbol{b}_r(x)p^{-r} \right|_{\mathscr{C}^l(K)} \leqslant C_{k,l,K}\, p^{-k-1}. \qquad (6.1.2)$$

Moreover, $\boldsymbol{b}_0 = \det\left(\frac{\dot{R}^L}{2\pi}\right)$ and $\boldsymbol{b}_1$ has the same form as in (4.1.9).

Let us remark that if $L = \det(T^{*(1,0)}X)$ is the canonical bundle K_X, the first two conditions in (6.1.1) are to be replaced by

$$h^L \text{ is induced by } \Theta \text{ and } \sqrt{-1}R^{\det} < -\varepsilon\Theta, \quad \sqrt{-1}R^E > -C\Theta\operatorname{Id}_E. \qquad (6.1.3)$$

Moreover, if (X, Θ) is Kähler, the condition on $\partial\Theta$ is trivially satisfied.

Proof. In general, on a non-compact manifold, we define a self-adjoint extension of $\Box_p$ by (3.1.5). By (3.2.4), the quadratic form associated to $\frac{1}{p}\Box_p$ is the form Q_p given by

$$\operatorname{Dom}(Q_p) := \operatorname{Dom}(\overline{\partial}^E_p) \cap \operatorname{Dom}(\overline{\partial}^{E,*}_p),$$
$$p\, Q_p(s_1, s_2) = \langle \overline{\partial}^E_p s_1, \overline{\partial}^E_p s_2 \rangle + \langle \overline{\partial}^{E,*}_p s_1, \overline{\partial}^{E,*}_p s_2 \rangle, \quad s_1, s_2 \in \operatorname{Dom}(Q_p). \qquad (6.1.4)$$

In the previous formulas, $\overline{\partial}^E_p$ is the maximal extension of $\overline{\partial}^E_p$ to L^2 forms and $\overline{\partial}^{E,*}_p$ is its Hilbert space adjoint.

Under hypothesis (6.1.1), there exists $C_1 > 0$ such that for p large enough

$$Q_p(s, s) \geqslant C_1 \|s\|_{L^2}^2, \quad \text{for } s \in \text{Dom}(Q_p) \cap L_{0,q}^2(X, L^p \otimes E), \; q > 0. \quad (6.1.5)$$

We apply the Nakano inequality (1.4.64) to prove (6.1.5). We denote by $i(\Theta) = (\Theta \wedge \cdot)^*$ the interior multiplication with Θ, and $\mathcal{T} = [i(\Theta), \partial\Theta]$ is the Hermitian torsion of Θ. By (1.4.64), for $s \in \Omega_0^{0,\bullet}(X, L^p \otimes E)$, we have

$$p\, Q_p(s, s) \geqslant \frac{2}{3} \big\langle (pR^L + R^E + R^{\det})(w_l, \overline{w}_k)\, \overline{w}^k \wedge i_{\overline{w}_l} s \, , \, s \big\rangle$$
$$- \frac{1}{3} \big(\|\mathcal{T}^* \widetilde{s}\|_{L^2}^2 + \|\overline{\mathcal{T}}\widetilde{s}\|_{L^2}^2 + \|\overline{\mathcal{T}}^* \widetilde{s}\|_{L^2}^2 \big), \quad (6.1.6)$$

where $\{w_k\}$ is an orthonormal frame of $T^{(1,0)}X$.

Relations (6.1.1) and (6.1.6) imply (6.1.5) for $s \in \Omega_0^{0,q}(X, L^p \otimes E)$ and $q \geqslant 1$. By the Andreotti–Vesentini density lemma 3.3.1, $\Omega_0^{0,\bullet}(X, L^p \otimes E)$ is dense in $\text{Dom}(Q_p)$ with respect to the graph norm (due to the completeness of the metric g^{TX}). Thus (6.1.5) holds in general.

Next, consider $f \in \text{Dom}(\Box_p) \cap L_{0,0}^2(X, L^p \otimes E)$ and set $s = \overline{\partial}_p^E f$. It follows from (6.1.4) and (6.1.5) that

$$\|\Box_p f\|_{L^2}^2 = \big\langle \overline{\partial}_p^{E,*} s \, , \, \overline{\partial}_p^{E,*} s \big\rangle = p\, Q_p(s, s) \geqslant C_1 p \|s\|_{L^2}^2 = C_1 p \big\langle \Box_p f, f \big\rangle. \quad (6.1.7)$$

(6.1.5) and (6.1.7) imply that for p large enough,

$$\begin{aligned} \text{Spec}(\Box_p) &\subset \{0\} \cup [pC_1, \infty[, \\ \text{Ker}(\Box_p) &= H_{(2)}^{0,0}(X, L^p \otimes E) = H_{(2)}^{0,\bullet}(X, L^p \otimes E). \end{aligned} \quad (6.1.8)$$

Certainly, under the condition (6.1.3), (6.1.5) thus (6.1.8) still holds.

As the Kodaira Laplacian $\Box_p = \frac{1}{2}D_p^2$ acting on sections of $L^p \otimes E$ has a spectral gap (6.1.5), by the argument in Section 4.1.2, we can localize the problem, and we get directly our theorem from Theorem 4.2.9, as in the proof of Theorem 4.2.1.

As for b_1, the argument leading to (4.1.112)–(4.1.114) in the proof of Theorem 4.1.3 still holds locally, thus we get b_1 from (4.1.9). □

For simplicity we consider now $\text{rk}(E) = 1$, with the important case $E = K_X$ in mind. Choose an orthonormal basis $(S_i^p)_{i \geqslant 1}$ of $H_{(2)}^0(X, L^p \otimes E)$. For each local holomorphic frame e_L and e_E of L and E, we have

$$S_i^p = f_i^p e_L^{\otimes p} \otimes e_E \quad (6.1.9)$$

for some local holomorphic functions f_i^p. Then

$$P_p(x, x) = \sum_{i \geqslant 1} |S_i^{\otimes p}(x)|^2 = \sum_{i \geqslant 1} |f_i^p(x)|^2 |e_L^{\otimes p}|_{h^{L^p}}^2 |e_E|_{h^E}^2 \quad (6.1.10)$$

is a smooth function. Observe that the quantity $\sum_{i \geqslant 1} |f_i^p(x)|^2$ is not globally defined, but the current

$$\omega_p = \frac{\sqrt{-1}}{2\pi} \partial\bar{\partial} \log \left(\sum_{i \geqslant 1} |f_i^p(x)|^2 \right) \tag{6.1.11}$$

is globally well defined on X. Indeed, by (1.5.8) and (6.1.10), we have

$$\frac{1}{p}\omega_p - \frac{\sqrt{-1}}{2\pi} R^L = -\frac{\sqrt{-1}}{2\pi p} \bar{\partial}\partial \log P_p(x,x) + \frac{\sqrt{-1}}{2\pi p} R^E . \tag{6.1.12}$$

If E is trivial of rank 1 and $\dim H^0_{(2)}(X, L^p) < \infty$, we have by (5.1.17) that $\omega_p = \Phi_p^*(\omega_{FS})$ where Φ_p is defined as in (2.2.10) with $H^0(X, L^p)$ replaced by $H^0_{(2)}(X, L^p)$.

Corollary 6.1.2. *Assume that* $\mathrm{rk}(E) = 1$ *and* (6.1.1) *or* (6.1.3) *holds true. Then:*

(a) *For any compact set* $K \subset X$, *the restriction* $\omega_p|_K$ *is a smooth* $(1,1)$-*form for sufficiently large* p; *moreover, for any* $l \in \mathbb{N}$, *there exists* $C_{l,K} > 0$ *such that*

$$\left| \frac{1}{p}\omega_p - \frac{\sqrt{-1}}{2\pi} R^L \right|_{\mathscr{C}^l(K)} \leqslant \frac{C_{l,K}}{p} ; \tag{6.1.13}$$

(b) *The Morse inequalities hold in bidegree* $(0,0)$:

$$\liminf_{p \longrightarrow \infty} p^{-n} \dim H^0_{(2)}(X, L^p \otimes E) \geqslant \frac{1}{n!} \int_X \left(\frac{\sqrt{-1}}{2\pi} R^L \right)^n . \tag{6.1.14}$$

Proof. Due to (6.1.2), $P_p(x,x)$ doesn't vanish on any given compact set K for p sufficiently large. Thus, (a) is a consequence of (6.1.2) and (6.1.12).

Part (b) follows from Fatou's lemma, applied on X with the measure $\Theta^n/n!$ to the sequence $p^{-n}P_p(x,x)$ which converges pointwise to $\left(\frac{\sqrt{-1}}{2\pi}R^L\right)^n/\Theta^n$ on X. □

Corollary 6.1.3. *Assume that* (6.1.1) *holds true with* E *trivial and suppose that* X *is Andreotti pseudoconcave. Then the graded ring* $\mathscr{A}(X, L) = \oplus_{p \geqslant 0} H^0(X, L^p)$ *separates points and gives local coordinates on* X. *For any compact set* $K \subset X$, *the restriction of the Kodaira map* $\Phi_p|_K$ *is an embedding for* p *large enough.*

Proof. Theorem 3.4.5 entails that $\dim H^0(X, L^p) < \infty$, for any $p \geqslant 0$, so we can consider the Kodaira map $\Phi_p : X \setminus \mathrm{Bl}_{|H^0(X,L^p)|} \to \mathbb{P}(H^0(X, L^p)^*)$. Fix a compact set $K \subset X$. Since $P_p(x,x)$ doesn't vanish on K for p sufficiently large, Φ_p is well defined on K. Using (6.1.13) as in Lemma 5.1.6, we see that $\Phi_p|_K$ is an immersion. Finally, we can define the notion of peak section (Definition 5.1.7) and show that $\Phi_p|_K$ is injective, following Lemma 5.1.8. □

6.1.2 Covering manifolds

Another generalization is a version of Theorem 4.1.1 for covering manifolds. Let $\widetilde{X}$ be a paracompact complex manifold, such that there is a discrete group Γ acting holomorphically and freely on $\widetilde{X}$ with a compact quotient $X = \widetilde{X}/\Gamma$. Let $\pi_\Gamma : \widetilde{X} \longrightarrow X$ be the projection. Let $\widetilde{\mathbf{J}}, g^{T\widetilde{X}}, \widetilde{\Theta}, \widetilde{\omega}, \widetilde{J}, \widetilde{L}, \widetilde{E}$ be the pull-back of the corresponding objects in Section 1.5.1 by the projection $\pi_\Gamma : \widetilde{X} \to X$. We suppose that the positivity condition (1.5.21) holds for R^L.

Let us consider the Kodaira Laplacian $\widetilde{\Box}_p = \Box^{\widetilde{L}^p \otimes \widetilde{E}}$ which is an essentially self-adjoint operator. It follows by the same argument as in Theorems 1.5.5 and 1.5.8 that with $\mu_0 > 0$ introduced in (1.5.26),

$$\mathrm{Spec}(\widetilde{\Box}_p) \subset \{0\} \cup \left]p\mu_0 - \tfrac{1}{2}C_L, +\infty\right[. \tag{6.1.15}$$

Finally, we define the Bergman kernel $\widetilde{P}_p(x, x')$ on $\widetilde{X}$ as in Section 4.1.1.

Theorem 6.1.4. *We fix* $0 < \varepsilon_0 < \mathrm{inj}^X := \inf_{x \in X}\{injectivity\ radius\ at\ x\}$. *For any* $k, l \in \mathbb{N}$, *there exists* $C > 0$ *such that for* $x, x' \in \widetilde{X}$, $p \in \mathbb{N}^*$,

$$\left|\widetilde{P}_p(x, x') - P_p(\pi_\Gamma(x), \pi_\Gamma(x'))\right|_{\mathscr{C}^l} \leqslant C\, p^{-k-1}, \quad if\ d(x, x') < \varepsilon_0,$$

$$\left|\widetilde{P}_p(x, x')\right|_{\mathscr{C}^l} \leqslant C\, p^{-k-1}, \qquad if\ d(x, x') \geqslant \varepsilon_0. \tag{6.1.16}$$

Especially, $\widetilde{P}_p(x, x)$ *has the same asymptotic expansion as* $P_p(\pi_\Gamma(x), \pi_\Gamma(x))$ *in Theorem 4.1.1 uniformly on* $\widetilde{X}$.

Proof. Let $\{\varphi_i\}$ be a partition of unity subordinate to $\{U_{x_i} = B^X(x_i, \varepsilon)\}$ as in Section 1.6.2. Then $\{\widetilde{\varphi}_{\gamma,i} = \varphi_i \circ \pi_\Gamma\}$ is a partition of unity subordinate to $\{\widetilde{U}_{\gamma, x_i}\}$ where $\pi_\Gamma^{-1}(U_{x_i}) = \cup_{\gamma \in \Gamma}\widetilde{U}_{\gamma, x_i}$ and $\widetilde{U}_{\gamma_1, x_i}$ and $\widetilde{U}_{\gamma_2, x_i}$ are disjoint for $\gamma_1 \neq \gamma_2$. The proof of Proposition 4.1.5 still holds for the pair $\{\widetilde{\varphi}_{\gamma,i}\}, \{\widetilde{U}_{\gamma, x_i}\}$, since we can apply the Sobolev embedding theorems A.1.6 and A.3.1 with uniform constant on $\widetilde{U}_{\gamma, x_i}$. Thus, the analogue of (4.1.12) holds uniformly on $\widetilde{X}$. Using the finite propagation speed as at the end of Section 4.1.2, we conclude. $\qquad\square$

Remark 6.1.5. Theorem 6.1.4 works well for coverings of non-compact manifolds. Let (X, J, Θ) be a complete Hermitian manifold, (L, h^L) be a holomorphic line bundle on X and let $\pi_\Gamma : \widetilde{X} \to X$ be a Galois covering of $X = \widetilde{X}/\Gamma$. If (X, Θ) and (L, h^L) satisfy one of the conditions (6.1.1) or (6.1.3), $(\widetilde{X}, \widetilde{\Theta})$ and $(\widetilde{L}, h^{\widetilde{L}})$ have the same properties. We obtain therefore as in (6.1.14) (by integrating over a fundamental domain):

$$\liminf_{p \longrightarrow \infty} p^{-n} \dim_\Gamma H^0_{(2)}(\widetilde{X}, \widetilde{L}^p) \geqslant \frac{1}{n!}\int_X \left(\frac{\sqrt{-1}}{2\pi}R^L\right)^n, \tag{6.1.17}$$

where $\dim_\Gamma$ is the von Neumann dimension of the Γ-module $H^0_{(2)}(X, L^p)$.

6.2   The Shiffman–Ji–Bonavero–Takayama criterion revisited

Let (L, h^L) be a singular Hermitian line bundle over a compact complex manifold X. We assume that the curvature current $\sqrt{-1}R^L$ is strictly positive and smooth on a non-empty Zariski open set. The main result of this section is the asymptotic expansion of the Bergman kernel of L^p on a Zariski open set endowed with the generalized Poincaré metric, which is the object of Theorem 6.2.3. One motivation is to give a new proof of Shiffman's conjecture studied in Section 2.3. Firstly, we will prove Theorem 2.3.7 without using the Demailly approximation Theorem 2.3.10. Secondly, if we apply the approximation Theorem 2.3.10, we get a new proof of Theorem 2.3.8.

We turn now to the proof of Theorem 2.3.7. In the course of the proof, we describe the Bergman kernel on the regular locus of the positive Kähler current.

Let X be a compact connected complex manifold of dimension n. Let Σ be a closed analytic subset of X.

Let $\pi : \widetilde{X} \longrightarrow X$ be a resolution of singularities (cf. Theorem 2.1.13) such that $\pi : \widetilde{X} \setminus \pi^{-1}(\Sigma) \longrightarrow X \setminus \Sigma$ is biholomorphic and $\pi^{-1}(\Sigma)$ is a divisor with normal crossings. More precisely, there exists a finite sequence of blow-ups

$$\widetilde{X} = X_m \xrightarrow{\tau_m} X_{m-1} \xrightarrow{\tau_{m-1}} \cdots \xrightarrow{\tau_2} X_1 \xrightarrow{\tau_1} X_0 = X \qquad (6.2.1)$$

such that

(a) τ_i is the blow-up along a non-singular center Y_{i-1} contained in the strict transform of Σ in X_{i-1}, $i \geqslant 1$,

(b) the strict transform of Σ in $\widetilde{X} = X_m$ through $\pi = \tau_1 \circ \tau_2 \circ \cdots \circ \tau_m$ is smooth and $\pi^{-1}(\Sigma)$ is a divisor with normal crossings.

Let $g_0^{T\widetilde{X}}$ be an arbitrary smooth J-invariant metric on $\widetilde{X}$ and $\Theta'(\cdot, \cdot) = g_0^{T\widetilde{X}}(J\cdot, \cdot)$ the corresponding $(1,1)$-form. The *generalized Poincaré metric* on $X \setminus \Sigma = \widetilde{X} \setminus \pi^{-1}(\Sigma)$ is defined by the Hermitian form

$$\Theta_{\varepsilon_0} = \Theta' + \varepsilon_0 \sqrt{-1} \sum_i \overline{\partial}\partial \log \left((-\log(\|\sigma_i\|_i^2))^2 \right), \quad 0 < \varepsilon_0 \ll 1 \text{ fixed}, \qquad (6.2.2)$$

where $\pi^{-1}(\Sigma) = \cup_i \Sigma_i$ is the decomposition into irreducible components Σ_i of $\pi^{-1}(\Sigma)$ and each Σ_i is non-singular; σ_i are holomorphic sections of the associated holomorphic line bundle $\mathscr{O}_{\widetilde{X}}(\Sigma_i)$ which vanish to first order on Σ_i, and $\|\sigma_i\|_i$ is the norm for a smooth Hermitian metric $\|\cdot\|_i$ on $\mathscr{O}_{\widetilde{X}}(\Sigma_i)$ such that $\|\sigma_i\|_i < 1$. Let $R^{\mathscr{O}_{\widetilde{X}}(\Sigma_i)}$ be the curvature of $(\mathscr{O}_{\widetilde{X}}(\Sigma_i), \|\cdot\|_i)$.

Lemma 6.2.1.

(i) *The generalized Poincaré metric* (6.2.2) *is a complete Hermitian metric of finite volume. Its Hermitian torsion* $\mathcal{T}_{\varepsilon_0} = [i(\Theta_{\varepsilon_0}), \partial\Theta_{\varepsilon_0}]$ *and the curvature* $R^{\det} = R^{K_{\widetilde{X}}^*}$ *is also bounded.*

(ii) *If (E, h^E) is a holomorphic Hermitian vector bundle over X, set*

$$H^0_{(2)}(X \smallsetminus \Sigma, E) = \{u \in L^2_{0,0}(X \smallsetminus \Sigma, E, \Theta_{\varepsilon_0}, h^E) : \overline{\partial}^E u = 0\}, \qquad (6.2.3)$$

then

$$H^0_{(2)}(X \smallsetminus \Sigma, E) = H^0(X, E). \qquad (6.2.4)$$

Proof. To describe the metric more precisely, we denote by $\mathbb{D}$ the unit disc in $\mathbb{C}$ and by $\mathbb{D}^* = \mathbb{D} \smallsetminus \{0\}$. On the product $(\mathbb{D}^*)^l \times \mathbb{D}^{n-l}$, we introduce the metric

$$\Omega_P = \frac{\sqrt{-1}}{2} \sum_{k=1}^{l} \frac{dz_k \wedge d\bar{z}_k}{|z_k|^2 (\log |z_k|^2)^2} + \frac{\sqrt{-1}}{2} \sum_{k=l+1}^{n} dz_k \wedge d\bar{z}_k. \qquad (6.2.5)$$

For any point $x \in \pi^{-1}(\Sigma)$, there exists a coordinate neighborhood U of x isomorphic to $\mathbb{D}^n$ in which $(\widetilde{X} \smallsetminus \pi^{-1}(\Sigma)) \cap U = \{z = (z_1, \ldots, z_n) : z_1 \neq 0, \ldots, z_l \neq 0\}$. Such coordinates are called special. We endow $(\widetilde{X} \smallsetminus \pi^{-1}(\Sigma)) \cap U \cong (\mathbb{D}^*)^l \times \mathbb{D}^{n-l}$ with the metric (6.2.5). We have

$$\sqrt{-1} \overline{\partial}\partial \log \left((-\log(\|\sigma_i\|_i^2))^2 \right)$$
$$= 2\sqrt{-1} \left(\frac{R^{\mathcal{O}_{\widetilde{X}}(\Sigma_i)}}{\log(\|\sigma_i\|_i^2)} + \frac{\partial \log(\|\sigma_i\|_i^2) \wedge \overline{\partial} \log(\|\sigma_i\|_i^2)}{(\log(\|\sigma_i\|_i^2))^2} \right). \qquad (6.2.6)$$

Since the terms $R^{\mathcal{O}_{\widetilde{X}}(\Sigma_i)} / \log(\|\sigma_i\|_i^2)$ tend to zero, as we approach Σ, for ε_0 small enough (cf. Definition B.2.8)

$$\Theta' + 2\sqrt{-1}\varepsilon_0 \sum_i \frac{R^{\mathcal{O}_{\widetilde{X}}(\Sigma_i)}}{\log(\|\sigma_i\|_i^2)} > 0. \qquad (6.2.7)$$

The last term in (6.2.6) is $\geqslant 0$, as $\sqrt{-1}\partial g \wedge \overline{\partial} g \geqslant 0$ for any real function g on $\widetilde{X}$. Thus Θ_{ε_0} is positive for ε_0 small enough.

We choose special coordinates in a neighborhood U of x_0 in which Σ_j has the equation $z_j = 0$ for $j = 1, \ldots, k$ and Σ_j, $j > k$, do not meet U. Then for $1 \leqslant i \leqslant k$, $\|\sigma_i\|_i^2 = \varphi_i |z_i|^2$ for some positive smooth function φ_i on U and

$$\frac{\partial \log(\|\sigma_i\|_i^2) \wedge \overline{\partial} \log(\|\sigma_i\|_i^2)}{(\log(\|\sigma_i\|_i^2))^2} = \frac{dz_i \wedge d\bar{z}_i + \psi_i}{|z_i|^2 (\log(\|\sigma_i\|_i^2))^2} \qquad (6.2.8)$$

where ψ_i is a smooth $(1,1)$-form on U such that $\psi_i|_{z_i=0} = 0$.

Using (6.2.6) and (6.2.8), we see that the metrics (6.2.2) and (6.2.5) are equivalent for $|z_i|$ small. Therefore we have to check the first assertion of (i) for the Poincaré metric on $\{z \in \mathbb{C} : 0 < |z| < c\}$,

$$\omega_P = \frac{\sqrt{-1}}{2} \frac{dz \wedge d\bar{z}}{|z|^2 (\log |z|^2)^2} \qquad \text{on } \{z \in \mathbb{C} : 0 < |z| < c < 1\}. \qquad (6.2.9)$$

The completeness follows from the fact that the length of the path $\{tz : t \in]0, c]\}$ ($|z| = 1$) is infinite:

$$\int_0^c \frac{dr}{r|\log r|} = -\log\log\frac{1}{r}\Big|_0^c = \infty.$$

Moreover, the volume calculated in polar coordinates equals

$$\int_0^{2\pi}\int_0^c \frac{r\,dr\,d\vartheta}{r^2(\log r)^2} = -2\pi(\log r)^{-1}\Big|_0^c < \infty.$$

From this the first assertion of (i) follows.

Recall that $R^{\det}$ is the curvature of the holomorphic Hermitian connection on $\det(T^{(1,0)}X)$ with respect to the Hermitian metric induced by Θ_{ε_0}. We wish to show that there exists a constant $C > 0$ such that

$$-C\Theta_{\varepsilon_0} < \sqrt{-1}R^{\det} < C\Theta_{\varepsilon_0},\ |\mathcal{T}_{\varepsilon_0}|_{\Theta_{\varepsilon_0}} < C \qquad (6.2.10)$$

where $|\mathcal{T}_{\varepsilon_0}|_{\Theta_{\varepsilon_0}}$ is its norm with respect to Θ_{ε_0}. Since $\partial\Theta_{\varepsilon_0} = \partial\Theta'$ by (6.2.2), $\partial\Theta_{\varepsilon_0}$ extends smoothly over $\widetilde{X}$, and thus we get the second relation of (6.2.10).

We turn now to the first condition of (6.2.10). By (6.2.2), (6.2.6) and (6.2.8), we know that

$$\Theta_{\varepsilon_0}^n = \frac{2^k\varepsilon_0^k + \beta(z)}{\prod_{i=1}^k |z_i|^2(\log\|\sigma_i\|_i^2)^2}\prod_{j=1}^n(\sqrt{-1}dz_j \wedge d\bar{z}_j) =: \gamma(z)\prod_{j=1}^n(\sqrt{-1}dz_j \wedge d\bar{z}_j).$$
$$(6.2.11)$$

Here $\beta(z)$ is a polynomial in the functions

$$a_{i\alpha}(z)|z_i|^2(\log\|\sigma_i\|_i^2)^2,\quad b_{i\alpha}(z)|z_i|^2\log\|\sigma_i\|_i^2 \quad\text{and}\quad c_{i\alpha}(z),\quad (1 \leqslant i \leqslant k),$$

with $a_{i\alpha}, b_{i\alpha}$ smooth functions on U and $c_{i\alpha}$ smooth functions on U such that $c_{i\alpha}(z)|_{z_i=0} = 0$. Moreover, $2^k\varepsilon_0^k + \beta(z)$ is positive on U as Θ_{ε_0} is positive. Since

$$|\tfrac{\partial}{\partial z_1} \wedge \cdots \wedge \tfrac{\partial}{\partial z_n}|^2_{\Theta_{\varepsilon_0}}\prod_{j=1}^n(\sqrt{-1}dz_j \wedge d\bar{z}_j) = \Theta_{\varepsilon_0}^n, \qquad (6.2.12)$$

we get from (6.2.11) and (6.2.12),

$$R^{\det} = \bar{\partial}\partial\log\gamma(z) = \bar{\partial}\partial\log\left(2^k\varepsilon_0^k + \beta(z)\right) + \sum_{i=1}^k\partial\bar{\partial}\log\left((\log\|\sigma_i\|_i^2)^2\right). \quad (6.2.13)$$

By (6.2.6), the last term of (6.2.13) is bounded with respect to Θ_{ε_0}. To examine the first term of the sum, we write

$$\bar{\partial}\partial\log(2^k\varepsilon_0^k + \beta(z)) = \frac{\bar{\partial}\partial\beta(z)}{2^k\varepsilon_0^k + \beta(z)} - \frac{\bar{\partial}\beta(z) \wedge \partial\beta(z)}{(2^k\varepsilon_0^k + \beta(z))^2}. \qquad (6.2.14)$$

Now we observe that for $W_i(z) = |z_i|^2 (\log \|\sigma_i\|_i^2)^2$ or $|z_i|^2 \log \|\sigma_i\|_i^2$, the terms $\partial \bar{\partial} W_i(z)$, $\partial W_i(z)$, $\bar{\partial} W_i(z)$ are bounded with respect to the Poincaré metric (6.2.5), thus with respect to Θ_{ε_0}. Combining with the form of β given after (6.2.11), this achieves the proof of (6.2.10).

Let us prove (ii). First observe that Θ_{ε_0} dominates the Euclidean metric in special coordinates near $\pi^{-1}(\Sigma)$, being equivalent with (6.2.5). Therefore it dominates some positive multiple of any smooth Hermitian metric on $\widetilde{X}$. We deduce that, given a smooth Hermitian form Θ'' on X, there exists $C > 0$ such that $\Theta_{\varepsilon_0} \geqslant C\Theta''$ on $X \smallsetminus \Sigma$. It follows that elements of $H^0_{(2)}(X \smallsetminus \Sigma, E)$ are L^2 integrable with respect to the smooth metrics Θ'' and h^E over X, which entails that they extend holomorphically to sections of $H^0(X, E)$ by Lemma 2.3.22. We have therefore $H^0_{(2)}(X \smallsetminus \Sigma, E) \subset H^0(X, E)$. The reverse of inclusion follows from the finiteness of the volume of $X \smallsetminus \Sigma$ in the Poincaré metric. $\qquad \square$

Let L be a holomorphic line bundle on X. Suppose that h^L is a singular Hermitian metric on L with curvature current R^L, smooth outside the proper analytic set Σ and $\sqrt{-1}R^L$ is a strictly positive current on X.

Lemma 6.2.2. *There exists a singular Hermitian line bundle $(\widetilde{L}, h^{\widetilde{L}})$ on $\widetilde{X}$ which is strictly positive and $\widetilde{L}|_{\widetilde{X} \smallsetminus \pi^{-1}(\Sigma)} \cong \pi^*(L^{k_0})$, for some $k_0 \in \mathbb{N}$.*

Proof. Let $\widetilde{Y}_0 = \tau_1^{-1}(Y_0)$, with $Y_0 = \Sigma$ in (6.2.1). By Proposition 2.1.11 (a), there exists a smooth Hermitian metric h_0 on the line bundle $\mathscr{O}_{X_1}(-\widetilde{Y}_0)$ whose curvature $R^{(\mathscr{O}_{X_1}(-\widetilde{Y}_0), h_0)}$ is strictly positive along $\widetilde{Y}_0$, and bounded on X_1, moreover, it vanishes outside a neighborhood of $\widetilde{Y}_0$.

On X_1, we consider the bundle $L_1 := \tau_1^*(L^{k_1}) \otimes \mathscr{O}_{X_1}(-\widetilde{Y}_0)$ endowed with the metric $h^{L_1} = (h^L)^{\otimes k_1} \otimes h_0$, for $k_1 \in \mathbb{N}$. The curvature current of (L_1, h^{L_1}) is

$$\sqrt{-1}R^{L_1} = k_1 \, \tau_1^*(\sqrt{-1}R^L) + \sqrt{-1}R^{(\mathscr{O}_{X_1}(-\widetilde{Y}_0), h_0)}. \qquad (6.2.15)$$

The current $\tau_1^*(\sqrt{-1}R^L)$ is positive on X_1 and strictly positive on any compact set disjoint from $\widetilde{Y}_0$. Hence, for k_1 sufficiently large, $\sqrt{-1}R^{L_1}$ is a strictly positive current on X_1 (cf. Proposition 2.1.8).

Continuing inductively, we construct a holomorphic line bundle $(\widetilde{L}, h^{\widetilde{L}})$ on $\widetilde{X}$ with curvature current $\sqrt{-1}R^{\widetilde{L}}$, smooth on $\widetilde{X} \smallsetminus \tau^{-1}(\Sigma)$ and strictly positive on $\widetilde{X}$. $\qquad \square$

We introduce on $L|_{X \smallsetminus \Sigma}$ the metric $(h^{\widetilde{L}})^{1/k_0}$ whose curvature extends to a strictly positive $(1,1)$-current on $\widetilde{X}$. Set

$$h_\varepsilon^L := (h^{\widetilde{L}})^{1/k_0} \prod_i (-\log(\|\sigma_i\|_i^2))^\varepsilon, \quad 0 < \varepsilon \ll 1, \qquad (6.2.16a)$$

$$H^0_{(2)}(X \smallsetminus \Sigma, L^p) := \{u \in L^2_{0,0}(X \smallsetminus \Sigma, L^p, \Theta_{\varepsilon_0}, h_\varepsilon^L) : \bar{\partial}^{L^p} u = 0\}. \qquad (6.2.16b)$$

The space $H^0_{(2)}(X \smallsetminus \Sigma, L^p)$ is the space of L^2-holomorphic sections relative to the metrics Θ_{ε_0} on $X \smallsetminus \Sigma$ and h^L_ε on $L|_{X\smallsetminus\Sigma}$. Since $(h^{\widetilde{L}})^{1/k_0}$ is bounded away from zero (having plurisubharmonic weights), as its curvature extends a strictly positive $(1,1)$-current on $\widetilde{X}$, the elements of this space are L^2 integrable with respect to the Poincaré metric and a smooth metric h^L_* of L over all of X. By the proof of Lemma 6.2.1 (ii), we have

$$H^0_{(2)}(X \smallsetminus \Sigma, L^p) \subset H^0(X, L^p). \tag{6.2.17}$$

(Here we cannot infer the other inclusion since h^L_ε might be infinity on Σ.) The space $H^0_{(2)}(X \smallsetminus \Sigma, L^p)$ is our space of polarized sections of L^p.

Theorem 6.2.3. *Let X be a compact complex manifold with an integral Kähler current ω. Let (L, h^L) be a singular polarization of $[\omega]$ with strictly positive curvature current having singular support contained in a proper analytic set Σ. Then the Bergman kernel of the space of polarized sections (6.2.16b) has the asymptotic expansion as in Theorem 6.1.1 for $X \smallsetminus \Sigma$.*

Proof. We will apply Theorem 6.1.1 to the non-Kähler Hermitian manifold $(X \smallsetminus \Sigma, \Theta_{\varepsilon_0})$ equipped with the Hermitian bundle $(L|_{X\smallsetminus\Sigma}, h^L_\varepsilon)$. Thus we have to show that there exist constants $\eta > 0$, $C > 0$ such that

$$\sqrt{-1}R^{(L|_{X\smallsetminus\Sigma}, h^L_\varepsilon)} > \eta\Theta_{\varepsilon_0}, \quad \sqrt{-1}R^{\det} > -C\Theta_{\varepsilon_0}, \quad |\mathcal{T}_{\varepsilon_0}| < C. \tag{6.2.18}$$

The first one results for all ε small enough from (6.2.2), (6.2.16a) and the fact that the curvature of $(h^{\widetilde{L}})^{1/k_0}$ extends to a strictly positive $(1,1)$-current on $\widetilde{X}$ (dominating a small positive multiple of Θ' on $\widetilde{X}$). The second and third relations were proved in (6.2.10). This achieves the proof of Theorem 6.2.3. $\square$

Proof of Theorem 2.3.7. Any Moishezon manifold possesses a strictly positive singular polarization (L, h^L) by Lemma 2.3.5 and 2.3.6.

Conversely, suppose X has such a polarization. Then by Theorem 6.2.3, as in (6.1.14), we have $\dim H^0_{(2)}(X \smallsetminus \Sigma, L^p) \geqslant Cp^n$ for some $C > 0$ and p large enough. By (6.2.17), it follows that L is big and X is Moishezon. $\square$

Corollary 6.2.4. *Let L be a holomorphic line bundle over a compact connected complex manifold X. L is big if and only if L possesses a singular Hermitian metric h^L with strictly positive curvature current $\sqrt{-1}R^L$ and smooth outside a proper analytic set.*

Proof. By Theorem 2.3.30, any big line bundle L on a compact manifold carries a singular Hermitian metric having strictly positive curvature current with singularities along a proper analytic set.

Conversely, by the above proof of Theorem 2.3.7, we know L is big. $\square$

Remark 6.2.5. The results of this section hold also for reduced compact complex spaces X possessing a holomorphic line bundle L with singular Hermitian metric h^L having positive curvature current. This is just a matter of desingularizing X. As a space of polarized sections, we obtain $H^0_{(2)}(X \setminus \Sigma, L^p)$ where Σ is an analytic set containing the singular set of X.

6.3 Compactification of manifolds

In this section, we apply the ideas developed so far about the spectral gap and the Morse inequalities to the compactification problem. We wish to find sufficient conditions under which a given non-compact complex manifold can be compactified – i.e., exhibited as a Zariski-open subset of a compact strongly pseudoconvex domain in a projective manifold. By considering such a problem, one hopes to reduce the study of certain non-compact complex manifolds to that of better understood manifolds.

This section is organized as follows. In Section 6.3.1, we review the known results about the compactification of strongly pseudoconcave ends. In Section 6.3.2, we prove Theorem 6.3.24 about the compactification of a complete Kähler manifold with pinched negative curvature, with a strongly pseudoconvex end and finite volume away from this end.

6.3.1 Filling strongly pseudoconcave ends

We give an overview here of the basic results about filling strongly pseudoconvex ends of complex manifolds.

Definition 6.3.1. For a compact subset K of a complex manifold X, an unbounded connected component of $X \setminus K$ is called an *end* of X (with respect to K). If $K_1 \subset K_2$ are two compact subsets, the number of ends with respect to K_1 is at most the number of ends with respect to K_2, so that we can define the number of ends of X. Namely, X is said to have *finitely many ends* if for some integer k, and for any $K \subset X$, the number of ends with respect to K is at most k. The smallest such k is called the number of ends of X, and then there exists $K_0 \subset X$ such that the number of ends with respect to K_0 is precisely the number of ends of X. If no such k exists, we say that X has infinitely many ends.

Definition 6.3.2. A manifold X with $\dim X \geqslant 2$ is said to be a *strongly pseudoconcave end* if there exist $u \in \mathbb{R} \cup \{-\infty\}$, $v \in \mathbb{R} \cup \{+\infty\}$ and a proper, smooth function $\varphi : X \longrightarrow]u, v[$, which is strictly plurisubharmonic on $\{\varphi < c_0\}$, for some $u < c_0 \leqslant v$. If $u = -\infty$, X is called a *hyperconcave end*. For $u < c < v$, we set $X_c = \{\varphi < c\}$. We call φ an *exhaustion function*. We say that a strongly pseudoconcave end can be *compactified* or *filled in* if there exists a complex space $\widehat{X}$ such that X is (biholomorphic to) an open set in $\widehat{X}$ and for any $c < v$, $(\widehat{X} \setminus X) \cup \{\varphi \leqslant c\}$

is a compact set. We will call $\widehat{X}$ (somewhat abusively) the *compactification* of X, although it is not necessarily compact.

Example 6.3.3. Let X be a Stein manifold and $A \subset X$ be a compact set. Assume that for any $x \in A$ there exists a neighborhood U_x and a strictly plurisubharmonic function $f_x : U_x \to [-\infty, \infty[$ such that $A \cap U_x = f_x^{-1}(-\infty)$. By a result of Colţoiu, there exists a neighborhood $A \subset U$ and a strictly plurisubharmonic function $f : U \to [-\infty, \infty[$ such that $A = f^{-1}(-\infty)$. Thus $X \smallsetminus A$ is a hyperconcave end.

Theorem 6.3.4 (Rossi–Andreotti–Siu).

(a) *Two normal Stein compactifications are biholomorphic by a map which is the identity on X.*

(b) *Any strongly pseudoconcave end X can be compactified provided $\dim X \geqslant 3$. If the exhaustion function is strictly plurisubharmonic on X, the compactification $\widehat{X}$ can be taken to be a normal Stein space with at worst isolated singularities.*

If $\dim X = 2$, then the previous theorem breaks down.

Example 6.3.5 (Grauert–Andreotti–Siu–Rossi). This example provides complex structures on $\mathbb{CP}^2 \smallsetminus \{[1,0,0]\}$, which are not fillable.

Let Q_ε be the family of quadrics in $\mathbb{CP}^3$ given in the homogeneous coordinates $[w_0, w_1, w_2, w_3]$ by the equation $w_3(w_3 + \varepsilon w_0) = w_1 w_2$. For $\varepsilon \neq 0$, they are non-singular. There exists an application $\Phi : \mathbb{CP}^2 \smallsetminus \{[1,0,0]\} \longrightarrow Q_\varepsilon \smallsetminus A$, where A is a real analytic sphere, such that Φ is a two-sheeted differentiable ramified covering. We can use Φ to pull back the complex structure of Q_ε on $\mathbb{CP}^2 \smallsetminus \{[1,0,0]\}$, so that Φ becomes holomorphic. Then $\mathbb{CP}^2 \smallsetminus \{[1,0,0]\}$ with the new structure cannot be compactified.

A large class of strongly pseudoconcave ends which can be compactified even in dimension 2 are the hyperconcave ends.

Example 6.3.6. Let (X, g^{TX}) be a complete Kähler manifold. For any linearly independent $U, V \in T_x X$, we define the *sectional curvature* of g^{TX} on the plane generated by U and V by

$$K(U, V) = R(U, V, U, V)/|U \wedge V|^2. \qquad (6.3.1)$$

Let us assume that g^{TX} has *pinched negative curvature*, that is, there are $C_1, C_2 > 0$ such that for any $x \in X$,

$$-C_1^2 \leqslant K(U, V) \leqslant -C_2^2 < 0, \quad \text{for any } U, V \in T_x X, \ U \wedge V \neq 0. \qquad (6.3.2)$$

A geodesic $\gamma : [0, \infty[\longrightarrow X$ is called a *ray* if $d(\gamma(t_1), \gamma(t_2)) = |t_1 - t_2|$ for all $t_1, t_2 \in [0, \infty[$, where $d(\cdot, \cdot)$ is the distance induced by g^{TX}. If X is non-compact, a ray emanating from any given point always exists. The *Busemann function* of the ray γ is defined by

$$r_\gamma : X \to \mathbb{R}, \quad r_\gamma(x) = \lim_{t \to \infty} (d(x, \gamma(t)) - t). \qquad (6.3.3)$$

Let N be an end of X. If N has finite volume, then N is called a *cusp*.

We give now two results about the structure of complete Kähler manifolds with negative curvature.

Theorem 6.3.7 (Wu). *Any simply connected complete Kähler manifold of nonpositive sectional curvature is Stein.*

Theorem 6.3.8 (Siu–Yau). *Let (X, g^{TX}) be a complete Kähler manifold of finite volume and pinched negative curvature. Then X is a hyperconcave manifold. In particular, any cusp N is a hyperconcave end.*

The first step in the proof is to show that, if $\tilde{\gamma}$ is the lift of γ to the universal cover of X, then $r_{\tilde{\gamma}}$ is strictly plurisubharmonic. Then, using results of Margulis and Gromov, Siu and Yau show that, for c sufficiently large, the restriction of $r_{\tilde{\gamma}}$ to $r_{\tilde{\gamma}} < -c$ descends to X. Moreover, the minimum of a finite number of such functions forms an exhaustion function on X.

The following result shows that hyperconcave ends can be always compactified.

Theorem 6.3.9.

(a) *Any hyperconcave end X can be compactified, i.e., there exists a complex space $\hat{X}$ such that X is (biholomorphic to) an open set in $\hat{X}$ and for any $c < v$, $(\hat{X} \smallsetminus X) \cup \{\varphi \leqslant c\}$ is a compact set. More specifically, if φ is strictly plurisubharmonic on the whole of X, $\hat{X}$ can be chosen a normal Stein space with at worst isolated singularities.*

(b) *Assume that X can be covered by Zariski-open sets whose universal coverings are Stein manifolds. Then $\hat{X} \smallsetminus X$ is the union of a finite set D' and an exceptional analytic set which can be blown down to a finite set D. Each connected component of X_c, for sufficiently small c, can be analytically compactified by one point from $D' \cup D$. If X itself has a Stein cover, $D' = \emptyset$ and D consists of the singular set of the Remmert reduction of $\hat{X}$.*

A strategy to fill in a strongly pseudoconcave end is to try to fill in the CR manifold $\{\varphi = \text{constant}\}$. Let us first review the notion of CR manifold of hypersurface type.

Definition 6.3.10. Let Y be a smooth orientable manifold of real dimension $(2n - 1)$. A *Cauchy-Riemann (CR) structure* on Y is an $(n - 1)$-dimensional complex subbundle $T^{(1,0)}Y$ of the complexified tangent bundle $TY \otimes_{\mathbb{R}} \mathbb{C}$ such that $T^{(1,0)}Y \cap \overline{T^{(1,0)}Y} = \{0\}$, and such that $T^{(1,0)}Y$ is integrable as a complex subbundle of $TY \otimes_{\mathbb{R}} \mathbb{C}$ (i.e., if U and X are sections of $T^{(1,0)}Y$, the Lie bracket $[U, X]$ is still a section of $T^{(1,0)}Y$).

Let X be a smooth domain in a complex manifold W. The complex structure of W induces the CR structure $T^{(1,0)}Y = T^{(1,0)}X \cap TY \otimes_{\mathbb{R}} \mathbb{C}$ on $Y = \partial X$.

If Y is a CR manifold, then its Levi distribution H is the real subbundle of TY defined by $H = \text{Re}\{T^{(1,0)}Y \oplus \overline{T^{(1,0)}Y}\}$. There exists on H a complex structure J given by $J(U + \overline{U}) = \sqrt{-1}(U - \overline{U})$, with $U \in T^{(1,0)}Y$. As Y is orientable, the real line bundle $H^\perp \subset T^*Y$, the annihilator of H, admits a global non-vanishing section θ.

Definition 6.3.11. The CR structure is said to be *strongly pseudoconvex* if $d\theta(\cdot, J\cdot)$ defines a positive definite metric on H. The *tangential Cauchy-Riemann operator* $\overline{\partial}_b : \mathscr{C}^\infty(Y) \to \mathscr{C}^\infty(Y, T^{*(1,0)}Y)$, associates to a function $f \in \mathscr{C}^1(Y)$ the projection on $T^{*(1,0)}Y$ of the exterior differential df. A function $f \in \mathscr{C}^1(Y)$ is called a CR function if $\overline{\partial}_b f = 0$. By a *CR embedding* of a manifold in a complex manifold, we mean an embedding whose components are CR functions.

When we say that a CR manifold is a submanifold of a complex manifold, we understand that the inclusion is a CR embedding, that is, the CR structure is induced from the ambient manifold.

If $Y = \partial X$, where X is a domain in a complex manifold W, then all restrictions of holomorphic functions on W to Y are CR functions. We have also the following converse which may be also seen as a form of Hartogs phenomenon for CR functions.

Theorem 6.3.12 (Kohn–Rossi). *Assume that X is a smooth, relatively compact connected domain in a complex manifold such that the Levi form of a defining function of X restricted to the holomorphic tangent space at ∂X has at least one positive eigenvalue everywhere. Then any CR function defined on ∂X extends to a holomorphic function in X. In particular, ∂X is connected.*

We also need the abstract notion of complex manifold with strongly pseudoconvex boundary. A priori, it is not a domain with boundary in a larger complex manifold.

Definition 6.3.13. A complex manifold X with strongly pseudoconvex boundary is a real manifold with boundary, of real dimension $2n$, satisfying the following conditions: (i) the interior $\mathrm{Int}(X) = X \smallsetminus \partial X$ has an integrable complex structure and (ii) for each point $x \in \partial X$, there exist a neighborhood U in X, a strongly pseudoconvex domain $M \subset \mathbb{C}^n$ with smooth boundary, and a diffeomorphism σ from U onto a relatively open subset $\sigma(U)$ such that $\sigma(\partial U) \subset \partial M$ and σ is biholomorphic from $\mathrm{Int}(U)$ to $\mathrm{Int}(\sigma(U))$.

From this definition, we infer:

Proposition 6.3.14. *There exists an induced integrable Cauchy-Riemann structure on the boundary ∂X, given locally by transporting the CR structure on ∂M. Moreover, if ∂X is compact, there exists a defining function $\varphi : X \to]-\infty, c]$ such that $\partial X = \{\varphi = c\}$, with the properties:*

(1) its Levi form is positive definite on the holomorphic tangent space of ∂X and

(2) there exists $c_0 < c$ such that φ is strictly plurisubharmonic on $\{c_0 < \varphi < c\}$.

Theorem 6.3.15 (Heunemann–Ohsawa). *Let X be a compact complex manifold with strongly pseudoconvex boundary. Then X can be realized as a domain with boundary in a larger complex manifold W: there exists a strongly pseudoconvex* *domain* $M \subset W$ *and a diffeomorphism* $\sigma : X \longrightarrow M$ *which is a biholomorphism between* $\mathrm{Int}(X)$ *and* $\sigma(\mathrm{Int}(X))$, $\sigma(\partial X) = \partial M$.

We give now the solution of the complex Plateau problem for strongly pseudoconvex CR submanifold of $\mathbb{C}^m$.

Theorem 6.3.16 (Harvey–Lawson). *Let* Y *be a compact strongly pseudoconvex CR submanifold of* $\mathbb{C}^m$. *Then there exists a normal Stein space* $S \subset \mathbb{C}^m$ *with boundary and at most isolated singularities, such that* $\partial S = Y$.

In this case, we say that Y *bounds* the Stein space S. In order to apply the Harvey–Lawson theorem, we need conditions for a strongly pseudoconvex CR manifold to be embeddable in the Euclidean space.

Theorem 6.3.17 (Boutet de Monvel). *Any compact CR manifold* Y *with* $\dim_{\mathbb{R}} Y > 3$ *admits a CR embedding in the Euclidean space.*

The proof is based on the Hodge decomposition for the *Kohn Laplacian*

$$\Box_b = \overline{\partial}_b \overline{\partial}_b^* + \overline{\partial}_b^* \overline{\partial}_b, \tag{6.3.4}$$

which is not an elliptic operator but has however as a parametrix a pseudodifferential operator of type $1/2$. If $\dim_{\mathbb{R}} Y = 3$, Boutet de Monvel's theorem breaks down. A counterexample is given by the boundary of the strongly pseudoconcave manifold constructed in Example 6.3.5 of Grauert–Andreotti–Rossi.

We are led to the following beautiful result which follows from the works of Boutet de Monvel–Sjöstrand, Harvey–Lawson, Burns and Kohn.

Theorem 6.3.18. *Let* Y *be a compact complex CR manifold,* $\dim_{\mathbb{R}} Y \geqslant 3$. *The following conditions are equivalent:*

(a) Y *is embeddable in the Euclidean space,*

(b) Y *bounds a strongly pseudoconvex complex manifold,*

(c) *The tangential Cauchy-Riemann operator* $\overline{\partial}_b$ *on functions of* Y *has closed range in* L^2.

In this context, we have even more:

Theorem 6.3.19 (Lempert). *Suppose a compact, strongly pseudoconvex CR manifold* Y, $\dim_{\mathbb{R}} Y \geqslant 3$, *bounds a strongly pseudoconvex Stein space (or, equivalently, a strongly pseudoconvex complex manifold). Then* Y *can be realized as a smooth real hypersurface in a complex projective manifold that* Y *divides into a strongly pseudoconvex and a strongly pseudoconcave part.*

The main ingredient is the following Nash-type approximation result, which will be also useful later.

Theorem 6.3.20 (Lempert). *Assume a reduced Stein space* X *has only isolated singularities, and* $K \subset X$ *is a compact subset. Then there are an affine algebraic variety* V, *and a neighborhood of* K *in* X *that is biholomorphic to an open set in* V.

It is thus interesting to have the following version of the Kodaira embedding theorem for 1-concave manifolds.

Theorem 6.3.21 (Andreotti–Tomassini). *Let X be a 1-concave manifold and let L be a holomorphic line bundle on X such that the ring $\mathscr{A}(X, L) = \oplus_{p \geqslant 0} H^0(X, L^p)$ gives local coordinates and separates points everywhere on X. Then X is biholomorphic to an open set of a projective algebraic variety.*

We finish with a result due to Epstein–Henkin.

Definition 6.3.22. A compact CR-hypersurface Y_0 is called *strictly CR-cobordant* to a compact CR-hypersurface Y_1 if there exists a complex space $\widetilde{X}$ with at most isolated singularities and a $\mathscr{C}^\infty$-strictly plurisubharmonic function ρ with at most isolated critical points on $\widetilde{X}$ such that the set $X = \{x \in \widetilde{X} : 0 < \rho(x) < 1\}$ is a relatively compact, complex subspace in $\widetilde{X}$ and $\partial X = Y_1 \cup Y_0$.

Theorem 6.3.23 (Epstein–Henkin). *Let Y_1 be an embeddable strictly pseudoconvex CR-hypersurface. Then any (not necessarily smooth) CR-hypersurface Y_0, strictly cobordant to Y_1, is also embeddable.*

6.3.2 The compactification theorem

Our goal is to prove the following result.

Theorem 6.3.24. *Let X be a connected complex manifold with compact strongly pseudoconvex boundary and of complex dimension $n \geqslant 2$. Assume that $\mathrm{Int}(X)$ is endowed with a complete Kähler metric g^{TX} with pinched negative curvature. The following assertions are equivalent:*

(a) *∂X is CR embeddable in some $\mathbb{C}^m$.*

(b) *X has finite volume away from a neighborhood of ∂X.*

If one of the equivalent conditions (a) or (b) holds true, there exists a compact strongly pseudoconvex domain M_1 in a smooth projective variety and an embedding $\sigma : X \longrightarrow M_1$ which is a biholomorphism between $\mathrm{Int}(X)$ and $\sigma(\mathrm{Int}(X))$, $\sigma(\partial X) = \partial M_1$, and $M_1 \setminus \sigma(X)$ is an exceptional analytic set which can be blown down to a finite set of singular points.

Proof. (a) $\Rightarrow$ (b). By Proposition 6.3.14, there exists a defining function $\varphi : X \to]-\infty, c]$ such that $\partial X = \{\varphi = c\}$, with the properties: (1) its Levi form is positive definite on the holomorphic tangent space of ∂X and (2) φ is strictly plurisubharmonic on $\{c_0 < \varphi < c\}$.

It follows that $\partial X = \{\varphi = c\}$ and $\{\varphi = d\}$ are strictly CR-cobordant, for $d \in]c_0, c[$. By assumption, $\partial X = \{\varphi = c\}$ is embeddable. We infer from Theorem 6.3.23 that also $\{\varphi = d\}$ is embeddable for $d \in]c_0, c[$. The Harvey–Lawson theorem 6.3.16 asserts there exists a normal Stein space with boundary and at most isolated singularities $S \subset \mathbb{C}^m$, such that $\partial S = \{\varphi = d\}$. We compactify the strip $\{d < \varphi <$

c} to the Stein space S, which can be realized as a Stein domain with boundary in a bigger Stein space S', by Theorem 6.3.15.

Lempert's approximation theorem 6.3.20 allows us to assume that the Stein space S' is an open set in an affine algebraic variety, hence also in a projective variety Y. Let $\delta > 0$ be sufficiently small. We set $W = \{d - \delta < \varphi \leqslant c\}$ and glue the manifolds X and $(Y \smallsetminus S) \cup W$ along W. The resulting manifold will be denoted by $\widehat{X}$. We have glued thus a pseudoconcave cap on the pseudoconvex end of X. The manifold $\widehat{X}$ contains a 1-concave open set and is thus Andreotti pseudoconcave. Hence by Theorem 3.4.5, there exists $C_1 > 0$ such that

$$\dim H^0(\widehat{X}, K_{\widehat{X}}^p) \leqslant C_1 p^n, \quad \text{for any } p \in \mathbb{N}^*. \tag{6.3.5}$$

A partition of unity argument delivers a Hermitian metric $g^{T\widehat{X}}$ on $\widehat{X}$ which agrees with the original metric g^{TX} of $\mathrm{Int}(X)$ on say $\{\varphi < \varepsilon\}$ with $\varepsilon < c$. We endow the canonical bundle $K_{\widehat{X}}$ with the induced metric. Since g^{TX} has pinched negative curvature, we deduce that there exists $C > 0$ such that on $\{\varphi < \varepsilon\}$,

$$\sqrt{-1}R^{K_{\widehat{X}}} = \sqrt{-1}R^{K_X} = -\sqrt{-1}R^{\det} \geqslant C\omega = C\widehat{\omega}, \tag{6.3.6}$$

where ω, $\widehat{\omega}$ are the associated $(1,1)$-forms to g^{TX}, $g^{T\widehat{X}}$. This means that $K_{\widehat{X}}$ is uniformly positive at infinity and we can apply the Morse inequalities (3.3.8) (or (3.3.9)) for $L = K_{\widehat{X}}$. We obtain that

$$\dim H^0(\widehat{X}, K_{\widehat{X}}^p) \geqslant \frac{p^n}{n!} \int_{\widehat{X}(\leqslant 1)} \left(\frac{\sqrt{-1}}{2\pi} R^{K_{\widehat{X}}}\right)^n + o(p^n), \quad p \to \infty. \tag{6.3.7}$$

From (6.3.5) and (6.3.7), we get

$$\int_{\widehat{X}(\leqslant 1)} \left(\frac{\sqrt{-1}}{2\pi} R^{K_{\widehat{X}}}\right)^n = \left(\int_{\{\varphi < \varepsilon\}} + \int_{\{\varphi > \varepsilon\}(\leqslant 1)}\right) \left(\frac{\sqrt{-1}}{2\pi} R^{K_{\widehat{X}}}\right)^n < +\infty.$$

The second term in the sum is an integral over a relatively compact set, therefore finite. We infer that the first term is also finite and from (6.3.6)

$$\int_{\{\varphi < \varepsilon\}} \frac{\omega^n}{n!} \leqslant \frac{C^{-n}}{n!} \int_{\{\varphi < \varepsilon\}} (\sqrt{-1}R^{K_{\widehat{X}}})^n < +\infty.$$

This means precisely that (b) is satisfied.

(b) $\Rightarrow$ (a). We show that all the ends of X, with the exception of the end corresponding to ∂X, are hyperconcave. Let U be a neighborhood of ∂X. Since $X \smallsetminus U$ has pinched negative curvature and finite volume, $X \smallsetminus U$ has finitely many ends, by a result of Eberlein. Let $N_1, \ldots, N_m$ be the cusps of $X \smallsetminus U$. We fix some end N_j. By Theorem 6.3.8, N_j is hyperconcave, so from the compactification Theorem 6.3.9, we deduce that there exists a Stein space S with boundary and an embedding of X as an open set in S, such that $\partial S = \partial X$.

Using Theorem 6.3.15, the space S can be embedded as a domain with boundary in a larger Stein space S' such that ∂S is a hypersurface in S'. By the Remmert–Bishop–Narasimhan embedding theorem B.3.6, S' admits a proper holomorphic embedding in $\mathbb{C}^q$ for some q. Restricting this embedding to $\partial S = \partial X$, we obtain (a).

We have thus proved that (a) $\Leftrightarrow$ (b). We assume now that (b) holds true and show that X can be compactified to a strongly pseudoconvex domain in a projective variety by adding an exceptional analytic set. Note that by applying the compactification theorem 6.3.9, we can deduce this assertion for some strongly pseudoconvex domain W, but we cannot say directly that this domain is an open set of a projective manifold. Therefore we use the manifold $\widehat{X}$ constructed in the proof of (a) $\Rightarrow$ (b), as well as the notation therein.

Since S' is an affine space in some $\mathbb{C}^q$, we can regard the embedding of $W' = \{d - \delta < \varphi < d\}$ in Y as a map with values in $\mathbb{C}^q$. Now X can be compactified to a compact strongly pseudoconvex domain, so the Kohn–Rossi extension theorem 6.3.12 , applied to the components of this embedding, shows that the embedding extends to a holomorphic map from X to $\mathbb{C}^q \subset \mathbb{C}\mathbb{P}^q$. It follows that the inclusion $\Psi : Y \hookrightarrow \mathbb{C}\mathbb{P}^q$ extends to a holomorphic map $\Psi : \widehat{X} \hookrightarrow \mathbb{C}\mathbb{P}^q$ which is an embedding on $(X \smallsetminus S') \cup W'$.

Pulling back the hyperplane line bundle of $\mathbb{C}\mathbb{P}^q$ through this map, we obtain a line bundle $F := \Psi^* \mathcal{O}_{\mathbb{C}\mathbb{P}^q}(1) \to \widehat{X}$ which is semi-positive on $\widehat{X}$ and positive on $(X \smallsetminus S') \cup W'$. Hence, the bundle $L = F^k \otimes K_{\widehat{X}}$ satisfies $\sqrt{-1}R^L > \varepsilon \omega$ (for some $\varepsilon > 0$) on $\widehat{X}$ for k sufficiently large. Moreover, the remaining conditions of (6.1.1) are fulfilled for $\widehat{\omega}$ (here E is trivial). Corollary 6.1.3 shows that the graded ring $\oplus_{p \geqslant 0} H^0(\widehat{X}, L^p)$ separates points and gives local coordinates on $\widehat{X}$.

On the other hand the manifold $\widehat{X}$ is hyper 1-concave (use again Theorem 6.3.8); we denote the exhaustion function by r. By Theorem 6.3.21, we find a smooth compact manifold $\widetilde{X} \subset \mathbb{C}\mathbb{P}^m$, such that $X \subset \widehat{X} \subset \widetilde{X}$ as open sets. The desired projective strongly pseudoconvex domain is $M = \widetilde{X} \smallsetminus (X \smallsetminus S')$. Theorem B.3.5 shows that there exists a Stein space M' and a Remmert reduction $\pi : M \to M'$, which blows down the exceptional set A of M to a discrete set of points. The set $M \smallsetminus X = \widetilde{X} \smallsetminus X$ is a pluripolar set, namely the set where the plurisubharmonic function r takes the value $-\infty$. By the maximum principle for plurisubharmonic functions, $A \subset M \smallsetminus X$. The function $\varphi = r \circ \pi^{-1} : M' \smallsetminus \pi(M \smallsetminus X) \to]-\infty, \infty[$ is strictly plurisubharmonic and proper.

By Theorem 6.3.7, the universal covering of X is Stein. We may thus apply Theorem 6.3.9 (b) for the hyperconcave end $M' \smallsetminus \pi(M \smallsetminus X)$ and deduce that the singular set M'_{sing} of M' equals $\pi(M \smallsetminus X)$.

Therefore $M \smallsetminus X = A$, so $M \smallsetminus X$ is the exceptional analytic set of M and by blowing down this exceptional set, we obtain M'_{sing}. Actually, each end $N_1, \ldots, N_m$ of M can be compactified with one point of the singular set $M'_{\mathrm{sing}} = \{x_1, \ldots, x_m\}$.

Moreover, by the uniqueness of the Stein completion from Theorem 6.3.4 (a), we see that M' and S' coincide.   □

The first application of Theorem 6.3.24 is the classical theorem of Siu–Yau, which is the particular case when $\partial X = \emptyset$. This provides a geometric proof of the Satake-Baily-Borel compactification of arithmetic quotients of rank 1.

Theorem 6.3.25 (Siu–Yau). *Let X be a complete Kähler manifold of finite volume and negative pinched sectional curvature. If $\dim X \geqslant 2$, X is biholomorphic to a quasiprojective manifold which can be compactified to a Moishezon space by adding finitely many singular points.*

Note that our proof of Theorem 6.3.25 is different from the original one in its second part. Siu and Yau use the Schwarz-Pick lemma of Yau, whereas we use Theorem 6.3.9, where one tool is Wermer's theorem.

As a second application, we study some quotients of the unit complex ball $\mathbb{B}^n$ in $\mathbb{C}^n$ which where considered by Burns and Napier–Ramachandran.

Corollary 6.3.26. *Let Γ be a torsion-free discrete group of automorphisms of the unit ball $\mathbb{B}^n$ in $\mathbb{C}^n$, $n \geqslant 2$, and let $X = \mathbb{B}^n/\Gamma$. Assume that the limit set Λ is a proper subset of $\partial\mathbb{B}^n$ and that the quotient $(\partial\mathbb{B}^n \setminus \Lambda)/\Gamma$ has a compact component Y. Let N be the end of X corresponding to Y. Then the following assertions are equivalent:*

(a) *$X \setminus N$ has finite volume.*
(b) *Y is CR embeddable in some $\mathbb{C}^m$.*

If one of (a) or (b) holds, X can be compactified to a strongly pseudoconvex domain in a projective variety by adding an exceptional analytic set.

Proof. As is well known, the *limit set* Λ is the set of accumulation points of any orbit $\Gamma \cdot x$, $x \in \mathbb{B}^n$, and is a closed Γ-invariant subset of the sphere at infinity $\partial\mathbb{B}^n$.

Lemma 6.3.27. *The complement $\partial\mathbb{B}^n \setminus \Lambda$ is precisely the set of points at which Γ acts properly discontinuously, and the space $X \cup (\partial\mathbb{B}^n \setminus \Lambda)/\Gamma$ is a manifold with boundary $(\partial\mathbb{B}^n \setminus \Lambda)/\Gamma$.*

Y is a compact subset of this boundary, hence there is a neighborhood N of Y in X which is diffeomorphic to the product $Y \times]0, 1[$. It follows that N is an end of X, because Y is compact and connected. Actually, N is a strongly pseudoconvex end, in the sense that its boundary Y at infinity is strictly pseudoconvex. Since $X = \mathbb{B}^n/\Gamma$ is a complete manifold with sectional curvature pinched between -4 and -1, Corollary 6.3.26 is an immediate consequence of Theorem 6.3.24.   □

Corollary 6.3.28 (Burns, Napier–Ramachandran). *Let Γ be a torsion-free discrete group of automorphisms of the unit ball $\mathbb{B}^n$ in $\mathbb{C}^n$ with $n \geqslant 3$ and let $X = \mathbb{B}^n/\Gamma$. Assume that the limit set Λ is a proper subset of $\partial\mathbb{B}^n$ and that the quotient $(\partial\mathbb{B}^n \setminus \Lambda)/\Gamma$ has a compact component Y. Then X has only finitely many ends, all of which, except for the unique end corresponding to Y, are cusps.*

Proof. Indeed, since Y is a strongly pseudoconvex manifold of dimension at least 5, it follows from Boutet de Monvel's theorem 6.3.17 that Y is embeddable. By Theorem 6.3.24, X has finite volume away from Y and finitely many ends, all of which, except for the unique end corresponding to Y, have finite volume, i.e., are cusps.   □

6.4 Weak Lefschetz theorems

Using Morse theory, Andreotti–Frankel and Bott gave the following formulation of the classical Lefschetz hyperplane theorem.

Theorem 6.4.1. *If Y is a smooth hypersurface in a compact projective manifold X of dimension n, such that the associated line bundle $\mathscr{O}_X(Y)$ is positive, then X is obtained from Y by attaching cells of real dimension $\geqslant n$. In particular, the natural map $\pi_1(Y) \to \pi_1(X)$ is surjective for $n \geqslant 2$ and an isomorphism for $n \geqslant 3$.*

Nori showed further:

Theorem 6.4.2. *If X and Y are connected projective manifolds with $\dim X = \dim Y + 1 > 1$ and $\iota : Y \to X$ is a holomorphic immersion with ample normal bundle, then the image of $\pi_1(Y) \to \pi_1(X)$ is of finite index.*

Napier and Ramachandran proved a generalization of Nori's Lefschetz type theorem by removing the codimension one condition. They use the L^2 estimates for $\bar\partial$ on complete Kähler manifolds to separate the sheets of appropriate coverings.

In the sequel, we use the Bergman kernel and consider not necessarily Kähler manifolds (see also Theorem 6.1.1).

First we introduce the notion of formal completion. Let Y be a complex analytic subspace of the complex manifold W and denote by $\mathcal{I}_Y$ the ideal sheaf of Y. The *formal completion* $\widehat{W}$ of W with respect to Y is the ringed space $(\widehat{W}, \mathscr{O}_{\widehat{W}}) = (Y, \operatorname{proj\,lim}_\nu \mathscr{O}_W/\mathcal{I}_Y^\nu)$. If $\mathcal{F}$ is an analytic sheaf on W, we denote by $\widehat{\mathcal{F}}$ the sheaf $\widehat{\mathcal{F}} = \operatorname{proj\,lim}_\nu \mathcal{F} \otimes (\mathscr{O}_W/\mathcal{I}_Y^\nu)$. If $\mathcal{F}$ is coherent then $\widehat{\mathcal{F}}$ is too.

Lemma 6.4.3. *The kernel of the canonical restriction mapping $H^0(W, \mathcal{F}) \longrightarrow H^0(\widehat{W}, \widehat{\mathcal{F}})$ consists of the sections of $\mathcal{F}$ which vanish on a neighborhood of Y. Hence for locally free $\mathcal{F}$, the map is injective.*

Theorem 6.4.4. *Let (X, Θ) be a complete connected Kähler manifold and let (L, h^L) be a holomorphic Hermitian line bundle such that for some $\varepsilon > 0$ we have $\sqrt{-1}R^L \geqslant \varepsilon\,\Theta$. Let moreover $Y \subset X$ be a connected compact complex subspace of X such that for any p,*

$$\dim H^0(\widehat{X}, \widehat{\mathcal{F}}_p) < \infty, \quad \text{for } \mathscr{F}_p = \mathscr{O}_X(L^p \otimes K_X). \tag{6.4.1}$$

Let $i : Y \hookrightarrow X$ be the inclusion and $G = i_*(\pi_1(Y))$ be the image of $\pi_1(Y)$ in $\pi_1(X)$. Then G is of finite index in $\pi_1(X)$.

Proof. We follow the strategy of Napier–Ramachandran. There exists a connected covering $\pi : \widetilde{X}_G \longrightarrow X$ such that $\pi_*(\pi_1(\widetilde{X}_G)) = G$ with degree $k = [\pi_1(X) : G]$. We fix a point $x \in X$ and neighborhood U of x such that $\widetilde{U} := \pi^{-1}(U)$ is a disjoint union of open sets $(U_i)_{1 \leqslant i \leqslant k}$ and $\pi|_{U_i} : U_i \to U$ is biholomorphic.

Let $(\widetilde{L}, h^{\widetilde{L}}) = \pi^*(L, h^L)$ and $\widetilde{\Theta} = \pi^*\Theta$. Then $\sqrt{-1}R^{\widetilde{L}} \geqslant \varepsilon\widetilde{\Theta}$ on $\widetilde{X}_G$. As usual set $n = \dim X$. By the holomorphic Morse inequalities (6.1.14) for $E = K_{\widetilde{X}}$, we have

$$\liminf_{p \longrightarrow \infty} \frac{n!}{p^n} \dim H_{(2)}^{n,0}(\widetilde{X}_G, \widetilde{L}^p) \geqslant \int_{\widetilde{X}_G} \left(\tfrac{\sqrt{-1}}{2\pi}R^{\widetilde{L}}\right)^n \geqslant \int_{\widetilde{U}} \left(\tfrac{\sqrt{-1}}{2\pi}R^{\widetilde{L}}\right)^n$$
$$= k \int_U \left(\tfrac{\sqrt{-1}}{2\pi}R^L\right)^n. \quad (6.4.2)$$

Therefore, for large p, we have $\dim H_{(2)}^{n,0}(\widetilde{X}_G, \widetilde{L}^p) \geqslant k$.

Let us choose a small open connected neighborhood V of Y such that $\pi_1(Y) \longrightarrow \pi_1(V)$ is an isomorphism; so the image of $\pi_1(V)$ in $\pi_1(X)$ is G. Hence, if we denote by $\jmath$ the inclusion of V in X, there exists a holomorphic lifting $\widetilde{\jmath} : V \longrightarrow \widetilde{X}, \pi \circ \widetilde{\jmath} = \jmath$. Since $\widetilde{\jmath}$ is locally biholomorphic, the pull-back map $\widetilde{\jmath}^* : H_{(2)}^{n,0}(\widetilde{X}_G, \widetilde{L}^p) \longrightarrow H^{n,0}(V, L^p)$ is injective. On the other hand, Lemma 6.4.3 delivers an injective map

$$H^0(V, \mathcal{F}_p) \hookrightarrow H^0(\widehat{V}, \widehat{\mathcal{F}}_p) = H^0(\widehat{X}, \widehat{\mathcal{F}}_p).$$

By hypothesis, the latter space is finite-dimensional so $k \leqslant \dim H_{(2)}^{n,0}(\widetilde{X}_G, \widetilde{L}^p) < \infty$. This finishes the proof. $\qquad\square$

Remark 6.4.5. (a) By a theorem of Grothendieck, the hypothesis (6.4.1) of Theorem 6.4.4 is fulfilled if Y is locally a complete intersection with ample normal bundle $N_{Y/X}$ (if Y is smooth cf. Definition B.3.12).

(b) We can replace (6.4.1) in Theorem 6.4.4 with the requirement that Y has a fundamental system of Andreotti pseudoconcave neighborhoods $\{V\}$. Then $\dim H^0(V, \mathcal{F}_p)$ is finite by Theorem 3.4.5. This happens for example if Y is a smooth hypersurface and the normal bundle $N_{Y/X}$ of Y in X admits a Hermitian metric $h^{N_{Y/X}}$ such that $\dot{R}^{N_{Y/X}}$ has at least one positive eigenvalue.

(c) If X contains a simply connected subvariety satisfying either (a) or (b), $\pi_1(X)$ is finite.

(d) We can slightly generalize Theorem 6.4.4, by assuming that Y is an analytic subset of a manifold V and there exists locally biholomorphic map $\psi : V \to X$, such that $\dim H^0(\widehat{V}, \widehat{\mathcal{F}}_p) < \infty$ holds for any $p \geqslant 1$, where $\mathcal{F}_p = \mathcal{O}_V(\psi^*L^p \otimes K_V)$. Then the image of the induced map $\pi_1(Y) \to \pi_1(X)$ has finite index in $\pi_1(X)$. The proof is the same as of Theorem 6.4.4, but we use the map ψ instead of the inclusion $\jmath$.

Corollary 6.4.6. *Let (X, Θ) be an n-dimensional complete Hermitian manifold and let (L, h^L) be a holomorphic Hermitian line bundle such that $\sqrt{-1}R^L > C\Theta$, with $C > 0$. Let $Y \subset X$ be a connected compact analytic subspace with $\dim Y \geqslant 1$. Suppose Y is locally a complete intersection and $N_{Y/X}$ is ample. Then the image G of $\pi_1(Y)$ in $\pi_1(X)$ has finite index in $\pi_1(X)$.*

Proof. Apply Theorem 6.4.4 and Remark 6.4.5. $\square$

Corollary 6.4.7. *Let X be a Zariski open set in a compact normal Moishezon space $\overline{X}$. Let $Y \subset X_{\mathrm{reg}}$ be a connected compact analytic subspace with $\dim Y \geqslant 1$. Suppose Y is locally a complete intersection and $N_{Y/X}$ is ample. Let G be the image of $\pi_1(Y)$ in $\pi_1(X)$. Then G has finite index in $\pi_1(X)$.*

Proof. Since $\overline{X}$ is normal, we have a surjective morphism $\pi_1(X_{\mathrm{reg}}) \longrightarrow \pi_1(X)$, so we can replace X with X_{reg}. We can thus resolve the singularities of $\overline{X}$ and assume that it is a manifold. We consider on $\overline{X}$ a singular Hermitian positive line bundle L. We modify then the proof of Theorem 6.2.3 in the following way. First we consider the singular support Σ of $\sqrt{-1}R^L$ and construct the generalized Poincaré metric on $X \smallsetminus \Sigma$. Then we consider a covering $\pi_\Gamma : \widetilde{X} \longrightarrow X$ such that $\pi_{\Gamma,*}(\pi_1(\widetilde{X})) = G$ with degree $k = [\pi_1(X) : G]$.

Then we apply the covering version of Theorem 6.2.3 on the covering $\widetilde{X} \smallsetminus \pi_\Gamma^{-1}(\Sigma)$ of $X \smallsetminus \Sigma$. We obtain in this way $(n, 0)$-forms on $\widetilde{X} \smallsetminus \pi_\Gamma^{-1}(\Sigma)$ which are L^2 with respect to the pull-back of the Poincaré metric on $\widetilde{X} \smallsetminus \pi_\Gamma^{-1}(\Sigma)$ and a metric on $\pi_\Gamma^{-1}(L)$ over $\widetilde{X} \smallsetminus \pi_\Gamma^{-1}(\Sigma)$ which is bounded below by a smooth metric on $\widetilde{X}$. But for $(n, 0)$-forms the L^2 condition does not depend on the metric on the base manifold (cf. the proof of Theorem 3.3.5 (i)), so we can take the L^2 condition with respect to smooth metrics on $\widetilde{X}$ and $\pi_\Gamma^{-1}(L)$. Hence these sections extend to $\widetilde{X}$ and we can apply the proof of Theorem 6.4.4. $\square$

Problems

Problem 6.1 (Theorem 6.1.1 for bounded geometry). Show that if in addition h^L, h^E, Θ, g^{TX} and their derivatives of order $\leqslant 2n+m+4$ (resp. $\leqslant 2n+m+m'+k+5$) are uniformly bounded on X in the norm induced by g^{TX}, then Proposition 4.1.5 (resp. Theorem 4.2.1) holds uniformly for $x, x' \in X$ (resp. $x_0 \in X$) with the norm $\mathscr{C}^k$. (Hint: Show that the injectivity radius of X is bounded by below by a positive constant and as in in the proof of Theorem 6.1.4, the constant in the Sobolev embedding theorem is uniformly bounded on X.)

Problem 6.2. Let X be a compact connected complex manifold. Assume that X admits a closed integral current ω with singular support contained in a proper analytic set Σ such that ω is positive in a neighborhood of Σ and $\int_{X(\leqslant 1,\omega)} \omega_{\mathrm{ac}}^n > 0$.

(a) Consider a singular Hermitian holomorphic line bundle (L, h^L) such that $\omega = c_1(L, h^L)$ (Lemma 2.3.5). Introduce on $X \smallsetminus \Sigma$ the metric (6.2.2) and on

$L|_{X \smallsetminus \Sigma}$ the metric (6.2.16a). Using that (L, h^L) has positive curvature near Σ, show that the fundamental estimate (3.2.2) holds for $(0, 1)$-forms on $X \smallsetminus \Sigma$. Apply Theorem 3.2.16 and deduce that $\dim H^0_{(2)}(X \smallsetminus \Sigma, L^p) \geqslant Cp^n$ for some $C > 0$ and p large enough, where $H^0_{(2)}(X \smallsetminus \Sigma, L^p)$ is defined in (6.2.16b).

(b) Infer that L is big and X is Moishezon.

This is a proof of (6) $\Rightarrow$ (1) in Theorems 2.3.28 and 2.3.30.

Problem 6.3. Let X be a connected complex manifold which is pseudoconcave in the sense of Andreotti. Assume that X is embedded in a projective space $\mathbb{CP}^n$. Using the same reasoning from Problem 5.5 deduce that X is contained in a projective variety of the same dimension as X.

6.5 Bibliographic notes

Sections 6.1 and 6.2 appear in [161]. The generalized Poincaré metric appeared in [59, §2]. See also [113, 114].

Section 6.3. Theorem 6.3.4 is due to Rossi [207, Th. 3, p. 245]. Andreotti–Siu [5, Prop. 3.2] improved the result in different directions, e.g., they showed that it holds for normal complex spaces. The uniqueness result comes from [5, Cor. 3.2]. Example 6.3.5 appeared in Rossi [207, p. 252–256] (being attributed to Andreotti), Andreotti–Siu [5, p. 262–270] (where credit is given to Grauert) and Grauert [117, p. 273].

The study of compactifications of manifolds with negative curvature was initiated in the paper of Siu–Yau [228], where Theorem 6.3.25 appeared. For more results on the compactification of complete Kähler–Einstein manifolds of finite volume and bounded curvature we refer to Mok [178] and the references therein.

Theorem 6.3.9 was proved in [172]. In this connection, note a result of Colţoiu-Tibăr [68], asserting that the universal cover of a small punctured neighborhood of an isolated singularity of dimension 2 is Stein, whenever the fundamental group of the link is infinite. The result from Example 6.3.3 is proved in [67].

Kohn–Rossi's theorem 6.3.12 stems from [145]. Theorem 6.3.15 is due to Heunemann [127, Theorem 0.2] (see also Ohsawa [188]). The Harvey–Lawson theorem 6.3.16 appeared in [125]. The embeddability theorem of Boutet de Monvel 6.3.17 comes from [51, p. 5]. A straightforward argument that the boundary of the Grauert–Andreotti–Rossi example is not embeddable can be found in Burns [57]. The author shows that the CR functions on S^3 equipped with the induced CR structure from the complex structure of Example 6.3.5 are equal at antipodal points. Therefore, CR functions cannot embed this structure in the Euclidean space. For Theorem 6.3.18, we refer to [53, 125, 57, 144].

L. Lempert introduced the idea of linking the deformations of the CR structures on Y to the deformations of the complex structure on a strongly pseudoconcave manifold Z bounding Y and proved in this context Theorems 6.3.19 and 6.3.20 in [151, 152, 153].

The result of Problem 6.3 was proved in [2, Théorème 6] and is a generalization of Chow's theorem from Problems 5.4, 5.5. Proofs of Theorem 6.3.21 are in [6, Theorem 3, p.97], [5, Theorem 4.1] (a generalization for complex normal spaces) and [181, Lemma 2.1]. For Theorem 6.3.23, we refer to [104, Theorem 2].

The use of holomorphic Morse inequalities in compactification questions appears in Nadel–Tsuji [181] and Napier–Ramachandran [183].

Theorem 6.3.24 and Corollary 6.3.26 appeared in [174]. The proof of the implication (a) ⇒ (b) in Theorem 6.3.24 follows [183]. The result of Eberlein alluded to in the proof of (a) ⇒ (b) can be found in [100]. Corollary 6.3.28 is [183, Theorem 4.2]. For Lemma 6.3.27 see [101, §10]. Wu's result (Theorem 6.3.7) was obtained in [255].

Section 6.4. The classical Lefschetz hyperplane theorem was proved by Lefschetz [150]. Andreotti–Frankel and Bott gave Morse theoretical proofs [3, 46, 176]. The weak Lefschetz theorems were studied by Nori [186], Campana [58], Kollár [147], Napier–Ramachandran [183]. The approach using the holomorphic Morse theory on covering stems from [244], from where the results of Section 6.4 are taken. Lemma 6.4.3 is [56, Prop. VI.2.7]; in [56] the reader can find out more about formal completions. For the definition of the ampleness of $N_{Y/X}$ for non-smooth Y, see [123, 124]. For the morphism $\pi_1(X_{\text{reg}}) \longrightarrow \pi_1(X)$, for a normal Zariski open set, see [110]. There are many references about Lefschetz theorems in algebraic geometry. We refer to Grothendieck [121], Fulton–Lazarsfeld [110] and Lazarsfeld [149].

Chapter 7

Toeplitz Operators

We show in this chapter how the asymptotic expansion of the Bergman kernel implies the semi-classical properties of Toeplitz operators acting on high tensor powers of a positive line bundle over a compact manifold. In particular we obtain a construction of a star-product (a deformation quantization) using this technique. Moreover, our approach works with some modifications on non-compact and symplectic manifolds.

This chapter is organized as follows: In Section 7.1, we explain the formal calculus on $\mathbb{C}^n$ for our model operator $\mathscr{L}$. In Section 7.2, we establish the asymptotic expansion for the kernel of Toeplitz operators. In Section 7.3, we establish that the asymptotic expansion is also a sufficient condition for a family of operators to be Toeplitz. In Section 7.4, we conclude finally that the Toeplitz operators form an algebra. In Section 7.5, we extend it to the non-compact case.

7.1 Kernel calculus on $\mathbb{C}^n$

In this section, we state the properties of the calculus of the kernels $(F\mathscr{P})(Z, Z')$, where $F \in \mathbb{C}[Z, Z']$ and $\mathscr{P}(Z, Z')$ is the kernel of the projection on the null space of the model operator $\mathscr{L}$. This calculus is the main ingredient of our approach.

Let us consider the canonical coordinates $(Z_1, \ldots, Z_{2n})$ on the real vector space $\mathbb{R}^{2n}$. On the complex vector space $\mathbb{C}^n$ we consider the complex coordinates $(z_1, \ldots, z_n)$. The two sets of coordinates are linked by the relation $z_j = Z_{2j-1} + \sqrt{-1}Z_{2j}$, $j = 1, \ldots, n$.

Let $0 < a_1 \leqslant a_2 \leqslant \cdots \leqslant a_n$. Following (4.1.74) and (4.1.76), we define the differential operators:

$$b_i = -2\frac{\partial}{\partial z_i} + \frac{1}{2}a_i\bar{z}_i, \quad b_i^+ = 2\frac{\partial}{\partial \bar{z}_i} + \frac{1}{2}a_i z_i,$$

$$b = (b_1, \ldots, b_n), \quad \mathscr{L} = \sum_i b_i b_i^+.$$

$$(7.1.1)$$

On $\mathbb{R}^{2n}$, we consider the L^2-norm $\| \cdot \|_{L^2} = \left(\int_{\mathbb{R}^{2n}} | \cdot |^2 \, dZ \right)^{1/2}$ where dZ is the standard Euclidean volume form. $\mathscr{L}$ acts as a densely defined self-adjoint operator on $(L^2(\mathbb{R}^{2n}), \| \cdot \|_{L^2})$ and its spectrum was determined in Theorem 4.1.20. From there we deduced the formula for the kernel of the orthogonal projection

$$\mathscr{P} : L^2(\mathbb{R}^{2n}) \longrightarrow \mathrm{Ker}(\mathscr{L}) = \mathrm{span}\Big\{ z^\beta \exp\Big(-\frac{1}{4} \sum_{j=1}^n a_j |z_j|^2 \Big), \ \beta \in \mathbb{N}^n \Big\}. \quad (7.1.2)$$

Namely, the kernel of $\mathscr{P}$ with respect to dZ is (cf. (4.1.84))

$$\mathscr{P}(Z, Z') = \prod_{i=1}^n \frac{a_i}{2\pi} \ \exp\Big(-\frac{1}{4} \sum_i a_i \big(|z_i|^2 + |z_i'|^2 - 2 z_i \bar{z}_i' \big) \Big). \quad (7.1.3)$$

We will repeat the use of Theorem 4.1.20 when we meet the operator $\mathscr{P}$, instead of doing the direct computation by using (7.1.3). This point of view will help us to simplify a lot of computations and to understand better our operations. As an example, if $\varphi(Z) = b^\alpha z^\beta \exp\big(-\frac{1}{4} \sum_{j=1}^n a_j |z_j|^2 \big)$ with $\alpha, \beta \in \mathbb{N}^n$, then

$$(\mathscr{P}\varphi)(Z) = \begin{cases} z^\beta \exp\Big(-\dfrac{1}{4} \sum_{j=1}^n a_j |z_j|^2 \Big) & \text{if } |\alpha| = 0, \\[4mm] 0 & \text{if } |\alpha| > 0. \end{cases} \quad (7.1.4)$$

In what follows, all operators are defined by their kernels with respect to dZ. In this way, if F is a polynomial on Z, Z', then $F\mathscr{P}$ is an operator on $L^2(\mathbb{R}^{2n})$ with kernel $F(Z, Z')\mathscr{P}(Z, Z')$ with respect to dZ.

We will add a subscript z or z' when we need to specify that the operator is acting on the variables Z or Z'.

Lemma 7.1.1. *For any polynomial* $F(Z, Z') \in \mathbb{C}[Z, Z']$, *there exist polynomials* $F_\alpha \in \mathbb{C}[z, Z']$ *and* $F_{\alpha,0} \in \mathbb{C}[z, \bar{z}']$, $(\alpha \in \mathbb{N}^n)$ *such that*

$$(F\mathscr{P})(Z, Z') = \sum_\alpha b_z^\alpha (F_\alpha \mathscr{P})(Z, Z'),$$

$$((F\mathscr{P}) \circ \mathscr{P})(Z, Z') = \sum_\alpha b_z^\alpha F_{\alpha,0}(z, \bar{z}') \mathscr{P}(Z, Z'). \quad (7.1.5)$$

Moreover, $|\alpha| + \deg F_\alpha$, $|\alpha| + \deg F_{\alpha,0}$ *have the same parity with the degree of* F *in* Z, Z'. *In particular,* $F_{0,0}(z, \bar{z}')$ *is a polynomial in* $z, \bar{z}'$ *and its degree has the same parity with* $\deg F$.

For any polynomials $F, G \in \mathbb{C}[Z, Z']$ *there exist polynomials* $\mathscr{K}[F, G] \in \mathbb{C}[Z, Z']$ *such that*

$$((F\mathscr{P}) \circ (G\mathscr{P}))(Z, Z') = \mathscr{K}[F, G](Z, Z')\mathscr{P}(Z, Z'). \quad (7.1.6)$$

Proof. Note that from (7.1.1) and (7.1.2), we get

$$b_{j,z}\,\mathscr{P}(Z,Z') = a_j(\overline{z}_j - \overline{z}'_j)\mathscr{P}(Z,Z'),$$

$$[g(z,\overline{z}), b_{j,z}] = 2\frac{\partial}{\partial z_j}g(z,\overline{z}),$$

(7.1.7)

for any polynomial $g(z,\overline{z}) \in \mathbb{C}[z,\overline{z}]$. Using repeatedly (7.1.7), we can replace $\overline{z}$ in $F(Z,Z')$ by a combination of $b_{j,z}$ and $\overline{z}'$ and the first equation of (7.1.5) follows.

From (7.1.4) and the first equation of (7.1.5), we get

$$(\mathscr{P} \circ (F\mathscr{P}))(Z,Z') = (F_0\mathscr{P})(Z,Z'),$$

(7.1.8)

with $F_0 \in \mathbb{C}[z,Z']$. When we take the adjoint of (7.1.8) for $\overline{F}$, as $\mathscr{P}$ is self-adjoint, we get $((F\mathscr{P}) \circ \mathscr{P})(Z,Z') = F'(Z,\overline{z}')\mathscr{P}(Z,Z')$, and F' is a polynomial on $Z,\overline{z}'$. Now using again the first equation of (7.1.5), we get the second equation of (7.1.5).

From (7.1.5), we get (7.1.6). $\qquad\square$

As an example of how Lemma 7.1.1 works, we calculate from (7.1.7)

$$\overline{z}_j\,\mathscr{P}(Z,Z') = \frac{b_{j,z}}{a_j}\,\mathscr{P}(Z,Z') + \overline{z}'_j\mathscr{P}(Z,Z'),$$

$$z_i\overline{z}_j\,\mathscr{P}(Z,Z') = \frac{1}{a_j}\,b_{j,z}\,z_i\mathscr{P}(Z,Z') + \frac{2}{a_j}\,\delta_{ij}\,\mathscr{P}(Z,Z') + z_i\overline{z}'_j\,\mathscr{P}(Z,Z').$$

(7.1.9)

We calculate some examples for $\mathscr{K}[F,G]$ in (7.1.6). From (7.1.4) and (7.1.9), we get

$$\mathscr{K}[1,\overline{z}_j]\mathscr{P} = \mathscr{P} \circ (\overline{z}_j\mathscr{P}) = \overline{z}'_j\mathscr{P}, \quad \mathscr{K}[1,z_j]\mathscr{P} = \mathscr{P} \circ (z_j\mathscr{P}) = z_j\mathscr{P},$$

$$\mathscr{K}[z_i,\overline{z}_j]\mathscr{P} = (z_i\mathscr{P}) \circ (\overline{z}_j\mathscr{P}) = z_i\mathscr{P} \circ (\overline{z}_j\mathscr{P}) = z_i\overline{z}'_j\mathscr{P},$$

$$\mathscr{K}[\overline{z}_i,z_j]\mathscr{P} = (\overline{z}_i\mathscr{P}) \circ (z_j\mathscr{P}) = \overline{z}_i\mathscr{P} \circ (z_j\mathscr{P}) = \overline{z}_iz_j\mathscr{P},$$

$$\mathscr{K}[z'_i,\overline{z}_j]\mathscr{P} = (z'_i\mathscr{P}) \circ (\overline{z}_j\mathscr{P}) = \mathscr{P} \circ (z_i\overline{z}_j\mathscr{P}) = \frac{2}{a_j}\delta_{ij}\mathscr{P} + z_i\overline{z}'_j\mathscr{P},$$

$$\mathscr{K}[\overline{z}'_i,z_j]\mathscr{P} = (\overline{z}'_i\mathscr{P}) \circ (z_j\mathscr{P}) = \mathscr{P} \circ (\overline{z}_iz_j\mathscr{P}) = \frac{2}{a_j}\delta_{ij}\mathscr{P} + \overline{z}'_iz_j\mathscr{P}.$$

(7.1.10)

Thus we get

$$\mathscr{K}[1,\overline{z}_j] = \overline{z}'_j, \quad \mathscr{K}[1,z_j] = z_j,$$

$$\mathscr{K}[z_i,\overline{z}_j] = z_i\overline{z}'_j, \quad \mathscr{K}[\overline{z}_i,z_j] = \overline{z}_iz_j,$$

$$\mathscr{K}[\overline{z}'_i,z_j] = \mathscr{K}[z'_j,\overline{z}_i] = \frac{2}{a_j}\delta_{ij} + \overline{z}'_iz_j.$$

(7.1.11)

For a polynomial $F \in \mathbb{C}[Z,Z']$, let $(F\mathscr{P})_p$ be the operator defined by the kernel $p^n(F\mathscr{P})(\sqrt{p}Z, \sqrt{p}Z')$. Then by changing the variable, we get for $F,G \in \mathbb{C}[Z,Z']$,

$$((F\mathscr{P})_p \circ (G\mathscr{P})_p)(Z,Z') = p^n((F\mathscr{P}) \circ (G\mathscr{P}))(\sqrt{p}Z, \sqrt{p}Z').$$

(7.1.12)

7.2   Asymptotic expansion of Toeplitz operators

We use the notation in Sections 1.6.1, 4.1.1.

Let (X, J) be an n-dimensional compact complex manifold with complex structure J. Let g^{TX} be a Riemannian metric on TX compatible with J. Let (L, h^L) and (E, h^E) be two holomorphic Hermitian vector bundles on X and $\mathrm{rk}(L) = 1$.

We suppose that the positivity condition (1.5.21) holds for R^L.

Recall that P_p is the orthogonal projection from $\Omega^{0, \bullet}(X, L^p \otimes E)$ onto $\mathrm{Ker}(D_p)$. As explained in Section 4.1.1, there exists $p_0 \in \mathbb{N}$ such that for any $p > p_0$, P_p is simply the orthogonal projection from $(L^2(X, L^p \otimes E), \langle\,,\,\rangle)$ onto $H^0(X, L^p \otimes E)$, where the Hermitian product $\langle\,,\,\rangle$ on $L^2(X, L^p \otimes E)$ is defined by (1.3.14) associated to g^{TX}, h^L, h^E. In what follows, we always assume $p > p_0$.

For simplicity we denote the linear operator $M_g : L^2(X, L^p \otimes E) \to L^2(X, L^p \otimes E)$ of multiplication with a bounded function g just by $g = M_g$.

Definition 7.2.1. A *Toeplitz operator* is a family $\{T_p\}$ of linear operators

$$T_p : L^2(X, L^p \otimes E) \longrightarrow L^2(X, L^p \otimes E), \tag{7.2.1}$$

with the properties:

(i) For any $p \in \mathbb{N}$, we have

$$T_p = P_p T_p P_p. \tag{7.2.2}$$

(ii) There exists a sequence $g_l \in \mathscr{C}^\infty(X, \mathrm{End}(E))$ such that for all $k \geqslant 0$ there exists $C_k > 0$ with

$$\left\| T_p - P_p \Big(\sum_{l=0}^{k} p^{-l} g_l \Big) P_p \right\| \leqslant C_k \, p^{-k-1}, \tag{7.2.3}$$

where $\|\cdot\|$ denotes the operator norm on the space of bounded operators.

The full symbol of $\{T_p\}$ is the formal series $\sum_{l=0}^{\infty} \hbar^l g_l \in \mathscr{C}^\infty(X, \mathrm{End}(E))[[\hbar]]$ and the *principal symbol* of $\{T_p\}$ is g_0. If each T_p is self-adjoint, $\{T_p\}$ is called self-adjoint.

As our notation, we will denote (7.2.3) by

$$T_p = P_p \Big(\sum_{l=0}^{k} p^{-l} g_l \Big) P_p + \mathcal{O}(p^{-k-1}). \tag{7.2.4}$$

If (7.2.3) holds for any $k \in \mathbb{N}$, then we denote it by

$$T_p = P_p \Big(\sum_{l=0}^{\infty} p^{-l} g_l \Big) P_p + \mathcal{O}(p^{-\infty}). \tag{7.2.5}$$

An important particular case is when $g_l = 0$ for all $l \geqslant 1$. We set $g_0 = f$. We denote then

$$T_{f,p} : L^2(X, L^p \otimes E) \longrightarrow L^2(X, L^p \otimes E), \quad T_{f,p} = P_p f P_p. \tag{7.2.6}$$

The Schwartz kernel of $T_{f,p}$ is given by

$$T_{f,p}(x, x') = \int_X P_p(x, x'') f(x'') P_p(x'', x') \, dv_X(x''). \tag{7.2.7}$$

Let us remark that if $f \in \mathscr{C}^\infty(X, \mathrm{End}(E))$ is self-adjoint, i.e., $f(x) = f(x)^*$ for all $x \in X$, then the operator of multiplication with f and therefore $T_{f,p}$ are self-adjoint. The map which associates to a section $f \in \mathscr{C}^\infty(X, \mathrm{End}(E))$ the bounded operator $T_{f,p}$ on $L^2(X, L^p \otimes E)$ is the famous *Berezin–Toeplitz quantization*.

We first note that outside the diagonal of $X \times X$ the kernel of the Toeplitz operators $T_{f,p}$ has the growth $\mathscr{O}(p^{-\infty})$.

Lemma 7.2.2. *For any $\varepsilon > 0$ and any $l, m \in \mathbb{N}$ there exists $C_{l,m} > 0$ such that*

$$|T_{f,p}(x, x')|_{\mathscr{C}^m(X \times X)} \leqslant C_{l,m} p^{-l} \tag{7.2.8}$$

for all $p \geqslant 1$ and all $(x, x') \in X \times X$ with $d(x, x') > \varepsilon$. Here $\mathscr{C}^m$-norm is induced by ∇^L, ∇^E and h^L, h^E, g^{TX}.

Proof. We know that from (4.1.12), (7.2.8) holds if we replace $T_{f,p}$ by P_p. Moreover, from (4.2.1), for any $m \in \mathbb{N}$ there exist $C_m > 0$, $M_m > 0$ such that $|P_p(x, x')|_{\mathscr{C}^m(X \times X)} < C_m p^{M_m}$ for all $(x, x') \in X \times X$. These two facts and formula (7.2.7) imply the lemma. $\qquad\square$

We concentrate next on a neighborhood of the diagonal in order to obtain the asymptotic expansion of the kernel $T_{f,p}(x, x')$.

As in Section 1.6.2, let inj^X be the injectivity radius of X and fix $0 < \varepsilon < \mathrm{inj}^X / 4$. For any $x \in X$, we identify the geodesic ball $B^X(x, 4\varepsilon)$ with the ball $B^{T_x X}(0, 4\varepsilon) \subset T_x X$ by means of the exponential map. We consider trivializations of L and E as in Section 4.1.3, i.e., we trivialize L by using unit frames $e_L(Z)$ and $e_E(Z)$ which are parallel with respect to ∇^L and ∇^E along $[0, 1] \ni u \to uZ$ for $Z \in B^{T_x X}(0, 4\varepsilon)$.

For $f \in \mathscr{C}^\infty(X, \mathrm{End}(E))$, we denote it as $f_{x_0}(Z) \in \mathrm{End}(E_{x_0})$ a family (with parameter $x_0 \in X$) of functions on Z in the normal coordinate near x_0.

Recall that dv_{TX} is the Riemannian volume form on $(T_{x_0}X, g^{T_{x_0}X})$, and $\kappa_{x_0}(Z)$ is the smooth positive function defined by (cf. (4.1.28))

$$dv_X(Z) = \kappa_{x_0}(Z) dv_{TX}(Z), \quad \kappa_{x_0}(0) = 1, \tag{7.2.9}$$

where we denote it by $\kappa_{x_0}(Z)$ to indicate the base point $x_0 \in X$.

Recall that for $x_0 \in X$, we choose $\{w_i\}_{i=1}^n$ as an orthonormal basis of $T_{x_0}^{(1,0)}X$, such that (1.5.18) holds. Then $e_{2j-1} = \frac{1}{\sqrt{2}}(w_j + \overline{w}_j)$ and $e_{2j} = \frac{\sqrt{-1}}{\sqrt{2}}(w_j - \overline{w}_j)$, $j = 1, \ldots, n$ forms an orthonormal basis of $T_{x_0}X$. We use the coordinates on $T_{x_0}X \simeq \mathbb{R}^{2n}$ induced by $\{e_i\}$ as in (1.6.22). For the functions on the normal coordinate, we will add a subscript x_0 to indicate the base point $x_0 \in X$.

Recall $\pi : TX \times_X TX \to X$ is the natural projection from the fiberwise product of TX on X.

Let $\Xi_p : L^2(X, L^p \otimes E) \longrightarrow L^2(X, L^p \otimes E)$ be a family of linear operators with smooth kernel $\Xi_p(x,y)$ with respect to $dv_X(y)$. Under our trivialization, $\Xi_p(x,y)$ induces a smooth section $\Xi_{p,x_0}(Z, Z')$ of $\pi^*(\mathrm{End}(E))$ over $TX \times_X TX$ with $Z, Z' \in T_{x_0}X$, $|Z|, |Z'| < 4\varepsilon$.

To study the asymptotic expansion of $\Xi_p(x,y)$ near the diagonal of $X \times X$, we denote

$$p^{-n}\Xi_{p,x_0}(Z, Z') \cong \sum_{r=0}^{k}(Q_{r,x_0}\mathscr{P}_{x_0})(\sqrt{p}Z, \sqrt{p}Z')p^{-\frac{r}{2}} + \mathcal{O}(p^{-\frac{k+1}{2}}), \qquad (7.2.10)$$

with $\mathscr{P}_{x_0}$ in (4.1.84) (cf. (7.1.3)), if $Q_{r,x_0} \in \mathrm{End}(E_{x_0})[Z, Z']$ are a smooth family on $x_0 \in X$ of polynomials on Z, Z' with values in $\mathrm{End}(E_{x_0})$, and there exist $0 < \varepsilon' \leq 4\varepsilon$, $C_0 > 0$ such that for any $l \in \mathbb{N}$, there exist $C_l > 0$, $M > 0$ such that for any $Z, Z' \in T_{x_0}X$, $|Z|, |Z'| < \varepsilon'$, $p > p_0$, we have

$$\left| p^{-n}\Xi_{p,x_0}(Z, Z')\kappa_{x_0}^{1/2}(Z)\kappa_{x_0}^{1/2}(Z') - \sum_{r=0}^{k}(Q_{r,x_0}\mathscr{P}_{x_0})(\sqrt{p}Z, \sqrt{p}Z')p^{-\frac{r}{2}} \right|_{\mathscr{C}^l(X)}$$
$$\leqslant C_{k,l}\, p^{-\frac{k+1}{2}}(1 + \sqrt{p}|Z| + \sqrt{p}|Z'|)^M \exp(-\sqrt{C_0 p}|Z - Z'|) + \mathcal{O}(p^{-\infty}). \qquad (7.2.11)$$

Recall that $\mathscr{C}^l(X)$ is the $\mathscr{C}^l$-norm for the parameter $x_0 \in X$, and the term $\mathcal{O}(p^{-\infty})$ means that for any $l, l_1 \in \mathbb{N}$, there exists $C_{l,l_1} > 0$ such that its $\mathscr{C}^{l_1}$-norm is dominated by $C_{l,l_1}p^{-l}$.

Recall that $J_r(Z, Z') \in \mathrm{End}(\Lambda(T^{*(0,1)}X) \otimes E)_{x_0}$, $(r \in \mathbb{N})$ defined in Theorem 4.1.21. By Theorem 4.1.21 and (4.1.98), we know $J_{r,x_0} \in \mathrm{End}(E_{x_0})[Z, Z']$ are polynomials on Z, Z' with the same parity as r and smooth on $x_0 \in X$, and

$$J_{0,x_0} = \mathrm{Id}_E. \qquad (7.2.12)$$

Lemma 7.2.3. *For any $k \in \mathbb{N}$, $Z, Z' \in T_{x_0}X$, $|Z|, |Z'| < 2\varepsilon$, we have*

$$p^{-n}P_{p,x_0}(Z, Z') \cong \sum_{r=0}^{k}(J_{r,x_0}\mathscr{P}_{x_0})(\sqrt{p}Z, \sqrt{p}Z')p^{-\frac{r}{2}} + \mathcal{O}(p^{-\frac{k+1}{2}}). \qquad (7.2.13)$$

Proof. By Theorem 4.2.1 we get: for any $k, m' \in \mathbb{N}$, there exist $M \in \mathbb{N}, C > 0$ such that for $Z, Z' \in T_{x_0}X$, $|Z|, |Z'| \leqslant 2\varepsilon$,

$$\left| p^{-n} P_{p,x_0}(Z, Z') \kappa_{x_0}^{\frac{1}{2}}(Z) \kappa_{x_0}^{\frac{1}{2}}(Z') - \sum_{r=0}^{k} (J_{r,x_0} \mathscr{P}_{x_0})(\sqrt{p}Z, \sqrt{p}Z') p^{-\frac{r}{2}} \right|_{\mathscr{C}^{m'}(X)}$$

$$\leqslant C p^{-(k+1)/2}(1 + \sqrt{p}|Z| + \sqrt{p}|Z'|)^M \exp(-\sqrt{C''\mu_0}\sqrt{p}|Z - Z'|) + \mathscr{O}(p^{-\infty}). \tag{7.2.14}$$

From (7.2.14), we get (7.2.13). $\qquad\qquad\square$

Lemma 7.2.4. *Let* $f \in \mathscr{C}^\infty(X, \mathrm{End}(E))$. *There exists a family of polynomials* $\{Q_{r,x_0}(f) \in \mathrm{End}(E_{x_0})[Z, Z']\}_{x_0 \in X}$ *with the same parity as* r, *and smooth in* $x_0 \in X$ *such that for any* $k \in \mathbb{N}$, $Z, Z' \in T_{x_0}X$, $|Z|, |Z'| < \varepsilon/2$,

$$p^{-n} T_{f,p,x_0}(Z, Z') \cong \sum_{r=0}^{k} (Q_{r,x_0}(f)\mathscr{P}_{x_0})(\sqrt{p}Z, \sqrt{p}Z') p^{-\frac{r}{2}} + \mathscr{O}(p^{-\frac{k+1}{2}}). \tag{7.2.15}$$

Moreover, under the notation in (7.1.6) and J_{r,x_0} *in (7.2.13), we have*

$$Q_{r,x_0}(f) = \sum_{r_1+r_2+|\alpha|=r} \mathscr{K}\left[J_{r_1,x_0}, \frac{\partial^\alpha f_{x_0}}{\partial Z^\alpha}(0) \frac{Z^\alpha}{\alpha!} J_{r_2,x_0} \right], \tag{7.2.16}$$

especially,

$$Q_{0,x_0}(f) = f(x_0). \tag{7.2.17}$$

Proof. From (7.2.7) and (7.2.8), we know that $T_{f,p,x_0}(Z, Z')$ for $|Z|, |Z'| < \varepsilon/2$, is determined up to terms of order $\mathscr{O}(p^{-\infty})$ by the behavior of f in $B^X(x_0, \varepsilon)$. For $|Z|, |Z'| < \varepsilon/2$, with ρ in (1.6.19), we get

$$T_{f,p,x_0}(Z, Z') = \int_{T_{x_0}X} P_{p,x_0}(Z, Z'') \rho(2|Z''|/\varepsilon) f_{x_0}(Z'') P_{p,x_0}(Z'', Z')$$

$$\times \kappa_{x_0}(Z'') \, dv_{TX}(Z'') + \mathscr{O}(p^{-\infty}). \tag{7.2.18}$$

We consider the Taylor expansion of f_{x_0}:

$$f_{x_0}(Z) = \sum_{|\alpha| \leqslant k} \frac{\partial^\alpha f_{x_0}}{\partial Z^\alpha}(0) \frac{Z^\alpha}{\alpha!} + \mathscr{O}(|Z|^{k+1})$$

$$= \sum_{|\alpha| \leqslant k} p^{-\alpha/2} \frac{\partial^\alpha f_{x_0}}{\partial Z^\alpha}(0) \frac{(\sqrt{p}Z)^\alpha}{\alpha!} + p^{-\frac{k+1}{2}} \mathscr{O}(|\sqrt{p}Z|^{k+1}). \tag{7.2.19}$$

We multiply now the expansions given in (7.2.19) and (7.2.14) and obtain the expansion of

$$\kappa_{x_0}^{1/2}(Z) P_{p,x_0}(Z, Z'') (\kappa_{x_0} f_{x_0})(Z'') P_{p,x_0}(Z'', Z') \kappa_{x_0}^{1/2}(Z')$$

which we substitute in (7.2.18). We integrate then on $T_{x_0}X$ by using the change of variable $\sqrt{p}Z'' = W$ and conclude (7.2.15) and (7.2.16) by using formulas (7.1.6) and (7.1.12).

From (7.2.12) and (7.2.16), we get

$$Q_{0,x_0}(f) = \mathscr{K}[1, f_{x_0}(0)] = f_{x_0}(0) = f(x_0). \tag{7.2.20}$$

The proof of Lemma 7.2.4 is complete. $\square$

As an example, we compute $Q_{1,x_0}(f)$.

Lemma 7.2.5. *For $Q_{1,x_0}(f)$ in (7.2.15), We have*

$$Q_{1,x_0}(f) = f(x_0)J_{1,x_0} + \mathscr{K}\left[J_{0,x_0}, \frac{\partial f_{x_0}}{\partial Z_j}(0)Z_j J_{0,x_0}\right]. \tag{7.2.21}$$

Proof. At first, by taking $f = 1$ in (7.2.16), we get

$$J_{1,x_0} = \mathscr{K}[J_{0,x_0}, J_{1,x_0}] + \mathscr{K}[J_{1,x_0}, J_{0,x_0}]. \tag{7.2.22}$$

From (1.2.30) and (4.1.34), we know that $\mathcal{O}_1$ defined in (4.1.31) (considered as a differential operator with coefficients in $\text{End}(\Lambda(T^{*(1,0)}X) \otimes E)_{x_0})$ acts as identity on the E-component and preserves the $\mathbb{Z}$-grading on $\Lambda(T^{*(1,0)}X)$. (Note that, if $g^{TX} = \omega(\cdot, J\cdot)$, its restriction on $\mathbb{C} \otimes E$ is zero by (4.1.100), thus $J_{1,x_0} = 0$.) Thus from (4.1.93) and (7.1.6), we obtain

$$\mathscr{K}[J_{1,x_0}, f(x_0)J_{0,x_0}] = f(x_0)\mathscr{K}[J_{1,x_0}, J_{0,x_0}]. \tag{7.2.23}$$

From (7.2.16), (7.2.22) and (7.2.23), we get (7.2.21). $\square$

7.3   A criterion for Toeplitz operators

We will prove next a useful criterion which ensures that a given family is a Toeplitz operator.

Theorem 7.3.1. *Let $\{T_p : L^2(X, L^p \otimes E) \longrightarrow L^2(X, L^p \otimes E)\}$ be a family of bounded linear operators which satisfies the following three conditions:*

(i) *For any $p \in \mathbb{N}$, $P_p T_p P_p = T_p$.*

(ii) *For any $\varepsilon_0 > 0$ and any $l, m \in \mathbb{N}$, there exists $C_{l,m} > 0$ such that for all $p \geqslant 1$ and all $(x, x') \in X \times X$ with $d(x, x') > \varepsilon_0$,*

$$|T_p(x, x')| \leqslant C_{l,m}p^{-l}. \tag{7.3.1}$$

(iii) *There exists a family of polynomials $\{\mathcal{Q}_{r,x_0} \in \text{End}(E_{x_0})[Z, Z']\}_{x_0 \in X}$ such that: (a) each $\mathcal{Q}_{r,x_0}$ has the same parity as r, (b) the family is smooth in*

$x_0 \in X$ and (c) there exists $0 < \varepsilon' < \varepsilon$ such that for any $x_0 \in X$ and $Z, Z' \in T_{x_0}X$, $|Z|, |Z'| < \varepsilon'$, in the sense of (7.2.10) and (7.2.11), we have

$$p^{-n} T_{p,x_0}(Z, Z') \cong \sum_{r=0}^{k} (\mathcal{Q}_{r,x_0} \mathscr{P}_{x_0})(\sqrt{p}Z, \sqrt{p}Z') p^{-\frac{r}{2}} + \mathcal{O}(p^{-\frac{k+1}{2}}). \quad (7.3.2)$$

Then $\{T_p\}$ is a Toeplitz operator.

We start the proof of Theorem 7.3.1. Let T_p^* be the adjoint of T_p. By writing

$$T_p = \frac{1}{2}(T_p + T_p^*) + \sqrt{-1} \frac{1}{2\sqrt{-1}}(T_p - T_p^*),$$

we may and we will assume from now on that T_p is self-adjoint.

We will define inductively the sequence $(g_l)_{l \geqslant 0}$, $g_l \in \mathscr{C}^\infty(X, \mathrm{End}(E))$ such that for any $m \geqslant 1$,

$$T_p = \sum_{l=0}^{m} P_p \, g_l \, p^{-l} \, P_p + \mathcal{O}(p^{-m-1}). \quad (7.3.3)$$

Moreover, we can make these g_l's to be self-adjoint. For $x_0 \in X$, we set

$$g_0(x_0) = \mathcal{Q}_{0,x_0}(0,0). \quad (7.3.4)$$

We will show that

$$T_p = P_p \, g_0 \, P_p + \mathcal{O}(p^{-1}). \quad (7.3.5)$$

Proposition 7.3.2. *In the conditions of Theorem 7.3.1 we have* $\mathcal{Q}_{0,x_0}(Z, Z') = \mathcal{Q}_{0,x_0}(0,0)$ *for all* $x_0 \in X$ *and all* $Z, Z' \in T_{x_0}X$.

Proof. Our first observation is as follows.

Lemma 7.3.3. $\mathcal{Q}_{0,x_0}$ *is a polynomial in* $z, \bar{z}'$.

Proof. Indeed, by (7.3.2)

$$p^{-n} T_{p,x_0}(Z, Z') \cong (\mathcal{Q}_{0,x_0} \mathscr{P}_{x_0})(\sqrt{p}Z, \sqrt{p}Z') + \mathcal{O}(p^{-1/2}). \quad (7.3.6)$$

By (7.2.12) and (7.2.13), we have

$$p^{-n}(P_p T_p P_p)_{x_0}(Z, Z') \cong (\mathscr{P} \circ (\mathcal{Q}_0 \mathscr{P}) \circ \mathscr{P})_{x_0}(\sqrt{p}Z, \sqrt{p}Z') + \mathcal{O}(p^{-1/2}). \quad (7.3.7)$$

Since $P_p T_p P_p = T_p$, we deduce from (7.3.6) and (7.3.7) that

$$\mathcal{Q}_{0,x_0} \mathscr{P}_{x_0} = \mathscr{P}_{x_0} \circ (\mathcal{Q}_{0,x_0} \mathscr{P}_{x_0}) \circ \mathscr{P}_{x_0}, \quad (7.3.8)$$

hence $\mathcal{Q}_{0,x_0} \in \mathrm{End}(E_{x_0})[z, \bar{z}']$ by (7.1.5) and (7.1.8). $\qquad \square$

For simplicity we denote in the rest of the proof $F_x = \mathcal{Q}_{0,x}$. Let $F_x = \sum_{i \geqslant 0} F_x^{(i)}$ be the decomposition of F_x in homogeneous polynomials $F_x^{(i)}$ of degree i. We will show, that $F_x^{(i)}$ vanish identically for $i > 0$, that is,

$$F_x^{(i)}(z, \overline{z}') = 0 \quad \text{for all } i > 0 \text{ and } z, z' \in \mathbb{C}^n . \tag{7.3.9}$$

The first step is to prove

$$F_x^{(i)}(0, \overline{z}') = 0 \quad \text{for all } i > 0 \text{ and all } z' \in \mathbb{C}^n . \tag{7.3.10}$$

Let us remark that since T_p are self-adjoint we have

$$F_x^{(i)}(z, \overline{z}') = (F_x^{(i)}(z', \overline{z}))^* . \tag{7.3.11}$$

For $y = \exp_x^X(Z')$, $Z' \in \mathbb{R}^{2n} \simeq T_{x_0}X$ as explained above (7.2.10), set

$$\begin{aligned}
F^{(i)}(x, y) &= F_x^{(i)}(0, \overline{z}') \in \text{End}(E_x), \\
\widetilde{F}^{(i)}(x, y) &= (F^{(i)}(y, x))^* \in \text{End}(E_y).
\end{aligned} \tag{7.3.12}$$

$F^{(i)}$ and $\widetilde{F}^{(i)}$ define smooth sections on a neighborhood of the diagonal of $X \times X$. Clearly, $\widetilde{F}^{(i)}(x, y)$'s need not be polynomials of z and $\overline{z}'$.

Since we wish to define global operators induced by these kernels, we use a cut-off function in the neighborhood of the diagonal. Pick a smooth function $\eta \in \mathscr{C}^\infty(\mathbb{R})$, such that $\eta(u) = 1$ for $|u| \leqslant \varepsilon'/2$ and $\eta(u) = 0$ for $|u| \geqslant \varepsilon'$.

We denote by $F^{(i)}P_p$ and $P_p\widetilde{F}^{(i)}$ the operators defined by the kernels

$$\eta(d(x, y))F^{(i)}(x, y)P_p(x, y) \quad \text{and} \quad \eta(d(x, y))P_p(x, y)\widetilde{F}^{(i)}(x, y)$$

with respect to $dv_X(y)$. Set

$$\mathscr{T}_p = T_p - \sum_{i \leqslant \deg F} (F^{(i)}P_p)\, p^{i/2}. \tag{7.3.13}$$

The operators $\mathscr{T}_p$ extend naturally to bounded operators on $L^2(X, L^p \otimes E)$.

From (7.3.2) and (7.3.13), in the sense of (7.2.11), we deduce that for any $k \geqslant 1$ and $|Z'| \leqslant \varepsilon'$, we have the following expansion in the normal coordinates around $x_0 \in X$:

$$p^{-n}\mathscr{T}_{p,x_0}(0, Z') \cong \sum_{r=1}^{k} (R_{r,x_0}\mathscr{P}_{x_0})(0, \sqrt{p}Z')p^{-r/2} + \mathcal{O}(p^{-(k+1)/2}), \tag{7.3.14}$$

for some polynomials R_{r,x_0} of the same parity as r. For simplicity we denote by $R_{r,p}$ the operator defined as in (7.3.12) by the kernel

$$R_{r,p}(x, y) = p^n (R_{r,x}\mathscr{P}_x)(0, \sqrt{p}Z')\kappa_x^{-1/2}(Z')\eta(d(x, y)), \tag{7.3.15}$$

where $y = \exp_x^X(Z')$.

Lemma 7.3.4. *There exists $C > 0$ such that for any $p > p_0$, $s \in L^2(X, L^p \otimes E)$,*

$$\|\mathscr{T}_p s\|_{L^2} \leqslant C p^{-1/2} \|s\|_{L^2}, \tag{7.3.16}$$

$$\|\mathscr{T}_p^* s\|_{L^2} \leqslant C p^{-1/2} \|s\|_{L^2}. \tag{7.3.17}$$

Proof. In order to use (7.3.14) we write

$$\|\mathscr{T}_p s\|_{L^2} \leqslant \left\|(\mathscr{T}_p - \sum_{r=1}^{k} p^{-r/2} R_{r,p}) s\right\|_{L^2} + \left\|\sum_{r=1}^{k} p^{-r/2} R_{r,p} s\right\|_{L^2}. \tag{7.3.18}$$

By the Cauchy-Schwarz inequality we have

$$\left\|(\mathscr{T}_p - \sum_{r=1}^{k} p^{-r/2} R_{r,p}) s\right\|_{L^2}^2 \leqslant \int_X \left(\int_X \left|(\mathscr{T}_p - \sum_{r=1}^{k} p^{-r/2} R_{r,p})(x,y)\right| dv_X(y)\right)$$

$$\times \left(\int_X \left|(\mathscr{T}_p - \sum_{r=1}^{k} p^{-r/2} R_{r,p})(x,y)\right| |s(y)|^2 dv_X(y)\right) dv_X(x). \tag{7.3.19}$$

We split then the inner integrals into integrals over $B^X(x, \varepsilon')$ and $X \smallsetminus B^X(x, \varepsilon')$ and use the fact that the kernel of $\mathscr{T}_p - \sum_{r=1}^{k} p^{-r/2} R_{r,p}$ has the growth $\mathcal{O}(p^{-\infty})$ outside the diagonal, by (7.3.1), the definition of the operators $F^{(i)} P_p$ in (7.3.13) (using the cut-off function η), and the definition (7.3.15) of $R_{r,p}$ (which involves $\mathscr{P}$). We get for example, uniformly in $x \in X$,

$$\int_X \left|(\mathscr{T}_p - \sum_{r=1}^{k} p^{-r/2} R_{r,p})(x,y)\right| |s(y)|^2 dv_X(y)$$

$$= \int_{B^X(x,\varepsilon')} \left|(\mathscr{T}_p - \sum_{r=1}^{k} p^{-r/2} R_{r,p})(x,y)\right| |s(y)|^2 dv_X(y)$$

$$+ \mathcal{O}(p^{-\infty}) \int_{X \smallsetminus B^X(x,\varepsilon')} |s(y)|^2 dv_X(y). \tag{7.3.20}$$

By (7.2.11), (7.3.14) for k sufficiently large, which we fix from now on, we obtain

$$\int_{B^X(x,\varepsilon')} \left|(\mathscr{T}_p - \sum_{r=1}^{k} p^{-r/2} R_{r,p})(x,y)\right| |s(y)|^2 dv_X(y)$$

$$= \mathcal{O}(p^{-1}) \int_{B^X(x,\varepsilon')} |s(y)|^2 dv_X(y). \tag{7.3.21}$$

In the same vein we obtain

$$\int_X \left|(\mathscr{T}_p - \sum_{r=1}^{k} p^{-r/2} R_{r,p})(x,y)\right| dv_X(y) = \mathcal{O}(p^{-1}) + \mathcal{O}(p^{-\infty}). \tag{7.3.22}$$

Using (7.3.19)–(7.3.22) finally gives

$$\left\|\Big(\mathscr{T}_p - \sum_{r=1}^k p^{-r/2} R_{r,p}\Big) s\right\|_{L^2} \leqslant C\, p^{-1/2} \|s\|_{L^2}, \quad s \in L^2(X, L^p \otimes E). \qquad (7.3.23)$$

A similar proof as for (7.3.23) delivers for $s \in L^2(X, L^p \otimes E)$,

$$\left\| R_{r,p}\, s \right\|_{L^2} \leqslant C \|s\|_{L^2}, \qquad (7.3.24)$$

which implies

$$\left\| \sum_{r=1}^k p^{-r/2} R_{r,p}\, s \right\|_{L^2} \leqslant C\, p^{-1/2} \|s\|_{L^2}, \quad \text{for } s \in L^2(X, L^p \otimes E), \qquad (7.3.25)$$

for some constant $C > 0$. Relations (7.3.23) and (7.3.25) entail (7.3.16), which is equivalent to (7.3.17), by taking the adjoint. □

Let us consider the Taylor development of $\widetilde{F}^{(i)}$ in normal coordinates around x with $y = \exp_x^X(Z')$:

$$\widetilde{F}^{(i)}(x,y) = \sum_{|\alpha| \leqslant k} \frac{\partial^\alpha \widetilde{F}^{(i)}}{\partial Z'^\alpha}(x,0) \frac{(\sqrt{p}Z')^\alpha}{\alpha!} p^{-|\alpha|/2} + \mathscr{O}(|Z'|^{k+1}). \qquad (7.3.26)$$

The next step in the proof of Proposition 7.3.2 is the following.

Lemma 7.3.5. *For any $j > 0$, we have*

$$\frac{\partial^\alpha \widetilde{F}^{(i)}}{\partial Z'^\alpha}(x,0) = 0, \quad \text{for } i - |\alpha| \geq j > 0. \qquad (7.3.27)$$

Proof. The definition (7.3.13) of $\mathscr{T}_p$ shows

$$\mathscr{T}_p^* = T_p - \sum_{i \leqslant \deg F} p^{i/2}(P_p \widetilde{F}^{(i)}). \qquad (7.3.28)$$

Let us develop the sum in the right-hand side. Combining the Taylor development (7.3.26) with the expansion (7.2.13) of the Bergman kernel we obtain:

$$p^{-n} \sum_i (P_p\, \widetilde{F}^{(i)})_{x_0}(0, Z') p^{i/2}$$

$$\cong \sum_i \sum_{|\alpha|, r \leqslant k} (J_{r,x_0} \mathscr{P}_{x_0})(0, \sqrt{p}Z') \frac{\partial^\alpha \widetilde{F}^{(i)}}{\partial Z'^\alpha}(x_0, 0) \frac{(\sqrt{p}Z')^\alpha}{\alpha!} p^{(i-|\alpha|-r)/2}$$

$$+ \mathscr{O}(p^{(\deg F - k - 1)/2}), \qquad (7.3.29)$$

where $k \geqslant \deg F + 1$. Having in mind (7.3.17), this is only possible if the coefficients of $p^{j/2}$, $j > 0$ in the right-hand side of (7.3.29) vanish. Thus we get for any $j > 0$,

$$\sum_{l=j}^{\deg F} \sum_{|\alpha|+r=l-j} J_{r,x_0}(0, \sqrt{p}Z') \frac{\partial^\alpha \widetilde{F}^{(l)}}{\partial Z'^\alpha}(x_0, 0) \frac{(\sqrt{p}Z')^\alpha}{\alpha!} = 0. \tag{7.3.30}$$

From (7.3.30), we will prove by recurrence that for any $j > 0$, (7.3.27) holds. As the first step of the recurrence, let us take $j = \deg F$ in (7.3.30). Since $J_{0,x_0} = \mathrm{Id}_E$ (see (7.2.12)), we get $\widetilde{F}^{(\deg F)}(x_0, 0) = 0$, thus (7.3.27) holds for $j = \deg F$.

Assume that (7.3.27) holds for $j > j_0 > 0$. Then for $j = j_0$, the coefficient with $r > 0$ in (7.3.30) is zero, thus by $J_{0,x_0} = \mathrm{Id}_E$, (7.3.30) reads

$$\sum_\alpha \frac{\partial^\alpha \widetilde{F}^{(j_0+|\alpha|)}}{\partial Z'^\alpha}(x_0, 0) \frac{(\sqrt{p}Z')^\alpha}{\alpha!} = 0. \tag{7.3.31}$$

From (7.3.31), we get (7.3.27) for $j = j_0$. The proof of (7.3.27) is complete. $\qquad\square$

Lemma 7.3.6. *We have*

$$\frac{\partial^\alpha F_x^{(i)}}{\partial \bar{z}'^\alpha}(0, 0) = 0, \quad |\alpha| \leqslant i. \tag{7.3.32}$$

Therefore $F_x^{(i)}(0, \bar{z}') = 0$ for all $i > 0$ and $z' \in \mathbb{C}^n$, i.e., (7.3.10) holds true. Moreover,

$$F_x^{(i)}(z, 0) = 0 \quad \text{for all } i > 0 \text{ and all } z \in \mathbb{C}^n. \tag{7.3.33}$$

Proof. Let us start with some preliminary observations.

By (7.3.17), (7.3.27) and (7.3.29), by comparing the coefficient of p^0 in (7.3.6) and (7.3.28), we get

$$\widetilde{F}^{(i)}(x, Z') = F_x^{(i)}(0, \bar{z}') + \mathcal{O}(|Z'|^{i+1}). \tag{7.3.34}$$

By the definition (7.3.12) of $\widetilde{F}$, we take the adjoint of (7.3.34) and get

$$F^{(i)}(Z', x) = (F_x^{(i)}(0, \bar{z}'))^* + \mathcal{O}(|Z'|^{i+1}) \tag{7.3.35}$$

which implies

$$\frac{\partial^\alpha}{\partial z^\alpha} F^{(i)}(\cdot, x)|_x = \left(\left(\frac{\partial^\alpha}{\partial \bar{z}'^\alpha} F_x^{(i)}\right)(0, \bar{z}')\right)^*, \quad \text{for } |\alpha| \leqslant i, \tag{7.3.36}$$

so in order to prove the lemma it suffices to show that

$$\frac{\partial^\alpha}{\partial z^\alpha} F^{(i)}(\cdot, x)|_x = 0, \quad \text{for } |\alpha| \leqslant i. \tag{7.3.37}$$

We prove this by induction over $|\alpha|$. For $|\alpha| = 0$ it is obvious that $F^{(i)}(0, x) = 0$ since we have a homogeneous polynomial of degree $i > 0$. For the induction step let $j_X : X \to X \times X$ be the diagonal injection. By Lemma 7.3.3 and the definition (7.3.12) of $F^{(i)}(x, y)$,

$$\frac{\partial}{\partial z'_j} F^{(i)}(x, y) = 0, \quad \text{near } j_X(X), \tag{7.3.38}$$

where $y = \exp_x^X(Z')$. Assume now that $\alpha \in \mathbb{N}^n$ and (7.3.37) holds for $|\alpha| - 1$. Consider j with $\alpha_j > 0$ and set $\alpha' = (\alpha_1, \ldots, \alpha_j - 1, \ldots, \alpha_n)$.

Taking the derivative of (7.3.12) and using the induction hypothesis and (7.3.38) we have

$$\frac{\partial^\alpha}{\partial z^\alpha} F^{(i)}(\cdot, x)|_x = \frac{\partial}{\partial z_j} j_X^* \left(\frac{\partial^{\alpha'}}{\partial z^{\alpha'}} F^{(i)} \right) \Big|_x - \frac{\partial^{\alpha'}}{\partial z^{\alpha'}} \frac{\partial}{\partial z'_j} F^{(i)}(\cdot, \cdot) \Big|_{0,0} = 0. \tag{7.3.39}$$

Thus (7.3.32) is proved and this is equivalent to (7.3.10). (7.3.33) follows from (7.3.10) and (7.3.11). This finishes the proof of Lemma 7.3.6. $\qquad \square$

Lemma 7.3.7. *We have* $F_x^{(i)}(z, \overline{z}') = 0$ *for all* $i > 0$ *and* $z, z' \in \mathbb{C}^n$.

Proof. Let us consider the operator

$$\frac{1}{\sqrt{p}} P_p \left(\nabla^{L^p \otimes E}_{\eta(d(x,y))(\frac{\partial}{\partial z_j} + \frac{\partial}{\partial \overline{z}_j})_x} T_p \right) P_p. \tag{7.3.40}$$

The leading term of its asymptotic expansion as in (7.3.2) is

$$\left(\frac{\partial}{\partial z_j} F_{x_0} \right) (\sqrt{p}\, z, \sqrt{p}\, \overline{z}') \mathscr{P}_{x_0}(\sqrt{p}\, Z, \sqrt{p}\, Z'). \tag{7.3.41}$$

By (7.3.10) and (7.3.33), $(\frac{\partial}{\partial z_j} F_{x_0})(z, \overline{z}')$ is an even degree polynomial on $z, \overline{z}'$ whose constant term vanishes. We reiterate the arguments from (7.3.13)–(7.3.36) by replacing the operator T_p with the operator (7.3.40); we get for $i > 0$,

$$\frac{\partial}{\partial z_j} F_x^{(i)}(0, \overline{z}') = 0. \tag{7.3.42}$$

By (7.3.11) and (7.3.42),

$$\frac{\partial}{\partial \overline{z}'_j} F_x^{(i)}(z, 0) = 0. \tag{7.3.43}$$

By continuing this process, we show that for all $i > 0, \alpha \in \mathbb{Z}^n, z, z' \in \mathbb{C}^n$,

$$\frac{\partial^\alpha}{\partial z^\alpha} F_x^{(i)}(0, \overline{z}') = \frac{\partial^\alpha}{\partial \overline{z}'^\alpha} F_x^{(i)}(z, 0) = 0. \tag{7.3.44}$$

Thus the lemma is proved and (7.3.9) holds true. $\qquad \square$

Lemma 7.3.7 finishes the proof of Proposition 7.3.2. $\qquad \square$

Proposition 7.3.8. *In the sense of notation* (7.2.4), *we have* (7.3.5).

Proof. Let us compare the asymptotic expansion of T_p and $T_{g_0,p} = P_p\, g_0\, P_p$. In the notation (7.2.10), the expansion (7.2.15) (for $k = 1$) reads

$$p^{-n} T_{g_0,p,x_0}(Z, Z') \cong (g_0(x_0)\mathscr{P}_{x_0} + Q_{1,x_0}(g_0)\mathscr{P}_{x_0}\, p^{-1/2})(\sqrt{p}Z, \sqrt{p}Z') + \mathcal{O}(p^{-1}),$$
(7.3.45)

since $\mathcal{Q}_{0,x_0}(g_0) = g_0(x_0)$ by (7.2.17). The expansion (7.3.2) (also for $k = 1$) takes the form

$$p^{-n} T_{p,x_0} \cong (g_0(x_0)\mathscr{P}_{x_0} + \mathcal{Q}_{1,x_0}\mathscr{P}_{x_0}\, p^{-1/2})(\sqrt{p}Z, \sqrt{p}Z') + \mathcal{O}(p^{-1})$$
(7.3.46)

where we have used Proposition 7.3.2 and the definition (7.3.4) of g_0. Thus, subtracting (7.3.45) from (7.3.46) we obtain

$$p^{-n}(T_p - T_{g_0,p})_{x_0}(Z, Z') \cong \big((\mathcal{Q}_{1,x_0} - Q_{1,x_0}(g_0))\mathscr{P}_{x_0}\big)(\sqrt{p}Z, \sqrt{p}Z')\, p^{-1/2} + \mathcal{O}(p^{-1}).$$
(7.3.47)

Thus it suffices to prove:

Lemma 7.3.9.

$$F_{1,x} := \mathcal{Q}_{1,x} - Q_{1,x}(g_0) \equiv 0.$$
(7.3.48)

Proof. As in Lemma 7.3.3 we show that $F_{1,x}$ is an odd polynomial in z, $\overline{z}'$. Let $F_{1,x} = \sum_{i \geqslant 0} F_{1,x}^{(i)}$ be the decomposition of $F_{1,x}$ in homogeneous polynomials $F_{1,x}^{(i)}$ of degree i. We will show that

$$F_{1,x}^{(i)}(z, \overline{z}') = 0 \quad \text{for all } i > 0 \text{ and } z, z' \in \mathbb{C}^n.$$
(7.3.49)

(7.3.49) together with the fact that $F_{1,x}$ is an odd polynomial, hence with vanishing constant term, will prove (7.3.48). The proof of (7.3.49) is similar to that of (7.3.9). Namely, we define as in (7.3.12) the operator $F_1^{(i)}$, by replacing $F_x^{(i)}(0, \overline{z}')$ by $F_{1,x}^{(i)}(0, \overline{z}')$, and we set (analogously to (7.3.13))

$$\mathscr{T}_{p,1} = T_p - P_p\, g_0\, P_p - \sum_{i \leqslant \deg F_1} (F_1^{(i)} P_p)\, p^{i/2}.$$
(7.3.50)

Then by (7.2.15) and (7.3.2), there exist odd polynomials $\widetilde{R}_{r,x_0} \in \mathbb{C}[Z, Z']$ such that the following expansion in the normal coordinates around $x_0 \in X$ holds:

$$p^{-n} \mathscr{T}_{p,1,x_0}(0, Z') \cong \sum_{r=2}^{k} (\widetilde{R}_{r,x_0}\mathscr{P}_{x_0})(0, \sqrt{p}Z')\, p^{-\frac{r}{2}} + \mathcal{O}(p^{-(k+1)/2}),$$
(7.3.51)

for $k \geqslant 2$ and $|Z'| \leqslant \varepsilon'/2$. This is the analogue of (7.3.14). Now we can repeat with obvious modifications the proof of (7.3.9) and obtain (7.3.9) for $F_1^{(i)}$. This achieves the proof of Lemma 7.3.9 and of Proposition 7.3.8. $\qquad\square$

By Lemma 7.2.4 and Proposition 7.3.8, $p(T_p - P_p g_0 P_p)$ verifies the condition of Theorem 7.3.1, thus we can continue our process to get (7.3.3), thus the proof of Theorem 7.3.1 is complete. $\qquad\square$

7.4   Algebra of Toeplitz operators

The Poisson bracket $\{\,,\,\}$ on $(X, 2\pi\omega)$ is defined by: for $f, g \in \mathscr{C}^\infty(X)$, if ξ_f is the Hamiltonian vector field generated by f which is defined by $2\pi i_{\xi_f}\omega = df$, then

$$\{f, g\} = \xi_f(dg). \tag{7.4.1}$$

One of our main goals is to show that the set of Toeplitz operators is closed under the composition of operators, so forms an algebra.

Theorem 7.4.1. *Let $f, g \in \mathscr{C}^\infty(X, \mathrm{End}(E))$. The product of the Toeplitz operators $T_{f,p}$ and $T_{g,p}$ is a Toeplitz operator; more precisely, it admits the asymptotic expansion in the sense of* (7.2.5),

$$T_{f,p} T_{g,p} = \sum_{r=0}^{\infty} p^{-r} T_{C_r(f,g),p} + \mathscr{O}(p^{-\infty}), \tag{7.4.2}$$

where C_r are bidifferential operators. In particular $C_0(f, g) = fg$.
  If $f, g \in \mathscr{C}^\infty(X)$, we have

$$C_1(f, g) - C_1(g, f) = \sqrt{-1}\{f, g\}, \tag{7.4.3}$$

and therefore

$$[T_{f,p}, T_{g,p}] = \frac{\sqrt{-1}}{p} T_{\{f,g\},p} + \mathscr{O}(p^{-2}). \tag{7.4.4}$$

Proof. Firstly, it is obvious that $P_p T_{f,p} T_{g,p} P_p = T_{f,p} T_{g,p}$. By Lemmas 7.2.2 and 7.2.4, we know $T_{f,p} T_{g,p}$ verifies (7.3.1). Now as in (7.2.18), for $Z, Z' \in T_{x_0}X$, $|Z|, |Z'| < \varepsilon/4$, we have

$$(T_{f,p} T_{g,p})_{x_0}(Z, Z') = \int_{T_{x_0}X} T_{f,p,x_0}(Z, Z'')\rho(4|Z''|/\varepsilon)T_{g,p,x_0}(Z'', Z')$$
$$\times \kappa_{x_0}(Z'') \, dv_{TX}(Z'') + \mathscr{O}(p^{-\infty}). \tag{7.4.5}$$

By Lemma 7.2.4 and (7.4.5), we know as in the proof of Lemma 7.2.4, for $Z, Z' \in T_{x_0}X$, $|Z|, |Z'| < \varepsilon/4$, we have

$$p^{-n}(T_{f,p} T_{g,p})_{x_0}(Z, Z') \cong \sum_{r=0}^{k} (Q_{r,x_0}(f, g)\mathscr{P}_{x_0})(\sqrt{p}Z, \sqrt{p}Z')p^{-\frac{r}{2}} + \mathscr{O}(p^{-\frac{k+1}{2}}), \tag{7.4.6}$$

and with the notation in (7.1.6),

$$Q_{r,x_0}(f, g) = \sum_{r_1+r_2=r} \mathscr{K}[Q_{r_1,x_0}(f), Q_{r_2,x_0}(g)]. \tag{7.4.7}$$

Thus $T_{f,p}T_{g,p}$ is a Toeplitz operator from Theorem 7.3.1. Moreover, it follows from the proofs of Lemma 7.2.4 and Theorem 7.3.1 that $g_l = C_l(f,g)$, where C_l are bidifferential operators.

From (7.1.6), (7.2.17) and (7.4.7), we get

$$C_0(f,g)(x) = Q_{0,x}(f,g) = \mathscr{K}[Q_{0,x}(f), Q_{0,x}(g)] = f(x)g(x). \tag{7.4.8}$$

By the proof of Theorem 7.3.1 (cf. Proposition 7.3.2, Lemma 7.3.9 and (7.3.4)), we get

$$\begin{aligned}
Q_{1,x}(f,g) &= Q_{1,x}(C_0(f,g)), \\
C_1(f,g) &= (Q_{2,x}(f,g) - Q_{2,x}(C_0(f,g)))(0,0).
\end{aligned} \tag{7.4.9}$$

Moreover, by (7.2.17) and (7.4.7), we get

$$Q_{2,x}(f,g) = \mathscr{K}[f(x), Q_{2,x}(g)] \\
+ \mathscr{K}[Q_{1,x}(f), Q_{1,x}(g)] + \mathscr{K}[Q_{2,x}(f), g(x)]. \tag{7.4.10}$$

Now $T_{f,p}P_p = P_p T_{f,p}$ implies $Q_{r,x}(f,1) = Q_{r,x}(1,f)$, we get from (7.4.10),

$$\mathscr{K}[J_{0,x}, Q_{2,x}(f)] - \mathscr{K}[Q_{2,x}(f), J_{0,x}] = \mathscr{K}[Q_{1,x}(f), J_{1,x}] \\
- \mathscr{K}[J_{1,x}, Q_{1,x}(f)] + \mathscr{K}[f(x)J_{0,x}, J_{2,x}] - \mathscr{K}[J_{2,x}, f(x)J_{0,x}]. \tag{7.4.11}$$

Assume now that $f, g \subset \mathscr{C}^\infty(X)$, by (7.4.9), (7.4.10) and (7.4.11), we get

$$\begin{aligned}
C_1(f,g)(x) - C_1(g,f)(x) &= \mathscr{K}[Q_{1,x}(f), Q_{1,x}(g)] - \mathscr{K}[Q_{1,x}(g), Q_{1,x}(f)] \\
&+ f(x)\Big(\mathscr{K}[Q_{1,x}(g), J_{1,x}] - \mathscr{K}[J_{1,x}, Q_{1,x}(g)]\Big) \\
&- g(x)\Big(\mathscr{K}[Q_{1,x}(f), J_{1,x}] - \mathscr{K}[J_{1,x}, Q_{1,x}(f)]\Big).
\end{aligned} \tag{7.4.12}$$

From (7.2.12), (7.2.17), (7.2.21) and (7.4.12), we get

$$\begin{aligned}
C_1(f,g)(x) - C_1(g,f)(x) &= \mathscr{K}\Big[\mathscr{K}[1, \tfrac{\partial f_x}{\partial Z_j}(0)Z_j], \mathscr{K}[1, \tfrac{\partial g_x}{\partial Z_j}(0)Z_j]\Big] \\
&- \mathscr{K}\Big[\mathscr{K}[1, \tfrac{\partial g_x}{\partial Z_j}(0)Z_j], \mathscr{K}[1, \tfrac{\partial f_x}{\partial Z_j}(0)Z_j]\Big].
\end{aligned} \tag{7.4.13}$$

From (7.1.11), we obtain

$$\mathscr{K}\Big[1, \tfrac{\partial f_x}{\partial Z_j}(0)Z_j\Big] = \tfrac{\partial f_x}{\partial z_i}(0)z_i + \tfrac{\partial f_x}{\partial \bar{z}_i}(0)\bar{z}_i'. \tag{7.4.14}$$

Plugging (7.1.11), (7.4.14) into (7.4.13),

$$\begin{aligned}
C_1(f,g)(x) - C_1(g,f)(x) &= \sum_{i=1}^{n} \frac{2}{a_i}\Big[\tfrac{\partial f_x}{\partial \bar{z}_i}(0)\tfrac{\partial g_x}{\partial z_i}(0) - \tfrac{\partial f_x}{\partial z_i}(0)\tfrac{\partial g_x}{\partial \bar{z}_i}(0)\Big]\mathrm{Id}_{E_x} \\
&= \sqrt{-1}\{f,g\}\mathrm{Id}_{E_x}.
\end{aligned} \tag{7.4.15}$$

This finishes the proof of Theorem 7.4.1. $\qquad\square$

The next result and Theorem 7.4.1 show that the Berezin–Toeplitz quantization has the correct semi-classical behavior.

Theorem 7.4.2. *For $f \in \mathscr{C}^\infty(X, \mathrm{End}(E))$, the norm of $T_{f,p}$ satisfies*

$$\lim_{p \to \infty} \|T_{f,p}\| = \|f\|_\infty := \sup_{0 \neq u \in E_x, x \in X} |f(x)(u)|_{h^E}/|u|_{h^E}. \tag{7.4.16}$$

Proof. Take a point $x_0 \in X$ and $u_0 \in E_{x_0}$ with $|u_0|_{h^E} = 1$ such that $|f(x_0)(u_0)| = \|f\|_\infty$. Recall that in Section 7.2, we trivialize the bundles L, E in our normal coordinates near x_0, and e_L is the unit frame of L which trivializes L. Moreover, in these normal coordinates, u_0 is a trivial section of E. Considering the sequence of sections $S_{x_0}^p = p^{-n/2} P_p(e_L^{\otimes p} \otimes u_0)$, we have by (4.2.1),

$$\|T_{f,p} S_{x_0}^p - f(x_0) S_{x_0}^p\|_{L^2} \leqslant \frac{C}{\sqrt{p}} \|S_{x_0}^p\|_{L^2}. \tag{7.4.17}$$

The proof of (7.4.16) is complete.   □

Remark 7.4.3. If $E = \mathbb{C}$, then to $f, g \in \mathscr{C}^\infty(X)$, we associated by Theorem 7.4.1 a formal power series $\sum_{l=0}^\infty \hbar^l C_l(f,g) \in \mathscr{C}^\infty(X)[[\hbar]]$, where C_l are bidifferential operators. Therefore, we have constructed in a canonical way an associative star-product $f * g = \sum_{l=0}^\infty \hbar^l C_l(f,g)$.

7.5   Toeplitz operators on non-compact manifolds

We assume that (X, Θ), (L, h^L) and (E, h^E) satisfy the same hypothesis as in Theorem 6.1.1 or (6.1.3). Especially, (X, Θ) is now a complete Hermitian manifold.

Let $\mathscr{C}_{\mathrm{const}}^\infty(X, \mathrm{End}(E))$ denote the algebra of smooth sections of X which are a constant map outside a compact set. For any $f \in \mathscr{C}_{\mathrm{const}}^\infty(X, \mathrm{End}(E))$, we consider the Toeplitz operator $(T_{f,p})_{p \geqslant 1}$ as in (7.2.6)

$$T_{f,p} : L^2(X, L^p \otimes E) \longrightarrow L^2(X, L^p \otimes E), \quad T_{f,p} = P_p f P_p. \tag{7.5.1}$$

The following result generalizes Theorem 7.4.1 to non-compact manifolds.

Theorem 7.5.1. *Let $f, g \in \mathscr{C}_{\mathrm{const}}^\infty(X, \mathrm{End}(E))$. The product of the two corresponding Toeplitz operators admits the asymptotic expansion (7.4.2) in the sense of (7.2.3). Where C_r are bidifferential operators, especially,*

$$\mathrm{supp}(C_r(f,g)) \subset \mathrm{supp}(f) \cap \mathrm{supp}(g), \quad \text{and } C_0(f,g) = fg.$$

If $f, g \in \mathscr{C}_{\mathrm{const}}^\infty(X)$, then (7.4.4) holds. Theorem 7.4.2 also holds for $f \in \mathscr{C}_{\mathrm{const}}^\infty(X)$.

Proof. The most important observation here is that for p large enough, by the spectral gap property, Theorem 1.5.5, as in (4.1.14), for any $s \in H_{(2)}^0(X, L^p \otimes E)$, we have

$$F(D_p)s = P_p s, \quad \|F(D_p) - P_p\| = \mathscr{O}(p^{-\infty}), \tag{7.5.2}$$

moreover, by the proof of Proposition 4.1.5, for any compact set K, $l, m \in \mathbb{N}$, $\varepsilon > 0$, there exists $C_{l,m,\varepsilon} > 0$ such that for $p \geq 1$, $x, x' \in K$,

$$|F(D_p)(x, x') - P_p(x, x')|_{\mathscr{C}^m(K \times K)} \leq C_{l,m,\varepsilon}\, p^{-l}. \tag{7.5.3}$$

As explained in Section 4.1.2, $F(D_p)(x, \cdot)$ only depends on the restriction of D_p to $B^X(x, \varepsilon)$ and is zero outside $B^X(x, \varepsilon)$.

For $g \in \mathscr{C}_0^\infty(X, \mathrm{End}(E))$, let $(F(D_p)gF(D_p))(x, x')$ be the smooth kernel of $F(D_p)gF(D_p)$ with respect to $dv_X(x')$. Then for any relative compact open set U in X such that $\mathrm{supp}(g) \subset U$, we have from (7.5.2) and (7.5.3),

$$\begin{aligned} T_{p,g} - F(D_p)gF(D_p) &= \mathcal{O}(p^{-\infty}), \\ T_{p,g}(x, x') - (F(D_p)gF(D_p))(x, x') &= \mathcal{O}(p^{-\infty}) \quad \text{on } U \times U. \end{aligned} \tag{7.5.4}$$

Now we fix $f, g \in \mathscr{C}_0^\infty(X, \mathrm{End}(E))$. Let $U \subset W$ be relative compact open sets in X such that $\mathrm{supp}(f) \cup \mathrm{supp}(g) \subset U$ and $d(x, y) > 2\varepsilon$ for any $x \in \mathrm{supp}(f) \cup \mathrm{supp}(g)$, $y \in X \setminus U$. From (7.5.2), we have

$$T_{f,p}T_{g,p} = P_p F(D_p) f P_p g F(D_p) P_p. \tag{7.5.5}$$

Let $(F(D_p)fP_p gF(D_p))(x, x')$, be the smooth kernel of $F(D_p)fP_p gF(D_p)$ with respect to $dv_X(x')$. Then the support of $(F(D_p)fP_p gF(D_p))(\cdot, \cdot)$ is contained in $U \times U$. If we fix $x_0 \in U$, the kernel of $F(D_p)fP_p gF(D_p)$ has exactly the same asymptotic expansion as in the compact case by (7.5.3). More precisely, as in (7.4.6), we have

$$\begin{aligned} &p^{-n}(F(D_p)fP_p gF(D_p))_{x_0}(Z, Z') \\ &\cong \sum_{r=0}^{k} (Q_{r,x_0}(f, g)\mathscr{P}_{x_0})(\sqrt{p}Z, \sqrt{p}Z')p^{-\frac{r}{2}} + \mathcal{O}(p^{-\frac{k+1}{2}}), \end{aligned} \tag{7.5.6}$$

with the same local formula for $Q_{r,x_0}(f, g)$ given in (7.4.7).

But since all formal computation is local, we have studied well $Q_{r,x_0}(f, g)$, which is a polynomial with coefficients as bidifferential operators acting on f and g. Thus we know from (7.5.4) that there exist $(g_l)_{l \geq 0}$, $g_l \in \mathscr{C}_0^\infty(X, \mathrm{End}(E))$, $\mathrm{supp}(g_l) \subset \mathrm{supp}(f) \cap \mathrm{supp}(g)$ such that for any $m \geq 1$, $s \in L^2(X, L^p \otimes E)$,

$$\left\| F(D_p)fP_p gF(D_p)s - \sum_{l=0}^{k} F(D_p)P_p\, g_l\, p^{-l}\, P_p F(D_p)s \right\|_{L^2} \leq \frac{C}{p^{k+1}} \|s\|_{L^2}. \tag{7.5.7}$$

(7.5.5) and (7.5.7) imply that

$$\left\| T_{f,p}T_{g,p} - \sum_{l=0}^{k} P_p\, g_l\, p^{-l}\, P_p \right\| \leq C_k\, p^{-k-1}. \tag{7.5.8}$$

For the last part of Theorem 7.5.1, we repeat the proof of Theorem 7.4.2. We conclude thus Theorem 7.5.1. $\qquad \square$

Remark 7.5.2. By Sections 7.2–7.4, we know that we can associate to any $f, g \in \mathscr{C}^\infty(X, \mathrm{End}(E))$ a formal power series $\sum_{l=0}^\infty \hbar^l C_l(f, g) \in \mathscr{C}^\infty(X, \mathrm{End}(E))[[\hbar]]$, where C_l are bidifferential operators. This follows from the fact that the construction in Section 7.4 is local. However, the problem here is which Hilbert space they act on. Theorem 7.5.1 claims that the space of holomorphic L^2-sections $H_{(2)}^0(X, L^p \otimes E)$ of $L^p \otimes E$, is a suitable Hilbert space which allows the Berezin–Toeplitz quantization of the algebra $\mathscr{C}^\infty_{\mathrm{const}}(X, \mathrm{End}(E))$.

Problems

Problem 7.1. Using (4.1.4) and (4.1.7), deduce that $|P_p(x, x')|_{\mathscr{C}^0(X \times X)} \leqslant Cp^n$, for some constant $C > 0$.

Problem 7.2. If $f \in \mathscr{C}^\infty(X)$, verify that in (7.2.16),

$$Q_{2,x}(f) = f(x)J_{2,x} + \mathscr{K}\left[J_{1,x}, \tfrac{\partial f_x}{\partial Z_j}(0)Z_j\right]$$
$$+ \mathscr{K}\left[1, \tfrac{\partial f_x}{\partial Z_j}(0)Z_j J_{1,x} + \sum_{|\alpha|=2} \tfrac{\partial^\alpha f_x}{\partial Z^\alpha}(0)\tfrac{Z^\alpha}{\alpha!}\right].$$

Assume moreover that $\omega = \frac{\sqrt{-1}}{2\pi} R^L$ is the Kähler form of (TX, g^{TX}), then

$$T_{f,p}(x, x) = f(x)p^n + \left(\boldsymbol{b}_1(x)f(x) - \frac{1}{4\pi}(\Delta f)(x)\right)p^{n-1} + \mathscr{O}(p^{n-2}),$$

with $\boldsymbol{b}_1$ in (4.1.8). (Hint: by (4.1.111), $J_{1,x} = 0$ and $J_{2,x}(0,0) = \boldsymbol{b}_1(x)$.)

7.6 Bibliographic notes

The Berezin–Toeplitz quantization was studied by Bordemann–Meinrenken–Schlichenmaier [42], [211] (cf. the references therein for earlier special cases). They consider the case when g^{TX} is the Kähler metric associated to $\omega = \frac{\sqrt{-1}}{2\pi} R^L$ and the twisting bundle E is trivial. Under these assumptions, they established Theorems 7.4.1 and 7.4.2 (cf. also [135], [63]). We also have the Berezin–Toeplitz quantization for a pseudoconvex domain by Englis [102, 103]. All the above works are based on the Boutet de Monvel–Sjöstrand parametrix for the Szegö kernel [53], [105], and the theory of Toeplitz operators of Boutet de Monvel–Guillemin [52].

The approach used here is taken from [162]. In contrast to the previous approaches, we establish directly the results as a consequence of our full asymptotic expansion of Bergman kernel (Theorem 4.2.1). The present approach and the results from Section 5.4.3 also imply the Berezin–Toeplitz quantization for complex orbifolds.

Chapter 8

Bergman Kernels on Symplectic Manifolds

In this chapter, we study the asymptotic expansion of the Bergman kernel associated to modified Dirac operators and renormalized Bochner Laplacians on symplectic manifolds. We will also explain some applications of the asymptotic expansion in the symplectic case. One is, for example, the extension of the Berezin–Toeplitz quantization studied in Chapter 7. We also find Donaldson's Hermitian scalar curvature as the second coefficient of the expansion.

This chapter is organized as follows: In Section 8.1, we study the asymptotic expansion of the Bergman kernel associated to modified Dirac operators and the corresponding Toeplitz operators. In Section 8.2, we obtain the corresponding results when we allow line bundles with mixed curvature (In Section 8.1, we suppose that the curvature of the line bundle is positive). In Section 8.3, we establish the asymptotic expansion for the renormalized Bochner Laplacian.

8.1 Bergman kernels of modified Dirac operators

Let (X, J) be a compact manifold with almost complex structure J and $\dim_{\mathbb{R}} X = 2n$. Let (L, h^L) be a Hermitian line bundle on X, and let (E, h^E) be a Hermitian vector bundle on X. Let ∇^E, ∇^L be Hermitian connections on (E, h^E), (L, h^L) with curvatures R^E, R^L. Let g^{TX} be any Riemannian metric on TX compatible with J.

We assume that the positivity condition (1.5.21) holds for R^L.

This section is organized as follows: In Section 8.1.1, we establish the asymptotic expansion of Bergman kernels of modified Dirac operators by adapting the arguments in Chapter 4; again the spectral gap property plays an essential role here. In Section 8.1.2, we establish the theory on Toeplitz operators associated to the modified Dirac operators in the symplectic case.

We use the notation from Section 1.5.2 and from (1.5.14)–(1.5.19).

8.1.1   Asymptotic expansion of the Bergman kernel

Let $\nabla^{\det}$ be a Hermitian connection on $\det(T^{*(0,1)}X)$ with curvature $R^{\det}$. We denote by D_p^c the spin^c Dirac operator associated to $L^p \otimes E$ and $\nabla^{\det}$ as in (1.3.15). Following (1.6.1), we denote

$$\mathbf{E}^j = \Lambda^j(T^{*(0,1)}X) \otimes E, \quad \mathbf{E}^- := \oplus_j \mathbf{E}^{2j+1}, \quad \mathbf{E}^+ := \oplus_j \mathbf{E}^{2j},$$

$$\mathbf{E} = \mathbf{E}^+ \oplus \mathbf{E}^-, \quad E_p^- = \mathbf{E}^- \otimes L^p, \quad E_p^+ = \mathbf{E}^+ \otimes L^p, \quad E_p = E_p^+ \oplus E_p^-. \tag{8.1.1}$$

For $A \in \Lambda^3(T^*X)$, let ∇^A be the Hermitian connection on E_p induced by ∇^{Cl}, A and ∇^{L^p}, ∇^E as in (1.3.33). Let $D_p^{c,A}$ be the modified Dirac operator on X defined in (1.5.27).

Definition 8.1.1. The *Bergman kernel* $P_p^A(x, x')$, $(x, x' \in X)$, is the smooth kernel of the orthogonal projection P_p^A, from $\Omega^{0,\bullet}(X, L^p \otimes E)$ onto $\mathrm{Ker}(D_p^{c,A})$, with respect to the Riemannian volume form $dv_X(x')$. Hence, $P_p^A(x, x') \in (E_p)_x \otimes (E_p)_{x'}^*$, and $P_p^A(x, x) \in \mathrm{End}(\Lambda(T^{*(0,1)}X) \otimes E)_x$.

By using the spectral gap property (Theorem 1.5.8), and the same proof of Proposition 4.1.5, we obtain:

Theorem 8.1.2. *For any* $l, m \in \mathbb{N}$, $\varepsilon > 0$, *there exists* $C_{l,m,\varepsilon} > 0$ *such that for any* $p \geqslant 1$ *the following estimate holds:*

$$|P_p^A(x, x')|_{\mathscr{C}^m} \leqslant C_{l,m,\varepsilon} p^{-l} \quad \text{for any } x, x' \in X \text{ with } d(x, x') \geqslant \varepsilon. \tag{8.1.2}$$

The $\mathscr{C}^m$ *norm used here is induced by* ∇^{TX}, ∇^L, ∇^E *and* h^L, h^E, g^{TX}.

We denote by $I_{\mathbb{C} \otimes E}$ the projection from $\Lambda(T^{*(0,1)}X) \otimes E$ onto $\mathbb{C} \otimes E$ corresponding to the decomposition $\Lambda(T^{*(0,1)}X) = \mathbb{C} \oplus \Lambda^{>0}(T^{*(0,1)}X)$.

Theorem 8.1.3. *There exist smooth coefficients* $\boldsymbol{b}_r(x) \in \mathrm{End}(\Lambda(T^{*(0,1)}X) \otimes E)_x$ *where*

$$\boldsymbol{b}_0 = \det(\dot{R}^L/(2\pi))I_{\mathbb{C} \otimes E}, \tag{8.1.3}$$

with the following property: for any $k, l \in \mathbb{N}$, *there exists* $C_{k,l} > 0$ *such that*

$$\left| P_p^A(x, x) - \sum_{r=0}^{k} \boldsymbol{b}_r(x)p^{n-r} \right|_{\mathscr{C}^l} \leqslant C_{k,l}p^{n-k-1}. \tag{8.1.4}$$

for any $x \in X$, $p \in \mathbb{N}^*$.

The coefficients $\boldsymbol{b}_r(x)$ are polynomials in A, R^{TX}, $R^{\det}$, R^E (and R^L) and their derivatives with order $\leqslant 2r - 1$ (resp. $2r$) and reciprocals of linear combinations of eigenvalues of $\dot{R}^L$ at x.

Moreover, the expansion (8.1.4) is uniform in the following sense: for any fixed $k, l \in \mathbb{N}$, assume that the derivatives of g^{TX}, h^L, ∇^L, h^E, ∇^E, J, with order $\leqslant 2n + 2k + l + 6$ run over a set bounded in the $\mathscr{C}^l$-norm taken with respect to the parameter $x \in X$ and, moreover, g^{TX} runs over a set bounded below. Then the constant $C_{k,l}$ is independent of g^{TX}; and the $\mathscr{C}^l$-norm in (4.1.7) includes also the derivatives on the parameters.

Let $x_0 \in X$. As in Section 4.1.3, we trivialize E_p on $B^X(x_0, 4\varepsilon) \simeq B^{T_{x_0}X}(0, 4\varepsilon)$ by using the parallel transport with respect to the connection ∇^A along the curve $[0,1] \ni u \to uZ$. Under this trivialization, we have $P_p^A(Z, Z') \in \mathrm{End}(E_{p,x_0}) = \mathrm{End}(\mathbf{E}_{x_0})$ for $Z, Z' \in T_{x_0}X$. Thus we can view $P_{p,x_0}^A(Z, Z') := P_p^A(Z, Z')$, $(Z, Z' \in T_{x_0}X, |Z|, |Z'| \leqslant 2\varepsilon)$, as a smooth section of $\pi^*(\mathrm{End}(\mathbf{E}))$ over $TX \times_X TX$ as in Section 4.1.5, and we denote by $|\cdot|_{\mathscr{C}^m(X)}$ the $\mathscr{C}^m$-norm with respect to the parameter $x_0 \in X$.

Recall that μ_0 and κ were defined in (1.5.26), (4.1.28). Recall also that $\mathscr{P}(Z, Z')$ is the smooth kernel of the orthogonal projection from $L^2(\mathbb{R}^{2n})$ onto $\mathrm{Ker}(\mathscr{L})$ defined in (4.1.84).

The following off-diagonal expansion of the Bergman kernel, which is an analogue of Theorem 4.2.1, is the main result of this section.

Theorem 8.1.4. *There exist $J_r(Z, Z') \in \mathrm{End}(\Lambda(T^{*(0,1)}X) \otimes E)_{x_0}$ polynomials in Z, Z' with the same parity as r and $\deg J_r(Z, Z') \leqslant 3r$, whose coefficients are polynomials in A, R^{TX}, $R^{\det}$, R^E (and R^L) and their derivatives of order $\leqslant r - 1$ (resp. r), and reciprocals of linear combinations of eigenvalues of $\dot{R}^L$ at x_0, such that*

$$\mathscr{F}_r(Z, Z') = J_r(Z, Z')\mathscr{P}(Z, Z'), \quad J_0(Z, Z') = I_{\mathbb{C} \otimes E}, \quad (8.1.5)$$

and such that there exists $C'' > 0$ such that for $k, m, m' \in \mathbb{N}$, there exist $C > 0$, $N \in \mathbb{N}$ such that if $p \geqslant 1$, $Z, Z' \in T_{x_0}X$, $|Z|, |Z'| \leqslant 2\varepsilon$, $\alpha, \alpha' \in \mathbb{N}^{2n}$, $|\alpha| + |\alpha'| \leqslant m$, then

$$\left| \frac{\partial^{|\alpha|+|\alpha'|}}{\partial Z^\alpha \partial Z'^{\alpha'}} \left(\frac{1}{p^n} P_p^A(Z, Z')\kappa^{1/2}(Z)\kappa^{1/2}(Z') - \sum_{r=0}^{k} \mathscr{F}_r(\sqrt{p}Z, \sqrt{p}Z')p^{-r/2} \right) \right|_{\mathscr{C}^{m'}(X)}$$

$$\leqslant Cp^{-(k+1-m)/2}(1 + |\sqrt{p}Z| + |\sqrt{p}Z'|)^N \exp(-\sqrt{C''\mu_0}\sqrt{p}|Z - Z'|) + \mathscr{O}(p^{-\infty}). \quad (8.1.6)$$

Remark 8.1.5. By Theorem 1.5.7 and because $(D_p^{c,A})^2$ preserves the $\mathbb{Z}_2$-grading of $\Omega^{0,\bullet}(X, L^p \otimes E)$, P_p^A is the orthogonal projection from $\mathscr{C}^\infty(X, E_p^+)$ onto $\mathrm{Ker}(D_p^{c,A})$ for p large enough. Thus $P_p^A(x, x), b_r(x) \in \mathrm{End}(\mathbf{E}^+)_x$ and $J_r(Z, Z') \in \mathrm{End}(\mathbf{E}^+)_{x_0}$.

As in Theorem 4.2.3, the following result relates the coefficients of the expansion of the Bergman kernel and the heat kernel.

Theorem 8.1.6. *There exist smooth sections $b_{r,u}$ of $\operatorname{End}(\Lambda(T^{*(0,1)}X) \otimes E)$ on X which are polynomials in A, R^{TX}, $R^{\det}$, R^E (and R^L) and their derivatives with order $\leqslant 2r - 1$ (resp. $2r$) and functions on the eigenvalues of $\dot{R}^L$ at x, and $b_{0,u}$ in (4.2.3) such that for each $u > 0$ fixed, we have the asymptotic expansion in the sense of (8.1.4) as $p \to \infty$,*

$$\exp(-\frac{u}{p}(D_p^{c,A})^2)(x,x) = \sum_{r=0}^{k} b_{r,u}(x)p^{n-r} + \mathscr{O}(p^{n-k-1}). \qquad (8.1.7)$$

Moreover, as $u \to +\infty$, with $\boldsymbol{b}_r$ in (8.1.4), we have

$$b_{r,u}(x) = \boldsymbol{b}_r(x) + \mathscr{O}(e^{-\frac{1}{8}\mu_0 u}). \qquad (8.1.8)$$

Proof of Theorems 8.1.3–8.1.6. Observe that in the construction from Section 4.1.3, we only need to replace the holomorphic connection $\nabla^{\det}$ on $\det(T^{(1,0)}X)$ by the Hermitian connection $\nabla^{\det}$ considered here, and $-\frac{1}{4}T_{as}$ by A. Then we get the modified Dirac operator D_p^{c,A_0} on $\mathbb{R}^{2n}$. Now in Theorem 4.1.7, we only need to replace $R^{B,\Lambda^{0,\bullet}}$, $-\frac{1}{4}T_{as}$ therein by R^{Cl} and A. Then the arguments in Section 4.1 go through until (4.1.96). In particular, by using (1.3.11), we obtain $\mathscr{F}_r$ from (8.1.5) in the same way as in Theorem 4.1.21.

Of course, here $\mathscr{F}_r$ will not preserve the $\mathbb{Z}$-grading on $\mathbf{E}_{x_0} = (\Lambda(T^{*(0,1)}X) \otimes E)_{x_0}$. Since $(D_p^{c,A})^2$ preserves the $\mathbb{Z}_2$-grading on $\mathscr{C}^\infty(X, E_p)$ and is invertible on $\mathscr{C}^\infty(X, E_p)$ for p large enough, then in view of our construction, $\mathscr{L}_2^t$ preserves the $\mathbb{Z}_2$-grading on $L^2(\mathbb{R}^{2n}, \mathbf{E}_{x_0})$ and is invertible on $L^2(\mathbb{R}^{2n}, \mathbf{E}_{x_0})$ for t small enough. Consequently, we know that for p large enough, $P_p^A(Z, Z')$ and $\mathscr{F}_r(Z, Z')$ from (8.1.6) are zero when we restrict to $\mathbf{E}_{x_0}^-$.

We run now the arguments from Section 4.2 without any change, concluding the proof of Theorems 8.1.4 and 8.1.6. $\qquad \square$

Remark 8.1.7. As in Remark 4.1.4, the formula (8.1.7) holds without the assumption (1.5.21) and $b_{0,u}$ is computed in Theorem 1.6.1.

8.1.2 Toeplitz operators on symplectic manifolds

We use the same notation as in Chapter 7. In view of Definition 7.3.1, it is natural to introduce the Toeplitz operator on symplectic manifolds.

Definition 8.1.8. A *Toeplitz operator* is a family $\{T_p\}$ of linear operators

$$T_p : L^2(X, E_p) \longrightarrow L^2(X, E_p), \qquad (8.1.9)$$

with the properties:

(i) For any $p \in \mathbb{N}$, we have
$$T_p = P_p^A T_p P_p^A . \tag{8.1.10}$$

(ii) There exist a sequence $g_l \in \mathscr{C}^\infty(X, \operatorname{End}(E))$ such that for all $k \geqslant 0$ there exists $C_k > 0$ with

$$\left\| T_p - P_p^A \left(\sum_{l=0}^k p^{-l} g_l \right) P_p^A \right\| \leqslant C_k \, p^{-k-1}, \tag{8.1.11}$$

where $\|\cdot\|$ denotes the operator norm on the space of bounded operators, and g_l acts on $\Lambda(T^{*(0,1)}X) \otimes L^p \otimes E$ as $\operatorname{Id}_{\Lambda(T^{*(0,1)}X) \otimes L^p} \otimes g_l$.

The full symbol of $\{T_p\}$ is the formal series $\sum_{l=0}^\infty \hbar^l g_l \in \mathscr{C}^\infty(X, \operatorname{End}(E))[[\hbar]]$ and the *principal symbol* of $\{T_p\}$ is g_0. If each T_p is self-adjoint, $\{T_p\}$ is called self-adjoint.

Certainly, for $f \in \mathscr{C}^\infty(X, \operatorname{End}(E))$, Lemma 7.2.2 still holds for the kernel $T_{f,p}(x, x')$ of the Toeplitz operator $T_{f,p}$ by Theorems 8.1.2 and 8.1.4.

Now let $\Xi_p : L^2(X, E_p) \longrightarrow L^2(X, E_p)$ be a family of linear operator with smooth kernel $\Xi_p(x, y)$. We use the same notation as in (7.2.10) except that here $Q_{r,x_0} \in \operatorname{End}(\Lambda(T^{*(0,1)}X) \otimes E)_{x_0}[Z, Z']$.

Then Lemmas 7.2.3 and 7.2.4 still hold and

$$J_{r,x_0}, Q_{r,x_0}(f) \in \operatorname{End}(\Lambda(T^{*(0,1)}X) \otimes E)_{x_0}[Z, Z']. \tag{8.1.12}$$

Now (7.2.12) and (7.2.17) need to be replaced by

$$J_{0,x_0} = I_{\mathbb{C} \otimes E}, \quad Q_{0,x_0}(f) = f(x_0) I_{\mathbb{C} \otimes E}. \tag{8.1.13}$$

We have still the following criterion for Toeplitz operators.

Theorem 8.1.9. *If $T_p : L^2(X, E_p) \longrightarrow L^2(X, E_p)$ be a family of bounded linear operators verifying* i), ii), iii) *of Theorem 7.3.1 by replacing P_p therein by P_p^A, and with $\{Q_{r,x_0} \in \operatorname{End}(\Lambda(T^{*(0,1)}X) \otimes E)_{x_0}[Z, Z']\}_{x_0 \in X}$; then $\{T_p\}$ is a Toeplitz operator.*

Proof. We claim first $Q_{0,x_0} \in \operatorname{End}(E_{x_0}) \circ I_{\mathbb{C} \otimes E}[Z, Z']$, and Q_{0,x_0} is a polynomial in $z, \overline{z}'$. In fact, by (7.2.13) and (8.1.13), analogous to (7.3.7), we have

$$p^{-n}(P_p^A T_p P_p^A)_{x_0}(Z, Z')$$
$$\cong ((\mathscr{P} J_0) \circ (Q_0 \mathscr{P}) \circ (\mathscr{P} J_0))_{x_0}(\sqrt{p} Z, \sqrt{p} Z') + \mathcal{O}(p^{-1/2}). \tag{8.1.14}$$

Since $P_p^A T_p P_p^A = T_p$, we deduce from (7.3.6), (8.1.13) and (8.1.14) that

$$Q_{0,x_0} \mathscr{P}_{x_0} = I_{\mathbb{C} \otimes E} \mathscr{P}_{x_0} \circ (Q_{0,x_0} \mathscr{P}_{x_0}) \circ \mathscr{P}_{x_0} I_{\mathbb{C} \otimes E}, \tag{8.1.15}$$

hence $Q_{0,x_0} \in \operatorname{End}(E_{x_0}) \circ I_{\mathbb{C} \otimes E}[z, \overline{z}']$ by (7.1.5) and (7.1.8).

By the same proof as of Proposition 7.3.2, we get

$$Q_{0,x_0}(Z, Z') = Q_{0,x_0}(0, 0) \in \operatorname{End}(E_{x_0}) \circ I_{\mathbb{C} \otimes E}. \tag{8.1.16}$$

Now we set

$$g_0(x_0) = \mathcal{Q}_{0,x_0}(0,0)|_{\mathbb{C}\otimes E} \in \mathrm{End}(E_{x_0}).\qquad(8.1.17)$$

Then as in the proof of Proposition 7.3.8, we get the analogue of (7.3.5):

$$T_p = P_p^A\, g_0\, P_p^A + \mathcal{O}(p^{-1}).\qquad(8.1.18)$$

Since $p(T_p - P_p^A g_0 P_p^A)$ verifies the condition of Theorem 8.1.9, we can continue the process to generate a sequence $(g_l)_{l\geqslant 0}$, $g_l \in \mathscr{C}^\infty(X, \mathrm{End}(E))$, such that (8.1.11) holds for any $k \geqslant 1$. This completes the proof of Theorem 8.1.9. $\qquad\square$

As in the holomorphic case, the set of Toeplitz operators is closed under the composition of operators, so forms an algebra.

Theorem 8.1.10. *Let $f, g \in \mathscr{C}^\infty(X, \mathrm{End}(E))$. The product of the Toeplitz operators $T_{f,p}$ and $T_{g,p}$ is a Toeplitz operator; more precisely, it admits the asymptotic expansion (7.4.2) where C_r are bidifferential operators, and $C_r(f,g) \in \mathscr{C}^\infty(X, \mathrm{End}(E))$. In particular $C_0(f,g) = fg$.*

If $f, g \in \mathscr{C}^\infty(X)$, then (7.4.4) holds.

The norm of $T_{f,p}$ satisfies (7.4.16), for any $f \in \mathscr{C}^\infty(X, \mathrm{End}(E))$.

Remark 7.4.3 about star-products still holds if $E = \mathbb{C}$.

Proof. The same proof as in Section 7.4 delivers Theorem 8.1.10. $\qquad\square$

8.2   Bergman kernel: mixed curvature case

In this section, we do not suppose that the almost complex structure polarizes the curvature of the line bundle L, that is we allow line bundles with mixed curvature (negative and positive eigenvalues). We will extend the results from Section 8.1 to this situation.

This section is organized as follows: In Section 8.2.1, we establish the spectral gap property for the modified Dirac operator in the mixed curvature case, by simply repeating the argument in the proof of Theorem 1.5.7. In Section 8.2.2 we explain that the asymptotic expansion from Section 8.1 still holds, and compute the first coefficients in the expansion.

We use the same notation as in Section 8.1. We choose the almost complex structure J such that ω is J-invariant, i.e., $\omega(\cdot,\cdot) = \omega(J\cdot, J\cdot)$. But we only suppose that ω is non-degenerate, and we do not suppose that $\omega(\cdot, J\cdot)$ is positive. This is the difference comparing with the assumption in Section 8.1.

8.2.1   Spectral gap

Under our assumption, $\dot{R}^L \in \mathrm{End}(T^{(1,0)}X)$ is still well defined by (1.5.15), and for any $x \in X$, we can diagonalize $\dot{R}_x^L$, i.e., find an orthonormal basis $\{w_j\}_{j=1}^n$ of $T^{(1,0)}X$ such that $\dot{R}_x^L w_j = a_j(x)w_j$ with $a_j(x) \in \mathbb{R}$. As ω is non-degenerate, the number of negative eigenvalues of $\dot{R}_x^L \in \mathrm{End}(T_x^{(1,0)} X)$ does not depend on x, and we denote it by q. (In Section 8.1, we suppose that $q = 0$.) From now on, we assume that

$$\dot{R}_x^L w_j = a_j(x) w_j, \quad a_j(x) < 0 \text{ for } j \leqslant q \text{ and } a_j(x) > 0 \text{ for } j > q. \tag{8.2.1}$$

Then the vectors $\{w_j\}_{j=1}^q$ span a sub-bundle W of $T^{(1,0)} X$. Set

$$\widetilde{\omega}_d(x) = -\sum_{j=1}^n a_j(x) \overline{w}^j \wedge i_{\overline{w}_j} + \sum_{j=1}^q a_j(x)$$

$$= \sum_{j=1}^q a_j(x) i_{\overline{w}_j} \wedge \overline{w}^j - \sum_{j=q+1}^n a_j(x) \overline{w}^j \wedge i_{\overline{w}_j}, \tag{8.2.2}$$

$$\widetilde{\tau}(x) = \sum_{j=1}^n |a_j(x)| = -\sum_{j=1}^q a_j(x) + \sum_{j=q+1}^n a_j(x),$$

$$\mu_0 = \inf_{x \in X, j} |a_j(x)|.$$

The following result extends Theorem 1.5.7 to the current situation. We denote $\Omega^{0, \neq q}(X, L^p \otimes E) := \bigoplus_{k \neq q} \Omega^{0,k}(X, L^p \otimes E)$.

Theorem 8.2.1. *There exists $C > 0$ such that for any $p \in \mathbb{N}$,*

$$\|D_p^{c,A} s\|_{L^2}^2 \geqslant (2p\mu_0 - C)\|s\|_{L^2}^2, \quad \text{for } s \in \Omega^{0, \neq q}(X, L^p \otimes E). \tag{8.2.3}$$

Proof. We consider now $s \in \mathscr{C}^\infty(X, L^p \otimes \mathbf{E})$, where $\mathbf{E} = \Lambda(T^{*(0,1)} X) \otimes E$. We will apply (1.5.30) to the almost complex structure $J' \in \mathrm{End}(TX)$ defined by

$$J' w_j = -\sqrt{-1} w_j \text{ for } j \leqslant q \text{ and } J' w_j = \sqrt{-1} w_j \text{ for } j > q,$$

then $\widetilde{\tau}$ is τ associated to J' in (1.5.16). By (1.5.30), there exists $C > 0$ such that for any $p \in \mathbb{N}$, $s \in \mathscr{C}^\infty(X, L^p \otimes \mathbf{E})$, we have

$$\|\nabla^{L^p \otimes \mathbf{E}} s\|_{L^2}^2 - p \langle \widetilde{\tau}(x) s, s \rangle \geqslant -C\|s\|_{L^2}^2. \tag{8.2.4}$$

By (1.3.35) and (8.2.2), comparing to (1.5.19) and (1.5.34), we have

$$\|D_p^{c,A} s\|^2 = \|\nabla^A s\|^2 - p\langle \widetilde{\tau} s, s \rangle - 2p\langle \widetilde{\omega}_d s, s \rangle$$

$$+ \left\langle \left(\frac{r^X}{4} + {}^c(R^E + \frac{1}{2} R^{\det}) + {}^c(dA) - 2|A|^2 \right) s, s \right\rangle. \tag{8.2.5}$$

If $s \in \Omega^{0, \neq q}(X, L^p \otimes E)$, the third term of (8.2.5), $-2p\langle \widetilde{\omega}_d s, s \rangle$ is bounded below by $2p\mu_0 \|s\|_{L^2}^2$, while the fourth term of (8.2.5) is $\mathscr{O}(\|s\|_{L^2}^2)$. The proof of (8.2.3) is completed. $\square$

From Theorem 8.2.1, the same proof as in Theorem 1.5.8 gives the following spectral gap property.

Theorem 8.2.2. *There exists $C_L > 0$ such that for $p \in \mathbb{N}$,*

$$\mathrm{Spec}((D_p^{c,A})^2) \subset \{0\} \cup [2p\mu_0 - C_L, +\infty[. \tag{8.2.6}$$

Set $o_q = -$ if q is even; $o_q = +$ if q is odd, then for p large enough, we have

$$\mathrm{Ker}(D_{o_q,p}^{c,A}) = \{0\}. \tag{8.2.7}$$

8.2.2　Asymptotic expansion of the Bergman kernel

The existence of the spectral gap expressed in Theorem 8.2.2 allows us to obtain immediately the analogue of (4.1.12). Namely, for any $l, m \in \mathbb{N}$ and $\varepsilon > 0$, there exists $C_{l,m,\varepsilon} > 0$ such that for $p \geqslant 1$, $x, x' \in X$, $d(x, x') > \varepsilon$,

$$|P_p^A(x, x')|_{\mathscr{C}^m} \leqslant C_{l,m,\varepsilon}\, p^{-l}. \tag{8.2.8}$$

Recall that W is the subbundle of $T^{(1,0)}X$ spanned by the eigenvectors of negative eigenvalues of $\dot{R}^L \in \mathrm{End}(T^{(1,0)}X)$. We denote by $(\det(\overline{W}^*))^\perp$ the orthogonal complement of $\det(\overline{W}^*)$ in $\Lambda(T^{*(0,1)}X)$. We denote by $I_{\det(\overline{W}^*)\otimes E}$ the orthogonal projection from $\mathbf{E} = \Lambda(T^{*(0,1)}X) \otimes E$ onto $\det(\overline{W}^*) \otimes E$.

We use the trivialization of E_p on the normal coordinate $B^X(x_0, 4\varepsilon) \simeq B^{T_{x_0}X}(0, 4\varepsilon)$ as in Section 8.1. We view $P_{p,x_0}^A(Z, Z') := P_p^A(Z, Z')$, $(Z, Z' \in T_{x_0}X, |Z|, |Z'| \leqslant 2\varepsilon)$, as a smooth section of $\pi^*(\mathrm{End}(\mathbf{E}))$ over $TX \times_X TX$.

We still use the notation in Section 4.1.6. Then with $\widetilde{\omega}_{d,x_0}, \widetilde{\tau}_{x_0}$ introduced in (8.2.2), we get that the limit operator $\mathscr{L}_2^0$ is

$$\widetilde{\mathscr{L}} = -\sum_i (\nabla_{0,e_i})^2 - \widetilde{\tau}_{x_0} = \sum_{j=1}^q b_j^+ b_j + \sum_{j=q+1}^n b_j b_j^+,$$
$$\mathscr{L}_2^0 = \widetilde{\mathscr{L}} - 2\widetilde{\omega}_{d,x_0}. \tag{8.2.9}$$

Note that by (8.2.2), we have for $\widetilde{\omega}_{d,x_0} \in \mathrm{End}(\Lambda(T^{*(0,1)}X))_{x_0}$,

$$\mathrm{Ker}(\widetilde{\omega}_{d,x_0}) = \det(\overline{W}^*)_{x_0},$$
$$\widetilde{\omega}_{d,x_0} \leqslant -\mu_0 \quad \text{on } (\det(\overline{W}^*))_{x_0}^\perp. \tag{8.2.10}$$

Set

$$\widetilde{z} = (\overline{z}_1, \ldots, \overline{z}_q, z_{q+1}, \ldots, z_n),$$
$$\widetilde{b} = (b_1^+, \ldots, b_q^+, b_{q+1}, \ldots, b_n). \tag{8.2.11}$$

Then by the same argument as the proof of Theorem 4.1.20, we have:

Theorem 8.2.3. *The spectrum of the restriction of $\widetilde{\mathscr{L}}$ to $L^2(\mathbb{R}^{2n})$ is given by*

$$\mathrm{Spec}(\widetilde{\mathscr{L}}|_{L^2(\mathbb{R}^{2n})}) = \Big\{ 2 \sum_{i=1}^{n} \alpha_i |a_i| \ : \ \alpha = (\alpha_1, \ldots, \alpha_n) \in \mathbb{N}^n \Big\} \qquad (8.2.12)$$

and an orthogonal basis of the eigenspace of $2\sum_{i=1}^{n} \alpha_i |a_i|$ is given by

$$\widetilde{b}^\alpha \Big(\widetilde{z}^\beta \exp\big(-\frac{1}{4} \sum_i |a_i| \, |z_i|^2 \big) \Big), \quad \text{with } \beta \in \mathbb{N}^n. \qquad (8.2.13)$$

Let $\widetilde{\mathscr{P}}(Z, Z')$ be the smooth kernel of the orthogonal projection $\widetilde{\mathscr{P}}$ from $(L^2(\mathbb{R}^{2n}), \|\cdot\|_{L^2})$ onto $\mathrm{Ker}(\widetilde{\mathscr{L}})$, calculated with respect to $dv_{TX}(Z')$. From (8.2.13) we get

$$\widetilde{\mathscr{P}}(Z, Z') = \prod_{i=1}^{n} \frac{|a_i|}{2\pi} \ \exp\Big(-\frac{1}{4} \sum_{i=1}^{n} |a_i|(|z_i|^2 + |z_i'|^2)$$

$$-\frac{1}{4} \Big(\sum_{i=1}^{q} |a_i| \, \overline{z}_i z_i' + \sum_{i=q+1}^{n} |a_i| \, z_i \overline{z}_i' \Big) \Big). \qquad (8.2.14)$$

Let P^N be the orthogonal projection from $(L^2(\mathbb{R}^{2n}), \mathbf{E}_{x_0}), \|\cdot\|_{L^2})$ onto $N = \mathrm{Ker}(\mathscr{L}_2^0)$, and $P^N(Z, Z')$ its smooth kernel with respect to $dv_{TX}(Z')$. Hence (8.2.10) yields

$$P^N(Z, Z') = \widetilde{\mathscr{P}}(Z, Z') I_{\det(\overline{W}^*) \otimes E}. \qquad (8.2.15)$$

From the above discussion we deduce:

Theorem 8.2.4. *Theorems 8.1.2, 8.1.3 and 8.1.6 hold with $b_{0,u}$ given by (4.2.3) and to b_0, J_0 from (8.1.3), (8.1.5) correspond now to*

$$b_0 = \big| \det(\dot{R}^L/(2\pi)) \big| \, I_{\det(\overline{W}^*) \otimes E}, \quad J_0(Z, Z') = I_{\det(\overline{W}^*) \otimes E}. \qquad (8.2.16)$$

The corresponding $\widetilde{\mathscr{P}}(Z, Z')$ is given by (8.2.14).

The Toeplitz operators are introduced as in Definition 8.1.8 by using the more general meaning of P_p^A considered in this section.

Theorem 8.2.5. *Theorem 8.1.10 holds under the assumption of this section.*

Proof. Observe that the kernel calculus for $\mathscr{P}$ presented in Section 7.1 has the corresponding version for $\widetilde{\mathscr{P}}$ (cf. (8.2.14)). Basically, we only need to replace b, z therein by $\widetilde{b}, \widetilde{z}$. If we denote by $\mathscr{K}[F, G] \in \mathbb{C}[Z, Z']$ the operation associated to

$\widetilde{\mathscr{P}}$ as in (7.1.6), then by exactly the same computation, we get the analogue of (7.4.13) by replacing $\mathscr{K}$ by $\widetilde{\mathscr{K}}$: for $f, g \in \mathscr{C}^\infty(X)$,

$$
C_1(f,g)(x) - C_1(g,f)(x) = \widetilde{\mathscr{K}}\left[\widetilde{\mathscr{K}}[1, \tfrac{\partial f_x}{\partial Z_j}(0)Z_j], \widetilde{\mathscr{K}}[1, \tfrac{\partial g_x}{\partial Z_j}(0)Z_j]\right]
$$
$$
- \widetilde{\mathscr{K}}\left[\widetilde{\mathscr{K}}[1, \tfrac{\partial g_x}{\partial Z_j}(0)Z_j], \widetilde{\mathscr{K}}[1, \tfrac{\partial f_x}{\partial Z_j}(0)Z_j]\right]. \quad (8.2.17)
$$

Now our theorem follows from Problem 8.2.   □

Remark 8.2.6. We specify now the preceding results to the complex case and we return to the notation from Section 1.5.1. Then our assumption here is that the curvature R^L is non-degenerate as a 2-form on X. D_p in (1.5.20) is a modified Dirac operator by (1.4.17). Thus all results in this section apply. As $D_p^2 = 2\square_p$ preserves the $\mathbb{Z}$-grading on $\Omega^{0,\bullet}(X, L^p \otimes E)$, we get from (8.2.3), that for p large enough,

$$
H^k(X, L^p \otimes E) = 0 \quad \text{for any } k \neq q. \tag{8.2.18}
$$

Moreover, for p large enough, the Bergman kernel $P_p(x, x')$ is in this case the kernel of the orthogonal projection from $\Omega^{0,q}(X, L^p \otimes E)$ on $\mathrm{Ker}(D_p^2) \simeq H^q(X, L^p \otimes E)$.

8.3   Generalized Bergman kernel

Let (X, J) be a compact manifold with almost complex structure J and $\dim_\mathbb{R} X = 2n$. Let (L, h^L) be a Hermitian line bundle on X, and let (E, h^E) be a Hermitian vector bundle on X. Let ∇^E, ∇^L be Hermitian connections on (E, h^E), (L, h^L). Let R^L, R^E be the curvature of ∇^L, ∇^E. Let g^{TX} be any Riemannian metric on TX compatible with J.

In this section, we assume that the positivity condition (1.5.21) holds for R^L. We still use the notation in Sections 1.5.2 and 8.1.1.

This section is organized as follows: In Section 8.3.1, we establish the spectral gap property for the renormalized Bochner Laplacian. In Section 8.3.2, we state the asymptotic expansion for the generalized Bergman kernel. In Section 8.3.3, we study the near diagonal asymptotic expansion. In particular, we obtain Theorem 8.3.3 as a special case of Theorem 8.3.8. In Section 8.3.4 we prove Theorem 8.3.4 and calculate the second coefficient of the expansion; one of the summands thereof is the Hermitian scalar curvature. In Section 8.3.5, we establish a symplectic version of the Kodaira embedding Theorem. Especially, this gives another analytic proof of the classical Kodaira embedding Theorem 5.1.12.

8.3.1   Spectral gap

Let $\Delta^{L^p \otimes E}$ be the Bochner Laplacian acting on $\mathscr{C}^\infty(X, L^p \otimes E)$ associated to ∇^L, ∇^E. We fix a smooth Hermitian section Φ of $\mathrm{End}(E)$ on X, i.e., $\Phi^* = \Phi$. The *renormalized Bochner Laplacian* $\Delta_{p,\Phi}$ on $L^p \otimes E$ is defined by

$$\Delta_{p,\Phi} = \Delta^{L^p \otimes E} - p\tau + \Phi. \tag{8.3.1}$$

Recall that μ_0 was defined in (1.5.26).

Theorem 8.3.1. *There exist* $\widetilde{C}_L, C_L > 0$ *such that for* $p \in \mathbb{N}$,

$$\mathrm{Spec}(\Delta_{p,\Phi}) \subset [-\widetilde{C}_L, \widetilde{C}_L] \cup [2p\mu_0 - C_L, +\infty[. \tag{8.3.2}$$

Let $\mathcal{H}_p$ *be the eigenspace of* $\Delta_{p,\Phi}$ *with the eigenvalues in* $[-\widetilde{C}_L, \widetilde{C}_L]$. *Then for* p *large enough,*

$$\dim \mathcal{H}_p = \int_X \mathrm{Td}(T^{(1,0)}X)\,\mathrm{ch}(L^p \otimes E). \tag{8.3.3}$$

Proof. Let $I_p : \Omega^{0,\bullet}(X, L^p \otimes F) \longrightarrow \mathscr{C}^\infty(X, L^p \otimes E)$ be the orthogonal projection. For $s \in \Omega^{0,\bullet}(X, L^p \otimes E)$, we will denote by $s_0 = I_p s$ its 0-degree component.

Recall that D_p^c is the spin^c Dirac operator as in Section 8.1, and $D_{+,p}^c$ is the restriction of D_p^c on $\mathscr{C}^\infty(X, E_p^+)$. We will estimate $\Delta_{p,\Phi}$ on $I_p(\mathrm{Ker}(D_{+,p}^c))$ and $(\mathrm{Ker}(D_{+,p}^c))^\perp \cap \mathscr{C}^\infty(X, L^p \otimes E)$.

In the sequel we denote with C all positive constants independent of p, although there may be different constants for different estimates. From (1.5.31) and (1.5.32) by taking $\nabla^{\det} = \nabla^{\det_1}$, $A = 0$, there exists $C > 0$ such that for $s \in \mathscr{C}^\infty(X, L^p \otimes E)$,

$$\left| \|D_p^c s\|_{L^2}^2 - \langle \Delta_{p,\Phi} s, s \rangle \right| \leqslant C\|s\|_{L^2}^2. \tag{8.3.4}$$

Theorem 1.5.8 and (8.3.4) show that there exists $C > 0$ such that for $p \in \mathbb{N}$,

$$\langle \Delta_{p,\Phi} s, s \rangle \geqslant (2p\mu_0 - C)\|s\|_{L^2}^2, \quad \text{for } s \in \mathscr{C}^\infty(X, L^p \otimes E) \cap (\mathrm{Ker}(D_{+,p}^c))^\perp. \tag{8.3.5}$$

We focus now on elements from $I_p(\mathrm{Ker}(D_{+,p}^c))$, and assume $s \in \mathrm{Ker}(D_p^c)$. Set $s' = s - s_0 \in \Omega^{0,>0}(X, L^p \otimes E) = \oplus_{q \geqslant 1}\Omega^{0,q}(X, L^p \otimes E)$. By (1.5.30) and (1.5.34),

$$-2p\langle \omega_d s, s \rangle \leqslant C\|s\|_{L^2}^2. \tag{8.3.6}$$

We obtain thus from (1.5.19), (1.5.21) and (8.3.6), for $p \gg 1$,

$$\|s'\|_{L^2} \leqslant Cp^{-1/2}\|s_0\|_{L^2}. \tag{8.3.7}$$

In view of (1.5.34) and (8.3.7),

$$\|\nabla^{\mathrm{Cl}} s\|_{L^2}^2 - p\langle \tau(x)s_0, s_0 \rangle \leqslant C\|s_0\|_{L^2}^2. \tag{8.3.8}$$

By (1.3.8) with $\nabla^{\det} = \nabla^{\det_1}$,

$$\nabla^{\mathrm{Cl}} s = \nabla^{L^p \otimes E} s_0 + A_2' s_2 + \alpha, \tag{8.3.9}$$

where s_2 is the component of degree 2 of s, A_2' is a contraction operator coming from the middle term of (1.3.8), and $\alpha \in \Omega^{0,>0}(X, L^p \otimes E)$. By (8.3.8) and (8.3.9), we know

$$\left\| \nabla^{L^p \otimes E} s_0 + A_2' s_2 \right\|_{L^2}^2 - p \langle \tau(x) s_0, s_0 \rangle \leqslant C \| s_0 \|_{L^2}^2, \tag{8.3.10}$$

and by (8.3.7) and (8.3.10),

$$\left\| \nabla^{L^p \otimes E} s_0 \right\|_{L^2}^2 \leqslant C p \| s_0 \|_{L^2}^2. \tag{8.3.11}$$

By (8.3.7) and (8.3.11), we get

$$\left\| \nabla^{L^p \otimes E} s_0 + A_2' s_2 \right\|_{L^2}^2 \geqslant \left\| \nabla^{L^p \otimes E} s_0 \right\|_{L^2}^2 - 2 \left\| \nabla^{L^p \otimes E} s_0 \right\|_{L^2} \left\| A_2' s_2 \right\|_{L^2}$$
$$\geqslant \left\| \nabla^{L^p \otimes E} s_0 \right\|_{L^2}^2 - C \| s_0 \|_{L^2}^2. \tag{8.3.12}$$

Thus, (8.3.10) and (8.3.12) yield

$$\left\| \nabla^{L^p \otimes E} s_0 \right\|_{L^2}^2 - p \langle \tau(x) s_0, s_0 \rangle \leqslant C \| s_0 \|_{L^2}^2. \tag{8.3.13}$$

By (1.5.30) and (8.3.13), there exists a constant $\widetilde{C}_L > 0$ such that

$$\left| \langle \Delta_{p,\Phi} s, s \rangle \right| \leqslant \widetilde{C}_L \| s \|_{L^2}^2, \qquad s \in I_p(\mathrm{Ker}(D_{+,p}^c)). \tag{8.3.14}$$

By (8.3.7), we know that $I_p : \mathrm{Ker}(D_{+,p}^c) \longrightarrow I_p(\mathrm{Ker}(D_{+,p}^c))$ is bijective for p large enough, and

$$\mathscr{C}^\infty(X, L^p \otimes E) = I_p(\mathrm{Ker}(D_{+,p}^c)) \oplus (\mathrm{Ker}(D_{+,p}^c))^\perp \cap \mathscr{C}^\infty(X, L^p \otimes E). \tag{8.3.15}$$

The proof is now reduced to a direct application of the minimax principle for the operator $\Delta_{p,\Phi}$. It is clear that (8.3.5) and (8.3.14) still hold for elements in the Sobolev space $\boldsymbol{H}^1(X, L^p \otimes E)$, which is the domain of the quadratic form $\mathbf{Q}_p(s) = \left\| \nabla^{L^p \otimes E} s \right\|_{L^2}^2 - p \langle \tau(x) s, s \rangle + \langle \Phi s, s \rangle$ associated to $\Delta_{p,\Phi}$. Let $\mu_1^p \leqslant \mu_2^p \leqslant \cdots \leqslant \mu_j^p \leqslant \cdots$, $(j \in \mathbb{N})$, be the eigenvalues of $\Delta_{p,\Phi}$. By the minimax principle (C.3.3),

$$\mu_j^p = \min_{V \subset \mathrm{Dom}(\mathbf{Q}_p)} \max_{s \in V, \|s\|_{L^2}=1} \mathbf{Q}_p(s) \tag{8.3.16}$$

where V runs over the subspaces of dimension j of $\mathrm{Dom}(\mathbf{Q}_p)$.

By (8.3.14) and (8.3.16), we know $\mu_j^p \leqslant \widetilde{C}_L$, for $j \leqslant \dim \mathrm{Ker}(D_{+,p}^c)$. Moreover, any subspace $V \subset \mathrm{Dom}(\mathbf{Q}_p)$ with $\dim V \geqslant \dim \mathrm{Ker}(D_{+,p}^c) + 1$ contains an element $0 \neq s \in V \cap (\mathrm{Ker}(D_{+,p}^c))^\perp$. By (8.3.5), (8.3.16), we obtain $\mu_j^p \geqslant 2p\mu_0 - C_L$, for $j \geqslant \dim \mathrm{Ker}(D_{+,p}^c) + 1$. Thus we get (8.3.2) and

$$\dim \mathcal{H}_p = \dim \mathrm{Ker}(D_{+,p}^c). \tag{8.3.17}$$

By Theorem 1.3.9 and Theorem 1.5.7, we obtain (8.3.3). The proof of our theorem is completed. $\qquad \square$

8.3.2 Generalized Bergman kernel

Definition 8.3.2. The smooth kernel of $P_{\mathcal{H}_p}$ with respect to the Riemannian volume form $dv_X(x')$ is denoted by $P_{\mathcal{H}_p}(x, x')$ $(x, x' \in X)$, and is called a *generalized Bergman kernel* of $\Delta_{p,\Phi}$. Especially, $P_{\mathcal{H}_p}(x, x') \in (L^p \otimes E)_x \otimes (L^p \otimes E)_{x'}^*$.

Recall that $\dot{R}^L \in \mathrm{End}(T^{(1,0)}X)$ is defined by (1.5.15) associated to R^L. Condition (1.5.21) means $a_j(x) \geqslant \mu_0 > 0$ in (1.5.18).

For ψ a tensor on X, set $\nabla^X \psi$ the covariant derivative of ψ induced by the Levi–Civita connection ∇^{TX}.

A corollary of Theorem 8.3.8 is one of our main results:

Theorem 8.3.3. *There exist smooth coefficients $\boldsymbol{b}_r(x) \in \mathrm{End}(E)_x$ where*

$$\boldsymbol{b}_0 = (2\pi)^{-n} \det(\dot{R}^L) \, \mathrm{Id}_E, \tag{8.3.18}$$

with the following property: for any $k, l \in \mathbb{N}$, there exists $C_{k,l} > 0$ such that

$$\left| \frac{1}{p^n} P_{\mathcal{H}_p}(x, x) - \sum_{r=0}^{k} \boldsymbol{b}_r(x) p^{-r} \right|_{\mathscr{C}^l} \leqslant C_{k,l} \, p^{-k-1}, \tag{8.3.19}$$

for any $x \in X$, $p \in \mathbb{N}^*$.

The coefficients $\boldsymbol{b}_r(x)$ are polynomials in R^{TX}, R^E, Φ (and R^L), their derivatives of order $\leqslant 2r - 2$ (resp. $2r$), and reciprocals of linear combinations of eigenvalues of $\dot{R}^L$ at x.

Moreover, the expansion is uniform in the following sense: for any fixed $k, l \in \mathbb{N}$, assume that the derivatives of g^{TX}, h^L, ∇^L, h^E, ∇^E, J and Φ with order $\leqslant 2n + 2k + l + 5$ run over a set bounded in the $\mathscr{C}^l$-norm taken with respect to the parameter $x \in X$ and, moreover, g^{TX} runs over a set bounded below; then the constant $C_{k,l}$ is independent of the data; and the $\mathscr{C}^l$-norm in (8.3.19) includes also the derivatives on the parameters.

The following result will be established in Section 8.3.4.

Theorem 8.3.4. *If $\dot{R}^L = 2\pi \, \mathrm{Id}$, we have*

$$\boldsymbol{b}_1 = \frac{1}{8\pi} \left[r^X + \frac{1}{4} |\nabla^X J|^2 + 4\sqrt{-1} \Lambda_\omega(R^E) \right], \tag{8.3.20}$$

where $|\nabla^X J|^2 = \sum_{ij} |(\nabla_{e_i}^X J) e_j|^2$, $\{e_j\}$ is an orthonormal basis of (TX, g^{TX}) and $\Lambda_\omega(R^E)$ was introduced in (4.1.5).

Recall that ∇^{TX} induces by projection a Hermitian connection $\nabla^{1,0}$ on $T^{(1,0)}X$ (cf. Section 1.3.1). By (1.3.45), the Chern–Weil representative of $c_1(T^{(1,0)}X)$ is

$$c_1(T^{(1,0)}X, \nabla^{1,0}) = \frac{\sqrt{-1}}{2\pi} \, \mathrm{Tr} \,|_{T^{(1,0)}X}(\nabla^{1,0})^2. \tag{8.3.21}$$

Recall that $\mathcal{R}, \kappa$ were defined in (4.1.18) and (4.1.28). Denote $t = \frac{1}{\sqrt{p}}$. For $s \in \mathscr{C}^\infty(\mathbb{R}^{2n}, E_{x_0})$ and $Z \in \mathbb{R}^{2n}$, as in (1.6.27), set

$$
\begin{aligned}
\nabla_t &= t S_t^{-1} \kappa^{\frac{1}{2}} \nabla^{L_0^p \otimes E_0} \kappa^{-\frac{1}{2}} S_t, \\
\mathscr{L}_t &= S_t^{-1} t^2 \kappa^{\frac{1}{2}} \Delta_{p,\Phi_0}^{X_0} \kappa^{-\frac{1}{2}} S_t.
\end{aligned}
\tag{8.3.33}
$$

Recall that the operators $\nabla_{0,\cdot}$ and $\mathscr{L}$ were defined in (4.1.72). Set with $\partial^\alpha := \frac{\partial^\alpha}{\partial Z^\alpha}$,

$$
\mathcal{I}(Z)(\cdot) = \sum_{|\alpha|=2} (\partial^\alpha R^L)_{x_0} \frac{Z^\alpha}{\alpha!} (\mathcal{R}, \cdot).
\tag{8.3.34}
$$

The operator $\mathscr{L}_t$ is the rescaled operator, which we now develop in Taylor series as in Theorem 4.1.7.

Theorem 8.3.7. *There exist polynomials $\mathcal{A}_{i,j,r}$ (resp. $\mathcal{B}_{i,r}$, $\mathcal{C}_r$) ($r \in \mathbb{N}, i, j \in \{1, \ldots, 2n\}$) in Z with the following properties:*

(i) *their coefficients are polynomials in R^{TX} (resp. R^{TX}, R^E, Φ, R^L) and their derivatives at x_0 up to order $r - 2$ (resp. $r - 2$, $r - 2$, $r - 2$, r),*

(ii) *$\mathcal{A}_{i,j,r}$ is a homogeneous polynomial in Z of degree r, the degree in Z of $\mathcal{B}_{i,r}$ is $\leqslant r + 1$ (resp. $\mathcal{C}_r$ is $\leqslant r + 2$), and has the same parity with $r - 1$ (resp. r),*

(iii) *if we denote $\mathbf{O}_r$ as in (4.1.30), then in the sense of (4.1.31), we have*

$$
\mathscr{L}_t = \mathscr{L} + \sum_{r=1}^{m} t^r \mathbf{O}_r + \mathscr{O}(t^{m+1}).
\tag{8.3.35}
$$

Moreover

$$
\mathbf{O}_1(Z) = -\frac{2}{3}(\partial_j R^L)_{x_0}(\mathcal{R}, e_i) Z_j \nabla_{0,e_i} - \frac{1}{3}(\partial_i R^L)_{x_0}(\mathcal{R}, e_i) - (\nabla_{\mathcal{R}}\tau)_{x_0},
$$

$$
\begin{aligned}
\mathbf{O}_2(Z) &= \frac{1}{3} \langle R_{x_0}^{TX}(\mathcal{R}, e_i)\mathcal{R}, e_j \rangle_{x_0} \nabla_{0,e_i} \nabla_{0,e_j} \\
&\quad + \left[\frac{1}{3}\langle R_{x_0}^{TX}(\mathcal{R}, e_j)e_j, e_i \rangle_{x_0} - \frac{1}{2}\mathcal{I}(Z)(e_i) - R_{x_0}^E(\mathcal{R}, e_i) \right] \nabla_{0,e_i} \\
&\quad - \frac{1}{4}\nabla_{e_i}(\mathcal{I}(Z)(e_i)) - \frac{1}{9}\sum_i \left[\sum_j (\partial_j R^L)_{x_0}(\mathcal{R}, e_i) Z_j \right]^2 \\
&\quad - \frac{r_{x_0}^X}{6} - \sum_{|\alpha|=2}(\partial^\alpha \tau)_{x_0} \frac{Z^\alpha}{\alpha!} + \Phi_{x_0}.
\end{aligned}
\tag{8.3.36}
$$

Proof. As in Lemma 1.2.3, set $g_{ij}(Z) = g^{TX}(e_i, e_j)(Z) = \langle e_i, e_j \rangle_Z$ and let $(g^{ij}(Z))$ be the inverse of the matrix $(g_{ij}(Z))$. Let $\nabla_{e_i}^{TX} e_j = \Gamma_{ij}^l(Z)e_l$. Now by (8.3.1),

$$
\Delta_{p,\Phi} = -g^{ij}(\nabla_{e_i}^{L^p \otimes E} \nabla_{e_j}^{L^p \otimes E} - \nabla_{\nabla_{e_i}^{TX} e_j}^{L^p \otimes E}) - p\tau + \Phi.
\tag{8.3.37}
$$

From (8.3.33) and (8.3.37) we infer the expression

$$\mathscr{L}_t = -g^{ij}(tZ)\Big[\nabla_{t,e_i}\nabla_{t,e_j} - t\Gamma^l_{ij}(t\cdot)\nabla_{t,e_l}\Big](Z) - \tau(tZ) + t^2\Phi(tZ). \qquad (8.3.38)$$

Let Γ^E, Γ^L be the connection forms of ∇^E and ∇^L with respect to any fixed frames for E, L which are parallel along the curve $\gamma_Z : [0,1] \ni u \to uZ$ under our trivializations on $B^{T_{x_0}X}(0,4\varepsilon)$. (8.3.33) yields, on $B^{T_{x_0}X}(0,2\varepsilon/t)$,

$$\nabla_{t,e_i}|_Z = \kappa^{\frac{1}{2}}(tZ)\Big(\nabla_{e_i} + \frac{1}{t}\Gamma^L(e_i)(tZ) + t\Gamma^E(e_i)(tZ)\Big)\kappa^{-\frac{1}{2}}(tZ). \qquad (8.3.39)$$

By (1.2.30), (4.1.101) and (8.3.39), we get

$$\nabla_{t,e_i}|_Z = \nabla_{0,e_i} + \frac{t}{3}(\partial_k R^L)_{x_0}(\mathcal{R},e_i)Z_k + \frac{t^2}{2}R^E_{x_0}(\mathcal{R},e_i)$$
$$+ \frac{t^2}{4}\mathcal{I}(Z)(e_i) + \frac{t^2}{6}\big\langle R^{TX}_{x_0}(\mathcal{R},e_k)e_k,e_i\big\rangle + \mathscr{O}(t^3). \qquad (8.3.40)$$

Relations (1.2.19), (4.1.102), (8.3.38) and (8.3.40), show that the first coefficient of the expansion (8.3.35) is $\mathscr{L}$ and prove (8.3.36). Lemmas 1.2.3, 1.2.4, (8.3.38) and (8.3.39) settle the rest of our theorem.   □

We denote by $\langle\cdot,\cdot\rangle_{0,L^2}$ and $\|\cdot\|_{0,L^2}$ the scalar product and the L^2 norm on $\mathscr{C}^\infty(X_0,E_{x_0})$ induced by g^{TX_0},h^{E_0} as in (1.3.14) and in Section 1.6.4. For $s \in \mathscr{C}^\infty(X_0,E_{x_0})$, set

$$\|s\|^2_{t,0} = \|s\|^2_0 = \int_{\mathbb{R}^{2n}} |s(Z)|^2_{h^{E_{x_0}}}\, dv_{TX}(Z), \qquad (8.3.41)$$

$$\|s\|^2_{t,m} = \sum_{l=0}^{m}\ \sum_{i_1,\cdots,i_l=1}^{2n} \|\nabla_{t,e_{i_1}}\cdots\nabla_{t,e_{i_l}}s\|^2_{t,0}.$$

Applying the same proof in Section 4.1.4, by using (8.3.31) and the corresponding Sobolev norms above, we get Theorems 4.1.9–4.1.14 for $\mathscr{L}_t$ defined in (8.3.38).

Now we view $P_{\mathcal{H}_p,x_0}(Z,Z') := P_{\mathcal{H}_p}(Z,Z')$ $(Z,Z' \in T_{x_0}X, |Z|,|Z'| < 2\varepsilon)$ as a smooth section of $\pi^*(\mathrm{End}(E))$ over $TX \times_X TX$ as in Section 4.1.5. The next step is to convert the estimates for the resolvent into estimates for the spectral projection $\mathcal{P}_{0,t} : (L^2(X_0,E_{x_0}),\|\cdot\|_0) \to (L^2(X_0,E_{x_0}),\|\cdot\|_0)$ of $\mathscr{L}_t$ corresponding to the interval $[-\widetilde{C}_{L_0}t^2,\widetilde{C}_{L_0}t^2]$. By the arguments in Sections 4.1.5 and 4.1.7, we finally get with $\mathscr{P}$ in (4.1.84):

Theorem 8.3.8. *There exist $J_r(Z,Z')$ polynomials in Z,Z' with the same parity as r and $\deg J_r(Z,Z') \leqslant 3r$, whose coefficients are polynomials in R^{TX}, R^E, Φ*

(and R^L) and their derivatives of order $\leqslant r-2$ (resp. r), and reciprocals of linear combinations of eigenvalues of $\dot{R}^L$ at x_0 , such that if we define

$$\mathscr{F}_r(Z,Z') = J_r(Z,Z')\mathscr{P}(Z,Z'), \quad J_0 = \mathrm{Id}_E, \qquad (8.3.42)$$

then for $k, m, m' \in \mathbb{N}$, $q > 0$, there exists $C > 0$ such that if $p \geqslant 1$, $Z, Z' \in T_{x_0}X$, $|Z|, |Z'| \leqslant q/\sqrt{p}$,

$$\sup_{|\alpha|+|\alpha'|\leqslant m} \left| \frac{\partial^{|\alpha|+|\alpha'|}}{\partial Z^\alpha \partial Z'^{\alpha'}} \left(\frac{1}{p^n} P_{\mathcal{H}_p}(Z,Z') \right. \right.$$

$$\left. \left. - \sum_{r=0}^k \mathscr{F}_r(\sqrt{p}Z, \sqrt{p}Z')\kappa^{-\frac{1}{2}}(Z)\kappa^{-\frac{1}{2}}(Z')p^{-\frac{r}{2}} \right) \right|_{\mathscr{C}^{m'}(X)} \leqslant C p^{-\frac{k-m+1}{2}}. \quad (8.3.43)$$

Moreover, $\mathscr{F}_1, \mathscr{F}_2$ were given by (4.1.93) by replacing $\mathscr{L}_2^0$, O_r therein by $\mathscr{L}$, $\boldsymbol{O}_r$ here.

Proof of Theorem 8.3.3. As in Section 4.1.7, from Theorem 8.3.8 with $Z = Z' = 0$, we get the first part of Theorem 8.3.3 and

$$b_r(x_0) = \mathscr{F}_{2r}(0,0). \qquad (8.3.44)$$

Now as the coefficients of $\mathscr{L}_t$ are functions of $g^{TX}, \nabla^L, \nabla^E$ their derivatives with order $\leqslant 1$, from the argument at the end of Section 4.1.7, we get the last part of Theorem 8.3.3. $\qquad \square$

Theorem 8.3.9 below gives a version of (4.1.94) in the current situation. We use the notation in Section 4.1.6 now. As in Section 1.2.2, we denote by $\langle \cdot, \cdot \rangle$ the $\mathbb{C}$-bilinear form on $TX \otimes_{\mathbb{R}} \mathbb{C}$ induced by g^{TX}.

Theorem 8.3.9. *We have the relation with $\boldsymbol{O}_1$ in (8.3.36),*

$$\mathscr{P}\boldsymbol{O}_1\mathscr{P} = 0. \qquad (8.3.45)$$

Proof. We define $\mathscr{J} \in \mathrm{End}(TX)$ by: for $U, V, W \in TX$,

$$\langle \mathscr{J}V, W \rangle = R^L(V,W). \qquad (8.3.46)$$

By (1.5.19) and (8.3.46),

$$(\nabla_U^X R^L)(V,W) = \langle (\nabla_U^X \mathscr{J})V, W \rangle,$$

$$\nabla_U \tau = -\frac{\sqrt{-1}}{2} \mathrm{Tr}\,|_{TX}[\nabla_U^X(J\mathscr{J})]. \qquad (8.3.47)$$

From (8.3.47), we obtain

$$\langle (\nabla_U^X \mathscr{J})V, W \rangle + \langle (\nabla_V^X \mathscr{J})W, U \rangle + \langle (\nabla_W^X \mathscr{J})U, V \rangle = dR^L(U,V,W) = 0. \quad (8.3.48)$$

As J, $\mathcal{J} \in \mathrm{End}(TX)$ are skew-adjoint and commute, $\nabla_U^X J$, $\nabla_U^X \mathcal{J}$ are skew-adjoint and $\nabla_U^X(J\mathcal{J})$ is symmetric. From $J^2 = -\mathrm{Id}$, we know that

$$J(\nabla^X J) + (\nabla^X J)J = 0, \tag{8.3.49}$$

thus $\nabla_U^X J$ exchanges $T^{(1,0)}X$ and $T^{(0,1)}X$. From (8.3.47) and (8.3.48), we have

$$(\nabla_{\mathcal{R}} \mathcal{T})_{x_0} = -2\sqrt{-1}\left\langle (\nabla_{\mathcal{R}}^X(J\mathcal{J}))\tfrac{\partial}{\partial z_i}, \tfrac{\partial}{\partial \bar{z}_i}\right\rangle = 2\left\langle (\nabla_{\mathcal{R}}^X \mathcal{J})\tfrac{\partial}{\partial z_i}, \tfrac{\partial}{\partial \bar{z}_i}\right\rangle,$$

$$(\partial_i R^L)_{x_0}(\mathcal{R}, e_i) = 2\left\langle (\nabla_{\frac{\partial}{\partial z_i}}^X \mathcal{J})\mathcal{R}, \tfrac{\partial}{\partial \bar{z}_i}\right\rangle + 2\left\langle (\nabla_{\frac{\partial}{\partial \bar{z}_i}}^X \mathcal{J})\mathcal{R}, \tfrac{\partial}{\partial z_i}\right\rangle \tag{8.3.50}$$

$$= 4\left\langle (\nabla_{\frac{\partial}{\partial z_i}}^X \mathcal{J})\mathcal{R}, \tfrac{\partial}{\partial \bar{z}_i}\right\rangle - 2\left\langle (\nabla_{\mathcal{R}}^X \mathcal{J})\tfrac{\partial}{\partial z_i}, \tfrac{\partial}{\partial \bar{z}_i}\right\rangle.$$

From (4.1.75), (8.3.36), (8.3.47) and (8.3.50), we infer

$$\boldsymbol{O}_1 = -\frac{2}{3}\Big[\left\langle (\nabla_{\mathcal{R}}^X \mathcal{J})\mathcal{R}, \tfrac{\partial}{\partial \bar{z}_i}\right\rangle b_i^+ - \left\langle (\nabla_{\mathcal{R}}^X \mathcal{J})\mathcal{R}, \tfrac{\partial}{\partial \bar{z}_i}\right\rangle b_i$$

$$+ 2\left\langle (\nabla_{\frac{\partial}{\partial z_i}}^X \mathcal{J})\mathcal{R}, \tfrac{\partial}{\partial \bar{z}_i}\right\rangle + 2\left\langle (\nabla_{\mathcal{R}}^X \mathcal{J})\tfrac{\partial}{\partial z_i}, \tfrac{\partial}{\partial \bar{z}_i}\right\rangle\Big] \tag{8.3.51}$$

$$= -\frac{2}{3}\Big[\left\langle (\nabla_{\mathcal{R}}^X \mathcal{J})\mathcal{R}, \tfrac{\partial}{\partial \bar{z}_i}\right\rangle b_i^+ - b_i \left\langle (\nabla_{\mathcal{R}}^X \mathcal{J})\mathcal{R}, \tfrac{\partial}{\partial \bar{z}_i}\right\rangle\Big].$$

Note that by (4.1.74) and (4.1.84),

$$(b_i^+ \mathscr{P})(Z, Z') = 0, \quad (b_i \mathscr{P})(Z, Z') = a_i(z_i - \bar{z}_i')\mathscr{P}(Z, Z'). \tag{8.3.52}$$

By Theorem 4.1.20 and (7.1.5), any polynomial $g(z, \bar{z})$ in $z, \bar{z}$ satisfies

$$\mathscr{P}b^\alpha g(z, \bar{z})\mathscr{P} = 0, \quad \text{for } |\alpha| > 0, \tag{8.3.53}$$

and relations (8.3.51)–(8.3.53) yield the desired relation (8.3.45). $\qquad\square$

8.3.4 The second coefficient b_1

In the rest of this section we assume that $g^{TX} = \omega(\cdot, J\cdot)$. A very useful observation is that (8.3.48), (8.3.49) imply

$$\mathcal{J} = -2\pi\sqrt{-1}J \text{ and } a_i = 2\pi \text{ in (1.5.18)}, \tau = 2\pi n. \nabla_U^X J \text{ is skew-adjoint}$$
$$\text{and the tensor } \left\langle (\nabla_\cdot^X J)\cdot, \cdot\right\rangle \text{ is of the type } (T^{*(1,0)}X)^{\otimes 3} \oplus (T^{*(0,1)}X)^{\otimes 3}. \tag{8.3.54}$$

Before computing b_1, we establish the relation between the scalar curvature r^X and $|\nabla^X J|^2$.

Lemma 8.3.10. *The following identity holds,*

$$r^X = 8\left\langle R^{TX}(\tfrac{\partial}{\partial z_i}, \tfrac{\partial}{\partial \bar{z}_i})\tfrac{\partial}{\partial z_j}, \tfrac{\partial}{\partial \bar{z}_i}\right\rangle - \frac{1}{4}|\nabla^X J|^2. \tag{8.3.55}$$

Proof. By (8.3.54),

$$|\nabla^X J|^2 = 4\left\langle (\nabla^X_{\frac{\partial}{\partial z_i}} J)e_j, (\nabla^X_{\frac{\partial}{\partial \overline{z}_i}} J)e_j \right\rangle = 8\left\langle (\nabla^X_{\frac{\partial}{\partial z_i}} J)\frac{\partial}{\partial z_j}, (\nabla^X_{\frac{\partial}{\partial \overline{z}_i}} J)\frac{\partial}{\partial \overline{z}_j} \right\rangle. \quad (8.3.56)$$

By (4.1.71), (8.3.48) and (8.3.54),

$$\left\langle (\nabla^X_{\frac{\partial}{\partial z_j}} J)\frac{\partial}{\partial z_i}, (\nabla^X_{\frac{\partial}{\partial \overline{z}_i}} J)\frac{\partial}{\partial \overline{z}_j} \right\rangle = 2\left\langle (\nabla^X_{\frac{\partial}{\partial z_j}} J)\frac{\partial}{\partial z_i}, \frac{\partial}{\partial z_k} \right\rangle \left\langle (\nabla^X_{\frac{\partial}{\partial \overline{z}_i}} J)\frac{\partial}{\partial \overline{z}_j}, \frac{\partial}{\partial \overline{z}_k} \right\rangle$$

$$= 2\left\langle (\nabla^X_{\frac{\partial}{\partial z_i}} J)\frac{\partial}{\partial z_k} - (\nabla^X_{\frac{\partial}{\partial z_k}} J)\frac{\partial}{\partial z_i}, \frac{\partial}{\partial z_j} \right\rangle \left\langle (\nabla^X_{\frac{\partial}{\partial \overline{z}_i}} J)\frac{\partial}{\partial \overline{z}_k}, \frac{\partial}{\partial \overline{z}_j} \right\rangle$$

$$= \left\langle (\nabla^X_{\frac{\partial}{\partial z_i}} J)\frac{\partial}{\partial z_k}, (\nabla^X_{\frac{\partial}{\partial \overline{z}_i}} J)\frac{\partial}{\partial \overline{z}_k} \right\rangle - \left\langle (\nabla^X_{\frac{\partial}{\partial z_k}} J)\frac{\partial}{\partial z_i}, (\nabla^X_{\frac{\partial}{\partial \overline{z}_i}} J)\frac{\partial}{\partial \overline{z}_k} \right\rangle. \quad (8.3.57)$$

By (8.3.56) and (8.3.57),

$$\left\langle (\nabla^X_{\frac{\partial}{\partial z_j}} J)\frac{\partial}{\partial z_i}, (\nabla^X_{\frac{\partial}{\partial \overline{z}_i}} J)\frac{\partial}{\partial \overline{z}_j} \right\rangle = \frac{1}{16}|\nabla^X J|^2. \quad (8.3.58)$$

The definition of $\nabla^X \nabla^X J$, R^{TX} and (8.3.48) imply, for $U, V, W, Y \in TX$,

$$(\nabla^X \nabla^X J)_{(U,V)} - (\nabla^X \nabla^X J)_{(V,U)} = [R^{TX}(U,V), J],$$

$$\left\langle (\nabla^X \nabla^X J)_{(Y,U)} V, W \right\rangle + \left\langle (\nabla^X \nabla^X J)_{(Y,V)} W, U \right\rangle \quad (8.3.59)$$

$$+ \left\langle (\nabla^X \nabla^X J)_{(Y,W)} U, V \right\rangle = 0.$$

Now, from (8.3.49), we get

$$(\nabla^X \nabla^X J)_{(U,V)} J + (\nabla^X_U J) \circ (\nabla^X_V J)$$

$$+ (\nabla^X_V J) \circ (\nabla^X_U J) + J(\nabla^X \nabla^X J)_{(U,V)} = 0. \quad (8.3.60)$$

Note that $\nabla^X \nabla^X J$ is a two tensor with values in the bundle of anti-symmetric endomorphisms of TX. Thus from (8.3.54), (8.3.59) and (8.3.60), for $u_1, u_2, u_3 \in T^{(1,0)}X$, $\overline{v}_1, \overline{v}_2 \in T^{(0,1)}X$,

$$(\nabla^X \nabla^X J)_{(u_1, u_2)} u_3, \ (\nabla^X \nabla^X J)_{(\overline{v}_1, \overline{v}_2)} u_3 \in T^{(0,1)}X,$$

$$(\nabla^X \nabla^X J)_{(u_1, \overline{v}_2)} u_3 \in T^{(1,0)}X, \quad (8.3.61)$$

$$2\sqrt{-1}\left\langle (\nabla^X \nabla^X J)_{(u_1, \overline{v}_1)} u_2, \overline{v}_2 \right\rangle = \left\langle (\nabla^X_{u_1} J)u_2, (\nabla^X_{\overline{v}_1} J)\overline{v}_2 \right\rangle.$$

Formulas (8.3.59) and (8.3.61) yield

$$\left\langle (\nabla^X \nabla^X J)_{(u_1, u_2)} \overline{v}_1, \overline{v}_2 \right\rangle$$

$$= -\left\langle (\nabla^X \nabla^X J)_{(u_1, \overline{v}_1)} \overline{v}_2, u_2 \right\rangle - \left\langle (\nabla^X \nabla^X J)_{(u_1, \overline{v}_2)} u_2, \overline{v}_1 \right\rangle$$

$$= \frac{1}{2\sqrt{-1}}\left\langle (\nabla^X_{u_1} J)u_2, (\nabla^X_{\overline{v}_1} J)\overline{v}_2 - (\nabla^X_{\overline{v}_2} J)\overline{v}_1 \right\rangle. \quad (8.3.62)$$

From (8.3.59), (8.3.58) and (8.3.62), we deduce

$$
\left\langle R^{TX}\left(\tfrac{\partial}{\partial z_i},\tfrac{\partial}{\partial z_j}\right)\tfrac{\partial}{\partial \bar{z}_i},\tfrac{\partial}{\partial \bar{z}_j}\right\rangle = \frac{\sqrt{-1}}{2}\left\langle [R^{TX}\left(\tfrac{\partial}{\partial z_i},\tfrac{\partial}{\partial z_j}\right),J]\tfrac{\partial}{\partial \bar{z}_i},\tfrac{\partial}{\partial \bar{z}_j}\right\rangle
$$

$$
= \frac{\sqrt{-1}}{2}\left\langle \left((\nabla^X\nabla^X J)_{(\tfrac{\partial}{\partial z_i},\tfrac{\partial}{\partial z_j})} - (\nabla^X\nabla^X J)_{(\tfrac{\partial}{\partial z_j},\tfrac{\partial}{\partial z_i})}\right)\tfrac{\partial}{\partial \bar{z}_i},\tfrac{\partial}{\partial \bar{z}_j}\right\rangle
$$

$$
= \frac{1}{4}\left\langle (\nabla^X_{\tfrac{\partial}{\partial z_i}} J)\tfrac{\partial}{\partial \bar{z}_j},(\nabla^X_{\tfrac{\partial}{\partial \bar{z}_i}} J)\tfrac{\partial}{\partial \bar{z}_j}\right\rangle = \frac{1}{32}|\nabla^X J|^2. \quad (8.3.63)
$$

The scalar curvature r^X of (X, g^{TX}) is

$$
r^X = -\left\langle R^{TX}(e_i, e_j)e_i, e_j\right\rangle = -4\left\langle R^{TX}\left(\tfrac{\partial}{\partial z_i}, e_j\right)\tfrac{\partial}{\partial \bar{z}_i}, e_j\right\rangle
$$

$$
= -8\left\langle R^{TX}\left(\tfrac{\partial}{\partial z_i},\tfrac{\partial}{\partial z_j}\right)\tfrac{\partial}{\partial \bar{z}_i},\tfrac{\partial}{\partial \bar{z}_j}\right\rangle - 8\left\langle R^{TX}\left(\tfrac{\partial}{\partial z_i},\tfrac{\partial}{\partial \bar{z}_j}\right)\tfrac{\partial}{\partial \bar{z}_i},\tfrac{\partial}{\partial z_j}\right\rangle. \quad (8.3.64)
$$

In conclusion, relations (8.3.63) and (8.3.64) imply (8.3.55). $\qquad\square$

Proof of Theorem 8.3.4. Recall that we do all our computations on $\mathscr{C}^\infty(\mathbb{R}^{2n}, E_{x_0})$. Thus here we still use $\mathscr{P}$ to denote the orthogonal projection from $L^2(\mathbb{R}^{2n}, E_{x_0})$ onto $\mathrm{Ker}(\mathscr{L})$ and $\mathscr{P}^\perp = 1 - \mathscr{P}$.

Recall that $\boldsymbol{b}_1(x_0) = \mathscr{F}_2(0,0)$. By (4.1.93) and (8.3.45), we get

$$
\mathscr{F}_2 = \mathscr{L}^{-1}\mathscr{P}^\perp \boldsymbol{O}_1 \mathscr{L}^{-1}\mathscr{P}^\perp \boldsymbol{O}_1 \mathscr{P} - \mathscr{L}^{-1}\mathscr{P}^\perp \boldsymbol{O}_2 \mathscr{P}
$$
$$
+ \mathscr{P}\boldsymbol{O}_1 \mathscr{L}^{-1}\mathscr{P}^\perp \boldsymbol{O}_1 \mathscr{L}^{-1}\mathscr{P}^\perp - \mathscr{P}\boldsymbol{O}_2 \mathscr{L}^{-1}\mathscr{P}^\perp \quad (8.3.65)
$$
$$
+ \mathscr{P}^\perp \mathscr{L}^{-1}\boldsymbol{O}_1 \mathscr{P}\boldsymbol{O}_1 \mathscr{L}^{-1}\mathscr{P}^\perp - \mathscr{P}\boldsymbol{O}_1 \mathscr{L}^{-2}\mathscr{P}^\perp \boldsymbol{O}_1 \mathscr{P}.
$$

First, by (8.3.51), (8.3.52) and (8.3.54), we obtain

$$
(\boldsymbol{O}_1\mathscr{P})(Z, Z') = \frac{2}{3}\left(b_i\left\langle (\nabla^X_{\bar{z}} J)\bar{z},\tfrac{\partial}{\partial \bar{z}_i}\right\rangle \mathscr{P}\right)(Z, Z')
$$

$$
= -\frac{4\sqrt{-1}\pi}{3}\left\{\left\langle \frac{b_i b_j}{2\pi}(\nabla^X_{\tfrac{\partial}{\partial \bar{z}_j}} J)\bar{z}' + b_i(\nabla^X_{\bar{z}'} J)\bar{z}',\tfrac{\partial}{\partial \bar{z}_i}\right\rangle \mathscr{P}\right\}(Z, Z'). \quad (8.3.66)
$$

Thus by Theorem 4.1.20 and $a_j = 2\pi$, we get

$$
(\mathscr{L}^{-1}\mathscr{P}^\perp \boldsymbol{O}_1 \mathscr{P})(Z, Z') = -\frac{4\sqrt{-1}\pi}{3}\left\{\left(\frac{b_i b_j}{16\pi^2}\left\langle (\nabla^X_{\tfrac{\partial}{\partial \bar{z}_j}} J)\bar{z}',\tfrac{\partial}{\partial \bar{z}_i}\right\rangle\right.\right.
$$
$$
\left.\left. + \frac{b_i}{4\pi}\left\langle (\nabla^X_{\bar{z}'} J)\bar{z}',\tfrac{\partial}{\partial \bar{z}_i}\right\rangle\right)\mathscr{P}\right\}(Z, Z'). \quad (8.3.67)
$$

From (8.3.52), (8.3.54), (8.3.66) and (8.3.67), we have

$$
(\boldsymbol{O}_1\mathscr{P})(Z, 0) = (\mathscr{L}^{-1}\mathscr{P}^\perp \boldsymbol{O}_1 \mathscr{P})(0, Z') = 0. \quad (8.3.68)
$$

In view of (8.3.68), the first and last two terms in (8.3.65) are zero at $(0,0)$. Thus we only need to compute $-(\mathscr{L}^{-1}\mathscr{P}^{\perp}\boldsymbol{O}_2\mathscr{P})(0,0)$, since the third and fourth terms in (8.3.65) are adjoint of the first two terms by the same remark as Remark 4.1.8.

Note that for ψ a 1-form, $\psi(e_j)\nabla_{0,e_j} = \psi(\frac{\partial}{\partial z_j})b_j^+ - \psi(\frac{\partial}{\partial \bar{z}_j})b_j$.

By (4.1.76), (8.3.36), (8.3.47) and (8.3.52), we get

$$
\begin{aligned}
(\boldsymbol{O}_2\mathscr{P})(Z,0) = \Big\{ &\Big(\frac{4\pi^2}{3}\big\langle R_{x_0}^{TX}(z,\bar{z})z,\bar{z}\big\rangle - \frac{4\pi}{3}\big\langle R^{TX}(\mathcal{R},\tfrac{\partial}{\partial z_j})\mathcal{R},\tfrac{\partial}{\partial \bar{z}_j}\big\rangle \\
&- \frac{2\pi}{3}\big\langle R_{x_0}^{TX}(\mathcal{R},e_j)e_j,\bar{z}\big\rangle + \pi\mathcal{I}(Z)(\bar{z}) - \frac{1}{4}\nabla_{e_l}\mathcal{I}(Z)(e_l) \\
&+ 2\pi R^E(z,\bar{z}) - \frac{r_{x_0}^X}{6} + \frac{4\pi^2}{9}|(\nabla_{\mathcal{R}}^X J)\mathcal{R}|^2 + \Phi_{x_0}\Big)\mathscr{P}\Big\}(Z,0). \quad (8.3.69)
\end{aligned}
$$

In normal coordinates, $(\nabla_{e_i}^{TX}e_j)_{x_0} = 0$, so from (4.1.102), we have at x_0,

$$
\begin{aligned}
\nabla_{e_j}\nabla_{e_i}\langle Je_k, e_l\rangle &= \big\langle (\nabla_{e_j}^X\nabla_{e_i}^X J)e_k + J(\nabla_{e_j}^{TX}\nabla_{e_i}^{TX}e_k), e_l\big\rangle + \big\langle Je_k, \nabla_{e_j}^{TX}\nabla_{e_i}^{TX}e_l\big\rangle \\
&= \big\langle (\nabla_{e_j}^X\nabla_{e_i}^X J)e_k, e_l\big\rangle - \frac{1}{3}\big\langle R^{TX}(e_j,e_i)e_k + R^{TX}(e_j,e_k)e_i, Je_l\big\rangle \\
&\quad + \frac{1}{3}\big\langle R^{TX}(e_j,e_i)e_l + R^{TX}(e_j,e_l)e_i, Je_k\big\rangle. \quad (8.3.70)
\end{aligned}
$$

From (8.3.34), (8.3.47), (8.3.54) and (8.3.70), we obtain

$$
\begin{aligned}
\mathcal{I}(Z)(e_l) &= -\sqrt{-1}\pi(\nabla_{e_j}\nabla_{e_i}\langle Je_k,e_l\rangle)_{x_0}Z_iZ_jZ_k \\
&= -\sqrt{-1}\pi\big\langle (\nabla^X\nabla^X J)_{(\mathcal{R},\mathcal{R})}\mathcal{R}, e_l\big\rangle - \frac{2\pi}{3}\big\langle R^{TX}(\mathcal{R},e_l)z,\bar{z}\big\rangle. \quad (8.3.71)
\end{aligned}
$$

From (8.3.59) and (8.3.61), we have

$$
\begin{aligned}
\big\langle (\nabla^X\nabla^X J)_{(\mathcal{R},\mathcal{R})}z,\bar{z}\big\rangle &= 2\big\langle (\nabla^X\nabla^X J)_{(z,\bar{z})}z,\bar{z}\big\rangle + \big\langle [R^{TX}(\bar{z},z),J]z,\bar{z}\big\rangle \\
&= -\sqrt{-1}\,|(\nabla_z^X J)z|^2. \quad (8.3.72)
\end{aligned}
$$

From (8.3.71), (8.3.72) and $(\nabla^X\nabla^X J)_{(Y,U)}$ is skew-adjoint, we get

$$
\mathcal{I}(Z)(\bar{z}) = -\pi\,|(\nabla_z^X J)z|^2 - \frac{2\pi}{3}\big\langle R^{TX}(z,\bar{z})z,\bar{z}\big\rangle. \quad (8.3.73)
$$

Note that $J, (\nabla^X\nabla^X J)_{(Y,U)}$ are skew-adjoint, thus,

$$
\begin{aligned}
&\frac{\partial}{\partial z_i}\big\langle (\nabla^X\nabla^X J)_{(\mathcal{R},\mathcal{R})}\mathcal{R}, \tfrac{\partial}{\partial \bar{z}_i}\big\rangle + \frac{\partial}{\partial \bar{z}_i}\big\langle (\nabla^X\nabla^X J)_{(\mathcal{R},\mathcal{R})}\mathcal{R}, \tfrac{\partial}{\partial z_i}\big\rangle \\
&= \big\langle (\nabla^X\nabla^X J)_{(\mathcal{R},\frac{\partial}{\partial z_i})}\mathcal{R} + (\nabla^X\nabla^X J)_{(\frac{\partial}{\partial z_i},\mathcal{R})}\mathcal{R}, \tfrac{\partial}{\partial \bar{z}_i}\big\rangle \\
&\quad + \big\langle (\nabla^X\nabla^X J)_{(\mathcal{R},\frac{\partial}{\partial \bar{z}_i})}\mathcal{R} + (\nabla^X\nabla^X J)_{(\frac{\partial}{\partial \bar{z}_i},\mathcal{R})}\mathcal{R}, \tfrac{\partial}{\partial z_i}\big\rangle, \quad (8.3.74)
\end{aligned}
$$

$$\frac{\partial}{\partial z_i}\left\langle R^{TX}(\mathcal{R},\frac{\partial}{\partial \bar{z}_i})z,\bar{z}\right\rangle + \frac{\partial}{\partial \bar{z}_i}\left\langle R^{TX}(\mathcal{R},\frac{\partial}{\partial z_i})z,\bar{z}\right\rangle$$

$$= \left\langle R^{TX}(\mathcal{R},\frac{\partial}{\partial \bar{z}_i})\frac{\partial}{\partial z_i},\bar{z}\right\rangle + \left\langle R^{TX}(\mathcal{R},\frac{\partial}{\partial z_i})z,\frac{\partial}{\partial \bar{z}_i}\right\rangle$$

$$= \left\langle R^{TX}(\bar{z},\frac{\partial}{\partial \bar{z}_i})\frac{\partial}{\partial z_i},\bar{z}\right\rangle + \left\langle R^{TX}(z,\frac{\partial}{\partial z_i})z,\frac{\partial}{\partial \bar{z}_i}\right\rangle.$$

Observe that for a polynomial H on $z,\bar{z}$, from (8.3.52), the contribution of $\mathscr{P}^\perp H \mathscr{P}$ at $(0,0)$ consists of the terms whose degree of $\bar{z}$ is the same as the degree of z. Thus for G a polynomial of degree 2 on $z,\bar{z}$, we get

$$(\mathscr{P}^\perp G \mathscr{P})(Z,0) = \left(\frac{b_j}{2\pi}\frac{\partial G}{\partial \bar{z}_j}\mathscr{P}\right)(Z,0),$$

$$(\mathscr{L}^{-1}\mathscr{P}^\perp G \mathscr{P})(0,0) = \left(\frac{b_j}{8\pi^2}\frac{\partial G}{\partial \bar{z}_j}\mathscr{P}\right)(0,0) = -\frac{1}{4\pi^2}\frac{\partial^2 G}{\partial z_j \partial \bar{z}_j}. \tag{8.3.75}$$

Thus from (8.3.62), (8.3.74) and (8.3.75),

$$\frac{1}{4}\left(\mathscr{L}^{-1}\mathscr{P}^\perp(\nabla_{e_l}\mathcal{I}(Z)(e_l))\mathscr{P}\right)(0,0) \tag{8.3.76}$$

$$= \frac{\sqrt{-1}}{4\pi}\mathrm{Re}\left[\left\langle (\nabla^X \nabla^X J)_{(\frac{\partial}{\partial z_j},\frac{\partial}{\partial z_i})}\frac{\partial}{\partial \bar{z}_j} + (\nabla^X \nabla^X J)_{(\frac{\partial}{\partial z_i},\frac{\partial}{\partial z_j})}\frac{\partial}{\partial \bar{z}_j},\frac{\partial}{\partial \bar{z}_i}\right\rangle\right] = 0,$$

$$\left(\mathscr{L}^{-1}\mathscr{P}^\perp\left(2\left\langle R^{TX}(\mathcal{R},\frac{\partial}{\partial \bar{z}_j})\mathcal{R},\frac{\partial}{\partial \bar{z}_j}\right\rangle + \left\langle R^{TX}_{x_0}(\mathcal{R},e_j)e_j,\bar{z}\right\rangle\right)\mathscr{P}\right)(0,0) = 0.$$

Let $f_{ij}(z)$, $(i,j = 1,\ldots,n)$ be arbitrary polynomials in z. By (4.1.75) and (8.3.52), we have

$$(f_{ij}\bar{z}_i\bar{z}_j\mathscr{P})(Z,0) = \left(f_{ij}\frac{b_i b_j}{4\pi^2}\mathscr{P}\right)(Z,0)$$

$$= \left\{\left(\frac{b_i b_j}{4\pi^2}f_{ij} + \frac{b_j}{2\pi^2}\frac{\partial f_{ij}}{\partial z_i} + \frac{b_i}{2\pi^2}\frac{\partial f_{ij}}{\partial z_j} + \frac{1}{\pi^2}\frac{\partial^2 f_{ij}}{\partial z_i \partial z_j}\right)\mathscr{P}\right\}(Z,0). \tag{8.3.77}$$

Thus from Theorem 4.1.20, (4.1.74) and (8.3.77), we get

$$(\mathscr{L}^{-1}\mathscr{P}^\perp f_{ij}\bar{z}_i\bar{z}_j\mathscr{P})(0,0) = \left\{\left(\frac{b_i b_j}{32\pi^3}f_{ij} + \frac{b_j}{8\pi^3}\frac{\partial f_{ij}}{\partial z_i} + \frac{b_i}{8\pi^3}\frac{\partial f_{ij}}{\partial z_j}\right)\mathscr{P}\right\}(0,0)$$

$$= -\frac{3}{8\pi^3}\frac{\partial^2 f_{ij}}{\partial z_i \partial z_j}. \tag{8.3.78}$$

By (8.3.63) and (8.3.78),

$$(\mathscr{L}^{-1}\mathscr{P}^\perp\left\langle R^{TX}_{x_0}(z,\bar{z})z,\bar{z}\right\rangle\mathscr{P})(0,0)$$

$$= -\frac{3}{8\pi^3}\left\langle R^{TX}_{x_0}(\frac{\partial}{\partial \bar{z}_i},\frac{\partial}{\partial \bar{z}_j})\frac{\partial}{\partial z_j} + R^{TX}_{x_0}(\frac{\partial}{\partial \bar{z}_j},\frac{\partial}{\partial \bar{z}_j})\frac{\partial}{\partial z_i},\frac{\partial}{\partial \bar{z}_i}\right\rangle$$

$$= -\frac{3}{8\pi^3}\left[2\left\langle R^{TX}_{x_0}(\frac{\partial}{\partial \bar{z}_i},\frac{\partial}{\partial \bar{z}_j})\frac{\partial}{\partial z_j},\frac{\partial}{\partial \bar{z}_i}\right\rangle + \frac{1}{2^5}|\nabla^X J|^2\right]. \tag{8.3.79}$$

and from (8.3.56), (8.3.58) and (8.3.78),

$$
\left(\mathscr{L}^{-1}\mathscr{P}^{\perp}|(\nabla_{z}^{X}J)z|^{2}\mathscr{P}\right)(0,0)
$$
$$
= -\frac{3}{8\pi^{3}}\left\langle(\nabla_{\frac{\partial}{\partial z_{i}}}^{X}J)\frac{\partial}{\partial z_{j}} + (\nabla_{\frac{\partial}{\partial z_{j}}}^{X}J)\frac{\partial}{\partial z_{i}}, (\nabla_{\frac{\partial}{\partial z_{j}}}^{X}J)\frac{\partial}{\partial \overline{z}_{j}}\right\rangle = -\frac{9}{27\pi^{3}}|\nabla^{X}J|^{2}. \quad (8.3.80)
$$

Thus by (8.3.73), (8.3.79) and (8.3.80),

$$
\left\{\mathscr{L}^{-1}\mathscr{P}^{\perp}\left(\frac{4\pi^{2}}{3}\left\langle R_{x_{0}}^{TX}(z,\overline{z})z,\overline{z}\right\rangle + \pi\mathcal{I}(Z)(\overline{z}) + \frac{4\pi^{2}}{9}|(\nabla_{\mathcal{R}}^{X}J)\mathcal{R}|^{2}\right)\mathscr{P}\right\}(0,0)
$$
$$
= \left\{\mathscr{L}^{-1}\mathscr{P}^{\perp}\left(\frac{2\pi^{2}}{3}\left\langle R_{x_{0}}^{TX}(z,\overline{z})z,\overline{z}\right\rangle - \frac{\pi^{2}}{9}|(\nabla_{z}^{X}J)z|^{2}\right)\mathscr{P}\right\}(0,0)
$$
$$
= -\frac{1}{2\pi}\left\langle R^{TX}(\tfrac{\partial}{\partial z_{i}}, \tfrac{\partial}{\partial \overline{z}_{j}})\tfrac{\partial}{\partial z_{j}}, \tfrac{\partial}{\partial \overline{z}_{i}}\right\rangle. \quad (8.3.81)
$$

Owing to (8.3.69), (8.3.75), (8.3.76) and (8.3.81), we get

$$
-\left(\mathscr{L}^{-1}\mathscr{P}^{\perp}\boldsymbol{O}_{2}\mathscr{P}\right)(0,0) = \frac{1}{2\pi}\left\langle R^{TX}(\tfrac{\partial}{\partial z_{i}}, \tfrac{\partial}{\partial \overline{z}_{j}})\tfrac{\partial}{\partial z_{j}}, \tfrac{\partial}{\partial \overline{z}_{i}}\right\rangle + \frac{1}{2\pi}R^{E}(\tfrac{\partial}{\partial z_{i}}, \tfrac{\partial}{\partial \overline{z}_{i}}).
$$
$$
\tag{8.3.82}
$$

Formulas (8.3.44), (8.3.55), (8.3.68) and (8.3.82), and the discussion after (8.3.68) yield finally

$$
\boldsymbol{b}_{1}(x_{0}) = -\left(\mathscr{L}^{-1}\mathscr{P}^{\perp}\boldsymbol{O}_{2}\mathscr{P}\right)(0,0) - \left(\left(\mathscr{L}^{-1}\mathscr{P}^{\perp}\boldsymbol{O}_{2}\mathscr{P}\right)(0,0)\right)^{*}
$$
$$
= \frac{1}{8\pi}\left[r_{x_{0}}^{X} + \frac{1}{4}|\nabla^{X}J|_{x_{0}}^{2} + 4\sqrt{-1}\Lambda_{\omega}(R^{E})\right]. \quad (8.3.83)
$$

The proof of Theorem 8.3.4 is complete.   □

8.3.5   Symplectic Kodaira embedding theorem

Recall that (X,ω) is a compact symplectic manifold of real dimension $2n$ and (L,∇^{L},h^{L}) is a Hermitian line bundle on X such that $\omega = \frac{\sqrt{-1}}{2\pi}R^{L}$, and g^{TX} is a Riemannian metric on X as at the beginning of Section 8.3.

Let $\mathcal{H}_{p}\subset\mathscr{C}^{\infty}(X,L^{p})$ be the span of those eigensections of $\Delta_{p} = \Delta^{L^{p}} - \tau p$ corresponding to eigenvalues from $[-\widetilde{C}_{L},\widetilde{C}_{L}]$. Note that $\mathcal{H}_{p}$ is endowed with the induced L^{2} product (1.3.14) so there is a well-defined Fubini–Study form ω_{FS} on $\mathbb{P}(\mathcal{H}_{p}^{*})$.

We have the following analogue of Theorem 5.1.3 in the symplectic case.

Theorem 8.3.11. *There exists $p_0 \in \mathbb{N}$ such that for $p > p_0$, $\Phi_p : X \longrightarrow \mathbb{P}(\mathcal{H}_p^*)$ defined by $\Phi_p(x) = \{s \in \mathcal{H}_p : s(x) = 0\}$ is well defined on X and the map*

$$\Psi_p : \Phi_p^* \mathscr{O}(1) \to L^p,$$
$$\Psi_p((\Phi_p^* \sigma_s)(x)) = s(x), \quad \text{for any } s \in \mathcal{H}_p$$

$$(8.3.84)$$

defines a canonical isomorphism from $\Phi_p^* \mathscr{O}(1)$ to L^p on X, and under this isomorphism, we have

$$h^{\Phi_p^* \mathscr{O}(1)}(x) = P_{\mathcal{H}_p}(x, x)^{-1} h^{L^p}(x).$$

$$(8.3.85)$$

Proof. By Theorem 8.3.4, as in Lemma 5.1.2, we know that there exists $p_0 > 0$ such that for $p > p_0$, Φ_p is defined on X. The remainder is the same proof of Theorem 5.1.3. $\qquad\square$

Theorem 8.3.12.

(i) *The induced Fubini–Study metric $\frac{1}{p}\Phi_p^*(\omega_{FS})$ converges in the $\mathscr{C}^\infty$ topology to ω; for any $l \geqslant 0$ there exists $C_l > 0$ such that*

$$\left| \frac{1}{p} \, \Phi_p^*(\omega_{FS}) - \omega \right|_{\mathscr{C}^l} \leqslant \frac{C_l}{p}.$$

$$(8.3.86)$$

(ii) *For large p the Kodaira maps Φ_p are embeddings.*

Proof. (i) Let us fix $x_0 \in X$. We identify a small geodesic ball $B^X(x_0, \varepsilon)$ to $B^{T_{x_0}X}(0, \varepsilon)$ by means of the exponential map and consider a trivialization of L as in Section 8.3.3, i.e., we trivialize L by using a unit frame $e_L(Z)$ which is parallel with respect to ∇^L along $[0, 1] \ni u \to uZ$ for $Z \in B^{T_{x_0}X}(0, \varepsilon)$. Let $\|w\|^2 = \sum_{j=1}^{d_p} |w_i|^2$. We can express the Fubini–Study metric in the homogeneous coordinate $[w_1, \ldots, w_{d_p}] \in \mathbb{P}(\mathcal{H}_p^*)$ as (cf. Problem 1.8)

$$\frac{\sqrt{-1}}{2\pi} \partial \bar{\partial} \log (\|w\|^2) = \frac{\sqrt{-1}}{2\pi} \Big[\frac{1}{\|w\|^2} \sum_{j=1}^{d_p} dw_j \wedge d\overline{w}_j - \frac{1}{\|w\|^4} \sum_{j,k=1}^{d_p} \overline{w}_j w_k \, dw_j \wedge d\overline{w}_k \Big].$$

Let $\{s_j\}$ be an orthonormal basis of $\mathcal{H}_p$, and $\{s^j\}$ be its dual basis. We write $s_j = f_j e_L^{\otimes p}$, then by (5.1.17),

$$\Phi_p^*(\omega_{FS})(x_0) = \frac{\sqrt{-1}}{2\pi} \Big[\frac{1}{|f^p|^2} \sum_{j=1}^{d_p} df_j \wedge d\overline{f}_j - \frac{1}{|f^p|^4} \sum_{j,k=1}^{d_p} \overline{f}_j f_k \, df_j \wedge d\overline{f}_k \Big](x_0)$$

$$= \frac{\sqrt{-1}}{2\pi} \Big[|f^p(x_0)|^{-2} d_x d_y f^p(x, y) - |f^p(x_0)|^{-4} d_x f^p(x, y) \wedge d_y f^p(x, y) \Big]\Big|_{x=y=x_0},$$

$$(8.3.87)$$

where $f^p(x, y) = \sum_{i=1}^{d_p} f_i(x)\overline{f}_i(y)$ and $|f^p(x)|^2 = f^p(x, x)$.

Since

$$P_{\mathcal{H}_p}(x,y) = f^p(x,y)e_L^p(x) \otimes e_L^p(y)^*, \tag{8.3.88}$$

thus $P_{\mathcal{H}_p}(x,y)$ is $f^p(x,y)$ under our trivialization of L. By (4.1.101), Theorem 8.3.8, we obtain

$$\frac{1}{p}\,\Phi_p^*(\omega_{FS})(x_0) = \frac{\sqrt{-1}}{2\pi}\Big[\frac{1}{\mathscr{F}_0}d_x d_y \mathscr{F}_0 - \frac{1}{\mathscr{F}_0^2}d_x\mathscr{F}_0 \wedge d_y\mathscr{F}_0\Big](0,0)$$

$$-\frac{\sqrt{-1}}{2\pi}\frac{1}{\sqrt{p}}\Big[\frac{1}{\mathscr{F}_0^2}(d_x\mathscr{F}_1 \wedge d_y\mathscr{F}_0 + d_x\mathscr{F}_0 \wedge d_y\mathscr{F}_1)\Big](0,0) + \mathscr{O}\Big(\frac{1}{p}\Big). \tag{8.3.89}$$

Using again (4.1.84), (8.3.42), we obtain

$$\frac{1}{p}\,\Phi_p^*(\omega_{FS})(x_0) = \frac{\sqrt{-1}}{4\pi}\sum_{j=1}^{n} a_j(x_0)dz_j \wedge d\bar{z}_j + \mathscr{O}\Big(\frac{1}{p}\Big) = \omega(x_0) + \mathscr{O}\Big(\frac{1}{p}\Big), \tag{8.3.90}$$

and the convergence takes place in the $\mathscr{C}^\infty$ topology with respect to $x_0 \in X$.

(ii) Since X is compact, we have to prove two things for p sufficiently large: (a) Φ_p are immersions and (b) Φ_p are injective. We note that (a) follows immediately from (8.3.86).

To prove (b) let us assume the contrary, namely that there exists a sequence of distinct points $x_p \neq y_p$ such that $\Phi_p(x_p) = \Phi_p(y_p)$. Since $\Phi_p(x_p) = \Phi_p(y_p)$ we can construct as in (5.1.24) the peak section $S_{x_p}^p = S_{y_p}^p$ as the unit norm generator of the orthogonal complement of $\Phi_p(x_p) = \Phi_p(y_p)$. We fix in the sequel such a section which peaks at both x_p and y_p.

We consider the distance $d(x_p, y_p)$ between the two points x_p and y_p. By passing to a subsequence we have two possibilities: either $\sqrt{p}d(x_p,y_p) \to \infty$ as $p \to \infty$ or there exists a constant $C > 0$ such that $d(x_p,y_p) \leqslant C/\sqrt{p}$ for all p.

Assume that the first possibility is true. For large p, we learn from relation (5.1.25) that the mass of $S_{x_p}^p = S_{y_p}^p$ (which is 1) concentrates both in neighborhoods $B(x_p, r_p)$ and $B(y_p, r_p)$ with $r_p = d(x_p, y_p)/2$ and approaches therefore 2 if $p \to \infty$. This is a contradiction which rules out the first possibility.

We identify as usual $B^X(x_p, \varepsilon)$ to $B^{T_{x_p}X}(0, \varepsilon)$ so the point y_p gets identified to $Z_p/\sqrt{p}$ where $Z_p \in B^{T_{x_p}X}(0, C)$. We define then

$$f_p : [0,1] \longrightarrow \mathbb{R}, \quad f_p(t) = \frac{|S_{x_p}^p(tZ_p/\sqrt{p})|^2}{P_{\mathcal{H}_p}(tZ_p/\sqrt{p}, tZ_p/\sqrt{p})}. \tag{8.3.91}$$

We have $f_p(0) = f_p(1) = 1$ (again because $S_{x_p}^p = S_{y_p}^p$) and $f_p(t) \leqslant 1$ by the definition of the generalized Bergman kernel. We deduce the existence of a point $t_p \in]0,1[$ such that $f_p''(t_p) = 0$. Equations (5.1.24), (8.3.43), (8.3.91) imply the estimate

$$f_p(t) = e^{-\frac{t^2}{4}\sum_j a_j|z_{p,j}|^2}\big(1 + g_p(tZ_p)/\sqrt{p}\big) \tag{8.3.92}$$

and the $\mathscr{C}^2$ norm of g_p over $B^{T_{x_p}X}(0, C)$ is uniformly bounded in p. From (8.3.92), we infer that $|Z_p|_0^2 := \frac{1}{4}\sum_j a_j |z_{p,j}|^2 = \mathscr{O}(1/\sqrt{p})$. Using a limited expansion $e^x = 1 + x + x^2\varphi(x)$ for $x = t^2|Z_p|_0^2$ in (8.3.92) and taking derivatives, we obtain

$$f_p''(t) = -2|Z_p|_0^2 + \mathscr{O}(|Z_p|_0^4) + \mathscr{O}(|Z_p|_0^2/\sqrt{p}) = (-2 + \mathscr{O}(1/\sqrt{p}))|Z_p|_0^2. \quad (8.3.93)$$

Evaluating at t_p we get $0 = f_p''(t_p) = (-2 + \mathscr{O}(1/\sqrt{p}))|Z_p|_0^2$, which is a contradiction since by assumption $Z_p \neq 0$. $\qquad\square$

Remark 8.3.13. The proof of Theorem 8.3.12 gives another analytic proof of the classical Kodaira embedding theorem, Theorem 5.1.12. Certainly, we can use also $\mathrm{Ker}(D_p^{c,A})$ in Section 8.1 instead $\mathcal{H}_p$ here to get again an analogue of Theorem 8.3.12.

Problems

Problem 8.1. Under the assumptions of Section 8.1.1, let G be a compact connected Lie group with Lie algebra $\mathfrak{g}$ and $\dim G = n_0$. Suppose that G acts on X and its action on X lifts on L and E. Moreover, we assume the G-action preserves the above connections and metrics on TX, L, E and J. Verify that $D_p^{c,A}$ commutes with the G-action. Thus $\mathrm{Ker}(D_p^{c,A})$ is a representation of G. Denote by $\mathrm{Ker}(D_p^{c,A})^G$ the G-trivial component of $\mathrm{Ker}(D_p^{c,A})$.

For $K \in \mathfrak{g}$, we denote by $K_x^X = \frac{\partial}{\partial t}e^{-tK}x|_{t=0}$ the corresponding vector field on X. Let $\mu : X \to \mathfrak{g}^*$ be defined by

$$2\pi\sqrt{-1}\mu(K) := \nabla_{K^X}^L - L_K, \quad K \in \mathfrak{g}.$$

Verify that μ is the *moment map*, associated to $\omega = \frac{\sqrt{-1}}{2\pi}R^L$, i.e., for any $K \in \mathfrak{g}$,

$$d\mu(K) = i_{K^X}\omega.$$

Let P_p^G be the orthogonal projection from $(\Omega^{0,\bullet}(X, L^p \otimes E), \langle\ \rangle)$ on $\mathrm{Ker}(D_p^{c,A})^G$. The *$G$-invariant Bergman kernel* is $P_p^G(x, x')$ $(x, x' \in X)$, the smooth kernel of P_p^G with respect to the Riemannian volume form $dv_X(x')$.

Assume that G acts freely on $\mu^{-1}(0)$. Denote by $\mathrm{vol}(Gx), (x \in \mu^{-1}(0))$ the volume of the orbit Gx equipped with the metric induced by g^{TX}. Prove that there exists an asymptotic expansion in the sense of (8.1.4) for $x_0 \in \mu^{-1}(0)$, $k \in \mathbb{N}$,

$$p^{-n+\frac{n_0}{2}}\,\mathrm{vol}(Gx_0)P_p^G(x_0, x_0) = \sum_{r=0}^{k} b_r^G(x_0)p^{-r} + \mathscr{O}(p^{-k-1}).$$

If $g^{TX} = \omega(\cdot, J\cdot)$, then $b_0^G(x_0) = 2^{n_0/2}I_{\mathbb{C}\otimes E}$. (Hint: For dg a Haar measure on G, we have $P_p^G(x, x) = \int_G (g, 1) \cdot P_p(g^{-1}x, x)dg$. Then use Theorem 8.1.4.)

Problem 8.2. In Theorem 8.2.5, verify that (7.1.11) holds for $\widetilde{\mathscr{K}}$ with $i,j > q$. While for $l,m \leqslant q$, we have

$$\widetilde{\mathscr{K}}[1,\overline{z}_l] = \overline{z}_l, \quad \widetilde{\mathscr{K}}[1,z_l] = z_l',$$

$$\widetilde{\mathscr{K}}[\overline{z}_l', z_m] = \widetilde{\mathscr{K}}[z_m', \overline{z}_l] = -\frac{2}{a_l}\delta_{lm} + \overline{z}_l z_m',$$

$$\widetilde{\mathscr{K}}[z_l, \overline{z}_m] = z_l \overline{z}_m, \quad \widetilde{\mathscr{K}}[\overline{z}_l, z_m] = \overline{z}_l z_m'.$$

Especially, analogue to (7.4.14), we have

$$\widetilde{\mathscr{K}}\Big[1, \tfrac{\partial f_x}{\partial Z_j}(0)Z_j\Big] = \sum_{i>q}\Big(\tfrac{\partial f_x}{\partial z_i}(0)z_i + \tfrac{\partial f_x}{\partial \overline{z}_i}(0)\overline{z}_i'\Big) + \sum_{l\leqslant q}\Big(\tfrac{\partial f_x}{\partial z_l}(0)z_l' + \tfrac{\partial f_x}{\partial \overline{z}_l}(0)\overline{z}_l\Big).$$

Problem 8.3. Let (X, J, g^{TX}) be a Kähler manifold and ∇^L, ∇^E be the holomorphic Hermitian connections on the holomorphic Hermitian bundles (L, h^L), (E, h^E). By using (1.4.31), (1.5.20) and (8.3.1), show that

$$\Delta_{p,0} = 2\square_p - R^E(w_j, \overline{w}_j), \quad \text{on } \mathscr{C}^\infty(X, L^p \otimes E).$$

Problem 8.4. Verify that in (8.3.21),

$$(\nabla^{1,0})^2 = P^{1,0}\Big[R^{TX} - \tfrac{1}{4}(\nabla^X J)\wedge(\nabla^X J)\Big]P^{1,0}.$$

Verify (8.3.22). (Hint: Use (8.3.49), (8.3.56), (8.3.63) and (8.3.64).)

Problem 8.5. We denote by $\mathcal{O}_r$ the coefficients of the Taylor expansion associated to the rescaled operator from $(D_p^{c,A})^2$ introduced in Section 8.1.1. As in (4.1.31), verify that

$$\mathcal{O}_1 = \boldsymbol{O}_1 + (\nabla_{\mathcal{R}}\mathcal{T})_{x_0} + \tfrac{1}{2}\langle(\nabla_{\mathcal{R}}^X J)_{x_0}e_l, e_m\rangle\, c(e_l)\,c(e_m).$$

with $\boldsymbol{O}_1, \mathcal{J}$ defined in (8.3.36), (8.3.46). Verify (4.1.94), i.e., $P^N\mathcal{O}_1 P^N = 0$. Finally prove that

$$\text{Tr}\,|_{\Lambda(T^{*(0,1)}X)}[\boldsymbol{b}_1(x)] = \frac{1}{8\pi}\Big[r^X + \tfrac{1}{4}|\nabla^X J|^2 + 4\sqrt{-1}\Lambda_\omega(R^E)\Big].$$

(Hint: Use (1.3.35) and compare the proof of Theorems 8.3.7, 8.3.9.)

8.4 Bibliographic notes

Our book is already quite long, thus we will not try to report exhaustively on results about the Bergman kernel on symplectic manifolds here.

Section 8.1. If $A = 0$, i.e., for the spin^c Dirac operator D_p^c, the results in Section 8.1.1 was obtained by Dai–Liu–Ma [69] by using the heat kernel as in Section 4.2. Section 8.1.2 is from [162].

Section 8.2 is taken from [163]. In the holomorphic case, i.e., in the situation of Remark 8.2.6, (8.2.18) is *Andreotti–Grauert's coarse vanishing theorem* [4, §23] (where it is proved by using the cohomology finiteness theorem for the disc bundle of L^*), the asymptotic expansion of $P_p(x, x')$ was studied independently by Berman and Sjöstrand [19].

Section 8.3 is from [160, 161]. In the case E is a trivial line bundle, (8.3.2) without knowing precise μ_0 is the main result of Guillemin and Uribe [122, Theorem 2], they apply the analysis of Toeplitz structures of Boutet de Monvel–Guillemin [52], cf. also [43, 45] for related topic. The Hermitian scalar curvature was used by Donaldson [88, 111] to define the moment map on the space of compatible almost-complex structures. This section is also related to Donaldson's work [87].

The proof of Theorem 8.3.12 is inspired by [219]. Borthwick and Uribe [44, Th. 1.1], Shiffman and Zelditch [219, Th. 2, 3] prove a different symplectic version of Theorem 8.3.12 when $g^{TX} = \omega(\cdot, J\cdot)$. Instead of $\mathcal{H}_p$, they use the space $H_J^0(X, L^p) := \mathrm{Im}(\Pi_p)$ (cf. [44, p. 601], [219, §2.3]) of 'almost holomorphic sections' proposed by Boutet de Monvel and Guillemin [52, 53] of a first order pseudodifferential operator D_b on the circle bundle of L^*. The associated Szegö kernels are well-defined modulo smooth operators on the associated circle bundle, even though D_b is neither canonically defined nor unique (Indeed, Boutet de Monvel–Guillemin define the Szegö kernels first, and construct the operator D_b from the Szegö kernels.)

Problem 8.1 is from Ma-Zhang [164, §6.3] where the full asymptotic expansion of P_p^G is obtained. Related results in the special case when (X, J, Θ) is Kähler, $\Theta = \omega$, $E = \mathbb{C}$, and ∇^L is the Chern connection can be found in [64, 190, 191]. Note, however, that the value $b_0^G(x_0) = 1$ in the main result of [191] is wrong. Problem 8.5 is from [163, §2].

Appendix A

Sobolev Spaces

In this appendix, we explain the Sobolev embedding theorem and the basic elliptic estimates. A good reference for the matters discussed here is [237, Chap. 4, 5].

A.1 Sobolev spaces on $\mathbb{R}^n$

We denote by $B(x, r)$ the ball with center x and radius r in $\mathbb{R}^n$. For $W \subset \mathbb{R}^n$, we denote by $\overline{W}$ its closure in $\mathbb{R}^n$.

Let $U \subset \mathbb{R}^n$ be an open set. For $k \in \mathbb{N}$, let $\mathscr{C}^k(U)$ be the space of complex-valued functions on U whose partial derivatives of order $\leqslant k$ exist and are continuous on U. The support $\mathrm{supp}(f)$ of a function $f \in \mathscr{C}^k(U)$ is defined as the closure in U of the set $\{x \in U : f(x) \neq 0\}$. Set $\mathscr{C}_0^k(U) := \{f \in \mathscr{C}^k(U) : \mathrm{supp}(f) \text{ is compact}\}$. Set

$$
\begin{aligned}
\mathscr{C}^\infty(U) &= \{f; f \in \mathscr{C}^k(U) \text{ for any } k \in \mathbb{N}\}, \\
\mathscr{C}_0^\infty(U) &= \{f; f \in \mathscr{C}_0^k(U) \text{ for any } k \in \mathbb{N}\}.
\end{aligned}
\tag{A.1.1}
$$

The space $\mathscr{C}^\infty(U)$ is called the space of smooth functions on U, and the space $\mathscr{C}_0^\infty(U)$ is called the space of smooth functions with compact support in U.

For $f \in \mathscr{C}^k(U)$, we define its $\mathscr{C}^k$-norm by

$$
|f|_{\mathscr{C}^k(U)} = \sum_{|\alpha| \leqslant k} \sup_{x \in U} \left| \frac{\partial^\alpha}{\partial x^\alpha} f(x) \right|.
\tag{A.1.2}
$$

Here $\frac{\partial^\alpha}{\partial x^\alpha} := \frac{\partial^{|\alpha|}}{\partial x^\alpha} := \left(\frac{\partial}{\partial x_1}\right)^{\alpha_1} \cdots \left(\frac{\partial}{\partial x_n}\right)^{\alpha_n}$ for $\alpha = (\alpha_1, \ldots, \alpha_n) \in \mathbb{N}^n$.

The L^2-space $L^2(U)$ is the space of square integrable functions on U, i.e., for a measurable function $f : U \to \mathbb{C}$, $f \in L^2(U)$ if and only if

$$
\|f\|_{L^2} = \left(\int_U |f|^2 dx \right)^{1/2} < +\infty,
\tag{A.1.3}
$$

where $dx = dx_1 \cdots dx_n$ is the canonical Euclidean volume form on $\mathbb{R}^n$.

Then $\|\cdot\|_{L^2}$ defines a norm on $L^2(U)$. The space $L^2(U)$ is the completion of $\mathscr{C}_0^\infty(U)$ with respect to the L^2-norm $\|\cdot\|_{L^2}$.

For $f, g \in L^2(U)$, set

$$\langle f, g \rangle = \int_U f \cdot \overline{g} \, dx. \tag{A.1.4}$$

Then $(L^2(U), \langle \cdot, \cdot \rangle)$ is a Hilbert space with inner product $\langle \cdot, \cdot \rangle$.

Theorem A.1.1 (Riesz representation theorem). *If a linear map $\phi : L^2(U) \to \mathbb{C}$ is continuous, i.e., there exists $C > 0$ such that for any $f \in L^2(U)$,*

$$|\phi(f)| \leqslant C\|f\|_{L^2}, \tag{A.1.5}$$

then there exists $u \in L^2(U)$ such that for any $f \in L^2(U)$,

$$\phi(f) = \langle f, u \rangle. \tag{A.1.6}$$

Definition A.1.2. For $u \in L^2(U)$, $\alpha \in \mathbb{N}^n$, we say that $\frac{\partial^\alpha}{\partial x^\alpha} u \in L^2(U)$ if there exists $C > 0$ such that for any $f \in \mathscr{C}_0^\infty(U)$,

$$\left| \left\langle \frac{\partial^\alpha}{\partial x^\alpha} f, u \right\rangle \right| \leqslant C\|f\|_{L^2}. \tag{A.1.7}$$

If (A.1.7) holds, by Theorem A.1.1, we can define $\frac{\partial^\alpha}{\partial x^\alpha} u \in L^2(U)$ as follows: for any $f \in \mathscr{C}_0^\infty(U)$,

$$\left\langle f, \frac{\partial^\alpha}{\partial x^\alpha} u \right\rangle = (-1)^{|\alpha|} \left\langle \frac{\partial^\alpha}{\partial x^\alpha} f, u \right\rangle. \tag{A.1.8}$$

Definition A.1.3. For $k \in \mathbb{N}$, the *Sobolev space* $\boldsymbol{H}^k(U)$ is defined as

$$\boldsymbol{H}^k(U) = \{u \in L^2(U); \frac{\partial^\alpha}{\partial x^\alpha} u \in L^2(U) \text{ for any } |\alpha| \leqslant k\}. \tag{A.1.9}$$

The Sobolev norm $\|\cdot\|_k$ on $\boldsymbol{H}^k(U)$ is defined by: for $u \in \boldsymbol{H}^k(U)$,

$$\|u\|_k^2 = \sum_{|\alpha| \leqslant k} \left\| \frac{\partial^\alpha}{\partial x^\alpha} u \right\|_{L^2}^2. \tag{A.1.10}$$

Definition A.1.4. For $k \in \mathbb{N}^*$, the Sobolev space $\boldsymbol{H}^{-k}(\mathbb{R}^n)$ is the dual space of $\boldsymbol{H}^k(\mathbb{R}^n)$, i.e., the continuous linear maps from $\boldsymbol{H}^k(\mathbb{R}^n)$ to $\mathbb{C}$.

Definition A.1.5. The *Schwartz space* $\mathcal{S}(\mathbb{R}^n)$ of rapidly decreasing functions is defined by

$$\mathcal{S}(\mathbb{R}^n) := \{u \in \mathscr{C}^\infty(\mathbb{R}^n); x^\beta \frac{\partial^\alpha}{\partial x^\alpha} u \text{ is bounded on } \mathbb{R}^n \text{ for all } \alpha, \beta \in \mathbb{N}^n\}. \tag{A.1.11}$$

Certainly $\mathscr{C}_0^\infty(\mathbb{R}^n) \subset \mathcal{S}(\mathbb{R}^n) \subset \boldsymbol{H}^k(\mathbb{R}^n)$ for any $k \in \mathbb{N}$.

Theorem A.1.6 (Sobolev embedding theorem). *If $m, k \in \mathbb{N}$, satisfy $m > \frac{n}{2} + k$, then*

$$\boldsymbol{H}^m(\mathbb{R}^n) \hookrightarrow \mathscr{C}^k(\mathbb{R}^n). \tag{A.1.12}$$

More precisely, there is $C > 0$ such that if $f \in \boldsymbol{H}^m(\mathbb{R}^n)$, then $f \in \mathscr{C}^k(\mathbb{R}^n)$ and

$$|f|_{\mathscr{C}^k(\mathbb{R}^n)} \leqslant C\|f\|_m. \tag{A.1.13}$$

Let $U_1 \subset U$ be a bounded open set such that $\overline{U_1} \subset U$. Let $\chi \in \mathscr{C}_0^\infty(U)$ such that $\chi = 1$ on U_1. Then there exists $C > 0$ such that for any $u \in \boldsymbol{H}^m(U)$, we have $\chi u \in \boldsymbol{H}^m(\mathbb{R}^n)$ and

$$\|\chi u\|_m \leqslant C\|u\|_m. \tag{A.1.14}$$

Assume $m > \frac{n}{2} + k$. Theorem A.1.6 and (A.1.14) imply the existence of a constant $C_{U_1} > 0$ such that for any $u \in \boldsymbol{H}^m(U)$, we have $u \in \mathscr{C}^k(U)$, and

$$|u|_{\mathscr{C}^k(U_1)} \leqslant C_{U_1}\|u\|_m. \tag{A.1.15}$$

In our book, we always apply (A.1.15) for $U = B(x, 1) \subset \mathbb{R}^n$ and $U_1 = B(x, \frac{1}{2})$.

We will denote by $\mathscr{C}^k(\mathbb{R}^n, \mathbb{C}^l), \cdots$, the corresponding spaces of $\mathbb{C}^l$-valued functions. Certainly, all above results hold for $\mathbb{C}^l$-valued functions.

Let $a_\alpha \in \mathscr{C}^\infty(\mathbb{R}^n, \mathrm{End}(\mathbb{C}^l))$ with $\alpha \in \mathbb{N}^n$ such that $m = \sup\{|\alpha| : a_\alpha \not\equiv 0\} < \infty$. We define a mth order differential operator H on $\mathbb{R}^n$ by

$$H = \sum_{|\alpha| \leqslant m} a_\alpha(x) \frac{\partial^\alpha}{\partial x^\alpha}. \tag{A.1.16}$$

The *principal symbol* of H is $\sigma(H)(x, \xi) = \sum_{|\alpha|=m} a_\alpha(x)(\sqrt{-1}\xi)^\alpha$ with $\xi \in \mathbb{R}^n$. If $\sigma(H)(x, \xi) = a(x, \xi) \mathrm{Id}_{\mathbb{C}^l}$ with $a(x, \xi) \in \mathscr{C}^\infty(\mathbb{R}^n)$, we say that H has scalar principal symbol.

We suppose that H is *elliptic* on $\mathbb{R}^n$, that is, for any $x \in \mathbb{R}^n$ and $\xi \in \mathbb{R}^n \setminus \{0\}$, $\sigma(H)(x, \xi) \in \mathrm{End}(\mathbb{C}^l)$ is invertible.

Theorem A.1.7 (Basic elliptic estimate). *Let $U \subset \mathbb{R}^n$ be a bounded open set. Then for any $k \in \mathbb{N}$, there exist $C_1, C_2 > 0$ such that for any $u \in \mathscr{C}_0^\infty(U, \mathbb{C}^l)$, we have*

$$\|u\|_{k+m} \leqslant C_1\|Hu\|_k + C_2\|u\|_{L^2}. \tag{A.1.17}$$

A.2   Sobolev spaces on $\mathbb{R}^n_+$

Let $\mathbb{R}^n_+ = \{x = (x', x_n) \in \mathbb{R}^n; x_n > 0\}$ be the half space with closure $\overline{\mathbb{R}^n_+}$.

Let $\mathcal{S}(\overline{\mathbb{R}^n_+})$ be the space of restrictions to $\mathbb{R}^n_+$ of elements of $\mathcal{S}(\mathbb{R}^n)$. For $k \in \mathbb{N}$, let $\mathscr{C}^k(\overline{\mathbb{R}^n_+})$, $\mathscr{C}^k_0(\overline{\mathbb{R}^n_+})$ be the space of restrictions to $\mathbb{R}^n_+$ of elements of $\mathscr{C}^k(\mathbb{R}^n)$, $\mathscr{C}^k_0(\mathbb{R}^n)$. Let $|\cdot|_{\mathscr{C}^k(\overline{\mathbb{R}^n_+})}$ be the $\mathscr{C}^k$-norm on $\mathscr{C}^k(\overline{\mathbb{R}^n_+})$ induced by (A.1.2). For $U \subset \mathbb{R}^n$, we define $\mathscr{C}^k(\overline{\mathbb{R}^n_+} \cap U)$, etc., in a similar way.

Now fix an integer $N > 0$, and for $u \in \mathcal{S}(\overline{\mathbb{R}^n_+})$, set

$$(\Psi u)(x) = \begin{cases} u(x), & \text{for } x_n \geqslant 0, \\ \sum_{j=1}^N c_j u(x', -jx_n), & \text{for } x_n < 0. \end{cases} \tag{A.2.1}$$

Theorem A.2.1. *The space $\mathcal{S}(\overline{\mathbb{R}^n_+})$ is dense in $\boldsymbol{H}^k(\mathbb{R}^n_+)$ for any $k \in \mathbb{N}$. For any $N \in \mathbb{N}^*$, there exist $\{c_1, \ldots, c_N\}$ such that the map Ψ has a unique continuous extension to*

$$\Psi : \boldsymbol{H}^k(\mathbb{R}^n_+) \to \boldsymbol{H}^k(\mathbb{R}^n), \text{ for } k \leqslant N - 1. \tag{A.2.2}$$

Especially, each $u \in \boldsymbol{H}^k(\mathbb{R}^n_+)$ is the restriction to $\mathbb{R}^n_+$ of an element of $\boldsymbol{H}^k(\mathbb{R}^n)$.

From Theorems A.1.6 and A.2.1, we get

Theorem A.2.2 (Sobolev embedding theorem). *If $m, k \in \mathbb{N}$, satisfy $m > \frac{n}{2} + k$ then*

$$\boldsymbol{H}^m(\mathbb{R}^n_+) \hookrightarrow \mathscr{C}^k(\overline{\mathbb{R}^n_+}). \tag{A.2.3}$$

More precisely, there is $C > 0$ such that if $f \in \boldsymbol{H}^m(\mathbb{R}^n_+)$, then $f \in \mathscr{C}^k(\overline{\mathbb{R}^n_+})$ and

$$|f|_{\mathscr{C}^k(\overline{\mathbb{R}^n_+})} \leqslant C\|f\|_m. \tag{A.2.4}$$

By using a cut-off function χ as in (A.1.14), we show that, if $m > \frac{n}{2} + k$, then there exists $C > 0$ such that if $u \in \boldsymbol{H}^m(\mathbb{R}^n_+ \cap B(0,1))$, then $u \in \mathscr{C}^k(\overline{\mathbb{R}^n_+} \cap B(0,1))$, and

$$|u|_{\mathscr{C}^k(\overline{\mathbb{R}^n_+} \cap B(0,\frac{1}{2}))} \leqslant C\|u\|_m. \tag{A.2.5}$$

A.3   Sobolev spaces on manifolds

Now let X be a Riemannian manifold without boundary endowed with the Riemannian metric g^{TX} on the tangent bundle TX, and $\dim_{\mathbb{R}} X = n$. Let ∇^{TX} be the Levi–Civita connection on (TX, g^{TX}). Let (E, h^E) be a Hermitian vector bundle on X with Hermitian connection ∇^E.

For $k \in \mathbb{N} \cup \{\infty\}$, we denote by $\mathscr{C}^k(X, E)$, (resp. $\mathscr{C}_0^k(X, E)$), the spaces of $\mathscr{C}^k$ (resp. $\mathscr{C}^k$ with compact support) sections of E on X. For a compact set $K \subset X$ we denote $\mathscr{C}_0^k(K, E) := \{s \in \mathscr{C}^k(X, E) : \mathrm{supp}(s) \subset K\}$.

For $l \in \mathbb{N}$, we denote by ∇^E the connection induced on $(T^*X)^{\otimes l} \otimes E$ by ∇^{TX} and ∇^E; and by $|\cdot|$ the pointwise norm induced by g^{TX} and h^E. Let $\|\cdot\|_{L^2}$ be the L^2-norm on $\mathscr{C}_0^\infty(X, (T^*X)^{\otimes l} \otimes E)$ introduced as in (1.3.14).

For $k \in \mathbb{N}$, the $\mathscr{C}^k(X)$-norm $|\cdot|_{\mathscr{C}^k}$ and Sobolev norm $\|\cdot\|_k$ are defined for $s \in \mathscr{C}_0^\infty(X, E)$ by

$$|s|_{\mathscr{C}^k(X)} = \sum_{l=0}^{k} \sup_X |\underbrace{\nabla^E \cdots \nabla^E}_{l \text{ times}} s|,$$

$$\|s\|_k^2 = \sum_{l=0}^{k} \|\underbrace{\nabla^E \cdots \nabla^E}_{l \text{ times}} s\|_{L^2}^2. \tag{A.3.1}$$

Let $K \subset X$ be a compact set. The *Sobolev space* $\boldsymbol{H}_0^k(K, E)$ is the completion of $\mathscr{C}_0^\infty(K, E)$ with respect to $\|\cdot\|_k$. If $K \subset Y$, where Y is a compact manifold, then $\boldsymbol{H}_0^k(K, E) = \{s \in \boldsymbol{H}^k(Y, E) : \mathrm{supp}(s) \subset K\}$. Assume moreover that $K = \overline{U}$, where U is a relatively open set; then $\boldsymbol{H}_0^k(K, E)$ coincides with the closure of $\mathscr{C}_0^\infty(U, E)$ with respect to $\|\cdot\|_k$, and we denote it by $\boldsymbol{H}_0^k(U, E)$. For these statements we refer to [237, p. 291].

Let $L_{\mathrm{loc}}^2(X, E)$ be the space of locally L^2-integrable sections of E on X. We also define $\boldsymbol{H}^k(X, E, loc) := \{s \in L_{\mathrm{loc}}^2(X, E) : \chi s \in \boldsymbol{H}_0^k(\mathrm{supp}(\chi), E) \text{ for any } \chi \in \mathscr{C}_0^\infty(X)\}$. If X is a compact manifold, we set $\boldsymbol{H}^k(X, E) := \boldsymbol{H}_0^k(X, E)$.

Now, we state the corresponding version of Theorems A.1.6 and A.1.7.

Theorem A.3.1. *Let $K \subset X$ be compact.*

a) (Rellich's theorem) *For any $k \in \mathbb{N}$, the inclusion map*

$$\jmath : \boldsymbol{H}_0^{k+1}(K, E) \to \boldsymbol{H}_0^k(K, E) \quad \text{is compact,}$$

i.e., $\jmath$ sends any bounded subset of $\boldsymbol{H}_0^{k+1}(K, E)$ to a relatively compact subset of $\boldsymbol{H}_0^k(K, E)$, equivalently, a set with compact closure.

b) (Sobolev embedding theorem) *If $m, k \in \mathbb{N}$ satisfy $m > \frac{n}{2} + k$, there is $C > 0$ such that if $s \in \boldsymbol{H}_0^m(K, E)$, then $s \in \mathscr{C}_0^k(K, E)$ and*

$$|s|_{\mathscr{C}^k(X)} \leqslant C \|s\|_m. \tag{A.3.2}$$

We call a *generalized Laplacian* associated to ∇^E an operator H of the form

$$H = \Delta^E + Q \tag{A.3.3}$$

where Q is a Hermitian section of $\mathrm{End}(E)$ on X, (i.e., $Q(x)^* = Q(x)$ for $x \in X$), and Δ^E is the Bochner Laplacian associated to ∇^E as in (1.3.19). We consider in the sequel a generalized Laplacian H.

Theorem A.3.2 (Basic elliptic estimate). *Let $K \subset X$ be compact. For any $k \in \mathbb{N}$, there exist $C_1, C_2 > 0$ such that for any $s \in \boldsymbol{H}_0^{k+2}(K, E)$, we have*

$$\|s\|_{k+2} \leqslant C_1 \|H\,s\|_k + C_2 \|s\|_{L^2}. \tag{A.3.4}$$

We have a variant of the basic elliptic estimate for the quadratic form (cf. [238, Chap.7, Th.6.1]).

Theorem A.3.3 (Gårding's inequality). *Let $K \subset X$ be compact. There exists $C > 0$ such that for any $s \in \boldsymbol{H}_0^1(K, E)$, we have*

$$\|s\|_1^2 \leqslant C(\langle Hs, s \rangle + \|s\|_{L^2}^2). \tag{A.3.5}$$

Theorem A.3.4 (Regularity). *Let $s \in L_{\mathrm{loc}}^2(X, E)$ with compact support, and let $K = \mathrm{supp}(s)$. Assume that $Hs \in \boldsymbol{H}_0^k(K, E)$ (in the sense that the current Hs is defined by an element of $\boldsymbol{H}_0^k(K, E)$). Then $s \in \boldsymbol{H}_0^{k+2}(K, E)$. In particular, if $Hs \in \mathscr{C}^\infty(X, E)$, then $s \in \mathscr{C}^\infty(X, E)$.*

Appendix B

Elements of Analytic and Hermitian Geometry

We collect in this appendix the necessary tools of complex analytic geometry that we used.

B.1 Analytic sets and complex spaces

A pair $(X, \mathscr{O}_X)$ of a topological space X and a subsheaf of rings $\mathscr{O}_X$ of the sheaf of continuous functions $\mathscr{C}_X$ such that each stalk $\mathscr{O}_{X,x}$ is a local $\mathbb{C}$-algebra is called a $\mathbb{C}$-*ringed space*.

Let X, Y be topological spaces, $f : X \longrightarrow Y$ be a continuous map and $\mathscr{A}$ be a sheaf of abelian groups on X. The presheaf on Y defined by $(f_*\mathscr{A})(U) = \mathscr{A}(f^{-1}(U))$ for any open set $U \subset Y$ is a sheaf, called a *direct image* and denoted $f_*\mathscr{A}$.

Let $f : X \longrightarrow Y$ be a continuous map. Then f induces a canonical comorphism $\tilde{f} : \mathscr{C}_Y \longrightarrow f_*\mathscr{C}_X$ by lifting the continuous functions:

$$\mathscr{C}_Y(U) \ni t \longrightarrow t \circ f \in \mathscr{C}_X(f^{-1}(U)) = (f_*\mathscr{C}_X)(U). \tag{B.1.1}$$

A morphism of $\mathbb{C}$-ringed spaces $f : (X, \mathscr{O}_X) \longrightarrow (Y, \mathscr{O}_Y)$ is a continuous map $f : X \longrightarrow Y$ such that the canonical morphism (B.1.1) induces a morphism $\tilde{f} : \mathscr{O}_Y \longrightarrow f_*\mathscr{O}_X$, i.e. pull-backs of sections of $\mathscr{O}_Y$ lie in $\mathscr{O}_X$.

Let $\mathscr{G}$ be a sheaf of $\mathscr{O}_Y$-modules over Y. The *topological inverse image* $f^{-1}(\mathscr{G})$ is a sheaf on X determined uniquely by the property $f^{-1}(\mathscr{G})_x = \mathscr{G}_{f(x)}$, for any $x \in X$. The *analytic inverse image sheaf* $\widetilde{\mathscr{G}}$ is defined by $\widetilde{\mathscr{G}}_x = f^{-1}(\mathscr{G})_x \otimes_{\mathscr{O}_{Y,f(x)}} \mathscr{O}_{X,x}$, $x \in X$. $\widetilde{\mathscr{G}}$ is a sheaf of $\mathscr{O}_X$-modules.

Definition B.1.1. Let X be a complex manifold and $\mathscr{O}_X$ be the sheaf of holomorphic functions on X. We say $A \subset X$ is an *analytic subset* of X if for any $x \in A$ there exist an open neighbourhood U_x in X and holomorphic functions $f_1, \ldots, f_k \in \mathscr{O}_X(U_x)$ on U_x such that $A \cap U_x = \{y \in U_x : f_1(y) = \cdots = f_k(y) = 0\}$. A point $x \in A$ is called a *regular point* if there exists an open neighbourhood U_x of x such that $A \cap U_x$ is a complex submanifold of X. The set of regular points of A is denoted by A_{reg}. A point $x \in A$ is a *singular point* of A if it is not regular; we denote $A_{\text{sing}} = A \setminus A_{\text{reg}}$. A is said to be *irreducible* if A can not be written as the union of two analytic subsets A_1, A_2 with $A_1, A_2 \neq A$.

$A \subset X$ is called an *analytic hypersurface* if A is locally the zero locus of a single non-zero holomorphic function g.

Lemma B.1.2 ([107, §2.15], [79, II.4.31]). *If $A \subset X$ is an analytic subset of a complex manifold X, then A_{reg}, A_{sing} are analytic subsets of X, and $\dim A_{\text{sing}} < \dim A_{\text{reg}}$.*

For the following basic result we refer to [79, III.7.6].

Definition and Theorem **B.1.3.** Let A be an analytic set of dimension k in a complex manifold X with $\dim X = n$. Let $x \in A$. For a generic choice of local coordinates $z' = (z_1, \ldots, z_k)$ and $z'' = (z_{k+1}, \ldots, z_n)$ around x there exists a neighborhood U of x in X and $C > 0$ such that $A \cap U \subset \{|z''| \leqslant C|z'|\}$. Denote by $B'(0, r)$ the ball of center 0 and radius r in $\mathbb{C}^k$ and $B''(0, Cr)$ the ball of center 0 and radius Cr in $\mathbb{C}^{n-k}$. For $r > 0$ sufficiently small the projection $\pi : A \cap (B'(0, r) \times B''(0, Cr)) \to B'(0, r)$ is a ramified covering with finite sheet number m, i.e., there exists a nowhere dense analytic set $S \subset B'(0, r)$ (called ramification locus) such that $\pi : A \setminus \pi^{-1}(S) \to B'(0, r) \setminus S$ is a topological covering with m sheets. The sheet number is independent on the choice of generic coordinates. It is called the *multiplicity of A at x* and is denoted by $m_x(A)$.

Definition B.1.4. Let $V \subset X$ be an irreducible analytic hypersurface of a complex manifold X, and g be a local defining function for V near $y \in V$. For any holomorphic function f defined near y, we define the *order* $\text{ord}_{V,y}(f)$ of f along V at y to be the largest integer k such that there exists $h \in \mathscr{O}_{X,y}$ such that $f = g^k \cdot h$ in the local ring $\mathscr{O}_{X,y}$. As relatively prime elements of $\mathscr{O}_{X,y}$ are still relatively prime in $\mathscr{O}_{X,x}$ for x near y, $\text{ord}_{V,y}(f)$ is independent of y and we denote it simply by $\text{ord}_V(f)$, the *order* of f along V.

If f is a meromorphic function on X, written locally as $f = f_1/f_2$ with f_1, f_2 holomorphic, then we define $\text{ord}_V(f) = \text{ord}_V(f_1) - \text{ord}_V(f_2)$. Certainly, there exist only a locally finite number of irreducible analytic hypersurfaces V of X such that $\text{ord}_V(f) \neq 0$. The *divisor* $\text{Div}(f)$ of a meromorphic function f on X is defined by

$$\text{Div}(f) = \sum_V \text{ord}_V(f) \cdot V. \tag{B.1.2}$$

Let U be an open set in $\mathbb{C}^n$ and let $\mathscr{I} \subset \mathscr{O}_U$ be a coherent ideal sheaf. The set $A := \operatorname{supp}(\mathscr{O}_U/\mathscr{I}) := \{x \in U \mid \mathscr{I}(x) \neq \mathscr{O}_{U,x}\}$ is an analytic subset of U and $(A, (\mathscr{O}_U/\mathscr{I})|_A)$ is a $\mathbb{C}$-ringed space called a *local model*.

Definition B.1.5 (complex spaces). A *complex space* $(X, \mathscr{O}_X)$ is a $\mathbb{C}$-ringed space which is Hausdorff and has the property that each $x \in X$ possesses an open neighbourhood W such that $(W, \mathscr{O}_W)$ is isomorphic to a local model (as $\mathbb{C}$-ringed spaces). A complex space is called *reduced* if its local models are of the form $(A, (\mathscr{O}_U/\mathscr{I}_A)|_A)$, where A is an analytic subset of U and $\mathscr{I}_A \subset \mathscr{O}_U$ is the sheaf of all holomorphic functions which vanish on A.

A section of the sheaf $\mathscr{O}_X$ is called a *holomorphic function*.

A complex space $(X, \mathscr{O}_X)$ is called *irreducible* if X cannot be represented as the union of two analytic subsets X_1, X_2 with $X_1, X_2 \neq X$.

When not otherwise stated, we consider only reduced complex spaces.

We will usually denote a complex space $(X, \mathscr{O}_X)$ simply by X. The property from the definition of being locally isomorphic to a model, means that there exists an open set $\widehat{W} \subset \mathbb{C}^N$, an analytic subset $W' \subset \widehat{W}$ and a homeomorphism $\imath : W \longrightarrow W'$, such that the canonical homomorphism $\widetilde{\imath} : \mathscr{O}_{W'} \longrightarrow \imath_* \mathscr{O}_W$ induces an isomorphism of sheaves of rings, where $\imath_* \mathscr{O}_W$ is the direct image sheaf. The map $\iota : W \longrightarrow \widehat{W}$, obtained by composing $\imath$ with the inclusion $W' \longhookrightarrow \widehat{W}$, is called a *local chart* on X. Note that we have induced isomorphisms $\widetilde{\imath}_x : \mathscr{O}_{W', \iota(x)} \longrightarrow \mathscr{O}_{X,x}$, $\widetilde{\imath}_x(t_x) = (t \circ \imath)_x$, $t_x \in \mathscr{O}_{W', \iota(x)}$.

We denote by X_{reg} the set of *regular points*, i.e. points for which a local chart $\iota : W \longrightarrow \widehat{W}$ can be found, such that ι is surjective. In other words, a local model is an open set in the Euclidean space. The other points of X are called *singular*, and the set of such points is denoted by X_{sing}. For any complex space the singular locus X_{sing} is a nowhere dense analytic subset (cf. [107, §2.15–16]). A reduced complex space is irreducible if and only if X_{reg} is connected.

We state now two useful extension theorems.

Theorem B.1.6 (Riemann's second extension theorem). *Let X be a complex space and A an analytic subset of codimension at least two at every point. Then every holomorphic function on $X \setminus A$ has a unique holomorphic extension to X.*

For the proof we refer to [107, p. 120], [79, I.6.4].

Theorem B.1.7 (Remmert-Stein extension theorem). *Let X be a complex space, A an analytic subset of X, Y an analytic subset of $X \setminus A$. Assume that there exists $k \in \mathbb{N}$ such that $\dim A \leqslant k$ and $\dim_y Y > k$ for all $y \in Y$. Then the closure $\overline{Y}$ of Y in X is an analytic set.*

For the proof see [120, p. 395-400], [79, II.8.6].

Theorem B.1.8 (Oka [79, II.3.19]). *Let E be a holomorphic vector bundle over a complex space X. The sheaf $\mathscr{O}_X(E)$ of holomorphic sections of E is a coherent sheaf.*

Definition B.1.9 (holomorphic map). If $(X, \mathcal{O}_X)$, $(Y, \mathcal{O}_Y)$ are arbitrary complex spaces, a morphism $\varphi : X \to Y$ of $\mathbb{C}$-ringed spaces is called *holomorphic map*. A holomorphic map is called *biholomorphic* if there exists a holomorphic map $\psi : Y \to X$ such that $\varphi \circ \psi = \mathrm{Id}_Y$, $\psi \circ \varphi = \mathrm{Id}_X$. A holomorphic map $\varphi : X \to Y$ is called *proper* if the preimage of any compact set is a compact set. A holomorphic map $\varphi : X \to Y$ is called *finite* if it is proper and each point $x \in X$ is isolated in $\varphi^{-1}(\varphi(x))$.

The following result shows that the fibers of a holomorphic map can be considered as a complex space. It is useful to determine if the fibers are connected.

Theorem B.1.10 (Stein factorization). *Let X, Y be complex spaces and $\varphi : X \to Y$ a proper surjective holomorphic map. Then there exists a complex space Z and surjective holomorphic maps $\psi : X \to Z$, $\tau : Z \to Y$ and a commutative diagram*

$$
\begin{array}{ccc}
X & \xrightarrow{\;\varphi\;} & Y \\
& \psi \searrow & \uparrow \tau \\
& & Z
\end{array}
\qquad\qquad (B.1.3)
$$

such that

(a) *The fibers of ψ are connected.*

(b) *The fibers of τ are finite, i.e., τ is a finite holomorphic map .*

(c) *For any $y \in Y$ there is a bijection between the points of $\tau^{-1}(y)$ and connected components of $\varphi^{-1}(y)$.*

For the proof we refer to [107, Th. 1.24, p. 70] and the references therein.

Theorem B.1.11 (Zariski's main theorem). *Let X, Y be irreducible complex spaces and $\varphi : X \to Y$ a finite holomorphic map which is a proper modification. Then φ is a biholomorphic map.*

For the proof we refer to [107, 4.9, p. 187].

Theorem B.1.12 (Remmert's proper mapping theorem). *Let $f : X \to Y$ be a proper holomorphic mapping between complex spaces. Then $f(X)$ is an analytic subset of Y.*

For the proof we refer to [107, 1.18, p. 64], [79, II.8.8].

B.2   Currents on complex manifolds

In this section, we collect the necessary facts about currents on complex manifolds. We refer the reader to the textbooks [120, 79] for more details.

If not otherwise stated, X will be a paracompact complex manifold of complex dimension n. We denote by TX the *real tangent bundle* of the underlying real manifold X.

Proposition B.2.1. *There exists a unique complex structure J on $T_x X$ which satisfies the intrinsic Cauchy-Riemann equations:*

$$df_x(Jv) = \sqrt{-1}df_x(v), \quad v \in T_x X$$

for all germs of holomorphic function $f \in \mathscr{O}_{X,x}$ at x.

If $(U, z_1, \ldots, z_n)$ is a holomorphic system of coordinates and $z_j := x_j + \sqrt{-1}y_j$, then $(U, x_1, y_1, \ldots, x_n, y_n)$ is a smooth system of coordinates for the real differentiable structure. The complex structure on $T_x X$ given by $J(\partial/\partial x_j) = \partial/\partial y_j$, $J(\partial/\partial y_j) = -\partial/\partial x_j$. We get in this way a bundle morphism $J : TX \longrightarrow TX$, $J^2 = -\,\mathrm{Id}$.

Definition B.2.2. The complex vector bundle $T_h X = (TX, J)$ is called the *holomorphic tangent bundle* of X.

There is a natural splitting of the bundle of $\mathbb{R}$-linear maps $\mathrm{Hom}_{\mathbb{R}}(TX, \mathbb{C}) = \mathrm{Hom}_{\mathbb{C}}(T_h X, \mathbb{C}) \oplus \mathrm{Hom}_{\mathbb{C}}(T_h X, \overline{\mathbb{C}})$ in $\mathbb{C}$-linear and $\mathbb{C}$-antilinear maps. We set

$$T^{*(1,0)} X = \mathrm{Hom}_{\mathbb{C}}(T_h X, \mathbb{C}), \quad T^{*(0,1)} X = \mathrm{Hom}_{\mathbb{C}}(T_h X, \overline{\mathbb{C}}). \tag{B.2.1}$$

Then $dz_j := dx_j + \sqrt{-1}dy_j$ (resp. $d\bar{z}_j := dx_j - \sqrt{-1}dy_j$) is a local frame of $T^{*(1,0)} X$ (resp. $T^{*(0,1)} X$). The duals, $T^{(1,0)} X$ and $T^{(0,1)} X$, are obtained similarly from the decomposition of $TX \otimes_{\mathbb{R}} \mathbb{C}$ in $\sqrt{-1}$ and $-\sqrt{-1}$ eigenspaces.

It is easy to check that $T^{(1,0)} X$ is a holomorphic vector bundle (i.e. the transition functions are holomorphic). The map $Y \in T_h X \to \frac{1}{2}(Y - \sqrt{-1}JY) \in T^{(1,0)} X$ induces the natural identification of $T_h X$ and $T^{(1,0)} X$ as complex vector bundles. Hence $T_h X$ has also the structure of a holomorphic vector bundle.

The bundle of (p, q)-*forms* is

$$\Lambda^{p,q}(T^* X) := \Lambda^p(T^{*(1,0)} X) \otimes \Lambda^q(T^{*(0,1)} X). \tag{B.2.2}$$

For any open set $U \subset X$ we denote by $\Omega^r(U)$ (resp. $\Omega^{p,q}(U)$) the space of smooth r-forms (resp. (p, q)-forms) on U. The correspondence $U \to \Omega^r(U)$ (resp. $U \to \Omega^{p,q}(U)$) defines the sheaf Ω^r (resp. $\Omega^{p,q}$) of smooth r-forms (resp. smooth (p, q)-forms). We have the direct sum decomposition $\Omega^r = \oplus_{p+q=r} \Omega^{p,q}$ and the differentials $d : \Omega^r \longrightarrow \Omega^{r+1}$ (resp. $\partial : \Omega^{p,q} \longrightarrow \Omega^{p+1,q}$ and $\bar{\partial} : \Omega^{p,q} \longrightarrow \Omega^{p,q+1}$) with $d = \partial + \bar{\partial}$.

We let $\Omega_0^r(X)$ (resp. $\Omega_0^{p,q}(X)$) denote the space of smooth r-forms (resp. (p, q)-forms) on X with compact support. In this context, forms with compact support are called *test forms*. For $K \subset X$ compact, let $\Omega_K^r(X)$ be the set of r-forms with support in K, endowed with $\mathscr{C}^\infty$-topology, which makes it a Fréchet space. We introduce the direct limit topology on $\Omega_0^r(X)$ by declaring a set $\mathcal{E} \subset \Omega_0^r(X)$ open if and only if $\mathcal{E} \cap \Omega_K^r(X)$ is open in $\Omega_K^r(X)$ for any compact set $K \subset X$. It is easy to see that for $\varphi_\nu, \varphi_0 \in \Omega_0^r(X)$, $\varphi_\nu \to \varphi_0$ in this topology, if and only if there exists a compact set K such that $\varphi_\nu, \varphi_0 \in \Omega_K^r(X)$ and $\varphi_\nu \to \varphi_0$ in $\Omega_K^r(X)$. In the same vein we introduce the direct limit topology on $\Omega_0^{p,q}(X)$.

Definition B.2.3. Let M be a *real* manifold of real dimension m. The space of *currents* of degree $m-r$ is the space of complex linear continuous forms on $\Omega_0^r(M)$.

This space is denoted by $\Omega'^{m-r}(M)$. We denote by $(T, \varphi) := T(\varphi)$ the pairing between a current $T \in \Omega'^{m-r}(M)$ and a test form $\varphi \in \Omega_0^r(M)$.

A current is said to be of order k if it is continuous in the $\mathscr{C}^k$-topology.

The space $\Omega'^0(M)$ dual to $\Omega_0^m(M)$ is called the space of *distributions* on M and is denoted by $\mathscr{D}'(M)$. If M is *orientable* we can identify $\Omega_0^m(M) \cong \Omega_0^0(M)$ using a volume element on M, so we can think $\mathscr{D}'(M) = \Omega'^0(M)$ as the dual of $\Omega_0^0(M)$.

In particular, let U be an open subset in $\mathbb{R}^m$. The space of distributions on U is denoted $\mathscr{D}'(U)$ and is the dual of the space $\mathscr{C}_0^\infty(U)$.

The Schwartz space of rapidly decreasing test functions $\mathcal{S}(\mathbb{R}^m)$ (cf. (A.1.11)) endowed with the seminorms $|f|_{\alpha,\beta} := \sup_{x \in \mathbb{R}^m} |x^\alpha \frac{\partial^\beta}{\partial x^\beta} f(x)|$ is a Fréchet space. The dual $\mathcal{S}'(\mathbb{R}^m)$ of $\mathcal{S}(\mathbb{R}^m)$ is called the space of *tempered distributions*.

Assume that E is a complex vector bundle on M. The dual of the space $\Omega_0^r(M, E^*)$, denoted by $\Omega'^{m-r}(M, E)$, is called the space of currents of degree $m - r$ with values in E.

Assume now that X is a complex manifold of complex dimension n. The space of currents of bidegree $(n-p, n-q)$ is the space of complex linear continuous forms on $\Omega_0^{p,q}(X)$. This space is denoted by $\Omega'^{n-p,n-q}(X)$.

The spaces of currents (in particular distributions) are endowed with the *weak topology*. For example, a sequence $\{T_k\}$ converges in $\Omega'^{m-r}(X)$ to T if and only if for any $\varphi \in \Omega_0^r(X)$ we have $\lim_{k \to \infty}(T_k, \varphi) = (T, \varphi)$.

Note that by the definition of the direct limit topology, a linear form T on $\Omega_0^r(X)$ (resp. $\Omega_0^{p,q}(X)$) is a current if and only if T is continuous on $\Omega_K^r(X)$ (resp. $\Omega_K^{p,q}(X)$) for any compact $K \subset X$. Equivalently, for any compact set $K \subset X$, there exist $k \in \mathbb{N}$ and $C > 0$ such that $|(T, \varphi)| \leqslant C|\varphi|_{\mathscr{C}^k(X)}$ for any $\varphi \in \Omega_K^r(X)$. We have the following direct sum decompositions:

$$\Omega^r(X) = \oplus_{p+q=r}\Omega^{p,q}(X), \quad \Omega'^r(X) = \oplus_{p+q=r}\Omega'^{p,q}(X). \qquad (B.2.3)$$

For two open sets $U \subset V \subset X$ we define the *restrictions* $\Omega'^r(V) \to \Omega'^r(U)$ and $\Omega'^{p,q}(V) \to \Omega'^{p,q}(U)$, $T \longmapsto T|_U$ simply by $(T|_U, \varphi) := (T, \varphi)$ for $\varphi \in \Omega_0^{n-r}(U)$ (resp. $\varphi \in \Omega_0^{n-p,n-q}(U)$). It is easy to see that this defines *sheaves* Ω'^r and $\Omega'^{p,q}$ on X. We define the *support* of T as $\operatorname{supp}(T) := X \smallsetminus \cup\{V \subset X : V \text{ open}, T|_V = 0\}$.

Definition B.2.4 (operations with currents).

1. For $T \in \Omega'^r(X)$, $\alpha \in \Omega^s(X)$ the wedge product $T \wedge \alpha$ is defined by

$$(T \wedge \alpha, \varphi) = (T, \alpha \wedge \varphi), \quad \varphi \in \Omega_0^{2n-r-s}(X). \qquad (B.2.4)$$

2. The operator $d: \Omega'^r(X) \longrightarrow \Omega'^{r+1}(X)$ is defined by

$$(dT, \varphi) = (-1)^{r+1}(T, d\varphi), \quad \varphi \in \Omega_0^{2n-r-1}(X). \qquad (B.2.5)$$

A current T is called *closed* if $dT = 0$.

3. The operator $\bar{\partial} : \Omega'^{p,q}(X) \longrightarrow \Omega'^{p,q+1}(X)$ is defined by

$$(\bar{\partial}T, \varphi) = (-1)^{p+q+1}(T, \bar{\partial}\varphi), \quad \varphi \in \Omega_0^{n-p,n-q-1}(X). \tag{B.2.6}$$

The operator $\partial : \Omega'^{p,q} \longrightarrow \Omega'^{p+1,q}(X)$ is defined analogously and $d = \partial + \bar{\partial}$.

4. Let $f : X \longrightarrow Y$ be a smooth map and $T \in \Omega'^r(X)$. If $f|_{\text{supp}(T)}$ is proper then

$$(f_*T, \varphi) = (T, f^*\varphi), \quad \varphi \in \Omega_0^{2n-r}(Y) \tag{B.2.7}$$

defines a current f_*T called the *push-forward* of T by f.

Example B.2.5. Let $u \in L^2_{\text{loc}}(X, \Lambda^r(T^*X) \otimes_{\mathbb{R}} \mathbb{C})$ be a locally L^2 integrable r-form. Then u defines a current $T_u \in \Omega'^r(X)$ by $(T_u, \varphi) = \int_X u \wedge \varphi$, for $\varphi \in \Omega_0^{2n-r}(X)$. The map $L^2_{\text{loc}}(X, \Lambda^r(T^*X) \otimes_{\mathbb{R}} \mathbb{C}) \to \Omega'^r(X)$, $u \mapsto T_u$ is injective, and we can identify T_u with u.

Definition B.2.6. A current $T \in \Omega'^r(X)$ is said to be smooth on an open set U if $T|_U = T_u$ for some $u \in \Omega^r(U)$. The *singular support* $\operatorname{sing\,supp}(T)$ of a current $T \in \Omega'^r(X)$ is defined as the smallest subset Y of X such that T is a smooth form on $X \smallsetminus Y$.

Let us recall the classical Schwartz kernel theorem [237, p. 296]. Let M be a real orientable manifold of dimension m and E, F be two vector bundles on M. We denote by $\mathscr{D}'_0(M, E)$ the space of E-valued distributions with compact support.

Let us fix a volume form $d\mu$ on M. Then $u \in L^1_{\text{loc}}(M, E)$ defines the distribution $T_u \in \mathscr{D}'(M, E)$ by setting $(T_u, \varphi) = \int_M u(y) \cdot \varphi(y) \, d\mu(y)$, for $\varphi \in \mathscr{C}_0^\infty(M, E^*)$, where $u(y) \cdot \varphi(y)$ is the pairing $E \times E^* \to \mathbb{C}$.

We denote by pr_1, pr_2 the projections from $M \times M$ on the first and second factor M respectively. We introduce the vector bundle $E \boxtimes F = \operatorname{pr}_1^* E \otimes \operatorname{pr}_2^* F$ on $M \times M$. For $u \in \mathscr{C}_0^\infty(M, E)$, $v \in \mathscr{C}_0^\infty(M, F^*)$ we define $v \otimes u \in \mathscr{C}_0^\infty(M \times M, F^* \boxtimes E)$ by $(v \otimes u)(x, y) := v(x) \otimes u(y)$.

Let $K \in \mathscr{D}'(M \times M, F \boxtimes E^*)$. For any fixed $u \in \mathscr{C}_0^\infty(M, E)$, the linear map $\mathscr{C}_0^\infty(M, F^*) \ni v \mapsto (K, v \otimes u) \in \mathbb{C}$ defines a distribution $A_K(u) \in \mathscr{D}'(M, F)$. The operator $A_K : \mathscr{C}_0^\infty(M, E) \to \mathscr{D}'(M, F)$, $u \mapsto A_K(u)$, is linear and continuous.

Theorem B.2.7 (Schwartz kernel theorem).

(a) *Let* $A : \mathscr{C}_0^\infty(M, E) \to \mathscr{D}'(M, F)$ *be a linear continuous operator. Then there exists a unique distribution* $K \in \mathscr{D}'(M \times M, F \boxtimes E^*)$, *called the* Schwartz *kernel distribution such that* $A = A_K$, *i.e.,*

$$(Au, v) = (K, v \otimes u) \tag{B.2.8}$$

for any $u \in \mathscr{C}_0^\infty(M, E)$, $v \in \mathscr{C}_0^\infty(M, F^*)$.

(b) *Assume* $A : \mathscr{C}_0^\infty(M, E) \longrightarrow \mathscr{C}^\infty(M, F)$ *is linear continuous. Then for any volume form* $d\mu$, *the Schwartz kernel of* A *is represented by a smooth kernel*

$K \in \mathscr{C}^\infty(M \times M, F \boxtimes E^*)$ *called the* Schwartz kernel *of A with respect to $d\mu$ such that*

$$(Au)(x) = \int_M K(x,y)u(y)\, d\mu(y), \quad \text{for any } u \in \mathscr{C}_0^\infty(M,E). \qquad (B.2.9)$$

Moreover, A can be extended as a linear continuous operator $A : \mathscr{D}_0'(M,E) \to \mathscr{C}^\infty(M,F)$ by setting

$$(Au)(x) = \big(u(\cdot), K(x,\cdot)\big), \quad x \in M, \qquad (B.2.10)$$

for any $u \in \mathscr{D}_0'(M,E)$.

Linear continuous operators $A : \mathscr{D}_0'(M,E) \to \mathscr{C}^\infty(M,F)$ are called *smoothing operators*.

For $I \subset \{1,2,\ldots,n\}$, we set $\complement I = \{1,2,\ldots,n\} \smallsetminus I$. If $I = \{i_1 < i_2 < \cdots < i_p\}$, $\complement I = \{j_1 < j_2 < \cdots < j_{n-p}\}$, we denote by $(I\complement I)$ the permutation $\{i_1, i_2, \ldots, i_p, j_1, j_2, \ldots, j_{n-p}\}$ of $\{1,2,\ldots,n\}$. Let $(z_1,\ldots,z_n)$ be local coordinates on an open set U. We can identify the space of currents $\Omega'^{p,q}(U)$ with the space of (p,q)-forms with distribution coefficients. For $T \in \Omega'^{p,q}(X)$ we have

$$T|_U = \sum_{|I|=p,|J|=q} T_{IJ} \wedge dz_I \wedge d\bar{z}_J, \quad T_{IJ} \in \Omega'^{0,0}(U), \qquad (B.2.11)$$

where $dz_I := dz_{i_1} \wedge \cdots \wedge dz_{i_p}$ and

$$T_{IJ}(dz_1 \wedge \cdots \wedge dz_n \wedge d\bar{z}_1 \wedge \cdots \wedge d\bar{z}_n)$$
$$= (-1)^{(n-p)q}\, \mathrm{sgn}(I\,\complement I)\, \mathrm{sgn}(J\,\complement J)\, T(dz_{\complement I} \wedge d\bar{z}_{\complement J}).$$

We can of course identify $T_{IJ} \in \Omega'^{0,0}(U)$ with a usual distribution $T_{IJ} \in \Omega'^{n,n}(U)$ due to the canonical isomorphism $\Omega^{0,0}(U) \cong \Omega^{n,n}(U)$.

We discuss now positivity notions for currents and start with the definition for $(1,1)$-forms. For a thorough discussion see [79, III.1].

Definition and Theorem B.2.8. Let $\omega : TX \times TX \to \mathbb{R}$ be a real $(1,1)$-form on the complex manifold X. Then the following assertions are equivalent:

1. For all $v \in TX$, $v \neq 0$, we have $\omega(v, Jv) > 0$ (resp. $\geqslant 0$).

2. If ω has the local form $\omega = \frac{\sqrt{-1}}{2} \sum_{jk} h_{jk}(z) dz_j \wedge d\bar{z}_k$, the Hermitian matrix $(h_{jk}(z))$ is positive definite (resp. semidefinite) for all z.

3. $\omega : T^{(1,0)}X \times T^{(0,1)}X \longrightarrow \mathbb{C}$ satisfies $-\sqrt{-1}\omega(u,\bar{u}) > 0$ (resp. $\geqslant 0$) for all $u \in T^{(1,0)}X$, $u \neq 0$.

If they are satisfied, ω is called *positive* (resp. *semipositive*). In the case of a positive form the conditions are also equivalent to

4. $\omega = -\operatorname{Im} h^{T_h X}$, where $h^{T_h X}$ is a Hermitian metric on $T_h X$.

Definition B.2.9.

(i) Let U be an open set in $\mathbb{C}$. A function $f : U \to [-\infty, \infty[$ is called *subharmonic* if it is upper semicontinuous and satisfies the mean value inequality on any closed ball $\overline{B}(a, r) \subset U$ with center a and radius r, i.e.,

$$u(a) \leqslant \frac{1}{2\pi} \int_0^{2\pi} u(a + re^{i\theta}) \, d\theta. \tag{B.2.12}$$

(ii) A function $f : X \longrightarrow [-\infty, \infty[$ is called *plurisubharmonic* if it is upper semicontinuous and for any holomorphic function $u : B(a, 1) \subset \mathbb{C} \to X$, $f \circ u$ is subharmonic.

Proposition B.2.10 ([79, I.Th. 5.8]). *Let* $\varphi : X \longrightarrow \mathbb{R}$ *be smooth. Then* φ *is plurisubharmonic if and only if* $\sqrt{-1}\partial\bar{\partial}\varphi$ *is a semipositive form.*

Definition B.2.11.

(i) A $(1,1)$-current T is called *real* if $T = \overline{T}$ in the sense that $\overline{T(\varphi)} = T(\bar{\varphi})$ for all $\varphi \in \Omega_0^{n-1,n-1}(X)$ and a real current T is *positive* when $(\sqrt{-1})^{(n-1)^2} T(\psi \wedge \bar{\psi}) \geqslant 0$ for all $\psi \in \Omega_0^{n-1,0}(X)$.

(ii) A real $(1,1)$-current T on X is called *strictly positive* if there exists a positive $\mathscr{C}^\infty$-$(1,1)$-form ω on X such that $T - \omega$ is a positive current on X.

Note that the terminology for forms and currents is not symmetric. We kept the traditional terminology of Lelong, although one might call "semi-positive" a current as in Definition B.2.11 (i). If a current T is positive we write $T \geqslant 0$. $T_1 \geqslant T_2$ means $T_1 - T_2 \geqslant 0$.

A current $T \in \Omega'^{1,1}(X)$ can be written in local coordinates $(z_1, \ldots, z_n)$ on an open set U as $T|_U = \sqrt{-1} \sum_{jk} T_{jk} \, dz_j \wedge d\bar{z}_k$ with distributions $T_{IJ} \in \Omega'^{n,n}(U)$ (cf. (B.2.11)). T is real if $T_{jk} = \overline{T}_{kj}$ and positive if and only if the current $\sum_{jk} T_{jk} \lambda_j \bar{\lambda}_k$ is a positive measure for every $(\lambda_1, \ldots \lambda_n) \in \mathbb{C}^n$. In particular T_{jk} are measures and T is of order zero.

Definition B.2.12. For a measure μ on a manifold M we denote by μ_{ac} and μ_{sing} the uniquely determined absolute continuous and singular measures (with respect to the Lebesgue measure on M) such that

$$\mu = \mu_{\text{ac}} + \mu_{\text{sing}} \tag{B.2.13}$$

which is called the Lebesgue decomposition of μ.

If T is a $(1,1)$-current of order 0 on X, written locally $T = \sqrt{-1} \sum T_{ij} \, dz_i \wedge d\bar{z}_j$, we define its *absolute continuous* and *singular* components by

$$T_{\text{ac}} = \sqrt{-1} \sum (T_{jk})_{\text{ac}} \, dz_j \wedge d\bar{z}_k,$$
$$T_{\text{sing}} = \sqrt{-1} \sum (T_{jk})_{\text{sing}} \, dz_j \wedge d\bar{z}_k. \tag{B.2.14}$$

The *Lebesgue decomposition* of T is then

$$T = T_{ac} + T_{sing}.\qquad\qquad(B.2.15)$$

Remark B.2.13. 1) If $T \geqslant 0$, it follows that $T_{ac} \geqslant 0$, $T_{sing} \geqslant 0$. Moreover, if $T \geqslant \alpha$ for a continuous $(1,1)$-form α, then $T_{ac} \geqslant \alpha$, $T_{sing} \geqslant 0$.

2) The Radon–Nikodym theorem insures that T_{ac} is (the current associated to) a $(1,1)$-form with L^1_{loc} coefficients. The form $T_{ac}(x)^n$ exists for almost all $x \in X$ and is denoted T^n_{ac}.

Proposition B.2.14 ([79, I,Th. 5.8]). *Let $\varphi : X \longrightarrow [-\infty, \infty[$ be a plurisubharmonic function such that $\varphi \in L^1_{loc}(X)$. Then the $(1,1)$-current $\sqrt{-1}\partial\bar\partial\varphi$ is positive. Conversely, if $\varphi \in L^1_{loc}(X)$ and $\sqrt{-1}\partial\bar\partial\varphi$ is positive, then there exists a plurisubharmonic function $\psi : X \longrightarrow [-\infty, \infty[$ such that $\varphi = \psi$ almost everywhere.*

Example B.2.15. If f is a holomorphic function on X, $\log|f|$ is plurisubharmonic.

Definition B.2.16. A plurisubharmonic function $\varphi : X \longrightarrow [-\infty, \infty[$ is called *strictly plurisubharmonic* if $\varphi \in L^1_{loc}(X)$ and $\sqrt{-1}\partial\bar\partial\varphi$ is strictly positive.

A function $\varphi : X \longrightarrow [-\infty, \infty[$ is called *quasi-plurisubharmonic* if φ can be written locally as the sum of a smooth function and a plurisubharmonic function.

Proposition B.2.17. *Let $\varphi : X \longrightarrow [-\infty, \infty[$ be a quasi-plurisubharmonic function such that $\varphi \in L^1_{loc}(X)$. Then there exists a continuous $(1,1)$ form α on X such that $\sqrt{-1}\partial\bar\partial\varphi \geqslant \alpha$ in the sense of currents. Conversely, if $\varphi \in L^1_{loc}(X)$ has the property that $\sqrt{-1}\partial\bar\partial\varphi \geqslant \alpha$ for some continuous $(1,1)$-form α on X, then φ equals a quasi-plurisubharmonic function almost everywhere.*

Example B.2.18 (cf. [79, III-1.20]). Let Y be an *analytic subset* of X of pure dimension m. The *current of integration on the analytic set Y*, denoted $[Y] \in \Omega'^{\,n-m,n-m}(X)$, is defined by

$$([Y], \varphi) = \int_{Y_{reg}} \varphi, \quad \varphi \in \Omega^{m,m}_0(X),\qquad\qquad(B.2.16)$$

where Y_{reg} is endowed with the canonical orientation.

P. Lelong showed that this is well defined (cf. [79, III-2.6], [120, p. 32]) which amounts to showing that Y_{reg} has finite volume locally near Y_{sing}. Moreover, a fundamental theorem of Lelong [79, III-2.7], [120, p. 33] states that $[Y]$ is closed.

Another important result due to Lelong is the following (see [79, III-2.15]). If $D = \sum c_j D_j$ with $c_j \in \mathbb{Z}$ is a divisor on X, where D_j are irreducible hypersurfaces, we define the current of integration on D by

$$[D] = \sum c_j [D_j].\qquad\qquad(B.2.17)$$

Theorem B.2.19 (Poincaré-Lelong formula). *Let f be a meromorphic function on X which is non-zero on any open set. Then*

$$\tfrac{\sqrt{-1}}{2\pi}\partial\bar\partial \log|f|^2 = [\operatorname{Div}(f)].\qquad\qquad(B.2.18)$$

Let T be a closed positive $(1,1)$-current in $\mathbb{C}^n$. We set

$$\nu(T, x, r) = \frac{1}{(\pi r^2)^{n-1}} \int_{B(x,r)} T \wedge \omega^{n-1}, \tag{B.2.19}$$

where $\omega = \frac{\sqrt{-1}}{2} \sum_j dz_j \wedge d\bar{z}_j$ is the Euclidean Kähler form on $\mathbb{C}^n$. Then $\nu(T, x, r)$ is an increasing function of r. We define the *Lelong number* of T at x by $\nu(T, x) = \lim_{r \to 0} \nu(T, x, r)$. By a result of Siu, the Lelong number is invariant by holomorphic changes of coordinates, and can be therefore defined for currents on manifolds. We refer the reader to [79] for detailed information about Lelong numbers.

We close this section with some remarks about cohomology of currents. It is easy to see that $d^2 = 0$ and $\bar{\partial}^2 = 0$ also at currents level. Thus we get complexes of sheaves X:

$$\begin{aligned}
(\Omega'^\bullet, d) &: \Omega'^0 \longrightarrow \Omega'^1 \longrightarrow \Omega'^2 \longrightarrow \cdots, \\
(\Omega'^{r,\bullet}, \bar{\partial}) &: \Omega'^{r,0} \longrightarrow \Omega'^{r,1} \longrightarrow \Omega'^{r,2} \longrightarrow \cdots.
\end{aligned} \tag{B.2.20}$$

Note that the sheaves Ω^r and Ω'^r, $(r \geq 0)$ are fine. Moreover, the Poincaré and Dolbeault-Grothendieck lemmas ($\bar{\partial}$-Lemma) are still valid at current level. Thus $(\Omega^\bullet, d)$ and $(\Omega'^\bullet, d)$ are resolutions of $\mathbb{C}$ (or $\mathbb{R}$) on X. There are natural homomorphisms of complexes of sheaves

$$\mathbb{Z} \longrightarrow \mathbb{R} \longrightarrow \mathbb{C} \longrightarrow (\Omega^\bullet, d) \longrightarrow (\Omega'^\bullet, d)$$

which induce homomorphisms of hypercohomology groups

$$H^*(X, \mathbb{Z}) \longrightarrow H^*(X, \mathbb{R}) \longrightarrow I\!I^*(X, \Omega^\bullet) \longrightarrow H^*(X, \Omega'^\bullet). \tag{B.2.21}$$

By the fineness of Ω^r and Ω'^r $(r \geq 0)$, the canonical edge homomorphisms

$$H^*(\Gamma(X, \Omega^\bullet)) \longrightarrow H^*(X, \Omega^\bullet), \quad I\!I^*(\Gamma(X, \Omega'^\bullet)) \longrightarrow H^*(X, \Omega'^\bullet)$$

are isomorphisms. We denoted by $\Gamma(X, \Omega'^\bullet)$ the space of sections of the sheaf $\Omega'^\bullet$ over X.

Lemma B.2.20 ($\partial\bar{\partial}$-Lemma for currents). *Let T be a closed, real, $(1,1)$-current of order 0. Then for every contractible open set U there exists a real function $\varphi \in L^1_{\text{loc}}(U)$, called a* local potential, *such that $T = \sqrt{-1}\partial\bar{\partial}\varphi$. If T is a (strictly) positive $(1,1)$-current, then φ is (strictly) plurisubharmonic.*

Indeed, since T is closed, as in [120, p. 387] and [79, III-1.18], the Poincaré and Dolbeault-Grothendieck lemmas for currents imply that $T = \sqrt{-1}\partial\bar{\partial}\varphi$ for some distribution φ. (Compare to the global version of the $\partial\bar{\partial}$-Lemma 1.5.1 given by Hodge Theory.) Then, locally $\Delta\varphi = -2\,\mathrm{Tr}(\partial\bar{\partial}\varphi)$ is of order 0 and by the regularity theory for elliptic equations, we have $\varphi \in L^1_{\text{loc}}$. If T is (strictly) positive then φ is (strictly) plurisubharmonic.

A real $(1,1)$-current ω on X is said to be a *Kähler current* if it is d-closed and strictly positive on X. A d-closed $(1,1)$-current or a d-closed $\mathscr{C}^\infty$-$(1,1)$-form is said to be *integral* if its cohomology class is in the image of $H^2(X, \mathbb{Z})$ under the map in (B.2.21).

B.3 q-convex and q-concave manifolds

We recall here briefly some notions linked to pseudoconvexity.

Definition B.3.1. Let X be a complex manifold, $\varphi \in \mathscr{C}^\infty(X)$. The *complex Hessian* of φ is the $(1,1)$-form $\partial\bar{\partial}\varphi$.

Let us introduce an auxiliary Riemannian metric on X; as in (1.5.15), we identify $\partial\bar{\partial}\varphi$ with a Hermitian matrix $H(\varphi) \in \mathscr{C}^\infty(X, \operatorname{End}(T^{(1,0)}X))$ and we will speak about the eigenvalues of $\partial\bar{\partial}\varphi$ at a point $x \in X$.

Definition B.3.2 (Andreotti-Grauert [4]). Let X be a complex manifold of complex dimension n. Let $1 \leqslant q \leqslant n$.

(i) X is called q-*convex* if there exists a smooth function $\varphi : X \longrightarrow [a, b[$, $a \in \mathbb{R}$, $b \in \mathbb{R} \cup \{+\infty\}$ such that the *sublevel set* $X_c = \{\varphi < c\} \Subset X$ for all $c \in [a, b[$ and the complex Hessian $\partial\bar{\partial}\varphi$ has at least $n - q + 1$ positive eigenvalues outside a compact set (exceptional set) K. Here $X_c = \{\varphi < c\} \Subset X$ means that the closure $\overline{X_c}$ of X_c is a compact set in X.

(ii) X is called q-*complete* if it is q-convex with $K = \emptyset$.

(iii) X is called q-*concave* if there exists a smooth function $\varphi : X \longrightarrow]a, b]$, $a \in \mathbb{R} \cup \{-\infty\}$, $b \in \mathbb{R}$ such that the *superlevel set* $X_c = \{\varphi > c\} \Subset X$ for all $c \in]a, b]$ and $\partial\bar{\partial}\varphi$ has at least $n - q + 1$ positive eigenvalues outside a compact set.

In all these cases we call φ an *exhaustion function* .

Immediate examples of 1-complete manifolds are $\mathbb{C}^n$ (resp. balls $B(a, r) = \{z \in \mathbb{C}^n : |z - a| < r\}$) with exhaustion function $\varphi = |z|^2$ (resp. $\varphi = |z - a|^2$).

Definition B.3.3. Let X be a complex space. We say that:

(a) X is *holomorphically separable* if for any $x, y \in X$, $x \neq y$, there exists $f \in \mathscr{O}_X(X)$ with $f(x) \neq f(y)$.

(b) X is *holomorphically regular* if for any $x \in X$ the set of germs $\{f_x : f \in \mathscr{O}_X(X)\}$ generates the maximal ideal $\mathbf{m}_x$ of $\mathscr{O}_{X,x}$.

(c) X is *holomorphically convex* if for any compact set $K \subset X$, $\widehat{K} = \{x \in X : |f(x)| \leqslant \sup_K |f|$, for all $f \in \mathscr{O}_X(X)\}$ is compact.

(d) X is called *Stein space* if X is holomorphically separable, regular and convex.

By definition, a complex manifold is 1-complete if it admits a smooth strictly plurisubharmonic function φ such that $\{\varphi < c\} \Subset X$ for any $c \in \mathbb{R}$. It is easy to see that a Stein manifold is 1-complete. The problem whether the converse is true was known as the Levi problem on complex manifolds. It was solved affirmatively by Hans Grauert [115] (see also [132, 108, 78]).

Theorem B.3.4 (Grauert). *A complex manifold is Stein if and only if it admits a smooth strictly plurisubharmonic exhaustion function.*

We need the solution of the Levi problem in a more general form.

Theorem B.3.5 (Grauert). *Let X be a 1-convex manifold. Then there exists*

(a) *a compact analytic set $A \subset X$ with $\dim_x A > 0$ for any $x \in A$,*
(b) *a Stein space X' with at worst isolated singularities, a finite set $D \subset X'$ and a proper holomorphic map $\pi : X \to X'$, such that $\pi(A) = D$, $\pi : X \setminus A \to X' \setminus D$ is a biholomorphism, and $\pi_* \mathcal{O}_X = \mathcal{O}_{X'}$.*

The Stein space X' is called the *Remmert reduction* of X and A the *exceptional analytic set* of X. We say that A can be *blown down to a finite set*.

The following result shows when Stein spaces can be embedded in the Euclidean space.

Theorem B.3.6 (Remmert-Bishop-Narasimhan [184]). *Assume that X is a Stein space of dimension n and of finite type $m > n$, that is, X can be locally realized as an analytic set in $\mathbb{C}^m$. Then the set of proper regular embeddings of X in $\mathbb{C}^{n+m}$ is dense in the set of all holomorphic mappings of X in $\mathbb{C}^{n+m}$ endowed with the topology of uniform convergence.*

The analytic convexity of a manifold is determined by the behavior of the complex Hessian of an exhaustion function on the analytic tangent space of sublevel sets. Let M be a relatively compact domain with smooth boundary in a complex manifold X. Let $\rho \in \mathscr{C}^\infty(U)$ be defined on an open neighbourhood U of $\overline{M}$ such that $M = \{x \subset X : \rho(x) < 0\}$ and $d\rho \neq 0$ on ∂M. We say that ρ is a *defining function* of M. The analytic tangent space to ∂M at $x \in \partial M$ is given by

$$T_x^{(1,0)} \partial M = \{v \in T_x^{(1,0)} X : \partial \rho(v) = 0\}. \tag{B.3.1}$$

The definition does not depend on the choice of ρ.

Definition B.3.7. The *Levi form* of ρ is the 2-form

$$\mathscr{L}_\rho \in \mathscr{C}^\infty(\partial M, T^{(1,0)} \partial M \otimes T^{(0,1)} \partial M), \quad \mathscr{L}_\rho(U, \overline{V}) = (\partial \overline{\partial} \rho)(U, \overline{V}), \tag{B.3.2}$$

for $U, V \in T_x^{(1,0)} \partial M$, $x \in \partial M$.

Lemma B.3.8. *The number of positive and negative eigenvalues of the Levi form is independent of the choice of the defining function ρ.*

Proof. Consider another defining function $\rho_1 \in \mathscr{C}^\infty(U)$. Let $x \in \partial M$. We can take ρ as part of a set of smooth coordinate functions (ρ, x'), $x' = (x_2, \dots, x_{2n})$ centered at $x = 0$. By applying the Taylor formula with integral rest we find $\rho_1(\rho, x') = \rho_1(\rho, x') - \rho_1(0, x') = \rho \cdot h(\rho, x')$, where h is smooth and $h(x) \neq 0$. Since $\rho_1 < 0$ and $\rho < 0$ on M we have $h(x) > 0$. Computing

$$\partial \overline{\partial}(\rho_1) = \partial \overline{\partial}(\rho h) = \rho \partial \overline{\partial} h + \partial \rho \wedge \overline{\partial} h + \partial h \wedge \overline{\partial} \rho + h \partial \overline{\partial} \rho \tag{B.3.3}$$

and taking into account that $\rho(x) = 0$ and $\partial \rho, \overline{\partial} \rho$ vanish on $T_x^{(1,0)} \partial M$ we obtain $\mathscr{L}_{\rho_1}(U, \overline{V}) = h(x) \mathscr{L}_\rho(U, \overline{V})$ for any $U, V \in T_x^{(1,0)} \partial M$. This implies immediately Lemma B.3.8. $\qquad \square$

Definition B.3.9. The domain M is called *strongly pseudoconvex* (resp. *(weakly) pseudoconvex*) if the Levi form is positive definite (resp. semidefinite).

Example B.3.10. Let X be a compact complex manifold and let (L, h^L) be a positive line bundle on X. We consider the Grauert tube $M = \{v \in L^* : |v|_{h^{L^*}} < 1\}$. If we set $\rho = |v|^2_{h^{L^*}} - 1$, then

$$\partial\overline{\partial}\rho|_{T^{(1,0)}\partial M} = \pi^*(R^L)|_{T^{(1,0)}\partial M}, \qquad (B.3.4)$$

where $\pi : M \longrightarrow X$ is the projection. Thus $\mathscr{L}_\rho$ is positive definite and M is a strongly pseudoconvex domain called the *Grauert tube*.

Example B.3.11. Assume that $M = \{\rho < 0\} \subset X$ is strongly pseudoconvex. By replacing ρ with $e^{C\rho} - 1$ for $C \gg 1$, we can achieve that $\partial\overline{\partial}\rho$ is positive definite on the whole tangent space $T_x^{(1,0)}X$, $x \in \partial M$, and therefore we can assume that ρ is strictly plurisubharmonic in a neighbourhood of ∂M. It follows that M is a 1-convex manifold. Indeed, the function $\varphi : M \longrightarrow \mathbb{R}$, $\varphi = \frac{1}{\rho^2}$ is a strictly plurisubharmonic exhaustion function. Conversely, if X is a 1-convex manifold and $X_c \supset K$ is smooth, then X_c is strongly pseudoconvex.

Let X be a compact complex manifold and E be a holomorphic vector bundle of rank m over X. Let us consider the projectivized bundle $\pi : \mathbb{P}(E^*) \longrightarrow X$ whose fiber $\pi^{-1}(x)$, $x \in X$, is the projective space $\mathbb{P}(E_x^*)$, i.e., the set of complex lines through the origin in E_x^*. The tautological line bundle $\mathscr{O}_{\mathbb{P}(E^*)}(-1) \subset \pi^*E^*$ over $\mathbb{P}(E^*)$ is the line bundle such that the fiber over $(y, [\xi])$ is $\mathbb{C}\xi \subset E_y^*$. Let $\mathscr{O}_{\mathbb{P}(E^*)}(1)$ be its dual line bundle.

Definition B.3.12. E is said to be *ample* if $\mathscr{O}_{\mathbb{P}(E^*)}(1)$ is ample on $\mathbb{P}(E^*)$ (cf. Definition 5.1.1). E is called *Grauert positive* if the zero section $Z(E^*)$ of the dual bundle E^* has a strongly pseudoconvex (equivalently, 1-convex) neighborhood.

The previous example shows that a positive line bundle is Grauert positive. The converse is proved in Problem 5.1. Thus, positivity, ampleness and Grauert positivity are equivalent for holomorphic line bundles.

Theorem B.3.13 (Grauert's ampleness criterion [116]). *A holomorphic vector bundle E on a compact complex manifold X is ample if and only if it is Grauert positive.*

In this case one can show that $Z(E^*)$ is the exceptional analytic set of its 1-convex neighborhood in E^*. Hence $Z(E^*)$ can be blown down to a finite set of points (cf. Theorem B.3.5).

The notion of weakly 1-complete manifold was introduced by S. Nakano [182] in order to solve the problem of the inverse of the monoidal transformation.

Definition B.3.14. A complex manifold X is called *weakly 1-complete* if there exists a smooth plurisubharmonic function $\varphi : X \to \mathbb{R}$ such that $\{\varphi < c\} \Subset X$ for any $c \in \mathbb{R}$. φ is called an exhaustion function.

Any 1-convex (and therefore any compact or Stein) manifold is weakly 1-complete. A proper modification (cf. Section 2.1) of a weakly 1-complete manifold is again weakly 1-complete.

Proposition B.3.15. *Every weakly 1-complete Kähler (in particular Stein) manifold carries a complete Kähler metric.*

Proof. By adding a constant we can consider a smooth exhaustion function $\varphi > 0$. Let $\chi : [0, +\infty[\to \mathbb{R}$ be a convex increasing function satisfying $\chi''(t) > 0$ and $\int_0^\infty \sqrt{\chi''(t)}dt = \infty$. Let ω be a Kähler form on X. Then

$$\omega_0 = \omega + \sqrt{-1}\partial\overline{\partial}\chi(\varphi) = \omega + \sqrt{-1}\chi''(\varphi)\partial\varphi \wedge \overline{\partial}\varphi + \sqrt{-1}\chi'(\varphi)\partial\overline{\partial}\varphi \qquad (B.3.5)$$

is the Kähler form of a complete metric on X. To prove this, let's observe that $|\partial\varphi|_{\omega_0} \leqslant 1/\sqrt{\chi''(\varphi)}$ by (B.3.5). Introduce the function $\lambda : X \to \mathbb{R}$, $\lambda(x) = \int_0^{\varphi(x)} \sqrt{\chi''(t)}dt$. Since φ is an exhaustion function, λ is proper. Moreover, $\partial\lambda = \sqrt{\chi''(\varphi)}\partial\varphi$, hence $|\partial\lambda|_{\omega_0} \leqslant 1$, so $|d\lambda|_{\omega_0} \leqslant 2$. By the same argument after (3.4.9) we conclude that ω_0 is complete. If X is 1-complete the above argument works with $\omega_0 = \sqrt{-1}\partial\overline{\partial}\chi(\varphi)$. $\qquad\square$

B.4 L^2 estimates for $\overline{\partial}$

We recall here the standard L^2 existence theorem of Hörmander–Andreotti–Vesentini [7, 131]. For excellent introductory books see [132, 79, 189].

Let (X, J, Θ) be a complex Hermitian manifold of dimension n as in Def. 1.2.7 and (E, h^E) be a holomorphic Hermitian vector bundle over X. We denote by dv_X the Riemannian volume form on (X, Θ).

We use the notation from Section 1.4.3 for $[\sqrt{-1}R^E, i(\Theta)]$. On the space of smooth forms with compact support $\Omega_0^{\bullet,\bullet}(X, E)$, we introduce the L^2 scalar product (1.3.14). The corresponding Hilbert space completion is denoted by $L^2_{\bullet,\bullet}(X, E)$.

Theorem B.4.1 (Hörmander-Andreotti-Vesentini). *Let (X, Θ) be a complete Kähler manifold of dimension n. Assume that for some (r, q), $q \geqslant 1$, there exists a continuous function $\psi : X \to [0, \infty[$ such that pointwise*

$$\langle[\sqrt{-1}R^E, i(\Theta)]s, s\rangle_{\Lambda^{r,q}\otimes E} \geqslant \psi\,|s|^2, \quad \text{for all } s \in \Omega_0^{r,q}(X, E). \qquad (B.4.1)$$

Then for any form $f \in L^2_{r,q}(X, E)$ satisfying $\overline{\partial}^E f = 0$ and $\int_X \psi^{-1}|f|^2\,dv_X < \infty$, there exists $g \in L^2_{r,q-1}(X, E)$ such that $\overline{\partial}^E g = f$ and

$$\int_X |g|^2\,dv_X \leqslant \int_X \psi^{-1}|f|^2\,dv_X. \qquad (B.4.2)$$

Moreover, if f is smooth, g can be chosen smooth.

Proof. Consider the complex of closed, densely defined operators analogue to (3.1.3)

$$L^2_{r,q-1}(X,E) \overset{T=\overline{\partial}^E}{\longrightarrow} L^2_{r,q}(X,E) \overset{S=\overline{\partial}^E}{\longrightarrow} L^2_{r,q+1}(X,E)\,, \qquad (B.4.3)$$

where T and S are the maximal extensions of $\overline{\partial}^E$. We apply the Nakano inequality (Theorem 1.4.14) and obtain for all $s \in \Omega^{r,q}_0(X,E)$

$$\|\overline{\partial}^E s\|^2_{L^2} + \|\overline{\partial}^{E,*} s\|^2_{L^2} \geqslant \langle [\sqrt{-1}R^E, i(\Theta)]s, s\rangle \geqslant \int_X \psi\,|s|^2 dv_X\,. \qquad (B.4.4)$$

Since the metric associated to Θ is complete, Lemma 3.3.1 shows that (B.4.4) holds for any $s \in \mathrm{Dom}(S) \cap \mathrm{Dom}(T^*)$. By the Cauchy–Schwarz inequality and (B.4.4) we have for any $s \in \mathrm{Dom}(S) \cap \mathrm{Dom}(T^*)$

$$\begin{aligned} |\langle f,s\rangle|^2 = |\langle \psi^{-1/2}f, \psi^{1/2}s\rangle|^2 &\leqslant \int_X \psi^{-1}|f|^2 dv_X \cdot \int_X \psi|s|^2 dv_X \\ &\leqslant \int_X \psi^{-1}|f|^2 dv_X \cdot (\|\overline{\partial}^E s\|^2_{L^2} + \|\overline{\partial}^{E,*} s\|^2_{L^2})\,. \end{aligned} \qquad (B.4.5)$$

Consider now $s \in \mathrm{Dom}(T^*)$ and write the decomposition $s = s_1 + s_2$, $s_1 \in \mathrm{Ker}(S)$, $s_2 \in \mathrm{Ker}(S)^\perp$. Since $f \in \mathrm{Ker}(S)$ by hypothesis, $|\langle f,s\rangle| = |\langle f,s_1\rangle|$. Applying (B.4.5) for $s_1 \in \mathrm{Ker}(S) \cap \mathrm{Dom}(T^*)$ we obtain

$$\begin{aligned} |\langle f,s\rangle|^2 = |\langle f,s_1\rangle|^2 \\ \leqslant \int_X \psi^{-1}|f|^2 dv_X \cdot \|T^*s_1\|^2_{L^2} = \int_X \psi^{-1}|f|^2 dv_X \cdot \|T^*s\|^2_{L^2}\,, \end{aligned} \qquad (B.4.6)$$

where we used that $s_2 \in \mathrm{Ker}(S)^\perp = [\mathrm{Im}(S^*)] \subset \mathrm{Ker}(T^*)$ (cf. (C.1.2)).

We consider the antilinear functional $\mathrm{Im}(T^*) \ni T^*s \mapsto \langle f,s\rangle \in \mathbb{C}$. By (B.4.6) it is well defined and bounded, with norm $\leqslant \left(\int_X \psi^{-1}|f|^2 dv_X\right)^{1/2}$. We can thus extend it to a bounded functional on $[\mathrm{Im}(T^*)]$ and then by zero on the complement of $[\mathrm{Im}(T^*)]$ to a bounded functional on $L^2_{r,q-1}(X,E)$, of norm $\leqslant \left(\int_X \psi^{-1}|f|^2 dv_X\right)^{1/2}$. By the Riesz representation theorem, there exists $g \in L^2_{r,q-1}(X,E)$ such that $\langle f,s\rangle = \langle g, T^*s\rangle$ for all $s \in \mathrm{Dom}(T^*)$ and $\|g\|^2_{L^2} \leqslant \int_X \psi^{-1}|f|^2 dv_X$. But this means that $g \in \mathrm{Dom}(T)$, $Tg = f$ and (B.4.2) is satisfied.

We wish to prove the regularity statement now. By replacing g with its projection on $\mathrm{Ker}(T)^\perp$ (which is still a solution of $Tg = f$) we can consider $g \in \mathrm{Ker}(T)^\perp$. This solution is sometimes called the *canonical solution* of $Tg = f$, since it is the unique solution of minimal L^2 norm. Let $L^2_{r,q-2}(X,E) \overset{U=\overline{\partial}^E}{\longrightarrow} L^2_{r,q-1}(X,E)$ be the maximal extension of $\overline{\partial}^E$ in bidegree $(r, q-2)$. Since $g \in \mathrm{Ker}(T)^\perp = [\mathrm{Im}(T^*)] \subset \mathrm{Ker}(U^*)$ (cf. (C.1.2)), we have

$$UU^*g = 0\,, \quad (UU^* + T^*T)g = T^*f \in \mathscr{C}^\infty(X, \Lambda^{r,q-1}(T^*X) \otimes E)\,. \qquad (B.4.7)$$

Applying the regularity theorem A.3.4 for the elliptic operator $UU^* + T^*T = \square^E$, we obtain that g is also smooth. The proof is complete. ☐

Using a weight function $\chi(\varphi)$ as in (3.5.8) and (B.3.5), with a convex rapidly increasing function $\chi : [0, \infty[\to \mathbb{R}$ and a positive exhaustion function φ, in order to temper the growth of smooth forms at infinity, we get:

Corollary B.4.2. *If X is a Stein manifold, then for any (r, q) with $q \geqslant 1$ and $f \in \Omega^{r,q}(X, E)$ satisfying $\overline{\partial}^E f = 0$, there exists $g \in \Omega^{r,q-1}(X, E)$ such that $\overline{\partial}^E g = f$.*

In particular, if $X = \mathbb{B}^n$, the unit ball in $\mathbb{C}^n$, we obtain:

Corollary B.4.3 (Dolbeault-Grothendieck Lemma or $\overline{\partial}$-Lemma). *If $q \geqslant 1$ and $f \in \Omega^{r,q}(\mathbb{B}^n)$ satisfying $\overline{\partial} f = 0$, then there exists $g \in \Omega^{r,q-1}(\mathbb{B}^n)$ such that $\overline{\partial} g = f$.*

We apply the results above to cohomology theory. For a sheaf $\mathscr{F}$ over X, we define the qth cohomology group of $\mathscr{F}$ by $H^q(X, \mathscr{F}) := H^q(\Gamma(X, \mathscr{S}^{\bullet}))$, where $0 \to \mathscr{F} \to \mathscr{S}^{\bullet}$ is the canonical (Godement) flasque resolution of $\mathscr{F}$ where we denote by $\Gamma(X, \mathscr{G})$ the space of sections of the sheaf $\mathscr{G}$ over X.

Consider the sheaves $\Omega_X^{\bullet,\bullet}(E)$ of smooth forms, defined by $U \mapsto \Omega^{\bullet,\bullet}(U, E)$, for any open set $U \subset X$. We are interested in computing the cohomology of the sheaf $\mathscr{O}_X^r(E)$ of *holomorphic r-forms* with values in E, i.e. holomorphic sections of $\Omega_X^{r,0}(E)$. For $r = 0$ we set $\mathscr{O}_X(E) := \mathscr{O}_X^0(E)$, the sheaf of holomorphic sections of E. We will denote $H^{\bullet}(X, E) := H^{\bullet}(X, \mathscr{O}_X(E))$.

By the Dolbeault–Grothendieck lemma B.4.3 applied to small coordinate balls, the following sequence of sheaves is exact (forms a resolution of $\mathscr{O}_X^r(E)$):

$$0 \longrightarrow \mathscr{O}_X^r(E) \longrightarrow \Omega_X^{r,0}(E) \xrightarrow{\overline{\partial}^E} \Omega_X^{r,1}(E) \xrightarrow{\overline{\partial}^E} \cdots \xrightarrow{\overline{\partial}^E} \Omega_X^{r,n}(E) \longrightarrow 0. \qquad \text{(B.4.8)}$$

The complex of global sections

$$0 \longrightarrow \Omega^{r,0}(X, E) \xrightarrow{\overline{\partial}^E} \Omega^{r,1}(X, E) \xrightarrow{\overline{\partial}^E} \cdots \xrightarrow{\overline{\partial}^E} \Omega^{r,n}(X, E) \longrightarrow 0, \qquad \text{(B.4.9)}$$

is called the *Dolbeault complex*.

The cohomology of (B.4.9) is denoted by $H^{\bullet,\bullet}(X, E)$ and called the *Dolbeault cohomology* of X with values in E. Since $\Omega_X^{\bullet,\bullet}(E)$ are fine sheaves, hence acyclic, the abstract de Rham theorem entails:

Theorem B.4.4 (Dolbeault isomorphism). *The canonical morphism*

$$H^{r,q}(X, E) \to H^q(X, \mathscr{O}_X^r(E)) \qquad \text{(B.4.10)}$$

is an isomorphism.

For a general complete Kähler manifold endowed with a semi-positive line bundle, the L^2 method applies to solve the $\overline{\partial}$-equation for (n, q)-forms.

Corollary B.4.5. *Let (L, h^L) be a semi-positive line bundle over a complete Kähler manifold (X, Θ) of dimension n. Let $a_1 \leqslant \cdots \leqslant a_n$ be the eigenvalues of $\dot{R}^L \in \operatorname{End}(T^{(1,0)}X)$ (cf. (1.5.15)) with respect to Θ. Then for any $q \geqslant 1$ and any form $f \in L^2_{n,q}(X, L)$ satisfying $\overline{\partial}^L f = 0$ and $\int_X (a_1 + \cdots + a_q)^{-1}|f|^2 dv_X < \infty$ there exists $g \in L^2_{n,q-1}(X, L)$ such that $\overline{\partial}^L g = f$ and*

$$\int_X |g|^2 dv_X \leqslant \int_X (a_1 + \cdots + a_q)^{-1}|f|^2 dv_X. \tag{B.4.11}$$

Proof. The Bochner–Kodaira–Nakano inequality (1.4.51) together with (1.4.61) (cf. also Problem 1.10) for $r = n$ delivers (B.4.1) with $\psi = a_1 + \cdots + a_q$. $\qquad\square$

In the seminal works of Bombieri [39] and Skoda [229] it has been already observed that Corollary B.4.5 still applies if h^L is singular; the following result was proved in full generality by Demailly [71, Th. 5.1] (cf. also [78, Cor. 5.3]). The idea is to reduce the problem to the case of smooth metrics by regularization (convolution of plurisubharmonic functions with smooth kernels). The solutions obtained by Theorem B.4.1 turn out to converge weakly to a solution for the singular metric, due to the uniformity of the constants in the L^2 estimates.

Let $L^2_{n,q}(X, L, \mathrm{loc})$ be the usual space of forms with locally L^2 coefficients with respect to smooth metrics on X and L. In the case of a singular Hermitian metric h^L in the sense of Definition 2.3.1, we set

$$L^2_{n,q}(X, L) := \{s \in L^2_{n,q}(X, L, \mathrm{loc}) : \int_X |s|^2_{h^L} dv_X < \infty\}. \tag{B.4.12}$$

Theorem B.4.6 (Demailly). *Let (X, Θ) be a complete Kähler manifold, $\dim X = n$, and let (L, h^L) be a holomorphic singular Hermitian line bundle on X such that $\sqrt{-1}R^L \geqslant \varepsilon\Theta$ in the sense of currents, for some constant $\varepsilon > 0$. Then for any $q \geqslant 1$ and any form $f \in L^2_{n,q}(X, L)$ satisfying $\overline{\partial}^L f = 0$, there exists $g \in L^2_{n,q-1}(X, L)$ such that $\overline{\partial}^L g = f$ and*

$$\int_X |g|^2_{h^L} dv_X \leqslant \frac{1}{q\varepsilon} \int_X |f|^2_{h^L} dv_X. \tag{B.4.13}$$

Let us consider the fine sheaf $\mathscr{L}^{n,q}$ given by

$$U \longrightarrow \mathscr{L}^{n,q}(U, L) = \{s \in L^2_{n,q}(U, L, \mathrm{loc}) : |s|_{h^L}, |\overline{\partial}^L s|_{h^L} \in L^2_{\mathrm{loc}}\}. \tag{B.4.14}$$

The sequence $0 \to \mathscr{O}_X(L \otimes K_X) \otimes \mathscr{I}(h^L) \to (\mathscr{L}^{n,\bullet}, \overline{\partial}^L)$ is exact, where $\mathscr{I}(h^L)$ is the Nadel ideal multiplier sheaf associated to h^L (cf. Definition 2.3.13). This follows by applying Theorem B.4.6 on a ball $B(0, r)$; we consider $h^L_1 = h^L e^{-\chi(\varphi)}$ and the complete Kähler form $\Theta = \sqrt{-1}\partial\overline{\partial}\chi(\varphi)$ on $B(0, r)$, constructed as in

(3.5.8) or Proposition B.3.15, where $\varphi(z) = -\log(r^2 - |z|^2)$ is the exhaustion function of $B(0, r)$. Therefore

$$H^q(X, L \otimes K_X \otimes \mathscr{I}(h^L)) = H^q(\Gamma(X, \mathscr{L}^{n,\bullet})). \tag{B.4.15}$$

As a consequence we have the singular variant of the Kodaira and Nakano vanishing theorem 1.5.4, cf. [182, 180, 77].

Theorem B.4.7 (Nadel). *Let (X, Θ) be a weakly 1-complete manifold endowed with a complete Kähler metric with associated $(1, 1)$-form Θ. Let (L, h^L) be a holomorphic singular Hermitian line bundle on X. Assume that $\sqrt{-1}R^L \geqslant \varepsilon \Theta$ for some $\varepsilon > 0$. Then*

$$H^q(X, L \otimes K_X \otimes \mathscr{I}(h^L)) = 0, \quad \text{for all} \quad q \geqslant 1. \tag{B.4.16}$$

Indeed, applying again Theorem B.4.6 globally we obtain $H^q(\Gamma(X, \mathscr{L}^{n,\bullet})) = 0$ for $q \geqslant 1$ and (B.4.15) entails (B.4.16).

Note that if h^L is smooth, we obtain as a special case the Kodaira vanishing theorem. Theorem B.4.7 is actually very strong: it contains also the Kawamata-Viehweg vanishing theorem.

The L^2 estimates for $\bar{\partial}$ imply also the following version of the finiteness theorem of Andreotti-Grauert.

Theorem B.4.8 (Andreotti-Grauert). *Let X be an n-dimensional complex manifold and E be a holomorphic vector bundle on X. If X is q-convex (resp. q-concave), then $\dim H^j(X, E) < \infty$ for $j \geqslant q$ (resp. $j \leqslant n - q - 1$).*

The proof of Andreotti and Grauert is sheaf-theoretic and makes use of the "bumping lemma" and it works on complex spaces too. On the analytic side, there are proofs based on L^2 estimates on complete manifolds (Andreotti-Vesentini [7], Hörmander [132], Ohsawa [187]), $\bar{\partial}$-Neumann problem (Kohn [108]), or integral representations (Henkin-Leiterer [126]).

B.5   Chern-Weil theory

Some good references for this section are [263], [31] and [15]. In [55], one can also find the relative Euler class for manifold with boundary. We will use the notation in Section 1.1.1.

Let X be a smooth manifold. For $R = \mathbb{C}$ or $\mathbb{R}$, let $\Omega^\bullet(X, R)$ be the space of smooth R-valued forms on X. Let $d : \Omega^\bullet(X, R) \to \Omega^{\bullet+1}(X, R)$ be the exterior differential. Then $d^2 = 0$. The elements of the kernel $\mathrm{Ker}(d)$ (resp. image $\mathrm{Im}(d)$) of d are called closed (resp. exact) forms. We will denote d also by d^X when we wish to emphasize the manifold X. The *de Rham cohomology* of X is defined by

$$H^j(X, \mathbb{R}) := \frac{\mathrm{Ker}(d) \cap \Omega^j(X, \mathbb{R})}{\mathrm{Im}(d) \cap \Omega^j(X, \mathbb{R})}, \quad H^\bullet(X, \mathbb{R}) := \bigoplus_j H^j(X, \mathbb{R}). \tag{B.5.1}$$

For a closed differential form η, we denote by $[\eta] \in H^\bullet(X, \mathbb{R})$ its cohomology class. In the same way we introduce the complex de Rham cohomology group $H^\bullet(X, \mathbb{C})$, by replacing $\mathbb{R}$ with $\mathbb{C}$ in the definition above.

Let E be a complex vector bundle on X. Let ∇^E be a connection on E, and $\nabla^E : \Omega^\bullet(X, E) \to \Omega^{\bullet+1}(X, E)$ be its unique extension verifying the Leibniz rule (1.1.4). Its curvature is $R^E = (\nabla^E)^2 \in \Omega^2(X, \operatorname{End}(E))$ and (1.1.5) holds.

The fiberwise trace $\operatorname{Tr} : \mathscr{C}^\infty(X, \operatorname{End}(E)) \to \mathscr{C}^\infty(X)$, $A \mapsto \operatorname{Tr}[A]$, extends to a map $\operatorname{Tr} : \Omega^\bullet(X, \operatorname{End}(E)) \to \Omega^\bullet(X, \mathbb{C})$, such that for $\eta \in \Omega^\bullet(X, \mathbb{C})$, $A \in \mathscr{C}^\infty(X, \operatorname{End}(E))$ we have $\operatorname{Tr}[\eta \otimes A] = \eta \operatorname{Tr}[A]$. For $A, B \in \mathscr{C}^\infty(X, \operatorname{End}(E))$ and $\eta, \gamma \in \Omega^\bullet(X, \mathbb{C})$ we have $(\eta \otimes A)(\gamma \otimes B) = (\eta \wedge \gamma) \otimes (AB)$. Let us extend the fiberwise commutator $[A, B]$ of two sections $A, B \in \mathscr{C}^\infty(X, \operatorname{End}(E))$ to a supercommutator $[\cdot, \cdot]$ on $\Omega^\bullet(X, \operatorname{End}(E))$ by setting

$$[\eta \otimes A, \gamma \otimes B] := (\eta \otimes A)(\gamma \otimes B) - (-1)^{\deg \eta \cdot \deg \gamma}(\gamma \otimes B)(\eta \otimes A)$$
$$= (\eta \wedge \gamma) \otimes [A, B]. \quad \text{(B.5.2)}$$

We have thus

$$\operatorname{Tr}\left[[A, B]\right] = 0, \quad \text{for any } A, B \in \mathscr{C}^\infty(X, \operatorname{End}(E)). \quad \text{(B.5.3)}$$

Let $f \in \mathbb{R}[z]$ be a real polynomial on z. We set

$$F(R^E) = \operatorname{Tr}\left[f\left(\tfrac{\sqrt{-1}}{2\pi} R^E\right)\right] \in \Omega^\bullet(X, \mathbb{C}). \quad \text{(B.5.4)}$$

Theorem B.5.1. $F(R^E)$ *is a closed differential form and its de Rham cohomology class* $[F(R^E)] \in H^{2\bullet}_{\mathrm{dR}}(X, \mathbb{C})$ *does not depend on the choice of* ∇^E. *If* ∇^E *is a Hermitian connection with respect to a Hermitian metric* h^E *on* E, *then* $F(R^E)$ *is a real differential form. Thus* $[F(R^E)] \in H^{2\bullet}(X, \mathbb{R})$.

Proof. Let us start by proving that for any $A \in \mathscr{C}^\infty(X, \operatorname{End}(E))$ we have

$$d \operatorname{Tr}[A] = \operatorname{Tr}\left[[\nabla^E, A]\right]. \quad \text{(B.5.5)}$$

We first observe that (B.5.5) does not depend on the choice of ∇^E. Indeed, if $\widetilde{\nabla}^E$ is another connection on E, the definition of a connection implies that $\widetilde{\nabla}^E - \nabla^E \in \Omega^1(X, \operatorname{End}(E))$, so by (B.5.3) we have $\operatorname{Tr}\left[[\widetilde{\nabla}^E - \nabla^E, A]\right] = 0$. It is also clear that (B.5.5) has local character, i.e., it suffices to prove it in a small neighborhood of each point. Moreover, in a neighborhood U such that $E|_U \cong U \times \mathbb{C}^m$ and for the trivial connection $\nabla^E(\varphi_1, \ldots, \varphi_m) = (d\varphi_1, \ldots, d\varphi_m)$, formula (B.5.5) clearly holds. Thus, the local character and the independence on the connection yield (B.5.5) in general.

Under the notation (1.1.4), the Bianchi identity holds

$$[\nabla^E, R^E] = [\nabla^E, (\nabla^E)^2] = 0, \tag{B.5.6}$$

which together with (B.5.5) yield

$$dF(R^E) = \mathrm{Tr}\left[[\nabla^E, f(\tfrac{\sqrt{-1}}{2\pi}R^E)]\right] = 0. \tag{B.5.7}$$

Thus $F(R^E)$ is closed.

If ∇_0^E, ∇_1^E are two connections on E. Let $\pi : X \times \mathbb{R} \to X$ be the natural projection. Let $\nabla^{\pi^* E}$ be a connection on $\pi^* E$ over $X \times \mathbb{R}$ with curvature $R^{\pi^* E}$ such that $\nabla_i^E = \nabla^{\pi^* E}|_{X \times \{i\}}$ for $i = 0, 1$. We denote $\nabla_t^E = \nabla^{\pi^* E}|_{X \times \{t\}}$ the connection on E with curvature R_t^E. Then we can write $F(R^{\pi^* E})$ as follows:

$$F(R^{\pi^* E}) = F(R_t^E) + dt \wedge Q_t, \tag{B.5.8}$$

with Q_t a differential form on X. Since $F(R^{\pi^* E})$ is closed on $X \times \mathbb{R}$ by (B.5.7), after considering the coefficient of dt in the relation $d^{X \times \mathbb{R}} F(R^{\pi^* E}) = 0$, we get

$$\frac{\partial}{\partial t} F(R_t^E) = dQ_t. \tag{B.5.9}$$

Thus

$$F(R_1^E) - F(R_0^E) = d \int_0^1 Q_t \, dt. \tag{B.5.10}$$

Hence $[F(R_1^E)] = [F(R_0^E)] \in H^{2\bullet}(X, \mathbb{C})$.

If ∇^E is a Hermitian connection on (E, h^E), then by (1.1.6), we know that for any $U, V \in TX$, $R^E(U, V) \in \mathrm{End}(E)$ is skew-adjoint, thus $\sqrt{-1}R^E$ is a Hermitian matrix, and $F(R^E)$ is a real form. But there always exist Hermitian metric h^E and Hermitian connections on (E, h^E); thus $[F(R^E)] \in H^{2\bullet}(X, \mathbb{R})$. $\square$

Remark B.5.2. Let ∇_t^E be any path connecting ∇_0^E and ∇_1^E, a natural choose of $\nabla^{\pi^* E}$ is given by

$$\nabla^{\pi^* E} = \nabla_t^E + dt \wedge \frac{\partial}{\partial t}. \tag{B.5.11}$$

Then $R^{\pi^* E} = R_t^E + dt \wedge \frac{\partial \nabla_t^E}{\partial t}$, and Q_t in (B.5.8) is

$$Q_t = \frac{\sqrt{-1}}{2\pi} \mathrm{Tr}\left[\frac{\partial \nabla_t^E}{\partial t} f'(\frac{\sqrt{-1}}{2\pi} R_t^E)\right] \in \Omega^{2\bullet - 1}(X, \mathbb{C}). \tag{B.5.12}$$

Definition B.5.3. Let ∇_0^E, ∇_1^E be two connections on E. For a connection $\nabla^{\pi^* E}$ on $X \times \mathbb{R}$ connecting ∇_0^E and ∇_1^E, with Q_t in (B.5.8), we define the *Chern-Simons form* of F by

$$\widetilde{F}(\nabla^{\pi^* E}) = \int_0^1 Q_t \, dt. \tag{B.5.13}$$

The *Chern-Simons class* of F associated to ∇_0^E and ∇_1^E is defined by

$$\widetilde{F}(\nabla_0^E, \nabla_1^E) := [\widetilde{F}(\nabla^{\pi^* E})] \in \Omega^\bullet(X, \mathbb{C})/d\Omega^\bullet(X, \mathbb{C}). \tag{B.5.14}$$

The following result is a special case of [31, Theorem 2.10].

Theorem B.5.4. *Modulo exact forms, $\widetilde{F}(\nabla^{\pi^* E})$ does not depend on the choice of the connection $\nabla^{\pi^* E}$. Thus $\widetilde{F}(\nabla_0^E, \nabla_1^E) \in \Omega^\bullet(X, \mathbb{C})/d\Omega^\bullet(X, \mathbb{C})$ is well defined, and*

$$d\widetilde{F}(\nabla_0^E, \nabla_1^E) = F(R_1^E) - F(R_0^E). \tag{B.5.15}$$

If ∇_i^E, $(i = 0, 1)$ are Hermitian connections on (E, h_i^E), then the imaginary part of $\widetilde{F}(\nabla_0^E, \nabla_1^E)$ is exact, thus $\widetilde{F}(\nabla_0^E, \nabla_1^E) \in \Omega^\bullet(X, \mathbb{R})/d\Omega^\bullet(X, \mathbb{R})$.

Proof. By (B.5.10), we get (B.5.15). Let $\nabla_0^{\pi^* E}$, $\nabla_1^{\pi^* E}$ be two connections on $X \times \mathbb{R}$ connecting ∇_0^E, ∇_1^E. Let us consider the trivial fibration $\rho : X \times \mathbb{R} \times \mathbb{R} \to X$. Let $\nabla^{\rho^* E}$ be a connection on $\rho^* E$ over $X \times \mathbb{R} \times \mathbb{R}$ defined by

$$\nabla_s^{\pi^* E} = (1 - s)\nabla_0^{\pi^* E} + s\nabla_1^{\pi^* E}, \quad \nabla^{\rho^* E} = \nabla_s^{\pi^* E} + ds \wedge \frac{\partial}{\partial s}. \tag{B.5.16}$$

Set $\nabla_{t,s}^E = \nabla^{\rho^* E}|_{X \times \{t\} \times \{s\}}$. Then we can decompose $F(R^{\rho^* E})$ as

$$F(R^{\rho^* E}) = F\big((\nabla_{t,s}^E)^2\big) + dt \wedge Q_{t,s} + ds \wedge T_{t,s} + dt \wedge ds \wedge \gamma, \tag{B.5.17}$$

where $Q_{t,s}, T_{t,s}, \gamma$ are forms on X. Again by considering the coefficient of $dt \wedge ds$ in $d^{X \times \mathbb{R} \times \mathbb{R}} F(R^{\rho^* E}) = 0$, we get

$$-\frac{\partial}{\partial s} Q_{t,s} + \frac{\partial}{\partial t} T_{t,s} + d^X \gamma = 0. \tag{B.5.18}$$

Integrating on $[0, 1]^2$ we obtain

$$\int_0^1 Q_{t,0} \, dt - \int_0^1 Q_{t,1} \, dt + \int_0^1 T_{1,s} \, ds - \int_0^1 T_{0,s} \, ds + d^X \int_0^1 \int_0^1 \gamma \, dt \, ds = 0. \tag{B.5.19}$$

Now we observe that $\nabla^{\rho^* E}|_{X \times \{i\} \times \mathbb{R}} = \nabla_{i,0}^E + ds \frac{\partial}{\partial s}$, thus $T_{1,s} = T_{0,s} = 0$ by (B.5.12). Therefore

$$\widetilde{F}(\nabla_1^{\pi^* E}) - \widetilde{F}(\nabla_0^{\pi^* E}) = d^X \int_0^1 \int_0^1 \gamma \, dt \, ds. \tag{B.5.20}$$

Assume that ∇_i^E, $(i = 0, 1)$ are Hermitian connections on (E, h_i^E). We define $G \in \mathscr{C}^\infty(X, \mathrm{End}(E))$ by $h_1^E(\xi_1, \xi_2) = h_0^E(G\xi_1, \xi_2)$. Then G is positive and self-adjoint, so $G^{1/2}$ is well defined. Now $\widetilde{\nabla}_1^E = G^{1/2}\nabla_1^E G^{-1/2}$ is a Hermitian connection on (E, h_0^E) and $\widetilde{\nabla}_t^E = (1-t)\nabla_0^E + t\,\widetilde{\nabla}_1^E$ are Hermitian connections on (E, h_0^E). If we take $g_t = (1-t)\,\mathrm{Id} + tG^{1/2}$, $h_t^E = h_0^E(g_t\cdot, g_t\cdot)$, then $\nabla_t^E = g_t^{-1}\widetilde{\nabla}_t^E g_t$ are Hermitian connections on (E, h_t^E). Now we choose the connection $\nabla^{\pi^* E}$ by

$$\nabla^{\pi^* E} = \nabla_t^E + dt \wedge \left(\frac{\partial}{\partial t} + \frac{1}{2}(h_t^E)^{-1}\frac{\partial}{\partial t}h_t^E \right). \tag{B.5.21}$$

Then $\nabla^{\pi^* E}$ is a Hermitian connection on $\pi^* E$ with the metric $h^{\pi^* E}$ induced by h_t^E. Now by the second part of Theorem B.5.1 and (B.5.8), the second part of Theorem B.5.4 follows. $\qquad\square$

Example B.5.5. Let ∇^E be a Hermitian connection on a Hermitian vector bundle (E, h^E). Taking $f(z) = z, e^{-z}, \log(\frac{z}{1-e^{-z}})$, we get from Theorem B.5.1 and (1.3.45) that $c_1(E, \nabla^F)$, $\mathrm{ch}(E, \nabla^E)$, $\mathrm{Td}(E, \nabla^E)$ are closed real differential forms on X and their cohomology classes in $H^{2\bullet}(X, \mathbb{R})$ do not depend on the choice of h^E and ∇^E. We define now

$$\det\left(1 + \frac{\sqrt{-1}}{2\pi}R^E \right) = 1 + \sum_{i=1}^{\mathrm{rk}(E)} c_i(E, \nabla^E), \text{ with } c_i(E, \nabla^E) \in \Omega^{2i}(X, \mathbb{R}). \tag{B.5.22}$$

Since for any positive matrix A, we have $\det A = \exp(\mathrm{Tr}[\log A])$, by taking $f(z) = \log(1 + z)$, we see that $c_i(E, \nabla^E)$ are closed real differential forms on X. The corresponding cohomology classes $c_i(E) \in H^{2\bullet}(X, \mathbb{R})$ are called ith Chern classes of E, and $c(E) = 1 + \sum_i c_i(E)$ is called the total Chern class of E.

Remark B.5.6. If E is a real vector bundle and ∇^E is a connection on E, then by applying the above construction to $E \otimes_\mathbb{R} \mathbb{C}$, we obtain the corresponding theory for E. Especially, if ∇^E preserves an Euclidean metric h^E, then $R^E(U, V)$ is antisymmetric, thus from (B.5.22), we get $c_{2i+1}(E, \nabla^E) = 0$ and $p_i(E) := (-1)^i c_{2i}(E)$ is the ith Pontryagin class of E.

Example B.5.7 (Chern-Simons functional [66]). We suppose that X is an oriented manifold of dimension 3. By a classical theorem of Stiefel (cf. N. Steenord, The topology of fiber bundles, Princeton Math. Series, Vol 14, 1951), the tangent bundle TX is trivial, and we fix a trivialization of TX. If ∇^{TX} is a connection on TX, there exists $A \in \mathscr{C}^\infty(X, T^*X \otimes \mathrm{End}(TX))$ such that

$$\nabla^{TX} = d + A. \tag{B.5.23}$$

Now we choose the special path $\nabla_t^{TX} = d + tA$ of connections from d to ∇^{TX}. Let $\pi : X \times \mathbb{R} \to X$ be the natural projection, and $E = TX$. We define a connection $\nabla^{\pi^* E}$ on $\pi^* E$ on $X \times \mathbb{R}$ as before: $\nabla^{\pi^* E} = \nabla_t^{TX} + dt \frac{\partial}{\partial t}$. Then

$$R^{\pi^* E} := \left(\nabla^{\pi^* E} \right)^2 = t\, dA + t^2 A \wedge A + dt \wedge A. \tag{B.5.24}$$

On $X \times \mathbb{R}$ we have

$$\mathrm{Tr}[(R^{\pi^* E})^2] = dt \wedge \mathrm{Tr}[A \wedge (t \, dA + t^2 A \wedge A)]. \tag{B.5.25}$$

Thus for the polynomial $f(z) = z^2$, from (B.5.13), we find

$$\begin{aligned} \widetilde{F}(\nabla^{\pi^* E}) &= \frac{-1}{4\pi^2} \int_0^t \mathrm{Tr}[A \wedge (t \, dA + t^2 A \wedge A)] \, dt \\ &= \frac{-1}{8\pi^2} \mathrm{Tr}[A \wedge dA + \frac{2}{3} A \wedge A \wedge A)]. \end{aligned} \tag{B.5.26}$$

$\mathrm{Tr}[A \wedge dA + \frac{2}{3} A \wedge A \wedge A)]$ is the classical *Chern-Simons functional*.

Appendix C

Spectral Analysis of Self-adjoint Operators

We collect in this appendix some of the properties related to self-adjoint extensions and their spectrum that we used. A short and general introduction on functional analysis for applications in partial differential equations is [237, Appendix A] which is very helpful to understand our book.

C.1 Quadratic forms and Friedrichs extension

Let H be a complex Hilbert space with inner product $\langle\,,\,\rangle$ and norm $\|\ \|$.

Let $A : \mathrm{Dom}(A) \subset H \longrightarrow H$ be a linear operator, where $\mathrm{Dom}(A)$ is a dense linear subspace of H. A is called *closed* if $\mathrm{Graph}(A) = \{(u, Au);\ u \in \mathrm{Dom}(A)\}$ is closed and *preclosed* if the closure $[\mathrm{Graph}(A)]$ is again the graph of a linear operator denoted $\overline{A}$ (the closure of A). An operator $B : \mathrm{Dom}(B) \subset H \longrightarrow H$ is called an extension of A if $\mathrm{Graph}(A) \subset \mathrm{Graph}(B)$, and we write $A \subset B$.

We define the *adjoint operator* of A by $\mathrm{Dom}(A^*) = \{u \in H;\ \text{there exists } C > 0 \text{ such that } |\langle Av, u\rangle| \leqslant C\|v\| \text{ for any } v \in \mathrm{Dom}(A)\}$, and for $u \in \mathrm{Dom}(A^*)$ we set A^*u to be the unique $w \in H$ such that $\langle Av, u\rangle = \langle v, w\rangle$ for any $v \in \mathrm{Dom}(A)$, by using the Riesz representation theorem. Clearly, $A \subset (A^*)^*$ and A^* is closed. If we denote $\Phi : H \times H \to H \times H$, $\Phi(u, v) = (v, -u)$, we have

$$\mathrm{Graph}(A)^\perp = \Phi(\mathrm{Graph}(A^*)). \tag{C.1.1}$$

We can also verify the important relations:

$$[\mathrm{Im}(A^*)] = \mathrm{Ker}(A)^\perp, \quad [\mathrm{Im}(A)] = \mathrm{Ker}(A^*)^\perp, \tag{C.1.2}$$

where $[V]$ denotes the closure of a linear space V, and $V^\perp$ denotes the orthogonal complement of V. We need a criterion for an operator to have closed range.

Lemma C.1.1. *Let* $T : \mathrm{Dom}(T) \subset H \longrightarrow H$ *be a linear, closed, densely defined operator. The following conditions on* T *are equivalent:*

1. $\mathrm{Im}(T)$ *is closed.*
2. *There is a constant* C *such that*

$$\|f\| \leqslant C\|Tf\| \quad \text{for all} \quad f \in \mathrm{Dom}(T) \cap [\mathrm{Im}(T^*)]. \tag{C.1.3}$$

3. $\mathrm{Im}(T^*)$ *is closed.*
4. *There is a constant* C *such that*

$$\|f\| \leqslant C\|T^*f\| \quad \text{for all} \quad f \in \mathrm{Dom}(T^*) \cap [\mathrm{Im}(T)]. \tag{C.1.4}$$

The best constants in (C.1.3) *and* (C.1.4) *are the same.*

Proof. We assume that (1) holds, hence $\mathrm{Im}(T)$ is a Hilbert space. From (C.1.2), $T : \mathrm{Dom}(T) \cap [\mathrm{Im}(T^*)] \longrightarrow \mathrm{Im}(T)$ is one-to-one, and its inverse $T^{-1} : \mathrm{Im}(T) \longrightarrow \mathrm{Dom}(T) \cap [\mathrm{Im}(T^*)]$ is a well-defined closed operator. Thus from the closed graph theorem, T^{-1} is continuous and this proves (2). It is obvious that (2) implies (1). Similarly, (3) and (4) are equivalent. To prove that (2) implies (4), notice that

$$|\langle g, Tf\rangle| = |\langle T^*g, f\rangle| \leqslant C\|T^*g\|\|Tf\|,$$

for $g \in \mathrm{Dom}(T^*)$ and $f \in \mathrm{Dom}(T) \cap [\mathrm{Im}(T^*)]$. Thus

$$|\langle g, h\rangle| \leqslant C\|T^*g\|\|h\|, \quad \text{for } g \in \mathrm{Dom}(T^*) \quad \text{and } h \in \mathrm{Im}(T),$$

which implies (4). Similarly, (4) implies (2). $\qquad\square$

We say that A is *self-adjoint* if $A = A^*$. We say A is *symmetric* if $A \subset A^*$. If A is symmetric, then A is preclosed and A is said to be *essentially self-adjoint* if $\overline{A} = A^*$. We say A is *positive* if $\langle Au, u\rangle \geqslant 0$ for any $u \in \mathrm{Dom}(A)$.

Lemma C.1.2. *Let* A *be an injective self-adjoint operator. Then* $\mathrm{Im}(A)$ *is dense and the operator* A^{-1} *with* $\mathrm{Dom}(A^{-1}) = \mathrm{Im}(A)$ *is self-adjoint.*

Proof. Since $\mathrm{Im}(A)^{\perp} = \mathrm{Ker}(A^*)$, $A = A^*$ and $\mathrm{Ker}(A) = 0$, we see that $\mathrm{Im}(A)$ is dense. Moreover, using (C.1.1) and $\mathrm{Graph}(A^{-1}) = \Phi(\mathrm{Graph}(-A))$, we deduce that $(A^{-1})^* = (A^*)^{-1}$, hence $(A^{-1})^* = A^{-1}$. $\qquad\square$

Lemma C.1.3 (von Neumann). *Let* A *be a closed operator. Then* A^*A *is self-adjoint and* $1 + A^*A$ *has a bounded inverse.*

Proof. Let $w \in H$. We write $(w, 0) = (u, v) + (u_0, v_0) \in \mathrm{Graph}(A) \oplus \mathrm{Graph}(A)^{\perp}$. But $\mathrm{Graph}(A)^{\perp} = \Phi(\mathrm{Graph}(A^*))$ so $u \in \mathrm{Dom}(A)$, $Au = v$, $v_0 \in \mathrm{Dom}(A^*)$, $A^*v_0 = -u_0$. Hence $w = u - A^*v_0$, $0 = Au + v_0$, thus $Au = -v_0 \in \mathrm{Dom}(A^*)$, $w = u + A^*Au = (1 + A^*A)u$. Thus $(1 + A^*A)\mathrm{Dom}(A^*A) = H$. Consequently

$1 + A^*A$ is bijective, with inverse $(1 + A^*A)^{-1}$ having range $\mathrm{Dom}(A^*A)$. Set $f = (1 + A^*A)^{-1}u$, $g = (1 + A^*A)^{-1}v$. Then

$$\langle u, (1 + A^*A)^{-1}v \rangle = \langle (1 + A^*A)f, g \rangle = \langle f, g \rangle + \langle Af, Ag \rangle = \langle (1 + A^*A^{-1})u, v \rangle,$$

so $(1 + A^*A)^{-1}$ is symmetric. Since its domain is H, $(1 + A^*A)^{-1}$ is bounded and self-adjoint. Lemma C.1.2 shows that $1 + A^*A$ and hence A^*A are self-adjoint. $\square$

We shall work in general with the quadratic form associated to an operator rather than with the operator directly.

A *quadratic form* is a sesquilinear map $Q : \mathrm{Dom}(Q) \times \mathrm{Dom}(Q) \longrightarrow \mathbb{C}$ where $\mathrm{Dom}(Q)$ is a dense linear subspace of H. Q is called *positive* if $Q(u, u) \geqslant 0$ for any $u \in \mathrm{Dom}(Q)$. A positive quadratic form Q is called *closed* if $(\mathrm{Dom}(Q), \|\cdot\|_Q)$ is complete, where

$$\|u\|_Q := (Q(u, u) + \|u\|^2)^{1/2}, \quad \text{for } u \in \mathrm{Dom}(Q). \tag{C.1.5}$$

There exists a basic correspondence between positive closed quadratic forms and positive self-adjoint operators.

Proposition C.1.4. *Let A be a positive self-adjoint operator. Then there exists a closed quadratic form Q_A such that*

$$\mathrm{Dom}(A) = \{u \in \mathrm{Dom}(Q_A) : \text{there exists } v \in H$$
$$\text{with } Q_A(u, w) = \langle v, w \rangle \text{ for any } w \in \mathrm{Dom}(Q_A)\}, \tag{C.1.6}$$
$$Au = v, \text{ for } u \in \mathrm{Dom}(A).$$

Q_A *is called the* associated quadratic form *to A*.

Proof. On $\mathrm{Dom}(A)$ we consider the scalar product $\langle u, v \rangle_A := \langle (1 + A)u, v \rangle$ with norm $\|u\|_A = \langle u, u \rangle_A^{1/2}$ and we consider the Hilbert space completion $(\mathscr{D}, \langle \cdot, \cdot \rangle_A)$ of $(\mathrm{Dom}(A), \langle \cdot, \cdot \rangle_A)$. This is the space of equivalence classes of Cauchy sequences $\{u_\nu\}$, where two Cauchy sequences $\{u_\nu\}$, $\{v_\nu\}$ are said to be equivalent if $u_\nu - v_\nu \to 0$ as $\nu \to \infty$. Since $\|u\|_A \geqslant \|u\|$ for $u \in \mathrm{Dom}(A)$, the canonical injection $\mathrm{Dom}(A) \hookrightarrow H$ extends to an injective bounded linear map $\mathscr{D} \hookrightarrow H$. We will identify $\mathscr{D}$ with its image in H:

$$\mathscr{D} = \{u \in H : \text{there exists } \{u_\nu\} \subset \mathrm{Dom}(A), \text{ such that } u_\nu \to u, \text{ and}$$
$$\|u_\nu - u_\mu\|_A \to 0, \text{ as } \mu, \nu \to +\infty\}. \tag{C.1.7}$$

Moreover

$$\|u\|_A \geqslant \|u\|, \quad u \in \mathscr{D},$$
$$\langle u, v \rangle_A = \langle (1 + A)u, v \rangle, \quad u \in \mathrm{Dom}(A), v \in \mathscr{D}. \tag{C.1.8}$$

We define $\mathrm{Dom}(Q_A) := \mathscr{D}$, $Q_A(u, v) := \langle u, v \rangle_A - \langle u, v \rangle$, for $u, v \in \mathscr{D}$. By construction, Q_A is a closed quadratic form. We prove now (C.1.6). Let us denote by

$\mathscr{D}_A$ the right-hand side of (C.1.6). From (C.1.8), $\text{Dom}(A) \subset \mathscr{D}_A$. Conversely, let $u \in \mathscr{D}_A$. It follows from (C.1.6), (C.1.8) that $\langle Aw, u \rangle = \langle w, v \rangle$ for all $w \in \text{Dom}(A)$. Thus $u \in \text{Dom}(A^*)$ and $A^* u = v$. Since $A = A^*$, $\mathscr{D}_A \subset \text{Dom}(A)$. $\square$

Proposition C.1.5. *Let Q be a closed positive quadratic form. Then there exists a positive self-adjoint operator A with $Q = Q_A$.*

Proof. Consider the scalar product $\langle u, v \rangle_Q := Q(u, v) + \langle u, v \rangle$ with norm $\|\cdot\|_Q$. By assumption $(\text{Dom}(Q), \langle \cdot, \cdot \rangle_Q)$ is a Hilbert space. For each $w \in H$ the mapping $u \mapsto \langle w, u \rangle$ is a bounded form on $(\text{Dom}(Q), \langle \cdot, \cdot \rangle_Q)$, since $\langle w, u \rangle \leqslant \|w\| \cdot \|u\| \leqslant \|w\| \cdot \|u\|_Q$. By the Riesz representation theorem, there exists a unique element $Tw \in \text{Dom}(Q)$ such that

$$\langle w, u \rangle = \langle Tw, u \rangle_Q \quad \text{for all } u \in \text{Dom}(Q). \tag{C.1.9}$$

It is easily seen that $T : H \to H$ is a bounded, injective, self-adjoint linear operator. Indeed, $\|Tw\|^2 \leqslant \langle Tw, Tw \rangle_Q = \langle w, Tw \rangle \leqslant \|w\| \cdot \|Tw\|$, so T is bounded. Moreover, $Tw = 0$ implies $\langle Tw, v \rangle_Q = \langle w, v \rangle = 0$ for all $v \in \text{Dom}(Q)$; since $\text{Dom}(Q)$ is dense, $w = 0$, and T is injective. Next,

$$\langle Tw, v \rangle = \overline{\langle v, Tw \rangle} = \overline{\langle Tv, Tw \rangle_Q} = \langle Tw, Tv \rangle_Q = \langle w, Tv \rangle,$$

so T is self-adjoint. By Lemma C.1.2, T^{-1} is self-adjoint. We define the operator A by

$$\text{Dom}(A) := \text{Dom}(T^{-1}), \quad A := T^{-1} - 1. \tag{C.1.10}$$

Then A is self-adjoint, positive (since $\langle Au, u \rangle = \|u\|_Q^2 - \|u\|^2 \geqslant 0$ for $u \in \text{Dom}(A)$) and by construction of T, we have $Q_A = Q$. $\square$

Let $A : \text{Dom}(A) \subset H \longrightarrow H$ be a positive operator. We define Q'_A by

$$\text{Dom}(Q'_A) := \text{Dom}(A), \quad Q'_A(u, v) := \langle Au, v \rangle, \quad \text{for } u, v \in \text{Dom}(A). \tag{C.1.11}$$

Proposition C.1.6. *Let A be a positive operator. Then Q'_A is closable, i.e., there exists the smallest positive closed form $\overline{Q}_A$ extending Q'_A.*

Proof. We repeat the construction given in Proposition C.1.4 word for word, and find a closed quadratic form $\overline{Q}_A$ (denoted there Q_A) such that $\text{Dom}(Q'_A)$ is dense in $\text{Dom}(\overline{Q}_A)$. $\square$

Definition C.1.7. Let A be a positive operator. The *Friedrichs extension* of A is the self-adjoint operator A_F such that $Q_{A_F} = \overline{Q}_A$.

The existence of A_F follows from Propositions C.1.4-C.1.6. Moreover, we deduce

$$\text{Dom}(A_F) = \big\{ u \in \text{Dom}(A^*) : \text{there exists } \{u_\nu\} \subset \text{Dom}(A),$$
$$\text{such that } u_\nu \to u, \text{ and } \langle A(u_\nu - u_\mu), u_\nu - u_\mu \rangle \to 0 \big\}, \tag{C.1.12}$$

and $A_F u = A^* u$ for $u \in \text{Dom}(A_F)$.

Definition C.1.8. A *one-parameter semi-group of operators* on a Hilbert space H is a set of bounded operators $P_t : H \longrightarrow H$, $t \in \mathbb{R}_+$, satisfying for $t_1, t_2 \in \mathbb{R}_+$,

$$P_{t_1+t_2} = P_{t_1} P_{t_2} \quad \text{and} \quad P_0 = \text{Id}, \tag{C.1.13}$$

and for any $v \in H$, $P_{t_j} v \to P_t v$ when $t_j \to t$.

We can find the following result in [237, Appendix A, §9].

Theorem C.1.9. *If $A : \text{Dom}(A) \subset H \longrightarrow H$ is positive self-adjoint, then $-A$ generates a semi-group $P_t = e^{-tA}$ consisting of positive self-adjoint operators of norm ≤ 1. P_t is characterized by : for any $v \in \text{Dom}(A)$, $t \in \mathbb{R}_+$,*

$$Av = \lim_{t \to 0} t^{-1}(P_t v - v),$$

$$P_t \text{Dom}(A) \subset \text{Dom}(A), \tag{C.1.14}$$

$$-AP_t v = -P_t A v = \frac{d}{dt} P_t v.$$

C.2 Spectral theorem

The point of extending an operator to a self-adjoint one is to study its spectral properties.

Let A be a closed operator on a Hilbert space H. We say that a complex number λ lies in the *resolvent set* of A if $\lambda - A$ is a bijection of $\text{Dom}(A)$ onto H with a bounded inverse. Note that by the closed-graph theorem if $\lambda - A : \text{Dom}(A) \longrightarrow H$ is a bijection, the inverse is automatically bounded. The *spectrum* of A, denoted $\sigma(A)$, is the complement in $\mathbb{C}$ of the resolvent set.

Let $\text{Bor}(\mathbb{R})$ be the family of Borel sets in $\mathbb{R}$. Let $\mathscr{L}(H)$ be the space of bounded operators of H. A map $E : \text{Bor}(\mathbb{R}) \longrightarrow \mathscr{L}(H)$ is called a *spectral measure* if a). $E(\Omega)$ is an orthogonal projection for any $\Omega \in \text{Bor}(\mathbb{R})$; b). $E(\emptyset) = 0$, $E(\mathbb{R}) = \text{Id}$; c). If $\Omega = \cup_{j \in I} \Omega_j$ with $I \subset \mathbb{N}$, and $\Omega_i \cap \Omega_j = \emptyset$ if $i \neq j$, then $E(\Omega) = $ s-$\lim_{k \to \infty} \sum_{j \in I \cap [0,k]} E(\Omega_j)$; d). $E(\Omega_1 \cap \Omega_2) = E(\Omega_1) E(\Omega_2)$. Here s-lim is the strong limit of operator.

Let $E : \text{Bor}(\mathbb{R}) \longrightarrow \mathscr{L}(H)$ be a spectral measure. If $f : \mathbb{R} \longrightarrow \mathbb{C}$ is a bounded Borel function we can define the integral

$$\int_{\mathbb{R}} f(t) \, dE(t) \in \mathscr{L}(H) \tag{C.2.1}$$

using the usual pattern of defining the integral for step-functions first and then writing the integral of a general bounded Borel function f as the limit of the integrals of a sequence of step-functions, which converge uniformly to f. If $E : \text{Bor}(\mathbb{R}) \longrightarrow \mathscr{L}(H)$ is a spectral measure we can define the associated scalar measures on $\mathbb{R}$, by setting $\text{Bor}(\mathbb{R}) \ni \Omega \longrightarrow \langle E(\Omega)u, v \rangle$, for each $(u, v) \in H \times H$. For any bounded Borel function $f : \mathbb{R} \longrightarrow \mathbb{C}$ we have then

$$\left\langle \left(\int_{\mathbb{R}} f(t) \, dE(t) \right) u, v \right\rangle = \int_{\mathbb{R}} f(t) \, d\langle E(t)u, v \rangle. \tag{C.2.2}$$

We have the following fundamental result [202, Theorem VIII.6], [70, Theorem 2.5.5].

Theorem C.2.1 (Spectral theorem). *Each self-adjoint operator A has a unique spectral measure E such that*

$$\mathrm{Dom}(A) = \left\{ u \in H : \int_{\mathbb{R}} t^2 \, d\langle E(t)u, u \rangle < \infty \right\} \tag{C.2.3}$$

and for $u \in \mathrm{Dom}(A)$

$$Au = \lim_{k \to \infty} \left(\int_{\mathbb{R}} 1_{[-k,k]}(t) t \, dE(t) \right) u =: \left(\int_{\mathbb{R}} t \, dE(t) \right) u. \tag{C.2.4}$$

From Theorem C.2.1 and (C.2.1), for any bounded Borel function $f : \mathbb{R} \longrightarrow \mathbb{C}$, we can define

$$f(A) = \int_{\mathbb{R}} f(t) \, dE(t) \in \mathscr{L}(H). \tag{C.2.5}$$

Especially, we see that the operator e^{-tA} defined in Theorem C.1.9 coincides with the definition in (C.2.5) for a positive self-adjoint operator A.

More generally, consider a Borel measurable function $f : [0, \infty[\to \mathbb{R}$. We define the self-adjoint operator $f(A)$ by setting

$$\mathrm{Dom}(f(A)) = \left\{ u \in H : \int_{\mathbb{R}} |f(t)|^2 \, d\langle E(t)u, u \rangle < \infty \right\}, \tag{C.2.6}$$

and for $u \in \mathrm{Dom}(f(A))$,

$$f(A)u = \lim_{k \to \infty} \left(\int_{\mathbb{R}} 1_{[-k,k]}(t) f(t) t \, dE(t) \right) u =: \left(\int_{\mathbb{R}} f(t) \, dE(t) \right) u. \tag{C.2.7}$$

$f(A)$ is positive if f is positive. Let A be a positive self-adjoint operator. For $f(t) = \sqrt{t}$ we obtain the positive operator $\sqrt{A} = A^{1/2}$; it is the unique positive self-adjoint operator B satisfying $B^2 = A$. Moreover $\mathrm{Dom}(Q_A) = \mathrm{Dom}(A^{1/2})$ and $Q_A(u, v) = (A^{1/2}u, A^{1/2}v)$ for any $u, v \in \mathrm{Dom}(Q_A)$. Therefore,

$$\mathrm{Dom}(Q_A) = \left\{ u \in H : \int_{\mathbb{R}} |t| \, d\langle E(t)u, u \rangle < \infty \right\},$$
$$Q_A(u, v) = \int_{\mathbb{R}} t \, d\langle E(t)u, v \rangle, \quad u, v \in \mathrm{Dom}(Q_A). \tag{C.2.8}$$

Let us define the *spectral resolution* associated to A by $(E_\lambda)_{\lambda \in \mathbb{R}}$ where $E_\lambda = E(]-\infty, \lambda])$. When we want to stress the dependence on A we note $E_\lambda(A)$. We set $\mathscr{E}(\lambda) = \mathscr{E}(\lambda, A) := \mathrm{Im}(E_\lambda(A))$. The *spectrum counting function* of A is defined as

$$N(\lambda) := \dim \mathrm{Im}(E_\lambda) = \dim \mathscr{E}(\lambda). \tag{C.2.9}$$

The *discrete spectrum* $\sigma_d(A)$ is the set of all eigenvalues λ of finite multiplicity which are isolated, in the sense that $]\lambda - \varepsilon, \lambda[$ and $]\lambda, \lambda + \varepsilon[$ are disjoint from the spectrum for some $\varepsilon > 0$. The *essential spectrum* is the complement of the discrete spectrum, $\sigma_{\mathrm{ess}}(A) = \sigma(A) \setminus \sigma_d(A)$.

Definition C.2.2. $B \in \mathscr{L}(H)$ is *compact*, if for any $U \subset H$ a bounded subset, the closure of $B(U)$ is compact in H.

Theorem C.2.3. *If $B \in \mathscr{L}(H)$ is compact and self-adjoint, then H has a complete orthonormal basis of eigenvectors $\{\varphi_j\}_{j=1}^\infty$ of B with corresponding eigenvalues λ_j which converge to 0 as $j \to \infty$.*

We have the following criterion for the non-existence of the essential spectrum [70, Cor. 4.2.3].

Theorem C.2.4. *Let A be a positive self-adjoint operator. Then the following conditions are equivalent:*

1. *The resolvent operator $(1 + A)^{-1}$ is compact.*
2. *The operator A has empty essential spectrum.*
3. *There exists a complete orthonormal set of eigenvectors $\{\varphi_j\}_{j=1}^\infty$ of A with corresponding eigenvalues $\lambda_j \geqslant 0$ which converge to $+\infty$ as $j \to \infty$.*

It follows from the proof of Proposition C.1.5 that A has compact resolvent $T = (1+A)^{-1}$ if the injection $(\mathrm{Dom}(Q_A), \|\cdot\|_{Q_A}) \hookrightarrow (H, \|\cdot\|)$ is a compact operator.

If the spectrum is discrete, we arrange the eigenvalues in increasing order and repeated according to multiplicity: $\lambda_1 \leqslant \lambda_2 \leqslant \cdots$. Then $N(\lambda) = \#\{j : \lambda_j \leqslant \lambda\}$.

C.3 Variational principle

The most important tool for estimating and for comparing the eigenvalues of different operators is the variational or minimax principle. The simplest but very useful form is the following.

Lemma C.3.1 (Glazman lemma). *The spectrum counting function of a positive self-adjoint operator A satisfies the variational formula*

$$N(\lambda) = \sup\{\dim F : F \text{ closed} \subset \mathrm{Dom}(Q_A)\,, \; Q_A(u, u) \leqslant \lambda \|u\|^2, \text{ for all } u \in F\}\,. \tag{C.3.1}$$

Proof. Let $u \in \mathrm{Im}(E_\lambda)$. Then

$$\langle E(\Omega)u, u \rangle = \langle E(\Omega)\, E(] - \infty, \lambda])u, u \rangle = \langle E(\Omega \cap] - \infty, \lambda])u, u \rangle$$

for any $\Omega \in \mathrm{Bor}(\mathbb{R})$, so formula (C.2.8) entails

$$Q_A(u, u) = \int_{-\infty}^{\lambda} t \, d\langle E(t)u, u \rangle \leqslant \lambda \langle E(] - \infty, \lambda])u, u \rangle = \lambda \|u\|^2. \qquad (\mathrm{C.3.2})$$

Hence $Q_A(u, u) \leqslant \lambda \|u\|^2$ for all $u \in \mathrm{Im}(E_\lambda)$ and $N(\lambda)$ does not exceed the right-hand side of (C.3.1). Consider a closed linear space $V \subset \mathrm{Dom}(Q_A)$ such that $Q_A(u, u) \leqslant \lambda \|u\|^2$ for all $u \in V$. We show that $E_\lambda : V \longrightarrow \mathrm{Im}(E_\lambda)$ is injective. If $u \in \mathrm{Ker}(E_\lambda) = \mathrm{Im}(E(]\lambda, +\infty[))$ we have

$$Q_A(u, u) = \int_{]\lambda, +\infty[} t \, d\langle E(t)u, u \rangle > \lambda \langle E(]\lambda, +\infty[)u, u \rangle = \lambda \|u\|^2$$

if $u \neq 0$. Thus any $u \in V \cap \mathrm{Ker}(E_\lambda)$ must vanish. We infer that $\dim V \leqslant \dim \mathrm{Im}(E_\lambda) = N(\lambda)$. Formula (C.3.1) is established. $\qquad \square$

Let A be a positive self-adjoint operator and let Q_A be the associated closed quadratic form. We consider the sequence $\lambda_1 \leqslant \lambda_2 \leqslant \cdots \leqslant \lambda_j \leqslant \cdots$ according to the formula

$$\lambda_j = \inf_{F \subset \mathrm{Dom}(Q_A)} \sup_{f \in F, \|f\|=1} Q_A(f, f) \qquad (\mathrm{C.3.3})$$

where F runs through the j-dimensional subspaces of $\mathrm{Dom}(Q_A)$.

We refer to [202, Vol IV, p.76-78], [70, Ch. 4.5] for the following result.

Theorem C.3.2 (Variational principle). *Let us define the bottom of the essential spectrum as* $\inf \sigma_{\mathrm{ess}}(A)$ *if* $\sigma_{\mathrm{ess}}(A) \neq \emptyset$ *and* $+\infty$ *if* $\sigma_{\mathrm{ess}}(A) = \emptyset$. *Then for each fixed j, either following a) or b) should be true:*

(a) *there are at least j eigenvalues (counted according to multiplicity) below the bottom of the essential spectrum and λ_j is the jth eigenvalue,*

(b) λ_j *is the bottom of the essential spectrum, in which case* $\lambda_j = \lambda_{j+1} = \lambda_{j+2} = \cdots$ *and there are at most $j - 1$ eigenvalues (counting multiplicity) below λ_j.*

Theorem C.3.3 ([70, Lemma 8.4.1]). *If A is a self–adjoint operator, then the following conditions are equivalent:*

(a) $\lambda \in \sigma_{\mathrm{ess}}(A)$,

(b) *there exists a noncompact sequence* $\{f_k\}_{k \in \mathbb{N}}$ *in* $\mathrm{Dom}(A)$ *with*

$$\|f_k\| = 1 \quad \text{for each} \quad k \in \mathbb{N} \quad \text{and} \quad \lim_{k \to \infty} \|(A - \lambda \,\mathrm{Id})f_k\| = 0, \qquad (\mathrm{C.3.4})$$

(c) *there exists a sequence* $\{f_k\}_{k \in \mathbb{N}}$ *in* $\mathrm{Dom}(A)$ *with* $\langle f_k, f_l \rangle = \delta_{kl}$ *and which satisfies (C.3.4).*

Appendix D

Heat Kernel and Finite Propagation Speed

The heat kernel and the finite propagation speed of solutions of hyperbolic equations play important roles in our book. Basically, we only need to know the existence of the heat kernel, then we always use the finite propagation speed to localize the problem at hand. To help the reader understand the finite propagation speed, we explain it here in detail.

This appendix is organized as follows. In Section D.1, we summarize the basic facts on heat kernels (cf. [148, Chap. 3], [15, Chap. 2]). In Section D.2, we explain the finite propagation speed of solutions of wave equations.

D.1   Heat kernel

Let (X, g^{TX}) be a smooth complete Riemannian manifold without boundary, endowed with the Riemannian metric g^{TX} on TX. Set $\dim_{\mathbb{R}} X = m$. Let $d(x, y)$ be the Riemannian distance on (X, g^{TX}). Let (E, h^E) be a Hermitian vector bundle on X with Hermitian connection ∇^E.

Since (X, g^{TX}) is complete, we know that the Bochner Laplacian Δ^E is essentially self-adjoint, i.e., the closure of Δ^E in $L^2(X, E)$, the L^2-sections of E on X with norm $\| \cdot \|_{L^2}$ as in (1.3.14), is self-adjoint (cf. Corollary 3.3.4, [9, Prop. 3.1], [148, Th. 2.5.7]).

Let $H = \Delta^E + Q$ be a generalized Laplacian on E defined in (A.3.3). We suppose that the Hermitian section $Q \in \mathscr{C}^{\infty}(X, \mathrm{End}(E))$ is bounded from below, i.e., there exists $C > 0$ such that $h^E(Q\xi, \xi) \geqslant -Ch^E(\xi, \xi)$ for any $x \in X$, $\xi \in E_x$. Then H is again essentially self-adjoint (cf. [238, §8.2]). Thus

Lemma D.1.1. *The operator* $\Delta^E + Q$ *is essentially self-adjoint if the Hermitian section* Q *is bounded from below.*

The *heat operator* e^{-tH} is defined in general by using operator theory (cf. Theorem C.1.9). For $t > 0$, the heat operator e^{-tH} is an operator from $L^2(X, E)$ to $L^2(X, E)$ which is $\mathscr{C}^1$ in t and which verifies the following properties: for $s \in L^2(X, E)$,

$$\left(\frac{\partial}{\partial t} + H \right) e^{-tH} s = 0, \tag{D.1.1a}$$

$$\lim_{t \to 0} e^{-tH} s = s \quad \text{in } L^2(X, E). \tag{D.1.1b}$$

The *heat kernel* $e^{-tH}(x, x') \in E_x \otimes E_{x'}^*$, $(x, x' \in X)$, is the Schwartz kernel of the heat operator e^{-tH} with respect to the Riemannian volume form $dv_X(x')$, i.e., for $s \in L^2(X, E)$, we have

$$(e^{-tH} s)(x) = \int_X e^{-tH}(x, x') s(x') dv_X(x'). \tag{D.1.2}$$

Theorem D.1.2. *The heat kernel $e^{-tH}(x, x')$ is smooth on $x, x' \in X$, $t \in]0, \infty[$, and*

$$e^{-tH}(x, x') = (e^{-tH}(x', x))^*, \tag{D.1.3}$$

where for $w \in E_x$, we denote $w^* \in E_x^*$, defined by $h^E(w_1, w) = (w^*, w_1)$ for $w_1 \in E_x$.

Proof. By (C.2.5), for any $t > 0, m_1, m_2 \in \mathbb{N}$, there exists $C_{m_1, m_2, t} > 0$ such that for any $s \in \mathscr{C}_0^\infty(X, E)$,

$$\|H^{m_1} e^{-tH} H^{m_2} s\|_{L^2} \leqslant C_{m_1, m_2, t} \|s\|_{L^2}. \tag{D.1.4}$$

(D.1.4) is equivalent to

$$\|H_x^{m_1} H_y^{m_2} e^{-tH}(x, y)\|_{L^2} \leqslant C_{m_1, m_2, t}. \tag{D.1.5}$$

Here H_y acts on E^* by identifying E^* to E by h^E. Now applying Theorems A.1.7 and A.3.2 to a compact subset K of $X \times X$, we know that $e^{-tH}(x, y)$ is $\mathscr{C}^\infty$ on K. From (D.1.1a), we get that $e^{-tH}(x, y)$ is also $\mathscr{C}^\infty$ on $t \in]0, \infty[$.

Finally, for any $s_1, s_2 \in \mathscr{C}_0^\infty(X, E)$, we have $\frac{\partial}{\partial t_1} \langle e^{-(t-t_1)H} s_1, e^{-t_1 H} s_2 \rangle = 0$. Thus after taking the integral in t_1, we have

$$\langle e^{-tH} s_1, s_2 \rangle = \langle s_1, e^{-tH} s_2 \rangle. \tag{D.1.6}$$

This means (D.1.3) holds. $\square$

In the rest of this appendix, we suppose that X is compact.

Theorem D.1.3. *If X is compact, the spectrum $\mathrm{Spec}(H)$ of H is discrete and the eigensections of H form a complete basis of $L^2(X, E)$. Let $\{\varphi_i\}_{i=1}^\infty$ be a complete orthogonal basis of $L^2(X, E)$ consisting of eigensections of H with $H\varphi_i = \lambda_i \varphi_i$*

and $-\infty < \lambda_1 \leqslant \lambda_2 \leqslant \cdots \to \infty;$ *then*

$$e^{-tH}(x, x') = \sum_{i=1}^{\infty} e^{-\lambda_i t} \varphi_i(x) \otimes \varphi_i(x')^* \in E_x \otimes E_{x'}^*. \qquad (D.1.7)$$

Proof. By adding a constant on H, we can assume that there exists $C > 0$ such that for any $s \in \mathscr{C}_0^{\infty}(X, E)$,

$$\langle Hs, s \rangle \geqslant C \|s\|_{L^2}^2. \qquad (D.1.8)$$

We claim that for any $k \in \mathbb{N}$,

$$H : \boldsymbol{H}^{k+2}(X, E) \to \boldsymbol{H}^k(X, E) \qquad (D.1.9)$$

is one-to-one and onto.

In fact, from (D.1.8), $\|Hs\|_{L^2} \geqslant C\|s\|_{L^2}$ for any $s \in \mathscr{C}_0^{\infty}(X, E)$. Thus for any $k \in \mathbb{N}$, H in (D.1.9) is injective and has closed rank in $\boldsymbol{H}^k(X, E)$. If it were not onto, then there would exist $0 \neq s_0 \in \boldsymbol{H}^k(X, E)$ that is orthogonal to $\mathrm{Im}(H)$, i.e.,

$$\langle Hs, s_0 \rangle = 0 \quad \text{for all } s \in \boldsymbol{H}^{k+2}(X, E). \qquad (D.1.10)$$

Note that $\boldsymbol{H}^{k+2}(X, E)$ (and $\mathscr{C}^{\infty}(X, E)$) is dense in $\boldsymbol{H}^k(X, E)$. Thus for $k \geqslant 2$, we can take $s = s_0$ in (D.1.10), which contradicts (D.1.8). Thus H in (D.1.9) is onto for $k \geqslant 2$. Since $\mathrm{Im}(H)$ is closed in $\boldsymbol{H}^k(X, E)$ for any k, we know that the closure of $\boldsymbol{H}^{k+2}(X, E)$ in $\boldsymbol{H}^k(X, E)$ (which is $\boldsymbol{H}^k(X, E)$ itself) is contained in $\mathrm{Im}(H)$.

Thus $H^{-1} : L^2(X, E) \to \boldsymbol{H}^2(X, E)$ is a bounded operator, by the open mapping theorem. Theorem A.3.1 implies that $H^{-1} : L^2(X, E) \to L^2(X, E)$ is a compact operator. From Theorem C.2.3, we get the first part of Theorem D.1.3.

Now for any $t > 0$, we have

$$\sum_{i=1}^{\infty} e^{-\lambda_i t} = \sum_{i=1}^{\infty} \langle e^{-tH} \varphi_i, \varphi_i \rangle = \sum_{i=1}^{\infty} \|e^{-tH/2} \varphi_i\|_{L^2}^2$$
$$= \int_{X \times X} |e^{-tH/2}(x, y)|^2 dv_X(x) dv_X(y) < +\infty. \qquad (D.1.11)$$

For $1 \leqslant k \leqslant l$, set $Q_{k,l}(x, y, t) = \sum_{i=k}^{l} e^{-\lambda_i t} \varphi_i(x) \otimes \varphi_i(x')^*$. Then from (D.1.11), we obtain that for any $m_1, m_2 \in \mathbb{N}$, there exists $C > 0$ such that for all k, l,

$$\|H_x^{m_1} H_y^{m_2} Q_{k,l}(x, y, t)\|_{L^2} \leqslant \sum_{i=k}^{l} e^{-\lambda_i t} \lambda_i^{m_1 + m_2} \qquad (D.1.12)$$
$$\leqslant Ct^{-m_1 - m_2} e^{-\lambda_k t/2} \sum_{i=1}^{\infty} e^{-\lambda_i t/4} < \infty.$$

Here we use the inequality $e^{-\lambda t/4} \lambda^{m_1 + m_2} \leqslant Ct^{-m_1 - m_2}$ for any $\lambda > 0$.

By using again the Sobolev embedding theorem we get that for any $r \in N$, the $\mathscr{C}^r$-norms of $Q_{k,l}(x, y)$ are uniformly bounded, and tend to 0 as $k \to \infty$. Thus $Q_{1,l}(x, y, t)$ converges uniformly to $Q_{1,\infty}(x, y, t)$. Certainly, $Q_{1,\infty}(x, y, t)$ verifies (D.1.1a) and (D.1.1b), so by the uniqueness of the heat kernel, we get (D.1.7). $\square$

We now recall some basic facts on trace class.

Definition D.1.4. Let $(\mathcal{H}, \| \; \|)$ be a Hilbert space. An operator $A \in \mathcal{L}(\mathcal{H})$ is a *Hilbert-Schmidt operator* if

$$\|A\|_{HS}^2 = \sum_j \|Ah_j\|^2 = \sum_{ij} |\langle Ah_i, h_j \rangle|^2 < \infty. \tag{D.1.13}$$

Here $\{h_j\}$ is an orthonormal basis of $\mathcal{H}$. The number $\|A\|_{HS}$ is called the Hilbert-Schmidt norm of A.

An operator $R \in \mathcal{L}(\mathcal{H})$ is called a *trace class* operator if there exist Hilbert-Schmidt operators A, B such that $R = AB$. In this case, the trace of R is defined by

$$\mathrm{Tr}[R] = \sum_j \langle Rh_j, h_j \rangle. \tag{D.1.14}$$

If $A, B \in \mathcal{L}(\mathcal{H})$ are Hilbert-Schmidt operators, then (D.1.13) implies that their adjoints A^*, B^* are also Hilbert-Schmidt and $\|A^*\|_{HS} = \|A\|_{HS}$. Thus $\sum_i |\langle ABh_i, h_i \rangle| \leqslant \|A\|_{HS} \|B\|_{HS}$, which means that (D.1.14) is well defined and

$$\mathrm{Tr}[AB] = \mathrm{Tr}[BA] = \sum_{ij} \langle Ah_i, h_j \rangle \langle Bh_j, h_i \rangle. \tag{D.1.15}$$

Recall that we assume that X is compact. Let $K(x, y)$ be the Schwartz kernel of an operator $K : L^2(X, E) \to L^2(X, E)$ with respect to $dv_X(y)$. If $K(x, y)$ is square integrable on $X \times X$, then by (D.1.13), K is Hilbert-Schmidt, and

$$\|K\|_{HS}^2 = \int_{X \times X} \mathrm{Tr}[K(x, y)^* K(x, y)] dv_X(x) \, dv_X(y). \tag{D.1.16}$$

Lemma D.1.5. *If the kernel $K(x, y)$ of K is smooth, then K is a trace class, and*

$$\mathrm{Tr}[K] = \int_X \mathrm{Tr}[K(x, x)] dv_X(x). \tag{D.1.17}$$

Proof. As in the proof of Theorem D.1.3, we assume that H verifies (D.1.8). Then by using the Sobolev embedding theorem as before, we see that the operator $H^{-(m+3)}$ has continuous kernel $H^{-(m+3)}(x, y)$. Then $K = H^{-(m+3)} \cdot H^{(m+3)} K$ and $H^{-(m+3)}$, $H^{(m+3)} K$ are Hilbert-Schmidt, thus K is a trace class.

Let $\{s_i\}$ be an orthonormal basis of $L^2(X, E)$. Then from (D.1.15), we have

$$\begin{aligned}
\mathrm{Tr}[K] &= \sum_{i,j} \langle H^{(m+3)} K s_i, s_j \rangle \langle s_j, H^{-(m+3)} s_i \rangle \\
&= \int_{X \times X} \langle (H^{(m+3)} K)(x, y), H^{-(m+3)}(x, y) \rangle dv_X(x) dv_X(y) \\
&= \int_{X \times X} ((H^{(m+3)} K)(x, y), (H^{-(m+3)}(x, y))^*) dv_X(x) dv_X(y) \\
&= \int_X \mathrm{Tr}[K(y, y)] dv_X(y).
\end{aligned} \tag{D.1.18}$$

Here we use the analogue of (D.1.6) and $H^{-(m+3)}(y, x) = (H^{-(m+3)}(x, y))^*$. $\qquad\square$

Theorem D.1.6. *If H_v is a smooth family of generalized Laplacians associated to a family of $g_v^{TX}, h_v^E, \nabla_v^E, Q_v$, then the heat operator e^{-tH_v} is smooth in v and*

$$\frac{\partial}{\partial v} e^{-tH_v} = -\int_0^t e^{-(t-t_1)H_v} \frac{\partial H_v}{\partial v} e^{-t_1 H_v} dt_1. \tag{D.1.19}$$

Proof. We will add a subscript v for the objects depending on v and work on $v \in [-1, 1]$. Then the Sobolev norms induced by $g_v^{TX}, h_v^E, \nabla_v^E$ are equivalent to the corresponding ones induced by g_0^{TX}, h_0^E. Thus from the proof of Theorem D.1.2, the heat kernel $e^{-tH_v}(x, x')$ of e^{-tH_v} with respect to $dv_{X,v}(x')$, is smooth in x, x', uniformly for $v \in [-1, 1]$.

We only need to prove (D.1.19) for $v = 0$. Set

$$dv_{X,v}(x) = \kappa_v(x) dv_{X,0}(x). \tag{D.1.20}$$

For $s \in \mathscr{C}^\infty(X, E^*)$, by (D.1.3) and (D.1.6), we get

$$\lim_{t\to 0} \int_X (s(z), e^{-tH_v}(z, x)) dv_{X,0}(z)$$

$$= \lim_{u\to 0} \int_X \left((e^{-tH_v}(x, z), (s(z))_v^*) \right)_v^* \kappa_v^{-1}(z) dv_{X,v}(z) = s(x)\kappa_v^{-1}(x). \tag{D.1.21}$$

By (D.1.1a), (D.1.1b), (D.1.21) and $H_0 e^{-tH_0} = e^{-tH_0} H_0$, we infer

$$e^{-tH_v}(x, w) - \kappa_v^{-1}(w) e^{-tH_0}(x, w)$$

$$= \int_0^t \frac{\partial}{\partial t_1} \int_X \left(e^{-(t-t_1)H_0}(x, z), e^{-t_1 H_v}(z, w) \right) dv_{X,0}(z) dt_1$$

$$= \int_0^t dt_1 \int_X \left(e^{-(t-t_1)H_0}(x, z), (H_0(z) - H_v(z)) e^{-t_1 H_v}(z, w) \right) dv_{X,0}(z). \tag{D.1.22}$$

From (D.1.22), we know that $e^{-tH_v}(x, w)$ is continuous in v, and we can take the differential in v, so we get at $v = 0$ (as $\kappa_0(x) = 1$),

$$\frac{\partial}{\partial v} (\kappa_v(w) e^{-tH_v}(x, w))$$

$$= -\int_0^t dt_1 \int_X \left(e^{-(t-t_1)H_0}(x, z), \frac{\partial H_v}{\partial v}(z) e^{-t_1 H_0}(z, w) \right) dt_1 dv_{X,0}(z). \tag{D.1.23}$$

But (D.1.20) and (D.1.23) means exactly (D.1.19).

By using (D.1.19), we conclude Theorem D.1.6. $\qquad\square$

As we suppose X is compact, we can also use the formal solution introduced by Minakshisundaram and Pleijel to construct the heat kernel (cf. [15, Chap. 2] for a detailed exposition). Especially, the following very useful asymptotic expansion for the heat kernel holds: There exist $h_j \in \mathscr{C}^\infty(X, \mathrm{End}(E))$ such that for any $k, l \in \mathbb{N}$, there exists $C_{k,l} > 0$ so that

$$\left| e^{-tH}(x, x) - \sum_{j=0}^{k} h_j(x) t^{j-\frac{m}{2}} \right|_{\mathscr{C}^l(X)} \leqslant C_{k,l}\, t^{k+1-\frac{m}{2}} \tag{D.1.24}$$

as $t \to 0$. The coefficients h_j depend smoothly on the geometric data and their derivatives at $x \in X$. Moreover, if we have a family H_v as in Theorem D.1.6, then in the expansion (D.1.24), the $\mathscr{C}^l$-norm includes also the parameter v (cf. also Section 5.5.6 for a proof in the spirit of the present book).

Assume now that the boundary ∂X of X is not empty. Then we need to fix boundary conditions in order to define a self-adjoint extension of H and the corresponding heat kernel. Usually, we impose Dirichlet (or Neumann) boundary conditions. Let e_n be the inward pointing unit normal at any boundary point of X. Then $s \in \mathscr{C}^\infty(X, E)$ satisfies the Dirichlet boundary conditions if

$$s = 0 \quad \text{on } \partial X. \tag{D.1.25}$$

Likewise $s \in \mathscr{C}^\infty(X, E)$ satisfies the Neumann boundary condition if

$$\nabla^E_{e_n} s = 0 \quad \text{on } \partial X. \tag{D.1.26}$$

D.2   Wave equation

Now we explain the finite propagation speed property for the wave equation. We suppose X is a compact manifold with boundary ∂X. Without loss of generality, we assume that $H = \Delta^E + Q$ in (A.3.3) is positive. If we consider $\mathrm{Dom}(H) = \{s \in \mathscr{C}^\infty(X, E) : s = 0 \text{ or } \nabla^E_{e_n} s = 0 \text{ on } \partial X\}$, and take the Friedrichs extension, the resulting operator, denoted still by H, is positive. The basic references for this part are [65, §7.8] and [237, §2.8].

Theorem D.2.1. *For $\omega = \omega(t, x)$, $t \in \mathbb{R}$, $x \in X$, we consider the wave equation*

$$\left(\frac{\partial^2}{\partial t^2} + H \right) \omega = 0 \tag{D.2.1}$$

with the Dirichlet (or Neumann) boundary condition.

Then for any $f_0, f_1 \in \mathscr{C}^\infty(X, E)$, verifying the corresponding boundary condition, there exists a unique solution ω for (D.2.1) with initial conditions $\omega(0, x) = f_0$, $\frac{\partial}{\partial t}\omega(0, x) = f_1$. The solution ω is given by the operator expression

$$\omega(t, x) = \cos(t\sqrt{H}) f_0 + \frac{\sin(t\sqrt{H})}{\sqrt{H}} f_1, \tag{D.2.2}$$

and satisfies

$$\mathrm{supp}(\omega(t, \cdot)) \subset K_t = \{x \in X; d(x, y) \leqslant t, \text{ for some } y \in \mathrm{supp}(f_0) \cup \mathrm{supp}(f_1)\}. \tag{D.2.3}$$

Proof. In fact, to get the finite propagation speed (D.2.3) and the uniqueness of the solution, we only need to obtain an energy estimate.

A vector $v = (v_t, v_x) \in T(\mathbb{R} \times X)$, is called timelike if $|v_x| < |v_t|$, and a hypersurface $\Sigma \subset [0, T] \times (X \setminus \partial X)$ is called spacelike, if the unit normal vector (N_t, N_x) is timelike.

Suppose that $\Omega \subset [0, T] \times X$ is bounded by two spacelike surfaces Σ_1, Σ_2, and $\Sigma_3 := \Omega \cap [0, T] \times \partial X$. We suppose that Ω is swept out by spacelike surfaces. Specifically, we assume that there is a smooth function h on a neighborhood of Ω such that $\mathrm{grad}(h)$ (defined by $\langle \mathrm{grad}(h), Y \rangle = Yh$ for $Y \in TX$) is timelike, and we set $\Omega(u) := \overline{\Omega} \cap \{h \leqslant u\}$, $\Sigma_2(u) := \Omega \cap \{h = u\}$. Suppose Ω is swept out by $\Sigma_2(u)$, $0 \leqslant u \leqslant u_1$ with $\Sigma_2 = \Sigma_2(u_1)$ and $\Omega(u) = \emptyset$ if $u < 0$. Also set $\Sigma_1^b(u) = \Sigma_1 \cap \{h \leqslant u\}$, $\Sigma_3^b(u) = \Sigma_3 \cap \{h \leqslant u\}$ (cf. Fig. D.1).

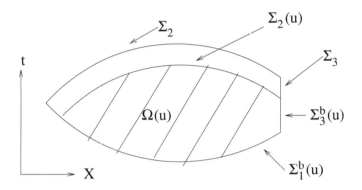

Figure D.1.

Denote by $\nu_2 = \mathrm{grad}h/|\mathrm{grad}h|$, then ν_2 is a timelike vector field. Set

$$E(\tfrac{\partial}{\partial t}\omega, \nabla^E \omega) = (|\tfrac{\partial}{\partial t}\omega|^2 + |\nabla^E \omega|^2)N_t - 2\left\langle \tfrac{\partial}{\partial t}\omega, \nabla_{\nu_2}^E \omega \right\rangle |N_x|, \qquad (\text{D.2.4})$$

on $\partial\Omega(u)$, then $\mathrm{sign}(N_t)E(\omega_t, \nabla^E \omega)$ is positive as $\partial\Omega(u)$ is spacelike.

For $s_1, s_2 \in \mathscr{C}^\infty(X, E)$, let α be the one form on X defined by $\alpha(Y)(x) = \langle \nabla_Y^E s_1, s_2 \rangle(x)$ for any $Y \in T_x X$. By (1.2.6), (1.2.9), (1.2.10) and Stokes' theorem (cf. (1.4.67)), we have

$$\int_X \mathrm{Tr}(\nabla\alpha)dv_X = \int_X d(i_W dv_X) = -\int_{\partial X} \alpha(e_{\mathbf{n}})dv_{\partial X}. \qquad (\text{D.2.5})$$

Thus by (1.3.20), (D.2.5), we get

$$\begin{aligned}
\int_X \langle \Delta^E s_1, s_2 \rangle dv_X &= \int_X \langle \nabla^E s_1, \nabla^E s_2 \rangle dv_X - \int_X \mathrm{Tr}(\nabla\alpha)dv_X \\
&= \int_X \langle \nabla^E s_1, \nabla^E s_2 \rangle dv_X + \int_{\partial X} \langle \nabla_{e_{\mathbf{n}}}^E s_1, s_2 \rangle dv_{\partial X}.
\end{aligned} \qquad (\text{D.2.6})$$

Now from $(\frac{\partial^2}{\partial t^2} + H)\omega = f_2$ on $[0, T] \times X$ with Dirichlet (or Neumann) boundary conditions and the initial conditions $\omega(0, x) = f_0, \frac{\partial}{\partial t}\omega(0, x) = f_1$, by applying (D.2.6) for $\Omega(u)$, we get

$$
\int_{\Omega(u)} \langle \tfrac{\partial}{\partial t}\omega, f_2 - Q\omega \rangle \, dv_X dt = \int_{\Omega(u)} \langle \tfrac{\partial}{\partial t}\omega, (\tfrac{\partial^2}{\partial t^2} + \Delta^E)\omega \rangle dv_X dt
$$

$$
= \frac{1}{2} \int_{\Omega(u)} \tfrac{\partial}{\partial t}(|\tfrac{\partial}{\partial t}\omega|^2 + |\nabla^E\omega|^2) dv_X \, dt + \int_{\partial\Omega(u)} \langle \tfrac{\partial}{\partial t}\omega, \nabla^E_{e_{n,u}}\omega \rangle dv_{\partial\Omega(u)} \quad \text{(D.2.7)}
$$

$$
= \frac{1}{2} \int_{\Sigma_1^b(u) \cup \Sigma_2(u)} E(\tfrac{\partial}{\partial t}\omega, \nabla^E\omega) dS + \int_{\Sigma_3^b(u)} \langle \tfrac{\partial}{\partial t}\omega, \nabla^E_{e_{n,u}}\omega \rangle |N_x| dS,
$$

where dS is the induced surface measure on $\partial\Omega(u) = \Sigma_1^b(u) \cup \Sigma_2(u) \cup \Sigma_3^b(u)$, and $e_{n,u}$ is the inward pointing unit normal at $\partial\Omega(u)$. On $\Sigma_3^b(u)$, the last term of the above equation is zero under the relative or absolute boundary condition (it can also be controlled under other suitable boundary conditions). Thus by using the Cauchy inequality and (D.2.7), we have

$$
\int_{\Sigma_2(u)} E(\tfrac{\partial}{\partial t}\omega, \nabla^E\omega) dS \leqslant \int_{\Sigma_1^b(u)} E(\tfrac{\partial}{\partial t}\omega, \nabla^E\omega) dS
$$
$$
+ C \int_{\Omega(u)} \left(2E(\tfrac{\partial}{\partial t}\omega, \nabla^E\omega) + |\omega|^2 + |f_2|^2 \right) dv_X dt. \tag{D.2.8}
$$

The following argument is quite standard. First, we have

$$
\int_{\Omega(u)} |\omega|^2 dv_X dt \leqslant C \int_{\Sigma_1(u)} |f_0|^2 dS + C \int_{\Omega(u)} E(\tfrac{\partial}{\partial t}\omega, \nabla^E\omega) dv_X dt. \tag{D.2.9}
$$

Now set

$$
E(u) = \int_{\Omega(u)} E(\tfrac{\partial}{\partial t}\omega, \nabla^E\omega) dv_X dt. \tag{D.2.10}
$$

Then by (D.2.4),

$$
\frac{dE}{du} \leqslant C \int_{\Sigma_2(u)} E(\tfrac{\partial}{\partial t}\omega, \nabla^E\omega) dS. \tag{D.2.11}
$$

By (D.2.8), (D.2.9) and (D.2.11), we have the estimate

$$
\frac{dE}{du} \leqslant CE(u) + F(u), \tag{D.2.12}
$$

where

$$
F(u) = C \int_{\Sigma_1} (E(\tfrac{\partial}{\partial t}\omega, \nabla^E\omega) + |f_0|^2) dS + C \int_{\Omega(u)} |f_2|^2 dv_X dt. \tag{D.2.13}
$$

(D.2.12) is equivalent to $\frac{d}{du}(e^{-Cu}E(u)) \leqslant e^{-Cu}F(u)$. Since $E(0) = 0$, we get

$$
e^{-Cu}E(u) \leqslant \int_0^u e^{-Cu_2}F(u_2)du_2. \tag{D.2.14}
$$

From (D.2.14), we get the following energy estimate, for $u \in [0, u_1]$,

$$\int_{\Omega(u)} E(\tfrac{\partial}{\partial t}\omega, \nabla^E \omega) dv_X dt \leqslant C u \int_{\Sigma_1} (|f_0|^2 + |\nabla^E f_0|^2 + |f_1|^2) dS$$
$$+ C \int_{\Omega(u)} |f_2|^2 dv_X dt. \qquad (D.2.15)$$

From (D.2.15), we get the uniqueness of the solution of (D.2.1) and the finite propagation speed property (D.2.3).

We turn now to the existence of the solution ω. One way to do it is as follows: Since X is compact, the heat kernel of H with boundary conditions exists, since H with corresponding boundary condition is essentially self-adjoint, thus the eigenfunctions of H span $L^2(X, E)$. Let $\varphi_j(x)$ be the eigenfunctions of H with eigenvalues λ_j. Then a solution of the Cauchy problem in the theorem is

$$\omega(t, x) = \sum_j \left[\langle f_0, \varphi_j \rangle \cos(\lambda_j^{1/2} t) + \langle f_1, \varphi_j \rangle \lambda_j^{-1/2} \sin(\lambda_j^{1/2} t) \right] \varphi_j(x), \qquad (D.2.16)$$

which is equivalent to $\omega(t, x) = \cos(t\sqrt{H}) f_0 + \frac{\sin(t\sqrt{H})}{\sqrt{H}} f_1$. Thus we conclude Theorem D.2.1. $\qquad \square$

Note that from the finite propagation speed and the uniqueness of the problem, it is clear now that the solution ω can be defined on a complete manifold M with boundary. Namely, let $\overline{\Omega}_j \subset M$ be compact subsets with smooth boundary such that $\overline{\Omega}_1 \subset \cdots \subset \overline{\Omega}_j \subset \nearrow M$. Then we can work on each $\overline{\Omega}_j$ to get the result.

From (D.2.2), we get the following important relation between the heat kernel and the wave equation for a non-negative operator H,

$$e^{-t^2 H/2} = \int_{\mathbb{R}} \cos(vt\sqrt{H}) e^{-v^2/2} \frac{dv}{\sqrt{2\pi}}. \qquad (D.2.17)$$

Problems

Problem D.1. Let Q_0 be a formally self-adjoint first order differential operator on E, and H be a generalized Laplacian on E associated to ∇^E. Verify that there exists a Hermitian connection $\widetilde{\nabla}^E$ such that $H + Q_0$ is a generalized Laplacian operator associated to $\widetilde{\nabla}^E$.

Problem D.2. Use the method of the proof of (D.1.6) to show that the heat operator e^{-tH} is unique.

Appendix E

Harmonic Oscillator

This appendix is organized as follows. In Section E.1, we explain in detail the spectrum and heat kernel of the harmonic oscillator on $\mathbb{R}$. In Section E.2, we explain Mehler's formula in the general case.

Some good references for this appendix are [112, §1.5], [238, §8.6], [27, §6].

E.1 Harmonic oscillator on $\mathbb{R}$

We consider the harmonic oscillator

$$H = \Delta + x^2 - 1 \quad \text{on } L^2(\mathbb{R}), \quad \text{with } \Delta = -\frac{\partial^2}{\partial x^2}. \tag{E.1.1}$$

The inner product on $L^2(\mathbb{R})$ is defined by $\langle f, g \rangle = \int_{\mathbb{R}} f \cdot \bar{g} dx$ with corresponding norm $\| \cdot \|_{L^2}$. We will establish the spectrum decomposition of H and Mehler's formula for the heat kernel e^{-tH}.

We define the creation and annihilation operators $\mathbf{a}$ and $\mathbf{a}^+$ by

$$\mathbf{a} = -\tfrac{\partial}{\partial x} + x, \quad \mathbf{a}^+ = \tfrac{\partial}{\partial x} + x. \tag{E.1.2}$$

Theorem E.1.1. *The spectrum of H is given by $\mathrm{Spec}(H) = \{2k \, : \, k \in \mathbb{N}\}$ and a basis of the eigenspace of $2k$ is given by $\mathbf{a}^k e^{-x^2/2}$.*

Proof. From (E.1.2), $\mathbf{a}^+$ is the adjoint of $\mathbf{a}$, and we have

$$[\mathbf{a}, \mathbf{a}^+] = -2. \tag{E.1.3}$$

From (E.1.2) and (E.1.3), we get

$$H = \mathbf{a}^+\mathbf{a} - 2 = \mathbf{a}\mathbf{a}^+, \quad [H, \mathbf{a}] = 2\mathbf{a}, \quad [H, \mathbf{a}^+] = -2\mathbf{a}^+. \tag{E.1.4}$$

Thus if $u_0 \in L^2(\mathbb{R})$ verifies $\mathbf{a}^+ u_0 = 0$ and $\|u_0\|_{L^2} = 1$, then u_0 is an eigenfunction of H with eigenvalue 0. But this means that $\frac{\partial}{\partial x} u_0 = -x u_0$, and we get

$$u_0 = \pi^{-1/4} e^{-x^2/2}. \tag{E.1.5}$$

Certainly, $e^{-x^2/2}$ belongs to $\mathscr{C}^\infty(\mathbb{R}) \cap L^2(\mathbb{R})$. Since $\mathbf{a}\mathbf{a}^+$ is a non-negative operator, we see that the smallest element of $\mathrm{Spec}(H)$ is 0, and the corresponding eigenfunction is $C e^{-x^2/2}$ with $C \in \mathbb{C}$.

Suppose that $u_j \in \mathscr{C}^\infty(\mathbb{R}) \cap L^2(\mathbb{R})$ is an eigenfunction of H with eigenvalue λ_j, i.e.,

$$H u_j = \lambda_j u_j. \tag{E.1.6}$$

By (E.1.4) and (E.1.6), we have

$$H(\mathbf{a} u_j) = (\lambda_j + 2)\mathbf{a} u_j, \quad H(\mathbf{a}^+ u_j) = (\lambda_j - 2)\mathbf{a}^+ u_j. \tag{E.1.7}$$

Thus $\mathbf{a} u_j, \mathbf{a}^+ u_j$ are eigenfunctions of H, provided $\mathbf{a} u_j, \mathbf{a}^+ u_j \in L^2(\mathbb{R})$. Especially $\mathbf{a}^k u_0$ is an eigenfunction of H with eigenvalue $2k$.

To complete the proof, we need to know that the eigenfunctions $\mathbf{a}^k u_0$ form a complete set in $L^2(\mathbb{R})$. This will be proved in Lemma E.1.3. □

We define the Hermite polynomials $H_k(x)$ by

$$H_k(x) e^{-x^2/2} = \mathbf{a}^k e^{-x^2/2}. \tag{E.1.8}$$

Thus $H_0 = 1$, $H_1(x) = 2x$. From (E.1.2), (E.1.8), we obtain

$$H_{k+1}(x) = (2x - \tfrac{\partial}{\partial x}) H_k(x). \tag{E.1.9}$$

Set

$$c_{k,j} = \frac{k!}{j!\, 2^j\, (k-2j)!}. \tag{E.1.10}$$

Then we show by recurrence that

$$H_k(x) = (-1)^k e^{x^2} (\tfrac{\partial}{\partial x})^k e^{-x^2} = \sum_{j=0}^{\lfloor k/2 \rfloor} (-1)^j c_{k,j} 2^{k-j} x^{k-2j}, \tag{E.1.11}$$

where $\lfloor k/2 \rfloor$ is the integer part of $k/2$.

Set

$$\varphi_k = [\pi^{1/2} 2^k (k!)]^{-1/2} H_k(x) e^{-x^2/2}. \tag{E.1.12}$$

Then φ_k is an eigenfunction of H with eigenvalue $2k$ and $\|\varphi_k\|_{L^2} = 1$. In fact, by (E.1.4), (E.1.8), if $\|\varphi_k\|_{L^2} = 1$, then $\|\varphi_{k+1}\|_{L^2}^2 = (2k+2)^{-1} \|\mathbf{a}\varphi_k\|_{L^2}^2 = (2k+2)^{-1} \langle \mathbf{a}^+ \mathbf{a}\varphi_k, \varphi_k \rangle = 1$.

Lemma E.1.2. *The inversion of* (E.1.11) *is*

$$x^k = \sum_{j=0}^{\lfloor k/2 \rfloor} c_{k,j} 2^{-(k-2j)/2} H_{k-2j}\left(\frac{x}{\sqrt{2}}\right). \tag{E.1.13}$$

Proof. We prove (E.1.13) by induction. At first (E.1.13) is true for $k = 0$. Suppose that (E.1.13) is true for $k \leqslant l$. Relation (E.1.9) yields

$$
\begin{aligned}
x^{l+1} &= \sum_{j=0}^{\lfloor l/2 \rfloor} c_{l,j} 2^{-(l-2j)/2} x H_{l-2j}\left(\frac{x}{\sqrt{2}}\right) \\
&= \sum_{j=0}^{\lfloor l/2 \rfloor} c_{l,j} 2^{-(l-2j+1)/2} \left(H_{l+1-2j} + \left(\tfrac{\partial}{\partial x} H_{l-2j}\right)\right)\left(\frac{x}{\sqrt{2}}\right).
\end{aligned}
\tag{E.1.14}
$$

By the induction hypothesis, the terms containing a derivative sum to $l x^{l-1}$. Using again the induction hypothesis for x^{l-1}, we infer from (E.1.14) that

$$x^{l+1} = \sum_{j=0}^{\lfloor l/2 \rfloor} c_{l,j} 2^{-(l-2j+1)/2} H_{l+1-2j} + \sum_{j=1}^{\lfloor (l+1)/2 \rfloor} l c_{l-1,j-1} 2^{-(l-2j+1)/2} H_{l+1-2j}\left(\frac{x}{\sqrt{2}}\right). \tag{E.1.15}$$

Note that $c_{l+1,j} = c_{l,j} + l c_{l-1,j-1}$, for $j \leqslant \lfloor l/2 \rfloor$, and $c_{l+1,j} = l c_{l-1,j-1}$ if l is odd and $j = (l+1)/2$. Thus we get (E.1.13) for $k = l+1$ from (E.1.15). $\qquad\square$

Lemma E.1.3. *The normalized Hermite functions* $\varphi_k(x)$ *introduced in* (E.1.12) *form a complete orthonormal set in* $L^2(\mathbb{R})$.

Proof. From (E.1.12), (E.1.13), we get

$$(\sqrt{2}x)^k \varphi_0 = \sum_{j=0}^{\lfloor k/2 \rfloor} c_{k,j} ((k-2j)!)^{1/2} \varphi_{k-2j}(x). \tag{E.1.16}$$

Thus

$$\|x^k \varphi_0\|_{L^2}^2 = \sum_{j=0}^{\lfloor k/2 \rfloor} c_{k,j}^2 2^{-k} (k-2j)! = \sum_{j=0}^{\lfloor k/2 \rfloor} \frac{(k!)^2}{(k-2j)!(j!)^2 2^{2j+k}} \leqslant \frac{(k!)^2}{\lfloor k/3 \rfloor! \, 2^{k-1}}. \tag{E.1.17}$$

In view of (E.1.17), we see that for any $\nu \in \mathbb{R}$,

$$e^{i\nu x} \varphi_0 = \sum_{k=0}^{\infty} (i\nu x)^k (k!)^{-1} \varphi_0 \tag{E.1.18}$$

converges in $L^2(\mathbb{R})$. Thus $e^{i\nu x} \varphi_0$ lies in the span of the $\{\varphi_k\}_k$. But $\{e^{i\nu x} \varphi_0\}_{\nu \in \mathbb{R}}$ span $L^2(\mathbb{R})$. In fact, if $v \in L^2(\mathbb{R})$ is orthogonal to $e^{i\nu x} \varphi_0$, then the Fourier transform $(v\varphi_0)\tilde{\ }(\nu)$ of $v\varphi_0$ is zero on $\mathbb{R}$. Thus $\|v\varphi_0\|_{L^2} = \|(v\varphi_0)\tilde{\ }\|_{L^2} = 0$, by Plancherel's theorem. This means that $v = 0$. The proof of Lemma E.1.3 is complete. $\square$

Let $e^{-tH}(x, x')$ be the kernel of the heat operator e^{-tH} with respect to dx'. Then we have

$$e^{-tH}(x, y) = \sum_{k=0}^{\infty} e^{-2kt}\varphi_k(x)\varphi_k(y). \qquad (E.1.19)$$

Recall that

$$\sinh(t) = \frac{1}{2}(e^t - e^{-t}), \quad \cosh(t) = \frac{1}{2}(e^t + e^{-t}), \quad \tanh(t) = \frac{\sinh(t)}{\cosh(t)}. \qquad (E.1.20)$$

Now we prove Mehler's formula.

Theorem E.1.4. *For $t > 0$, we have*

$$e^{-tH}(x, y) = (\pi(1 - e^{-4t}))^{-1/2} \exp\left\{-\frac{x^2 + y^2}{2\tanh(2t)} + \frac{xy}{\sinh(2t)}\right\}. \qquad (E.1.21)$$

Proof. We can prove (E.1.21) from (E.1.19) by Fourier transformation, but here we take a different route. We denote by $P_t(x, y)$ the right side of (E.1.21) and for $u \in L^2(\mathbb{R})$, set

$$v(t, x) = \int_{\mathbb{R}} P_t(x, y)u(y)dy. \qquad (E.1.22)$$

Since $P_t(x, y)$ is rapidly decreasing for $|x| + |y| \to \infty$, and $t > 0$, we deduce that $v \in \mathscr{C}^{\infty}(]0, \infty[, \mathcal{S}(\mathbb{R}))$, provided $u \in L^2(\mathbb{R})$. Here $\mathcal{S}(\mathbb{R})$ denotes the Schwartz space of rapidly decreasing functions. We verify by direct computation that

$$\frac{\partial v}{\partial t} = -Hv. \qquad (E.1.23)$$

We claim now that for each $\lambda > 0$,

$$v(t + \lambda, \cdot) - e^{-tH}v(\lambda, \cdot) = 0. \qquad (E.1.24)$$

In fact, if we denote the left-hand side of (E.1.24) by $w(t, \cdot)$, then we have $w(0, \cdot) = 0$, $\frac{\partial w}{\partial t} \in \mathscr{C}^{\infty}(]0, \infty[, L^2(\mathbb{R}))$, and by (E.1.4),

$$\frac{\partial}{\partial t}\|w(t, \cdot)\|_{L^2}^2 = -2\langle Hw, w\rangle \leqslant 0. \qquad (E.1.25)$$

Thus $w(t, \cdot) = 0$ for any $t > 0$.

Finally, from (E.1.21), we know that $v(t, \cdot) \to u$ in $L^2(\mathbb{R})$ as $t \to 0$. Thus letting $\lambda \to 0$ in (E.1.24), we get (E.1.21). $\square$

E.2   Harmonic oscillator on vector spaces

Let V be a real vector space with complex structure J and $\dim_{\mathbb{R}} V = 2n$. We have the decomposition $V \otimes_{\mathbb{R}} \mathbb{C} = V^{(1,0)} \oplus V^{(0,1)}$ where $V^{(1,0)}$ and $V^{(0,1)}$ are the eigenspaces of J corresponding to the eigenvalues $\sqrt{-1}$ and $-\sqrt{-1}$ respectively.

Let g^V be a Euclidean metric on V which is compatible with J. Then we consider $V_h = (V, J)$ as a complex vector space with Hermitian metric $h^V(\cdot, \cdot) = g^V(\cdot, \cdot) - \sqrt{-1} g^V(J\cdot, \cdot)$; moreover, $Y \to \frac{1}{2}(Y - \sqrt{-1}JY)$ induces a natural identification of V_h with $V^{(1,0)}$.

Let A be an invertible self-adjoint complex endomorphism of the complex vector space $V^{(1,0)}$. We extend A to $V \otimes_{\mathbb{R}} \mathbb{C}$ by defining $A\bar{v} = -\overline{Av}$ for any $v \in V^{(1,0)}$, then $\sqrt{-1}A$ induces an anti-symmetric endomorphism on the real vector space (V, g^V). We will denote by $\det$ the determinant on $V^{(1,0)}$.

Let $\{e_i\}_{i=1}^{2n}$ be an orthonormal basis of (V, g^V). For $Y \in V$, let ∇_Y be the ordinary differential operator on V in the direction Y. Consider the differential operators on V,

$$
\begin{aligned}
\mathcal{H} &= -\sum_i (\nabla_{e_i})^2 + \frac{1}{4}\langle A^2 Z, Z\rangle - \mathrm{Tr}_{V^{(1,0)}}[A], \quad Z \in V; \\
\mathscr{H} &= -\sum_i \left(\nabla_{e_i} + \frac{1}{2}\langle AZ, e_i\rangle\right)^2 - \mathrm{Tr}_{V^{(1,0)}}[A] = \mathcal{H} - \nabla_{AZ}, \quad Z \in V.
\end{aligned}
\tag{E.2.1}
$$

Let $e^{-t\mathcal{H}}(Z, Z')$ and $e^{-t\mathscr{H}}(Z, Z')$, $(Z, Z' \in V)$, be the smooth kernels of the heat operators $e^{-t\mathcal{H}}$ and $e^{-t\mathscr{H}}$ associated to the volume form dZ. Then by Theorem E.1.4, we know that

$$
\begin{aligned}
e^{-t\mathcal{H}}(Z, Z') = (2\pi)^{-n}\det\left(\frac{A}{1 - \exp(-2tA)}\right)\exp\Big\{ &-\frac{1}{2}\Big\langle \frac{A/2}{\tanh(tA)}Z, Z\Big\rangle \\
&-\frac{1}{2}\Big\langle \frac{A/2}{\tanh(tA)}Z', Z'\Big\rangle + \Big\langle \frac{A/2}{\sinh(tA)}Z, Z'\Big\rangle\Big\}.
\end{aligned}
\tag{E.2.2}
$$

Since $\sqrt{-1}A \in \mathrm{End}(V)$ is anti-symmetric, thus

$$
[\mathcal{H}, \nabla_{AZ}] = 0.
\tag{E.2.3}
$$

From (E.2.2) and (E.2.3), we get that

$$
\begin{aligned}
e^{-t\mathscr{H}}(Z, Z') &= e^{-t\mathcal{H}}(e^{tA}Z, Z') \\
&= (2\pi)^{-n}\det\left(\frac{A}{1 - \exp(-2tA)}\right)\exp\Big\{ -\frac{1}{2}\Big\langle \frac{A/2}{\tanh(tA)}Z, Z\Big\rangle \\
&\qquad -\frac{1}{2}\Big\langle \frac{A/2}{\tanh(tA)}Z', Z'\Big\rangle + \Big\langle \frac{A/2}{\sinh(tA)}e^{tA}Z, Z'\Big\rangle\Big\}.
\end{aligned}
\tag{E.2.4}
$$

If $A = 0$, then $\mathscr{H}$ is the Laplace operator $\Delta = -\sum_i (\nabla_{e_i})^2$, hence we have classically

$$e^{-t\Delta}(Z, Z') = (4\pi t)^{-n} \exp\left(-\frac{1}{4t}|Z - Z'|^2\right).$$
(E.2.5)

Bibliography

[1] A. Aeppli, *Modifikation von reellen und komplexen Mannigfaltigkeiten*, Comment. Math. Helv. **31** (1957), 219–301.

[2] A. Andreotti, *Théorèmes de dépendance algébrique sur les espaces complexes pseudo-concaves*, Bull. Soc. Math. France **91** (1963), 1–38.

[3] A. Andreotti and T. Frankel, *The Lefschetz theorem on hyperplane sections*, Ann. of Math. **69** (1959), 713–717.

[4] A. Andreotti and H. Grauert, *Théorème de finitude pour la cohomologie des espaces complexes*, Bull. Soc. Math. France **90** (1962), 193–259.

[5] A. Andreotti and Y.T. Siu, *Projective embeddings of pseudoconcave spaces*, Ann. Sc. Norm. Sup. Pisa **24** (1970), 231–278.

[6] A. Andreotti and G. Tomassini, *Some remarks on pseudoconcave manifolds*, Essays on Topology and Related Topics dedicated to G. de Rham (A. Haefinger and R. Narasimhan, eds.), Springer-Verlag, 1970, pp. 85–104.

[7] A. Andreotti and E. Vesentini, *Carleman estimates for the Laplace-Beltrami equation on complex manifolds*, Inst. Hautes Etudes Sci. Publ. Math. **25** (1965), 81–130; Erratum: **27** (1965), 153–155.

[8] F. Angelini, *An algebraic version of Demailly's asymptotic Morse inequalities*, Proc. Amer. Math. Soc. **124** (1996), no. 11, 3265–3269.

[9] M.F. Atiyah, *Elliptic operators, discrete groups and von Neumann algebras*, Astérisque **32–33** (1976), 43–72.

[10] M.F. Atiyah, R. Bott, and V.K. Patodi, *On the heat equation and the index theorem*, Invent. Math. **19** (1973), 279–330, Erratum **28** (1975), 277–280.

[11] M.F. Atiyah and I.M. Singer, *Index of elliptic operators. III*, Ann. of Math. **87** (1968), 546–604.

[12] W. Baily, *The decomposition theorem for V-manifolds*, Amer. J. Math. **78** (1956), 862–888.

[13] ——, *On the imbedding of V-manifolds in projective space*, Amer. J. Math. **79** (1957), 403–430.

[14] ——, *Satake's compactification of V_n*, Amer. J. Math. **80** (1958), 348–364.

[15] N. Berline, E. Getzler, and M. Vergne, *Heat kernels and Dirac operators*, Grundl. Math. Wiss. Band 298, Springer-Verlag, Berlin, 1992.

[16] R. Berman, *Bergman kernels and local holomorphic Morse inequalities*, Math. Z. **248** (2004), no. 2, 325–344.

[17] _____, *Holomorphic Morse inequalities on manifolds with boundary*, Ann. Inst. Fourier (Grenoble) **55** (2005), no. 4, 1055–1103.

[18] R. Berman, B. Berndtsson, and J. Sjöstrand, *Asymptotics of Bergman kernels*, Preprint available at arXiv:math.CV/0506367, 2005.

[19] R. Berman and J. Sjöstrand, *Asymptotics for Bergman-Hodge kernels for high powers of complex line bundles*, math.CV/0511158 (2005).

[20] B. Berndtsson, *An eigenvalue estimate for the $\bar{\partial}$-Laplacian*, J. Differential Geom. **60** (2002), no. 2, 295–313.

[21] _____, *Positivity of direct image bundles and convexity on the space of Kähler metrics*, (2006), math.CV/0608385.

[22] E. Bierstone and P. Milman, *Canonical desingularization in characteristic zero by blowing up the maximum strata of a local invariant*, Invent. Math. **128** (1997), no. 2, 207–302.

[23] O. Biquard, *Métriques kählériennes à courbure scalaire constante*, Astérisque (2006), no. 307, Exp. No. 938, Séminaire Bourbaki. Vol. 2004/2005.

[24] J.-M. Bismut, *The Witten complex and degenerate Morse inequalities*, J. Differential Geom. **23** (1986), 207–240.

[25] _____, *Demailly's asymptotic inequalities: a heat equation proof*, J. Funct. Anal. **72** (1987), 263–278.

[26] _____, *A local index theorem for non-Kähler manifolds*, Math. Ann. **284** (1989), no. 4, 681–699.

[27] _____, *Koszul complexes, harmonic oscillators, and the Todd class*, J. Amer. Math. Soc. **3** (1990), no. 1, 159–256, With an appendix by the author and C. Soulé.

[28] _____, *From Quillen metrics to Reidemeister metrics: some aspects of the Ray-Singer analytic torsion*, Topological methods in modern mathematics (Stony Brook, NY, 1991), Publish or Perish, Houston, TX, 1993, pp. 273–324.

[29] _____, *Equivariant immersions and Quillen metrics*, J. Differential Geom. **41** (1995), no. 1, 53–157.

[30] _____, *Local index theory and higher analytic torsion*, Proceedings of the International Congress of Mathematicians, Vol. I (Berlin, 1998), no. Extra Vol. I, 1998, pp. 143–162 (electronic).

[31] _____, *Local index theory, eta invariants and holomorphic torsion: a survey*, Surveys in differential geometry, Vol. III (Cambridge, MA, 1996), Int. Press, Boston, MA, 1998, pp. 1–76.

[32] J.-M. Bismut, H. Gillet, and C. Soulé, *Analytic torsion and holomorphic determinant bundles. III. Quillen metrics on holomorphic determinants*, Comm. Math. Phys. **115** (1988), no. 2, 301–351.

[33] J.-M. Bismut and G. Lebeau, *Complex immersions and Quillen metrics*, Inst. Hautes Études Sci. Publ. Math. (1991), no. 74, ii+298 pp. (1992).

[34] J.-M. Bismut and X. Ma, *Holomorphic immersions and equivariant torsion forms*, J. Reine Angew. Math. **575** (2004), 189–235.

[35] J.-M. Bismut and E. Vasserot, *The asymptotics of the Ray–Singer analytic torsion associated with high powers of a positive line bundle*, Comm. Math. Phys. **125** (1989), 355–367.

[36] ———, *The asymptotics of the Ray-Singer analytic torsion of the symmetric powers of a positive vector bundle*, Ann. Inst. Fourier (Grenoble) **40** (1990), no. 4, 835–848.

[37] P. Bleher, B. Shiffman, and S. Zelditch, *Poincaré-Lelong approach to universality and scaling of correlations between zeros*, Comm. Math. Phys. **208** (2000), no. 3, 771–785.

[38] ———, *Universality and scaling of correlations between zeros on complex manifolds*, Invent. Math. **142** (2000), no. 2, 351–395.

[39] E. Bombieri, *Algebraic values of meromorphic maps*, Invent. Math. **10** (1970), 267–287.

[40] L. Bonavero, *Inégalités de Morse holomorphes singulières*, Ph.D. thesis, 1995, Institut Fourier, Grenoble.

[41] ———, *Inégalités de Morse holomorphes singulières*, J. Geom. Anal. **8** (1998), 409–425; announced in C. R. Acad. Sci. Paris **317** (1993),1163–1166.

[42] M. Bordemann, E. Meinrenken, and M. Schlichenmaier, *Toeplitz quantization of Kähler manifolds and gl(N), N $\longrightarrow \infty$ limits*, Comm. Math. Phys. **165** (1994), 281–296.

[43] D. Borthwick and A. Uribe, *Almost complex structures and geometric quantization*, Math. Res. Lett. **3** (1996), 845–861. Erratum: Math. Res. Lett. **5** (1998), 211–212.

[44] ———, *Nearly Kählerian Embeddings of Symplectic Manifolds*, Asian J. Math. **4** (2000), no. 3, 599–620.

[45] ———, *The semiclassical structure of low-energy states in the presence of a magnetic field*, Trans. Amer. Math. Soc. **359** (2007), 1875–1888.

[46] R. Bott, *On a theorem of Lefschetz*, Mich. Math. J. **6** (1959), 211–216.

[47] Th. Bouche, *Inegalités de Morse pour la d″-cohomologie sur une variété non–compacte*, Ann. Sci. Ecole Norm.Sup. **22** (1989), 501–513.

[48] ———, *Convergence de la métrique de Fubini-Study d'un fibré linéare positif*, Ann. Inst. Fourier (Grenoble) **40** (1990), 117–130.

[49] _____, *Asymptotic results for Hermitian line bundles over complex manifolds: the heat kernel approach*, Higher-dimensional complex varieties (Trento, 1994), de Gruyter, Berlin, 1996, pp. 67–81.

[50] S. Boucksom, *On the volume of a line bundle*, Internat. J. Math. **13** (2002), no. 10, 1043–1063.

[51] L. Boutet de Monvel, *Intégration des équations de Cauchy-Riemann induites formelles*, Séminaire Goulaouic-Lions-Schwartz 1974–1975; Équations aux derivées partielles linéaires et non linéaires, Centre Math., École Polytech., Paris, 1975, pp. Exp. No. 9, 14.

[52] L. Boutet de Monvel and V. Guillemin, *The spectral theory of Toeplitz operators*, Annals of Math. Studies, no. 99, Princeton Univ. Press, Princeton, NJ, 1981.

[53] L. Boutet de Monvel and J. Sjöstrand, *Sur la singularité des noyaux de Bergman et de Szegö*, Journées: Équations aux Dérivées Partielles de Rennes (1975), Soc. Math. France, Paris, 1976, pp. 123–164. Astérisque, No. 34–35.

[54] M. Braverman, *Vanishing theorems on covering manifolds*, Contemp. Math. **213** (1999), 1–23.

[55] J. Brüning and X. Ma, *An anomaly formula for Ray-Singer metrics on manifolds with boundary*, Geom. Funct. Anal. **16** (2006), 767–837, announced in C. R. Math. Acad. Sci. Paris **335** (2002), no. 7, 603–608.

[56] C. Bănică and O. Stănăşilă, *Algebraic methods in the global theory of complex spaces*, Wiley, New York, 1976.

[57] D.M. Burns, *Global behavior of some tangential Cauchy-Riemann equations*, Partial differential equations and geometry (Proc. Conf., Park City, Utah, 1977), Lecture Notes in Pure and Appl. Math., vol. 48, Dekker, New York, 1979, pp. 51–56.

[58] F. Campana, *Remarques sur le revêtement universel des variétés Kählériennes compactes*, Bull. Soc. Math. France **122** (1994), 255–284.

[59] J. Carlson and P. Griffiths, *A defect relation for equidimensional holomorphic mappings between algebraic varieties*, Ann. of Math. **95** (1972), 557–584.

[60] H. Cartan, *Séminaires de H. Cartan*, 1953–1954.

[61] _____, *Quotient d'un espace analytique par un groupe d'automorphismes*, Algebraic geometry and topology., Princeton University Press, Princeton, N. J., 1957, A symposium in honor of S. Lefschetz,, pp. 90–102.

[62] D. Catlin, *The Bergman kernel and a theorem of Tian*, Analysis and geometry in several complex variables (Katata, 1997), Trends Math., Birkhäuser Boston, Boston, MA, 1999, pp. 1–23.

[63] L. Charles, *Berezin-Toeplitz operators, a semi-classical approach*, Comm. Math. Phys. **239** (2003), 1–28.

[64] _____, *Toeplitz operators and hamiltonian torus action*, J. Funct. Anal. **236** (2006), 299–350.

[65] J. Chazarain and A. Piriou, *Introduction à la théorie des équations aux dérivées partielles*, Gautier–Villars, Paris, 1981.

[66] S.S. Chern and J. Simons, *Characteristic forms and geometric invariants*, Ann. of Math. (2) **99** (1974), 48–69.

[67] M. Colţoiu, *Complete locally pluripolar sets*, J. Reine Angew. Math. **412** (1990), 108–112.

[68] M. Colţoiu and M. Tibăr, *Steinness of the universal covering of the complement of a 2-dimensional complex singularity*, Math. Ann. **326** (2003), no. 1, 95–104.

[69] X. Dai, K. Liu, and X. Ma, *On the asymptotic expansion of Bergman kernel*, J. Differential Geom. **72** (2006), no. 1, 1–41; announced in C. R. Math. Acad. Sci. Paris **339** (2004), no. 3, 193–198.

[70] E.B. Davies, *Spectral Theory and Differential Operators*, Cambridge Stud. Adv. Math., vol. 42, Cambridge Univ. Press, 1995.

[71] J.-P. Demailly, *Estimations L^2 pour l'opérateur $\bar{\partial}$ d'un fibré holomorphe semipositif au-dessus d'une variété kählérienne complète*, Ann. Sci. École Norm. Sup. **15** (1982), 457–511.

[72] _____, *Champs magnétiques et inegalités de Morse pour la d''-cohomologie*, Ann. Inst. Fourier (Grenoble) **35** (1985), 189–229.

[73] _____, *Sur l'identité de Bochner-Kodaira-Nakano en géométrie hermitienne*, Lecture Notes in Math., vol. 1198, pp. 88–97, Springer Verlag, 1985.

[74] _____, *Holomorphic Morse inequalities*, Several complex variables and complex geometry, Part 2 (Santa Cruz, CA, 1989), Proc. Sympos. Pure Math., vol. 52, Amer. Math. Soc., Providence, RI, 1991, pp. 93–114.

[75] _____, *Regularization of closed positive currents and Intersection Theory*, J. Alg. Geom. **1** (1992), 361–409.

[76] _____, *Singular hermitian metrics on positive line bundles*, Proc. Conf. Complex algebraic varieties (Bayreuth, April 2–6,1990), vol. 1507, Springer–Verlag, Berlin, 1992.

[77] _____, *A numerical criterion for very ample line bundles*, J. Differential Geom. **37** (1993), no. 2, 323–374.

[78] _____, *L^2 vanishing theorems for positive line bundles and adjunction theory*, Transcendental methods in algebraic geometry (Cetraro, 1994), Lecture Notes in Math., vol. 1646, Springer, Berlin, 1996, pp. 1–97.

[79] _____, *Complex analytic and differential geometry*, 2001, published online at www-fourier.ujf-grenoble.fr/~demailly/lectures.html.

[80] J.-P. Demailly, L. Ein, and R. Lazarsfeld, *A subadditivity property of multiplier ideals*, Michigan Math. J. **48** (2000), 137–156.

[81] J.-P. Demailly and M. Paun, *Numerical characterization of the Kähler cone of a compact Kähler manifold*, Ann. of Math. (2) **159** (2004), no. 3, 1247–1274.

[82] J.-P. Demailly, Th. Peternell, and M. Schneider, *Compact complex manifolds with numerically effective tangent bundles*, J. Algebraic Geometry **3** (1994), 295–345.

[83] P. Dingoyan, *Monge-Ampère currents over pseudoconcave spaces.*, Math. Ann. **320** (2001), no. 2, 211–238.

[84] T.-C. Dinh and N. Sibony, *Distribution des valeurs de transformations méromorphes et applications*, Comment. Math. Helv. **81** (2006), no. 1, 221–258.

[85] M. do Carmo, *Riemannian geometry*, Mathematics: Theory & Applications, Birkhäuser Boston Inc., Boston, MA, 1992.

[86] S.K. Donaldson, *Infinite determinants, stable bundles and curvature*, Duke Math. J. **54** (1987), no. 1, 231–247.

[87] ———, *Symplectic submanifolds and almost complex geometry*, J. Differential Geom. **44** (1996), 666–705.

[88] ———, *Remarks on gauge theory, complex geometry and 4-manifold topology*, Fields Medallists' lectures, World Sci. Ser. 20th Century Math., vol. 5, World Sci. Publishing, River Edge, NJ, 1997, pp. 384–403.

[89] ———, *Planck's constant in complex and almost-complex geometry*, XIIIth International Congress on Mathematical Physics (London, 2000), Int. Press, Boston, MA, 2001, pp. 63–72.

[90] ———, *Scalar curvature and projective embeddings*, J. Differential Geom. **59** (2001), no. 2, 479–522.

[91] ———, *Scalar curvature and stability of toric varieties*, J. Differential Geom. **62** (2002), no. 2, 289–349.

[92] ———, *Conjectures in Kähler geometry*, Strings and geometry, Clay Math. Proc., vol. 3, Amer. Math. Soc., Providence, RI, 2004, pp. 71–78.

[93] ———, *Lower bounds on the Calabi functional*, J. Differential Geom. **70** (2005), no. 3, 453–472.

[94] ———, *Scalar curvature and projective embeddings. II*, Q. J. Math. **56** (2005), no. 3, 345–356.

[95] ———, *Some numerical results in complex differential geometry*, (2006), math.DG/0512625.

[96] H. Donnelly, *Spectral theory for tensor products of Hermitian holomorphic line bundles*, Math. Z. **245** (2003), no. 1, 31–35.

[97] M. Douglas, B. Shiffman, and S. Zelditch, *Critical points and supersymmetric vacua. I*, Comm. Math. Phys. **252** (2004), no. 1-3, 325–358.

[98] ———, *Critical points and supersymmetric vacua. II. Asymptotics and extremal metrics*, J. Differential Geom. **72** (2006), no. 3, 381–427.

[99] ———, *Critical points and supersymmetric vacua. III. String/M models*, Comm. Math. Phys. **265** (2006), no. 3, 617–671.

[100] P. Eberlein, *Lattices in spaces of nonpositive curvature*, Ann. of Math. (2) **111** (1980), no. 3, 435–476.

[101] P. Eberlein and B. O'Neill, *Visibility manifolds*, Pacific J. Math. **46** (1973), 45–109.

[102] M. Engliš, *A Forelli-Rudin construction and asymptotics of weighted Bergman kernels*, J. Funct. Anal. **177** (2000), no. 2, 257–281.

[103] _____, *Weighted Bergman kernels and quantization*, Comm. Math. Phys. **227** (2002), no. 2, 211–241.

[104] C.L. Epstein and G.M. Henkin, *Can a good manifold come to a bad end?*, Tr. Mat. Inst. Steklova **235** (2001), no. Anal. i Geom. Vopr. Kompleks. Analiza, 71–93.

[105] C. Fefferman, *The Bergman kernel and biholomorphic mappings of pseudoconvex domains*, Invent. Math. **26** (1974), 1–65.

[106] W. Feller, *An introduction to probability theory and its applications. Vol. II.*, Second edition, John Wiley & Sons Inc., New York, 1971.

[107] G. Fischer, *Complex analytic geometry*, Lect. Notes, vol. 538, Springer-Verlag, 1976.

[108] G.B. Folland and J.J. Kohn, *The Neumann problem for the Cauchy-Riemann complex*, Princeton University Press, Princeton, N.J., 1972, Annals of Mathematics Studies, No. 75.

[109] T. Fujita, *Approximating Zariski decomposition of big line bundles*, Kodai Math. J. **17** (1994), 1–3.

[110] W. Fulton and R. Lazarsfeld, *Connectivity and its applications in algebraic geometry*, Algebraic geometry (Chicago, Ill., 1980), Lecture Notes in Math., vol. 862, Springer, Berlin, 1981, pp. 26–92.

[111] P. Gauduchon, *Calabi's extremal Kähler metrics: an elementary introduction*, 2005, book in preparation.

[112] G.J. Glimm and A. Jaffe, *Quantum Physics*, Springer-Verlag, 1987.

[113] C. Grant and P. Milman, *Metrics for singular analytic spaces*, Pacific J. Math. **168** (1995), no. 1, 61–156.

[114] C. Grant Melles and P. Milman, *Classical Poincaré metric pulled back off singularities using a Chow-type theorem and desingularization*, Ann. Fac. Sci. Toulouse Math. (6) **15** (2006), no. 4, 689–771.

[115] H. Grauert, *On Levi's problem and the imbedding of real analytic manifolds*, Ann. Math. **68** (1958), 460–472.

[116] _____, *Über Modifikationen und exzeptionelle analytische Mengen*, Math. Ann. **146** (1962), 331–368.

[117] _____, *Theory of q-convexity and q-concavity*, Several Complex Variables VII (H. Grauert, Th. Peternell, and R. Remmert, eds.), Encyclopedia of mathematical sciences, vol. 74, Springer Verlag, 1994.

[118] H. Grauert and O. Riemenschneider, *Verschwindungssätze für analytische Kohomologiegruppen auf komplexen Räumen*, Invent. Math. **11** (1970), 263–292.

[119] P. Griffiths, *The extension problem in complex analysis; embedding with positive normal bundle*, Amer. J. Math. **88** (1966), 366–446.

[120] P. Griffiths and J. Harris, *Principles of Algebraic Geometry*, John Wiley and Sons, New York, 1978.

[121] A. Grothendieck, *Cohomologie locale des faisceaux cohérents et théorèmes de Lefschetz locaux et globaux*, North-Holland, Amsterdam, 1968.

[122] V. Guillemin and A. Uribe, *The Laplace operator on the n-th tensor power of a line bundle: eigenvalues which are bounded uniformly in n*, Asymptotic Anal. **1** (1988), 105–113.

[123] R. Hartshorne, *Ample subvarieties of algebraic varieties*, Lecture Notes in Math., vol. 156, Springer-Verlag, Berlin, 1970.

[124] _____, *Algebraic Geometry*, Springer-Verlag, Berlin, 1977.

[125] R. Harvey and B. Lawson, *On boundaries of complex varieties I*, Ann. of Math.. **102** (1975), 233–290.

[126] G.M. Henkin and J. Leiterer, *Andreotti–Grauert theory by integrals formulas*, Birkhäuser, 1986.

[127] D. Heunemann, *Extension of the complex structure from Stein manifolds with strictly pseudoconvex boundary*, Math. Nachr. **128** (1986), 57–64.

[128] H. Hironaka, *Resolution of singularities of an algebraic variety over a field of characteristic zero*, Ann. of Math. **79** (1964), 109–326.

[129] _____, *Flattening theorem in complex-analytic geometry*, Am. J. Math. **97** (1975), 503–547.

[130] F. Hirzebruch, *Neue topologische Methoden in der algebraischen Geometrie*, Ergebnisse der Mathematik und ihrer Grenzgebiete (N.F.), Heft 9, Springer-Verlag, Berlin, 1956.

[131] L. Hörmander, L^2-*estimates and existence theorem for the $\overline{\partial}$-operator*, Acta Math. **113** (1965), 89–152.

[132] _____, *An introduction to complex analysis in several variables*, 1966, third ed., North-Holland Mathematical Library, vol. 7, North-Holland Publishing Co., Amsterdam, 1990.

[133] D. Huybrechts, *Compact hyper-Kähler manifolds: basic results*, Invent. Math. **135** (1999), no. 1, 63–113; Erratum: Invent. Math. **152** (2003), 209–212.

[134] S. Ji and B. Shiffman, *Properties of compact complex manifolds carrying closed positive currents*, J. Geom. Anal. **3** (1993), no. 1, 37–61.

[135] A.V. Karabegov and M. Schlichenmaier, *Identification of Berezin-Toeplitz deformation quantization*, J. Reine Angew. Math. **540** (2001), 49–76.

[136] T. Kawasaki, *The Riemann-Roch theorem for complex V-manifolds*, Osaka J. Math. **16** (1979), no. 1, 151–159.

[137] _____, *The index of elliptic operators over V-manifolds*, Nagoya Math. J. **84** (1981), 135–157.

[138] J. Keller, *Vortex type equations and canonical metrics*, math.DG/0601485, 2006.

[139] S.L. Kleiman, *Toward a numerical theory of ampleness*, Annals of Math. **84** (1966), 293–344.

[140] S. Kobayashi, *Differential geometry of complex vector bundles*, Publications of the Mathematical Society of Japan, vol. 15, Princeton University Press, Princeton, NJ, 1987, Kanô Memorial Lectures, 5.

[141] K. Kodaira, *On Kähler varieties of restricted type*, Ann. of Math. **60** (1954), 28–48.

[142] K. Köhler and D. Roessler, *A fixed point formula of Lefschetz type in Arakelov geometry I: statement and proof*, Invent. Math. **145** (2001), no. 2, 333–396.

[143] J.J. Kohn, *Harmonic integrals on strongly pseudoconvex manifolds I*, Ann. Math. **78** (1963), 112–148.

[144] _____, *The range of the tangential Cauchy–Riemann operators*, Duke Math. **53** (1986), 525–545.

[145] J.J. Kohn and H. Rossi, *On the extension of holomorphic functions from the boundary of a complex manifold*, Ann. Math. **81** (1965), 451–472.

[146] J. Kollár, *Flips, flops, minimal models*, Surveys in Diff. Geom. **1** (1991), 113–199.

[147] _____, *Shafarevich maps and automorphic forms*, Princeton Univ. Press, Princeton, NJ, 1995.

[148] H.B. Lawson and M.-L. Michelson, *Spin geometry*, Princeton Mathematical Series, vol. 38, Princeton Univ. Press, Princeton, NJ, 1989.

[149] R. Lazarsfeld, *Positivity in algebraic geometry. I, II*, Ergebnisse der Mathematik und ihrer Grenzgebiete. 3. Folge., vol. 48–49, Springer-Verlag, Berlin, 2004.

[150] S. Lefschetz, *L'analysis situs et la géométrie algébrique*, Dunod, Paris, 1924.

[151] L. Lempert, *On three-dimensional Cauchy-Riemann manifolds*, J. Amer. Math. Soc. **5** (1992), no. 4, 923–969.

[152] _____, *Embeddings of three-dimensional Cauchy-Riemann manifolds*, Math. Ann. **300** (1994), no. 1, 1–15.

[153] _____, *Algebraic approximations in analytic geometry*, Invent. Math. **121** (1995), no. 2, 335–353.

[154] N. Leung, *Einstein type metrics and stability on vector bundles*, J. Differential Geom. **45** (1997), no. 3, 514–546.

[155] K. Liu and X. Ma, *A remark on 'some numerical results in complex differential geometry'*, Math. Res. Lett. **14** (2007), 165–171.

[156] Z. Lu, *On the lower order terms of the asymptotic expansion of Tian-Yau-Zelditch*, Amer. J. Math. **122** (2000), no. 2, 235–273.

[157] Z. Lu and G. Tian, *The log term of the Szegő kernel*, Duke Math. J. **125** (2004), no. 2, 351–387.

[158] H. Luo, *Geometric criterion for Gieseker-Mumford stability of polarized manifolds*, J. Differential Geom. **49** (1998), no. 3, 577–599.

[159] X. Ma, *Orbifolds and analytic torsions*, Trans. Amer. Math. Soc. **357** (2005), no. 6, 2205–2233.

[160] X. Ma and G. Marinescu, *The Spin^c Dirac operator on high tensor powers of a line bundle*, Math. Z. **240** (2002), no. 3, 651–664.

[161] _____, *Generalized Bergman kernels on symplectic manifolds*, C. R. Acad. Sci. Paris **339** (2004), no. 7, 493–498, The full version: math.DG/0411559, Adv. in Math.

[162] _____, *Toeplitz operators on symplectic manifolds*, Preprint, 2005.

[163] _____, *The first coefficients of the asymptotic expansion of the Bergman kernel of the spin^c Dirac operator*, Internat. J. Math. **17** (2006), no. 6, 737–759.

[164] X. Ma and W. Zhang, *Bergman kernels and symplectic reduction*, C. R. Math. Acad. Sci. Paris **341** (2005), 297–302, see also *Toeplitz quantization and symplectic reduction*, Nankai Tracts in Mathematics Vol. **10**, World Scientific, 2006, 343–349. The full version: math.DG/0607605.

[165] _____, *Superconnection and family Bergman kernels*, C. R. Math. Acad. Sci. Paris **344** (2007), 41–44.

[166] T. Mabuchi, *An energy-theoretic approach to the Hitchin-Kobayashi correspondence for manifolds. I*, Invent. Math. **159** (2005), no. 2, 225–243.

[167] _____, *The Chow-stability and Hilbert-stability in Mumford's geometric invariant theory*, (2006), math.DG/0607590.

[168] _____, *An energy-theoretic approach to the Hitchin-Kobayashi correspondence for manifolds. II*, J. Differential Geom. (2006), to appear.

[169] G. Marinescu, *Morse inequalities for q-positive line bundles over weakly 1-complete manifolds*, C. R. Acad. Sci. Paris Sér. I Math. **315** (1992), no. 8, 895–899.

[170] _____, *Asymptotic Morse Inequalities for Pseudoconcave Manifolds*, Ann. Scuola Norm. Sup. Pisa Cl. Sci. **23** (1996), no. 1, 27–55.

[171] _____, *A criterion for Moishezon spaces with isolated singularities*, Annali di Matematica Pura ed Applicata **185** (2005), no. 1, 1–16.

[172] G. Marinescu and T.-C. Dinh, *On the compactification of hyperconcave ends and the theorems of Siu-Yau and Nadel*, Invent. Math. **164** (2006), 233–248.

[173] G. Marinescu, R. Todor, and I. Chiose, L^2 holomorphic sections of bundles over weakly pseudoconvex coverings, Geom. Dedicata **91** (2002), 23–43.

[174] G. Marinescu and N. Yeganefar, Embeddability of some strongly pseudoconvex CR manifolds, to appear in Trans. Amer. Math. Soc., Preprint available at arXiv:math.CV/0403044, 2004.

[175] V. Mathai and S. Wu, Equivariant holomorphic Morse inequalities. I. Heat kernel proof, J. Differential Geom. **46** (1997), no. 1, 78–98.

[176] J. Milnor, Morse Theory, Ann. Math. Studies, vol. 51, Princeton University Press, 1963.

[177] B. G. Moishezon, On n-dimensional compact complex manifolds having n algebraically independent meromorphic functions. I, II, III, Izv. Akad. Nauk SSSR Ser. Mat. **30** (1966), 133–174, 345–386, 621–656.

[178] N. Mok, Compactification of complete Kähler surfaces of finite volume satisfying certain curvature conditions, Ann. Math. **129** (1989), 383–425.

[179] J. Morrow and K. Kodaira, Complex manifolds, AMS Chelsea Publishing, Providence, RI, 2006, Reprint of the 1971 edition with errata.

[180] A. Nadel, Multiplier ideal sheaves and Kähler-Einstein metrics of positive scalar curvature, Ann. of Math. (2) **132** (1990), no. 3, 549–596.

[181] A. Nadel and H. Tsuji, Compactification of complete Kähler manifolds of negative Ricci curvature, J. Differential Geom. **28** (1988), no. 3, 503–512.

[182] S. Nakano, On the inverse of monoidal transformation, Publ. Res. Inst. Math. Sci. **6** (1970/71), 483–502.

[183] T. Napier and M. Ramachandran, The L^2-method, weak Lefschetz theorems and the topology of Kähler manifolds, JAMS **11** (1998), no. 2, 375–396.

[184] R. Narasimhan, Imbedding of holomorphically complete complex spaces, Amer. J. Math. **82** (1960), 917–934.

[185] Y. Nohara, Projective embeddings and Lagrangian fibrations of Kummer varieties, (2006), math.DG/0604329.

[186] M.V. Nori, Zariski's conjecture and related problems, Ann. Sci. École Norm. Sup. **16** (1983), 305–344.

[187] T. Ohsawa, Isomorphism theorems for cohomology groups of weakly 1-complete manifolds, Publ. Res. Inst. Math. Sci. **18** (1982), 191–232.

[188] _____, Holomorphic embedding of compact s.p.c. manifolds into complex manifolds as real hypersurfaces, Differential geometry of submanifolds (Kyoto, 1984), Lecture Notes in Math., vol. 1090, Springer, Berlin, 1984, pp. 64–76.

[189] _____, Analysis of several complex variables, Translations of Mathematical Monographs, vol. 211, American Mathematical Society, Providence, RI, 2002.

[190] R. Paoletti, *Moment maps and equivariant Szegö kernels*, J. Symplectic Geom. **2** (2003), 133–175.

[191] _____, *The Szegö kernel of a symplectic quotient*, Adv. Math. **197** (2005), 523–553.

[192] S.T. Paul and G. Tian, *Algebraic and analytic K-stability*, math.DG/0405530.

[193] _____, *CM stability and the generalised Futaki invariant*, math.AG/0605278.

[194] _____, *Analysis of geometric stability*, Int. Math. Res. Not. (2004), no. 48, 2555–2591.

[195] Th. Peternell, *Algebraicity criteria for compact complex manifolds*, Math. Ann. **275** (1986), no. 4, 653–672.

[196] _____, *Moishezon manifolds and rigidity theorems*, Bayreuth. Math. Schr. **54** (1998), 1–108.

[197] D. Phong and J. Sturm, *Scalar curvature, moment maps, and the Deligne pairing*, Amer. J. Math. **126** (2004), no. 3, 693–712.

[198] _____, *The Monge-Ampère operator and geodesics in the space of Kähler potentials*, Invent. Math. **166** (2006), no. 1, 125–149.

[199] _____, *Test configurations for K-stability and geodesic rays*, math.DG/0606423 (2006).

[200] D. Quillen, *Determinants of Cauchy-Riemann operators on Riemann surfaces*, Functional Anal. Appl. **19** (1985), no. 1, 31–34.

[201] D.B. Ray and I.M. Singer, *Analytic torsion for complex manifolds*, Ann. of Math. (2) **98** (1973), 154–177.

[202] M. Reed and B. Simon, *Methods of modern mathematical physics*, vol. IV, Academic Press, New York, 1978.

[203] R. Remmert, *Meromorphe Funktionen in kompakten komplexen Räumen*, Math. Ann. **132** (1956), 277–288.

[204] _____, *Holomorphe und meromorphe Abbildungen komplexer Räume*, Math. Ann. **133** (1957), 328–370.

[205] J. Ross and R. Thomas, *An obstruction to the existence of constant scalar curvature Kähler metrics*, J. Differential Geom. **72** (2006), no. 3, 429–466.

[206] _____, *A study of the Hilbert-Mumford criterion for the stability of projective varieties*, J. Algebraic Geom. **16** (2007), no. 2, 201–255.

[207] H. Rossi, *Attaching analytic spaces to an analytic space along a pseudoconcave boundary*, Proc. Conf. Complex. Manifolds (Minneapolis), Springer-Verlag, New York, 1965, pp. 242–256.

[208] W.-D. Ruan, *Canonical coordinates and Bergmann metrics*, Comm. Anal. Geom. **6** (1998), no. 3, 589–631.

[209] L. Saper, L^2-cohomology and intersection homology of certain algebraic varieties with isolated singularities, Invent. Math. **82** (1985), no. 2, 207–255.

[210] I. Satake, The Gauss-Bonnet theorem for V-manifolds, J. Math. Soc. Japan **9** (1957), 464–492.

[211] M. Schlichenmaier, Berezin-Toeplitz quantization and Berezin transform, Long time behaviour of classical and quantum systems (Bologna, 1999), Ser. Concr. Appl. Math., vol. 1, World Sci. Publ., River Edge, NJ, 2001, pp. 271–287.

[212] L. Schwartz, Théorie des distributions, Actualités Sci. Ind., no. 1091, Hermann & Cie., Paris, 1950.

[213] R.T. Seeley, Extension of C^∞ functions defined in a half space, Proc. Amer. Math. Soc. **15** (1964), 625–626.

[214] J.P. Serre, Faisceaux algébriques cohérents, Ann. of Math. **61** (1955), 197–278.

[215] _____, Géométrie algébrique et géométrie analytique, Ann. Inst. Fourier **6** (1956), 1–42.

[216] I.R. Shafarevich, Basic Algebraic geometry, Grundl. math. Wiss., vol. 213, Springer-Verlag, Berlin, 1974.

[217] B. Shiffman and A.J. Sommese, Vanishing theorems on complex manifolds, Progress in Math., vol. 56, Birkhäuser, 1985.

[218] B. Shiffman and S. Zelditch, Distribution of zeros of random and quantum chaotic sections of positive line bundles, Comm. Math. Phys. **200** (1999), no. 3, 661–683.

[219] _____, Asymptotics of almost holomorphic sections of ample line bundles on symplectic manifolds, J. Reine Angew. Math. **544** (2002), 181–222.

[220] _____, Equilibrium distribution of zeros of random polynomials, Int. Math. Res. Not. (2003), no. 1, 25–49.

[221] _____, Random polynomials with prescribed Newton polytope, J. Amer. Math. Soc. **17** (2004), no. 1, 49–108.

[222] M. Shubin, Semi-classical asymptotics on covering manifolds and Morse inequalities, Geom. Funct. Analysis **6** (1996), 370–409.

[223] C.L. Siegel, Meromorphe Funktionen auf kompakten analytischen Mannigfaltigkeiten, Nachr. Akad. Wiss. Göttingen. Math. Phis. Kl. IIa (1955), 71–77.

[224] Y.T. Siu, A vanishing theorem for semipositive line bundles over non-Kähler manifolds, J. Differential Geom. **20** (1984), 431–452.

[225] _____, Some recent results in complex manifold theory related to vanishing theorems for the semipositive case, Lecture Notes in Math. (Berlin – New York), vol. 1111, Springer Verlag, 1985, Arbeitstagung Bonn 1984, pp. 169–192.

[226] _____, *An effective Matsusaka big theorem*, Ann. Inst. Fourier (Grenoble) **43** (1993), no. 5, 1387–1405.

[227] Y.T. Siu and S.T. Yau, *Complete Kähler manifolds with nonpositive curvature of faster than quadratic decay*, Ann. of Math. (2) **105** (1977), no. 2, 225–264.

[228] _____, *Compactification of negatively curved complete Kähler manifolds of finite volume*, Ann. Math. Stud., vol. 102, Princeton Univ. Press, 1982, pp. 363–380.

[229] H. Skoda, *Application des techniques L^2 à la théorie des idéaux d'une algèbre de fonctions holomorphes avec poids*, Ann. Sci. École Norm. Sup. (4) **5** (1972), 545–579.

[230] J. Song, *The Szegö kernel on an orbifold circle bundle*, (2004), math.DG/0405071.

[231] _____, *The α-invariant on toric Fano manifolds*, Amer. J. Math. **127** (2005), no. 6, 1247–1259.

[232] C. Soulé, *Lectures on Arakelov geometry*, Cambridge Studies in Advanced Mathematics, vol. 33, Cambridge University Press, Cambridge, 1992.

[233] _____, *Genres de Todd et valeurs aux entiers des dérivées de fonctions L*, Astérisque (2007), no. 311, Exp. No. 955,, Séminaire Bourbaki, Vol. 2005/2006.

[234] S. Takayama, *A differential geometric property of big line nbundles*, Tôhoku Math. J. **46** (1994), no. 2, 281–291.

[235] _____, *Adjoint linear series on weakly 1-complete manifolds I: Global projective embedding*, Math.Ann. **311** (1998), 501–531.

[236] K. Takegoshi, *Global regularity and spectra of Laplace-Beltrami operators on pseudoconvex domains*, Publ. Res. Inst. Math. Sci. **19** (1983), no. 1, 275–304.

[237] M.E. Taylor, *Partial differential equations. 1: Basic theory*, Applied Mathematical Sciences, vol. 115, Springer-Verlag, Berlin, 1996.

[238] _____, *Partial differential equations. 2: Qualitative studies of linear equations*, Applied Mathematical Sciences, vol. 116, Springer-Verlag, Berlin, 1996.

[239] W. Thimm, *Meromorphe Abbildungen von Riemannschen Bereichen*, Math. Z. **60** (1954), 435–457.

[240] R. Thomas, *Notes on GIT and symplectic reduction for bundles and varieties*, Survey in Differential Geometry, **X** (2006), 221–273.

[241] G. Tian, *On a set of polarized Kähler metrics on algebraic manifolds*, J. Differential Geom. **32** (1990), 99–130.

[242] _____, *Canonical metrics in Kähler geometry*, Lectures in Mathematics ETH Zürich, Birkhäuser Verlag, Basel, 2000, Notes taken by Meike Akveld.

[243] _____ , *Extremal metrics and geometric stability*, Houston J. Math. **28** (2002), no. 2, 411–432, Special issue for S. S. Chern.

[244] R. Todor, I. Chiose, and G. Marinescu, *Morse inequalities for covering manifolds*, Nagoya Math. J. **163** (2001), 145–165.

[245] S. Trapani, *Numerical criteria for the positivity of the difference of ample divisors*, Math. Z. **219** (1995), no. 3, 387–401.

[246] K. Uhlenbeck and S.-T. Yau, *On the existence of Hermitian-Yang-Mills connections in stable vector bundles*, Comm. Pure Appl. Math. **39** (1986), no. S, suppl., S257–S293.

[247] L. Wang, *Bergman kernel and stability of holomorphic vector bundles with sections*, MIT Ph.D. Thesis (2003), 85 pages.

[248] X. Wang, *Balance point and stability of vector bundles over a projective manifold*, Math. Res. Lett. **9** (2002), no. 2-3, 393–411.

[249] _____ , *Moment map, Futaki invariant and stability of projective manifolds*, Comm. Anal. Geom. **12** (2004), no. 5, 1009–1037.

[250] _____ , *Canonical metrics on stable vector bundles*, Comm. Anal. Geom. **13** (2005), no. 2, 253–285.

[251] A. Weil, *Introduction a l'étude des variétés kählériennes*, Actualités scientifiques et industrielles, vol. 1267, Hermann, Paris, 1958.

[252] R.O. Wells, *Differential analysis on complex manifolds*, second ed., GTM, vol. 65, Springer-Verlag, New York, 1980.

[253] E. Witten, *Supersymmetry and Morse theory*, J. Differential Geom. **17** (1982), 661–692.

[254] _____ , *Holomorphic Morse inequalities*, Algebraic and differential topology – global differential geometry, Teubner-Texte Math., vol. 70, Teubner, Leipzig, 1984, pp. 318–333.

[255] H. Wu, *Negatively curved Kähler manifolds*, Notices. Amer. Math. Soc. **14** (1967), 515.

[256] S. Wu and W. Zhang, *Equivariant holomorphic Morse inequalities. III. Non-isolated fixed points*, Geom. Funct. Anal. **8** (1998), no. 1, 149–178.

[257] S.-T. Yau, *On the Ricci curvature of a compact Kähler manifold and the complex Monge-Ampère equation. I*, Comm. Pure Appl. Math. **31** (1978), no. 3, 339–411.

[258] _____ , *Nonlinear analysis in geometry*, Enseign. Math. (2) **33** (1987), no. 1-2, 109–158.

[259] _____ , *Perspectives on geometric analysis*, Survey in Differential Geometry, **X** (2006), 275–379.

[260] K.-I. Yoshikawa, *K3 surfaces with involution, equivariant analytic torsion, and automorphic forms on the moduli space*, Invent. Math. **156** (2004), no. 1, 53–117.

[261] S. Zelditch, *Szegő kernels and a theorem of Tian*, Internat. Math. Res. Notices (1998), no. 6, 317–331.

[262] S. Zhang, *Heights and reductions of semi-stable varieties*, Compositio Math. **104** (1996), no. 1, 77–105.

[263] W. Zhang, *Lectures on Chern–Weil Theory and Witten Deformations*, Nankai Tracts in Mathematics, vol. 4, World Scientific, Hong Kong, 2001.

Index

DISCARDED

CONCORDIA UNIVERSITY LIBRARIES
MONTREAL

CONCORDIA UNIV. LIBRARY